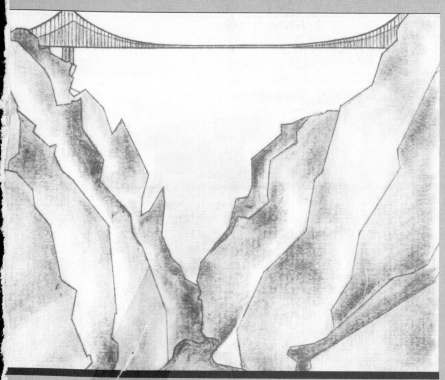

Over 1200 ft long, and 1000 ft above the Arkansas River, the Royal Gorge Suspension Bridge is not only a miracle of construction but a symbol of technical achievement that enhances and crowns the landscape.

DAVID E. STEVENS
Wentworth Institute of Technology

DISCOVERING TECHNICAL MATHEMATICS
AN INTRODUCTION

Addison-Wesley Publishing Company
Reading, Massachusetts · Menlo Park, California
Don Mills, Ontario · Wokingham, England
Amsterdam · Sydney · Singapore · Tokyo
Mexico City · Bogotá · Santiago · San Juan

Library of Congress Cataloging in Publication Data

Stevens, David E.

   Discovering technical mathematics.

   Includes index.

   1. Mathematics—1961–   . I. Title.

QA39.2.S746 1985   510   84–9347

ISBN 0-201-16365-9

8 9 10 DO 959493

PREFACE

AUDIENCE

Discovering Technical Mathematics: An Introduction is intended for students who are beginning a career in the industrial and engineering technology programs at technical institutes and two-year colleges. Chapters 1 to 5 are designed for students who have never taken high school algebra or for students who may be weak in the basic arithmetic and basic algebra skills. Chapters 6 to 12 are designed for students who have mastered these basic skills; they introduce the intermediate algebra and trigonometry topics necessary for today's technical student.

APPROACH

The text is written to encourage the student to become *involved* in the development of the mathematical ideas presented. By using carefully selected examples, the student is allowed to *discover* many of the rules and laws of algebra and trigonometry. Throughout the text, the scientific calculator is used as an instructional tool for verifying new concepts and for reducing the time spent on tedious calculations. Never is the calculator allowed to become a replacement for the understanding of any algebraic or trigonometric concept.

FEATURES

Every effort has been made to make this a book from which the student can *learn* and *succeed*. The following features make this possible.

Objectives

At the beginning of each section is a clearly defined list of objectives to be mastered. The material presented in each section is then keyed to these objectives by using a dice symbol ($\boxdot$, $\boxdot$, $\boxdot$, etc.).

Illustrative Examples

Each illustrative example has been carefully laid out. Every statement is supported by a *reason*, and the use of *arrows* allows the student to flow smoothly from one step to another. For steps that are more difficult, special side *notes* are given. Finally, *comments* at the end of many examples provide further insight into the mathematical idea being presented.

Readability

The text is written at an elementary level. Rules and formulas are boxed in for easy reference, and common student errors are clearly pointed out by using the symbol ⬡CAUTION. The drawings and illustrations are carefully detailed and given more attention than usual.

Applications

There is no doubt that student motivation is at its highest when relevant and interesting applications are directly integrated into the mathematics curriculum. For this reason, separate material has been set aside to explain carefully and use some of the more important formulas and concepts needed by the technical student in his or her area of concentration. A glance at the table of contents will reveal some of the outstanding applications that appear in this text.

Exercises

Each section contains a large set of exercises. Each set begins with routine problems that test the basic concepts being presented and concludes with more challenging problems designed to test the student's thorough understanding of the subject matter. Whenever possible, application problems have been integrated throughout each set of exercises. Answers to all *odd* exercises appear at the end of the book.

To help prepare students for exams, each chapter concludes with a *two*-part *Chapter Review* consisting of *Review Exercises* and a practice *Chapter Test*. The Review Exercises are grouped according to the objective that should have been mastered. If difficulty is encountered with any of these problems, the student can refer to and read the material keyed by the dice symbol to that objective. The practice Chapter Tests provide still further opportunity for the student to correct any weaknesses. Each Chapter Test contains a *time* element and a *score* category and each has been class-tested for the time and score listed. Answers to *all* the problems in the Review Exercises and Chapter Tests appear at the end of the book.

Flowchart

The following flowchart suggests a natural flow from one chapter to another. Chapter 11 can be covered immediately after Chapter 5 if an earlier introduction to right-triangle trigonometry is desired. The text has been made as flexible as possible.

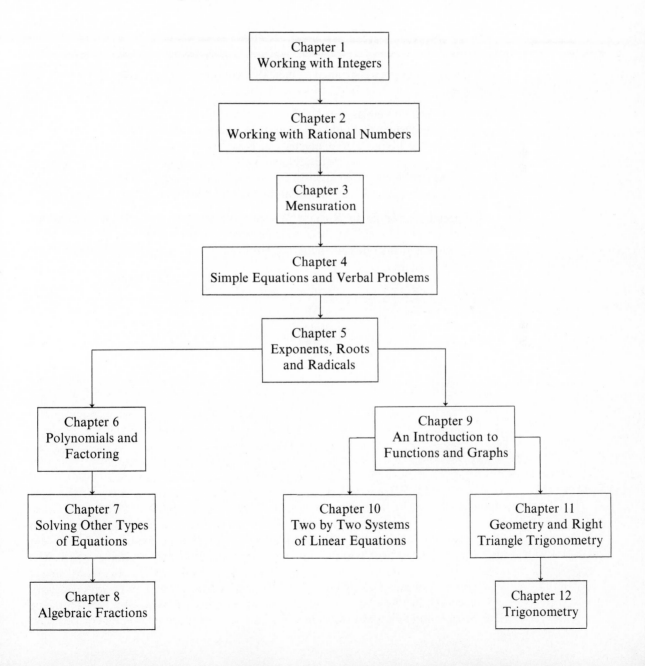

Supplementary Materials

An *Instructor's Manual* with answers to even-numbered exercises, general comments and suggestions on the use of the material in each chapter, and additional chapter tests is also available from the publisher.

ACKNOWLEDGEMENTS

This text has greatly profited from helpful comments and critiques by the following people:

Grace DeVelbias, Sinclair Community College
John Drury, Columbus Technical Institute
Gerald Flynn, State University of New York,
                    Agricultural and Technical College
Marcia Kemen, Wentworth Institute of Technology
Arlene Starwalt-Jeskey, Oklahoma City, Oklahoma

Special thanks go to Richard Wheeler (Wentworth Institute of Technology), whose work in checking the answers is greatly appreciated.

I would also like to thank the editorial group at Addison-Wesley for their kind assistance.

D.E.S.

CONTENTS

2

WORKING WITH RATIONAL NUMBERS   65

3

MENSURATION   115

4

SIMPLE EQUATIONS AND VERBAL PROBLEMS   175

5

EXPONENTS, ROOTS, AND RADICALS   241

6

POLYNOMIALS AND FACTORING   307

7

SOLVING OTHER TYPES OF EQUATIONS   343

8

ALGEBRAIC FRACTIONS   383

9

AN INTRODUCTION TO FUNCTIONS AND GRAPHS   437

10

TWO BY TWO SYSTEMS OF LINEAR EQUATIONS   493

11

GEOMETRY AND RIGHT TRIANGLE TRIGONOMETRY   539

12

TRIGONOMETRY 625

ANSWERS TO SELECTED PROBLEMS A-1

INDEX I-1

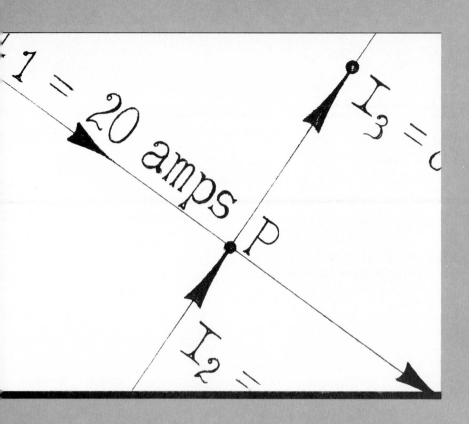

WORKING
WITH
INTEGERS

1.1 SIMPLE ALGEBRAIC EXPRESSIONS

Objectives

⊡ To write a simple algebraic expression that contains one operation.

⊡ To write an algebraic expression that contains more than one operation.

⊡ To evaluate an algebraic expression that contains one operation.

⊡ To simplify an algebraic expression that contains repeated addition.

⊡ Algebraic Expressions that Contain One Operation

Let's look at the following additions.

$$10 + 0, \text{ read "ten plus zero"}$$
$$10 + 1, \text{ read "ten plus one"}$$
$$10 + 2, \text{ read "ten plus two"}$$
$$10 + 3$$
$$10 + 4$$
$$10 + 5$$
$$\vdots$$

Suppose we wish to write a general expression for "ten plus *some number*." The symbols that mathematicians most often use for such generalities as *some number* are the letters of the alphabet.

We can write

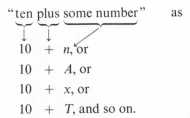

$$10 + n, \text{ or}$$
$$10 + A, \text{ or}$$
$$10 + x, \text{ or}$$
$$10 + T, \text{ and so on.}$$

Mathematicians call expressions like "$10 + n$" an **algebraic expression**. The 10 and the n are called the **terms** of the algebraic expression, and the $+$ symbolizes the **operation** of addition.

EXAMPLE 1

Write an algebraic expression for "some number minus eight."

Solution:

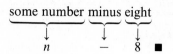

Note: The *n* and the 8 are the terms of the algebraic expression, and the −
symbolizes the operation of subtraction.

EXAMPLE 2

Write an algebraic expression for "some number plus some *different* number."

Solution:

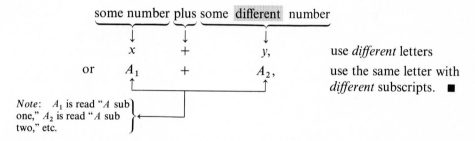

Note: A_1 is read "*A* sub one," A_2 is read "*A* sub two," etc.

EXAMPLE 3

Write an algebraic expression for "some number plus the *same* number."

Solution:

some number plus the   same   number

$$x \quad + \quad x,$$ use the *same* letter

or $A_1 \quad + \quad A_1,$ use the same letter with the *same* subscript.   ∎

In arithmetic, the cross × symbolizes multiplication. In technical mathematics,
however, the × is seldom used since it would be easily confused with the letter *x*.
The dot · or parentheses ( ) are used to show the multiplication of two numbers.

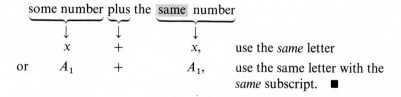

Thus, to denote 5 times 4 write $5 \cdot 4, 5(4), (5)4,$ or $(5)(4)$.

The dot · or parentheses ( ) are not needed to show the multiplication of two or more letters, or to show the multiplication of a number and one or more letters.

Thus, to denote 5 times x times y simply write $5xy$.

For products, always write the number first, then the letters.

EXAMPLE 4

Write an algebraic expression for "two times some number."

Solution:

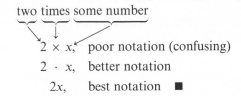

$2 \times x$, poor notation (confusing)

$2 \cdot x$, better notation

$2x$, best notation ∎

Note: "$2x$" is a one-term algebraic expression. The 2 is called the numerical *coefficient* of the letter x.

In arithmetic, the ÷ symbolizes division. To write 15 divided by 3, we could write $15 \div 3$. However, in technical mathematics, we usually write 15 divided by 3 as $\frac{15}{3}$.

EXAMPLE 5

Write an algebraic expression for "some number divided by five."

Solution:

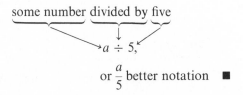

$a \div 5$,

or $\frac{a}{5}$ better notation ∎

⬚ Algebraic Expressions that Contain More than One Operation

It is extremely important to become skillful in reading words and changing these words into algebraic expressions. Some frequently occurring words and their meanings are given in Table 1.1.

TABLE 1.1

The expression	The operation
plus, sum, increased by, more than	addition
minus, difference, less, less than, take away, decreased by	subtraction
times, product, twice, doubled, tripled	multiplication
divided by, quotient	division

EXAMPLE 6

Write an algebraic expression for "three times a number increased by four."

Solution:

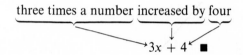

Note: Another possible solution is $4 + 3x$. The order in which you add does not matter.

EXAMPLE 7

Write an algebraic expression for "six *less* twice some number."

Solution:

six less twice some number
$6 - 2x$ ■

EXAMPLE 8

Write an algebraic expression for "six *less than* twice some number."

Solution:

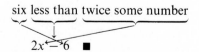

Note: Be sure you can distinguish the difference between Example 7 and Example 8. The order in which you subtract does matter.

⸬ Evaluating Simple Algebraic Expressions

If we replace the *n* in an algebraic expression like $10 + n$ with a specific number, say 3, the algebraic expression becomes

$$10 + n$$
$$10 + (3) \quad \text{replace } n \text{ with } 3$$
$$13. \quad \text{evaluate (add)}$$

When using *n* in this manner, *n* is called a **variable**.

DEFINITION
A **variable** is a symbol (usually a letter of the alphabet) that can be replaced by a certain number.

EXAMPLE 9

Evaluate $12n$, if $n = 15$.

Solution:

$$12n$$
$$12 \,(15) \qquad \text{replace } n \text{ with } 15$$
$$180 \quad \blacksquare \qquad \text{evaluate (multiply)} \rightarrow$$

$$\begin{array}{r} 12 \\ \times 15 \\ \hline 60 \\ 12 \\ \hline 180 \end{array}$$

Therefore, $12n = 180$ when $n = 15$.

EXAMPLE 10

Evaluate $182 - P$, if $P = 94$.

Solution:

$$182 - P$$
$$182 - (94) \qquad \text{replace } P \text{ with } 94$$
$$88 \quad \blacksquare \qquad \text{evaluate (subtract)} \rightarrow$$

$$\begin{array}{r} 182 \\ -94 \\ \hline 88 \end{array}$$

Therefore, $182 - P = 88$ when $P = 94$.

TABLE 1.2

	$x + x$	$2x$
$x = 3$	$(3) + (3) = 6$	$2(3) = 6$
$x = 4$	$(4) + (4) = 8$	$2(4) = 8$
$x = 7$	$(7) + (7) = 14$	$2(7) = 14$

⠢ Simplifying Algebraic Expressions

In Example 3, we wrote the algebraic expression $x + x$ for "some number plus the same number." In Example 4, we wrote the algebraic expression $2x$ for "two times some number." Let's compare these two algebraic expressions (see Table 1.2). Do you see a special relationship between $x + x$ and $2x$? Since $x + x$ and $2x$ are equal for *every* numerical replacement of the variable, the two expressions are said to be *equivalent*. Since $2x$ is a more compact form than $x + x$, $2x$ is more **simplified**.

As you can see from Table 1.2, any number added to itself is two times that number.

$$\boxed{\text{For any number } a,\ a + a = 2a.}$$

$$\text{Does } \underbrace{5 + 5 + 5}_{3 \text{ times}} = (3)5?$$

$$\text{Does } \underbrace{5 + 5 + 5 + 5}_{4 \text{ times}} = (4)5?$$

You can see that repeated addition is nothing more than multiplication.

$$\boxed{\begin{array}{c}\text{For any number } a \text{ added } n \text{ times} \\[4pt] \underbrace{a + a + a + \cdots + a}_{n \text{ times}} = na.\end{array}}$$

EXAMPLE 11

Simplify $T + T + T + T + T$.

Solution:

$$\underbrace{T + T + T + T + T}_{5 \text{ times}} = 5T \quad \blacksquare$$

EXAMPLE 12

Simplify $w + w + z + z + z$.

Solution:

$$w + w + z + z + z = 2w + 3z \quad \blacksquare$$

EXAMPLE 13

Simplify $1 + 2 + T + T$.

Solution:

$$1 + 2 + T + T = 3 + 2T \quad \blacksquare$$

 Do not write $3 + 2T$ as $5T$. The 3 and the $2T$ cannot be added together to get $5T$. The meaning of $5T$ is given in Example 11.

EXERCISES 1.1

Write an algebraic expression for each of the following:

1. Some voltage v decreased by eight.
2. Eight less some current I.
3. Some horsepower H divided by three.
4. The product of some force F_1 and five.
5. Twice some temperature T_2 increased by six.
6. Five more than some temperature T_1.
7. Twice some weight w less four.
8. The sum of some weight w and six, divided by 5.
9. Five less than some pressure P_1, increased by pressure P_2.
10. Some area A_1 less nine, take away four times area A_2.

Evaluate the algebraic expressions.

11. $3x; x = 71$
12. $55x; x = 13$
13. $118 - b; b = 38$
14. $16 - c; c = 9$
15. $n + 181; n = 96$
16. $172 + a; a = 1012$

17. $\dfrac{448}{p}$; $p = 8$ **18.** $\dfrac{n}{17}$; $n = 3434$

19. $x + y$; $x = 21$, $y = 711$ **20.** xy; $x = 191$, $y = 40$

Write the following in a more compact form. Simplify.

21. $q + q + q$ **22.** $f + f + f + f$

23. $p + p + p + 8 + 8$ **24.** $6 + 6 + m + m + m$

25. $M + M + M + M + R + R + 9 + 5$

Write exercises 26–30 entirely as sums (no products).

26. $5a$

27. $4h$

28. $3a + 2d$

29. $3K + 2S$

30. $4x + y$

31. Are $x + y$ and $y + x$ *equivalent* algebraic expressions?

32. Are xy and yx *equivalent* algebraic expressions?

1.2   A FIRST LOOK AT EQUATIONS AND FORMULAS

Objectives

☐ To write a simple equation from words.

☐ To guess the solution to a simple equation.

☐ To write a formula that describes a special relationship between the dimensions of a technical drawing.

☐ From Words to Equations

When the equal symbol $=$ is used in conjunction with algebraic expressions, we have what is called an **equation**.

EXAMPLE 1

Write an equation for "the product of five and some force F equals fifteen."

Solution:

The product of five and some force F equals fifteen.

$$5F = 15 \quad \blacksquare$$

EXAMPLE 2

Write an equation for "twice some current I less six equals four."

Solution:

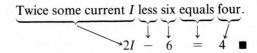

Twice some current I less six equals four.

$$2I - 6 = 4 \quad \blacksquare$$

⠒ Guessing the Solution to a Simple Equation

Any number that will make an equation true, when it replaces the variable in the equation, is called a **solution** to the equation. If the solution to a simple equation is a counting number (1, 2, 3, ...), then it is quite possible to guess its solution.

EXAMPLE 3

Guess the solution to the equation $5F = 15$.

Solution: We are looking for a number to replace F so that the equation becomes true. If $5F = 15$, then F must equal 3,

since $5\,(3) = 15$ is a true statement.

Thus, $5F = 15$ has the solution $F = 3$. $\blacksquare$

EXAMPLE 4

Guess the solution to the equation $2I - 6 = 4$.

Solution: If $2I - 6 = 4$, then $2I$ must equal 10

since $10 - 6 = 4$.

Now if $2I = 10$, then I must equal 5,

since $2\,(5) = 10$ is a true statement.

Thus, $2I - 6 = 4$ has the solution $I = 5$. $\blacksquare$

EXAMPLE 5

Guess the solution to the equation $8 + n = 3$.

Solution: We're looking for some number n that can be added to eight that will equal three. If we're already at eight, how can we *add* on anything to get to three?

If we consider replacements for n to be the *counting numbers* (1, 2, 3, 4, 5, ...), then $8 + n = 3$ has *no solution.* ■

Note: There are many types of numbers other than the counting numbers that might replace n to make the equation true. In Section 1.3 we'll introduce numbers called **integers**. In Section 1.4 you'll discover that $8 + n = 3$ has the solution -5 (read "negative five").

In Chapter 4, the addition and multiplication properties of equality will be introduced. These properties will allow you to solve some fairly complicated equations that you never could solve by guessing. Perhaps you will discover these properties yourself, before they are actually introduced to you.

∴ Formulas

In many technical applications we use equations that tell us about special relationships among several different variables. Such equations are called **formulas**.

DEFINITION

A **formula** is an equation that describes a special relationship among variables.

EXAMPLE 6

Write a formula that describes the distance y (see Fig. 1.1).

Solution:

$$y = \underbrace{x + x}_{} + 5$$

$$y = 2x + 5 \text{ or } y = 5 + 2x \quad ■$$

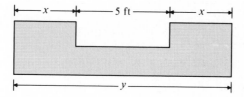

Figure 1.1

EXAMPLE 7

Use the formula from Example 6 to solve for x when $y = 11$ ft.

Solution: First, replace y with 11.

$$y \;\; = 2x + 5$$
$$\downarrow$$
$$11 \;\; = \underline{2x} + 5. \qquad \text{Now } 2x \text{ must equal } \boxed{6},$$
$$\downarrow$$
$$\text{since } 11 \;\; = \boxed{6} + 5.$$

Of course, if $2x \;\; = 6$, then x must equal $\boxed{3}$
$$\downarrow$$
$$\text{since } 2\,\boxed{(3)} \;\; = 6 \text{ is a true statement.}$$

Thus, if $y = 11$ ft, then $x = 3$ ft.   ■

Application

ELECTRICAL RESISTANCES IN SERIES

In electronics, when resistors are connected in series (see Fig. 1.2), the total resistance R_t between the points A and B can be found by adding the individual resistances. The name of the resistance unit is the ohm, and its symbol is Ω. (Ω is the Greek letter omega.) —✀— is the electrical symbol for a resistor.

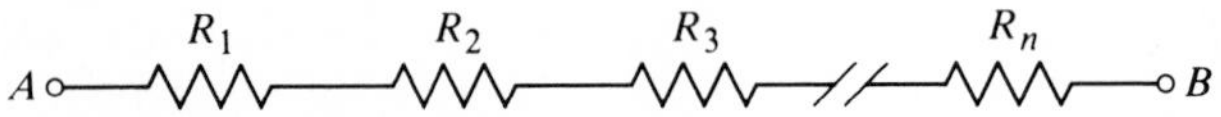

Figure 1.2

FORMULA: Total Series Resistance

$$R_t = R_1 + R_2 + R_3 + \cdots + R_n$$

EXAMPLE 8

Write a formula that describes the total resistance between points A and B (see Fig. 1.3).

Solution:

$$R_t = \underbrace{R_1 + R_1 + R_1} + 8$$

$$R_t = 3R_1 + 8 \qquad \text{or} \quad R_t = 8 + 3R_1 \quad ■$$

Figure 1.3

EXAMPLE 9

Use the formula from Example 8 to solve for R_1 when $R_t = 20\,\Omega$.

Solution: First, replace R_t with 20.

$$R_t = 3R_1 + 8$$
$$20 = 3R_1 + 8. \qquad \text{Now } 3R_1 \text{ must equal } \boxed{12},$$
$$\text{since } 20 = \boxed{12} + 8.$$

Of course, if $3R_1 = 12$, then R_1 must equal $\boxed{4}$,

since $3\,\boxed{(4)} = 12$ is a true statement.

Thus, if $R_t = 20\,\Omega$, then $R_1 = 4\,\Omega$.   ■

EXERCISES 1.2

Guess the solution to the following equations. Use the counting numbers as possible replacements for the variable.

1. $7 + n = 13$

2. $n + 9 = 21$

3. $18 - m = 2$

4. $v - 19 = 10$

5. $15A = 75$

6. $100w = 1000$

7. $\dfrac{x}{17} = 2$

8. $\dfrac{48}{y} = 6$

9. $7x - 1 = 20$

10. $15 + 2x = 29$

Write an equation for each of the following. Then guess the solution to each equation.

11. Some voltage v decreased by six equals four.

12. Twice some weight w equals 180.

13. Some temperature T increased by sixteen equals fifty.

14. Some pressure P_1 divided by fifteen equals three.

15. Four less than twice some horsepower H equals thirty.

16. Some force F tripled less eight equals nineteen.

Given the L-shaped bracket in Fig. 1.4, fill in the table below.

	EF	CD	AF	BC	AB	DE
17.	2	8	6	1	?	?
18.	x	y	t	v	?	?
19.	$3a$	$2b$	$4m$	$5n$	?	?
20.	2	$2w$	7	$2t$	?	?

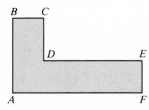

Figure 1.4

Note: AB represents the distance from A to B. BC represents the distance from B to C, and so on.

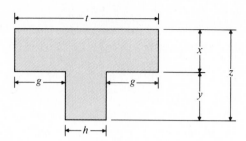

Figure 1.5

For exercises 21–24, refer to Fig. 1.5.

21. Write a formula that describes the distance x.

22. Write a formula that describes the distance t.

23. Using the formula from exercise 21, find x when $z = 48$ in. and $y = 9$ in.

24. Using the formula from exercise 22, solve for g when $t = 18$ in. and $h = 4$ in.

25. Write a formula that describes the total resistance between points A and B (see Fig. 1.6).

26. Suppose the total resistance between points A and B is $120\,\Omega$ (see Fig. 1.7). Write an equation describing this fact, and then solve for R_1.

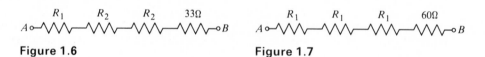

Figure 1.6   **Figure 1.7**

1.3 THE NUMBER LINE AND THE INTEGERS

Objectives

- ⊡ To represent certain words with integers.
- ⊡ To find the opposite of an integer.
- ⊡ To find the magnitude of an integer.
- ⊡ To place $>$, $<$, or $=$ between a pair of integers.

⊡ The Integers

The first numbers that you learned about were the **counting numbers**: 1, 2, 3, 4, 5, ... Another name for the counting numbers is the **positive integers**: $+1$, $+2$, $+3$, $+4$, $+5$, ... The positive sign $(+)$ in front of each number is usually not written when designating a positive integer. Thus, 4 and $+4$ are exactly the same.

You could think of the positive integers as being arranged on a **number line** as shown in Fig. 1.8. When constructing a number line, it is extremely important

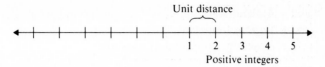

Figure 1.8

to keep the distances between 1 and 2, 2 and 3, 3 and 4, and so on, exactly the same. This common distance is called the *unit distance*.

To the left of 1 on the number line is the number 0, and to the left of 0 are the **negative integers**: -1, -2, -3, -4, -5, ... (read "negative one, negative two, negative three," etc.; see Fig. 1.9).

All of the positive integers combined with zero and all of the negative integers form what are simply called the **integers**.

DEFINITION
The numbers $\{\ldots -5, -4, -3, -2, -1, 0, 1, 2, 3, 4, 5, \ldots\}$ are called the **integers**.

You could think of the number line in Fig. 1.9 as a temperature scale with the numbers 1, 2, 3, 4, 5, ... representing degrees *above* zero, and the numbers -1, -2, -3, -4, -5, ... representing degrees *below* zero. Thus, $+5$ or 5 represents a temperature of five degrees *above* zero, while -5 represents a temperature of five degrees *below* zero. Certain other words can represent positive and negative values as well.

EXAMPLE 1

If a *deposit* into a savings account is designated as a positive integer, then what integer would describe a *withdrawal* of 6 dollars?

Solution: -6 ∎

Opposites

If you look at the number line in Fig. 1.9, you will notice that 5 and -5 are points that are the same distance from 0 on the number line. The same could be said of

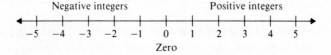

Figure 1.9

4 and −4, 3 and −3, and so on. Two such numbers are called *opposites*. Instead of writing the word opposite, mathematicians use the dash − along with parentheses ( ) to represent the opposite of a number.

EXAMPLE 2

Find the opposite of 7.

Solution: The opposite of 7 is written −(7), and −(7) = −7. ■

EXAMPLE 3

Find the opposite of −3.

Solution: The opposite of −3 is written −(−3), and −(−3) = 3. ■

Note: The opposite of a positive number is negative, and the opposite of a negative number is positive.

⠔ Magnitude

Technicians and engineers would say that 5 and −5 have *opposite direction* but the *same magnitude*. Both 5 and −5 have a magnitude of 5, since both integers are 5 units away from zero. Likewise, 4 and −4 have a magnitude of 4, since both are 4 units away from zero. Mathematicians refer to the **magnitude** of a number as its **absolute value**. The symbol used to designate magnitude or absolute value is | |.

EXAMPLE 4

Find the magnitude of −7.

Solution: The magnitude of −7 is written |−7|, and |−7| = 7 since −7 is seven units away from zero. ■

EXAMPLE 5

Find the magnitude of 10.

Solution: The magnitude of 10 is written |10|, and |10| = 10 since 10 is ten units away from zero. ■

Note: The magnitude or absolute value of a number can *never* be negative.

⠒ **Comparing Integers**

To compare two integers, mathematicians use the following symbols:

> is greater than

< is less than

= is equal to

For $a > b$, read "a is greater than b"; this means that on the number line, the number a is to the *right* of the number b. For $a < b$, read "a is less than b"; this means that on the number line, the number a is to the *left* of the number b. Of course, for $a = b$, read "a equals b"; this means that the numbers a and b occupy the same place on the number line.

EXAMPLE 6

Place $>$, $<$, or $=$ between -5___-3.

Solution: -5 is to the *left* of -3; therefore $-5 < -3$. ■

EXAMPLE 7

Place $>$, $<$, or $=$ between -3___-5.

Solution: -5 is to the left of -3 (see example 6), so certainly -3 is to the *right* of -5. Thus, $-3 > -5$. ■

Note: For any numbers a and b, if $a < b$ then $b > a$.

EXAMPLE 8

Place $>$, $<$, or $=$ between $|-6|$___$-(-6)$.

Solution: $|-6| = 6$ and $-(-6) = 6$. Therefore, $|-6| = -(-6)$. ■

EXERCISES 1.3

  1. If a distance *above* sea level is designated as a positive integer, then what integer would describe 100 feet *below* sea level?

  2. If a current flowing *toward* point A is positive, then what integer describes a current of 3 amperes flowing *away* from point A?

3. Suppose a force that is exerted *upward* is positive. What integer describes a *downward* force of 500 pounds?

4. Suppose a *gain* on the New York Stock Market is positive. What integer describes a *loss* of 8 points?

Give a simpler name for each of the following.

5. $-(16)$ **6.** $-(29)$ **7.** $-(-16)$

8. $-(-9)$ **9.** $-(x)$ **10.** $-(-x)$

Give a simpler name for each of the following.

11. $|4|$ **12.** $|17|$

13. $|-22|$ **14.** $|-35|$

15. $-|7|$ **16.** $-|-2|$

17. $-|-19|$ **18.** $|-3| + |0|$

19. $|-2| + |-7|$ **20.** $(|7|)(|-6|)$

Place $>$, $<$, or $=$ between the following pairs of numbers.

21. $-3\_\_7$ **22.** $-8\_\_2$

23. $-8\_\_-10$ **24.** $-4\_\_-9$

25. $+9\_\_9$ **26.** $-9\_\_9$

27. $-|-4|\_\_4$ **28.** $-|-4|\_\_-4$

29. $|-2| + |3|\_\_5$ **30.** $(|-4|)(|-3|)\_\_-12$

True or false. If false, give an example to verify the falsehood.

31. If a is a negative integer, then $|a| = a$.

32. If a is a positive integer, then $a > -a$.

33. If a is a negative integer, then $a < -a$.

34. Every integer has a different integer as its opposite.

35. Every integer has exactly one opposite.

Guess the solution to each of the following equations. Use the integers as possible replacements for the variable.

36. $-x = 3$ **37.** $-x = -8$ **38.** $-(-a) = 5$

39. $-(-t) = -15$ **40.** $-(-p) = -|-10|$

1.4 ADDING INTEGERS

Objectives

☐ To add two negative integers.

☐ To add a negative and a positive integer.

☐ To simplify an algebraic expression containing addition.

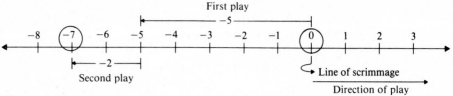

Figure 1.10

· Adding Two Negative Integers

Now that we know what integers are, let's see if we can *add* some integers. Think of a football game with *gains* being positive and *losses* being negative. The team is moving from left to right. It *loses* 5 yards on the first play, and *loses* another 2 yards on the second play (see Fig. 1.10).

The net yardage of the two plays is a *loss* of 7 yards. Therefore,

a loss of 5 yards plus a loss of 2 yards equals a loss of 7 yards.

$$(-5) + (-2) = (-7)$$

Two losses always yield a loss, and therefore *the sum of two negative numbers is always negative.*

ADDITION RULE 1:   Adding Two Negative Numbers
Add their magnitudes and retain the negative − sign.

EXAMPLE 1

Find the sum $(-19) + (-60)$.

Solution:

$$|-19| = 19$$
$$|-60| = 60$$
$$\overline{79}\ \text{adding their magnitudes}$$

Therefore, $(-19) + (-60) = -79$. ∎

·. Adding a Negative and Positive Integer

Again, think of a football game with *gains* being positive and *losses* being negative. The team is moving from left to right. It *loses* 5 yards on the first play, but *gains* 2 yards on the second play (see Fig. 1.11).

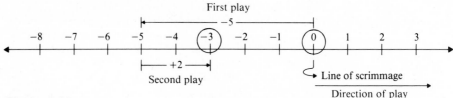

Figure 1.11

The net yardage of the two plays is a *loss* of 3 yards. Therefore,

a loss of 5 yards plus a gain of 2 yards equals a loss of 3 yards.

$$(-5) + (+2) = (-3)$$

However, suppose you *gain* 5 yards on the first play, then *lose* 2 yards on the second play (see Fig. 1.12).

The net yardage of the two plays is a *gain* of 3 yards. Therefore,

a gain of 5 yards plus a loss of 2 yards equals a gain of 3 yards.

$$(+5) + (-2) = (+3)$$

As you can see, *when adding a negative and a positive number, the answer could be either positive or negative.*

ADDITION RULE 2:   Adding a Positive and a Negative Number
Subtract the smaller magnitude from the larger magnitude, and retain the original sign of the number whose magnitude is larger.

EXAMPLE 2

Find the sum $32 + (-15)$.

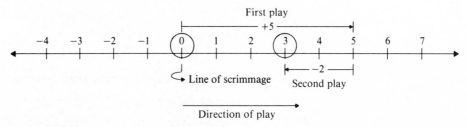

Figure 1.12

Solution:

$$|32| = 32$$
$$|-15| = \underline{15}$$
$$\overline{17} \text{ subtracting their magnitudes}$$

Since $32 > 15$ and the original sign of 32 is $+$, retain the $+$. Therefore, $32 + (-15) = 17$. ∎

EXAMPLE 3

Find the sum $-32 + 15$.

Solution:

$$|-32| = 32$$
$$|15| = \underline{15}$$
$$\overline{17} \text{ subtracting their magnitudes}$$

Since $32 > 15$ and the original sign of 32 is $-$, retain the $-$. Therefore, $-32 + 15 = -17$. ∎

⊡ Some Properties of Addition

Does the order in which you add two numbers together matter? $(-5) + (-2) = -7$, but also $(-2) + (-5) = -7$. Similarly, $(-5) + 2 = -3$, but also $2 + (-5) = -3$. As you can see, the order in which you add is immaterial.

COMMUTATIVE LAW OF ADDITION
For any numbers a and b, $a + b = b + a$.

EXAMPLE 4

Simplify $t + 8 + t + (-2)$.

Solution:

$$t + 8 + t + (-2) = \underbrace{t + t} + \underbrace{8 + (-2)} \qquad \text{commutative law of addition}$$
$$= 2t + 6 \quad ∎$$

What do you think always happens when you add a number and its opposite? $2 + (-2) = 0$, $-8 + 8 = 0$, and so on. As you can see, any number added to its opposite is zero.

For any number a, $a + (-a) = (-a) + a = 0.$

Any number added to zero remains unchanged: $-5 + 0 = -5, 0 + (-8) = -8,$ and so on.

For any number a, $a + 0 = 0 + a = a.$

EXAMPLE 5

Simplify $-x + (-3) + (-4) + x + 3$.

Solution:

$$-x + (-3) + (-4) + x + 3 = \underbrace{-x + x} + \underbrace{(-3) + 3} + (-4) \qquad \text{commutative law of addition}$$

$$= \underbrace{0 \quad + \quad 0} \quad + (-4) \qquad -a + a = 0$$

$$= -4 \quad \blacksquare \qquad 0 + a = a$$

Application

THE FORCE LAW OF EQUILIBRIUM

In mechanics, the **force law of equilibrium** states that if an object is at rest (in equilibrium), then the *algebraic sum* of all forces acting on that object must be zero. Assume that *downward* forces are negative and *upward* forces are positive. The name of the force unit is pounds (lb) or Newtons (N).

EXAMPLE 6

Does the object in Fig. 1.13 conform to the force law of equilibrium?

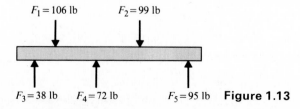

$F_1 = 106$ lb $F_2 = 99$ lb

$F_3 = 38$ lb $F_4 = 72$ lb $F_5 = 95$ lb   **Figure 1.13**

Solution: According to the force law of equilibrium,

$$(-F_1) + (-F_2) + F_3 + F_4 + F_5 = 0.$$ So, does

$$\underbrace{(-106) + (-99)}_{\downarrow} + \underbrace{(38) + (72)}_{\downarrow} + (95) = 0?$$

$$\underbrace{-205 \qquad + \qquad \underbrace{110 \quad + \; 95}_{\downarrow} = 0?}$$

$$\underbrace{-205 \qquad + \qquad \qquad 205}_{} \qquad = 0?$$

$$\longrightarrow 0 = 0?\ \text{Yes.}$$

Therefore, the forces do conform to the force law of equilibrium, and the object is at rest.   ■

Application

KIRCHHOFF'S CURRENT LAW

In electronics, **Kirchhoff's current law** states that the *algebraic sum* of all currents at a junction point P in a circuit must be zero. The name of the current unit is the ampere (A). Assume that if a current flows *away* from point P it is negative, and if it flows *toward* point P it is positive.

EXAMPLE 7

Does the circuit in Fig. 1.14 conform to Kirchhoff's current law?

Solution: According to the Kirchhoff's current law,

$$I_1 + I_2 + (-I_3) + (-I_4) = 0.$$ So, does

$$\underbrace{(20) + (5)}_{\downarrow} + \underbrace{(-6) + (-19)}_{\downarrow} = 0?$$

$$\underbrace{25 \qquad + \qquad -25}_{} \qquad = 0?$$

$$\longrightarrow 0 = 0?\ \text{Yes.}$$

Therefore, the circuit does conform to Kirchhoff's current law.   ■

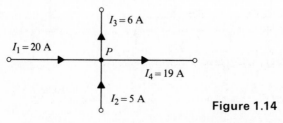

Figure 1.14

EXERCISES 1.4

Find the following sums.

1. $19 + (-19)$
2. $-32 + 32$
3. $17 + (-9)$
4. $-8 + 36$
5. $-16 + (-14)$
6. $-35 + (-11)$
7. $-385 + (-52)$
8. $-602 + (-101)$
9. $-21 + 7$
10. $9 + (-16)$
11. $67 + (-14) + 12$
12. $-3 + (-17) + (-10)$
13. $-3 + (-2) + (-11)$
14. $10 + (-8) + (-9) + 0$
15. $15 + (-4) + (-11)$
16. $-7 + (-4) + 11 + 0$

Simplify the following algebraic expressions.

17. $4 + (-1) + n + n$
18. $p + p + p + 7 + (-4)$
19. $x + (-x) + 15 + (-7)$
20. $8 + y + (-y) + (-8)$
21. $6 + t + (-9) + d$
22. $5 + v + (-5) + v$
23. $-x + 7 + (-3) + x$
24. $-3 + 6 + m + (-1) + m$
25. $a + (-15) + (-3) + a + a + 18$
26. $-5 + n + n + 6 + (-n) + (-1) + n + 3$

Using integers as the replacements for the variables, guess the solution to the following equations.

27. $8 + n = 3$
28. $7 + a = -19$
29. $R + (-9) = -15$
30. $-1 + x = 14$
31. $2F_1 + (-10) = 0$
32. $2I_1 + (-16) = 0$
33. $-D + 3 = 4$
34. $7 + (-w) = 15$

35. Does the object in Fig. 1.15 conform to the force law of equilibrium?
36. Find the magnitude and direction of F_7 for the object in Fig. 1.16 to conform to the force law of equilibrium.
37. Does the circuit in Fig. 1.17 conform to the Kirchhoff's current law?
38. Find the magnitude and direction of I_4 for the circuit in Fig. 1.18 to conform to Kirchhoff's current law.

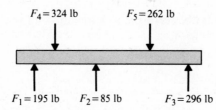

Figure 1.15

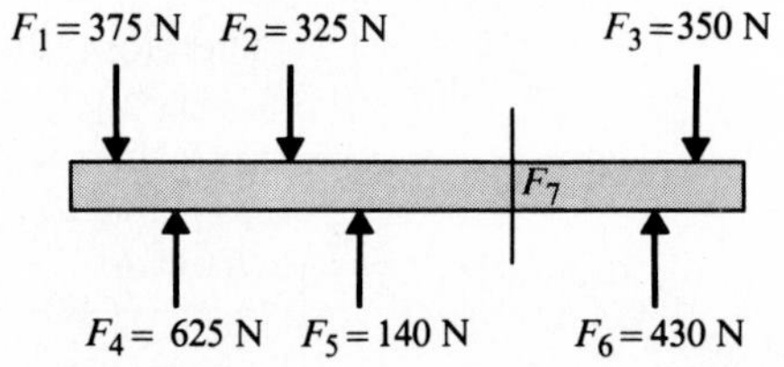

Figure 1.16

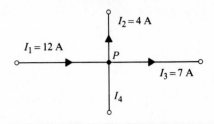

Figure 1.17   **Figure 1.18**

39. Does $|a| + |b| = |a + b|$ for all values of a and b? Fill in the table below and state your conclusion.

a	b	$|a|$	$|b|$	$|a| + |b|$	$|a + b|$
3	2				
-3	2				
3	-2				
-3	-2				

1.5 SUBTRACTING INTEGERS

Objectives

⊡ To subtract two integers.

⊡ To simplify an algebraic expression containing addition and subtraction.

⊡ The Subtraction Rule

If we could define *subtraction* in terms of addition, then we could apply one of the addition rules (from Section 1.4) to find the *difference* between two integers.

Consider this problem: Suppose you *gained* five pounds one week, but *lost* two pounds the next week. What was your overall change in weight? One approach to this problem would be to *add*.

gain of 5 pounds plus a loss of 2 pounds equals a gain of 3 pounds

$$(5) + (-2) = (3)$$

However, another approach would be to simply *subtract*.

gain of 5 pounds minus 2 pounds equals a gain of 3 pounds

$(5) - (2) = (3)$

The subtraction problem, $5 - (2) = 3$, is exactly the same as the addition problem, $5 + (-2) = 3$. Carefully look at

$5 - (2)$ and $5 + (-2)$.

What's happening to the minus (subtraction) sign?

What's happening to the 2?

As you can see, the minus (subtraction) sign changes to an addition sign, and the 2 switches to its opposite, -2. Therefore, to subtract a minus b, add a to the opposite of b.

SUBTRACTION RULE
For any numbers a and b, $a - b = a + (-b)$.

EXAMPLE 1

Find the difference $-8 - 10$.

Solution:

$$-8 - 10 = -8 + (-10) \qquad \text{subtraction rule}$$

$$= -18 \quad \blacksquare \qquad \text{addition rule 1}$$

EXAMPLE 2

Find the difference $-8 - (-10)$.

Solution:

$$-8 - (-10) = -8 + (10) \qquad \text{subtraction rule}$$

$$= 2 \quad \blacksquare \qquad \text{addition rule 2}$$

Note: Notice that subtracting a negative actually means to add a positive. $a - (-b) = a + b$.

⊡ **Some Properties of Subtraction**

Using the *subtraction rule in reverse*, we could say that adding a negative means to subtract a positive.

$$a + (-b) = a - b.$$

EXAMPLE 3

Rewrite $-x - (-y) + (-8)$ without using parentheses and double signs. Simplify.

Solution:

$$
\begin{aligned}
-x - (-y) + (-8) &= -x \oplus y + (-8) && \text{subtraction rule} \\
&= -x + y \ominus 8 \quad \blacksquare && \text{subtraction rule} \\
& && \text{in reverse}
\end{aligned}
$$

Does $a - b = b - a$ for all integers a and b? Is subtraction commutative?
    Suppose $a = 16$ and $b = 10$. Then

$$a - b = 16 - 10 = 6. \text{ However,}$$
$$b - a = 10 - 16 = 10 + (-16) = -6.$$

As you can see, $a - b$ and $b - a$ are not quite the same. Notice that $a - b$ is the *opposite* of $b - a$. Therefore, subtraction is *not* commutative.
    To simplify an algebraic expression containing subtraction, first apply the subtraction rule in order to obtain addition. Once the algebraic expression is written in terms of addition, the commutative law of addition can be applied.

EXAMPLE 4

Simplify $4 - a - 3 - (-a) - 8$.

Solution:

$$
\begin{aligned}
4 - a - 3 - (-a) - 8 &= 4 + (-a) + (-3) + a + (-8) && \text{subtraction rule} \\
&= \underbrace{4 + (-3)} + (-8) + \underbrace{a + (-a)} && \text{commutative law} \\
& && \text{of addition} \\
&= \quad\underbrace{1 \quad + (-8) +} \quad 0 \\
&= \quad -7 \quad \blacksquare
\end{aligned}
$$

EXAMPLE 5

Simplify $-4 - (-5) + y - 6 - (-y)$.

Solution:

$$-4 - (-5) + y - 6 - (-y) = -4 + 5 + y + (-6) + y \qquad \text{subtraction rule}$$

$$= \underbrace{y + y} + \underbrace{(-4) + 5} + (-6) \qquad \text{commutative law}$$
$$\text{of addition}$$

$$= \quad 2y \quad + \quad \underbrace{1 \quad + (-6)}$$

$$= \quad 2y \ + (-5)$$

$$= \quad 2y \ - 5 \quad \blacksquare \qquad \text{subtraction rule}$$
$$\text{in reverse}$$

EXERCISES 1.5

Find the following differences.

1. $18 - 8$

2. $92 - 27$

3. $6 - 17$

4. $15 - 57$

5. $-6 - 15$

6. $-9 - 18$

7. $-9 - (-8)$

8. $-11 - (-4)$

9. $-34 - 34$

10. $-17 - 18$

11. $0 - (-19)$

12. $-16 - 0$

13. $15 - 18 - 7$

14. $-13 - 8 - (-3)$

Rewrite the following without using parentheses and double signs. Simplify.

15. $-y - (-a)$

16. $x - (-y) + (-t)$

17. $-n - (-f) + (-3)$

18. $p + (-n) - (-2)$

19. $-(-x) - (-y) - (-9)$

20. $(-f) - (-p) + (-t)$

Simplify.

21. $14 - 9 - x$

22. $-a - 4 - (-9)$

23. $n - n - (-8)$

24. $q - 6 - q$

25. $y + 4 - y - 9 - (-5)$

26. $R - 8 - R - (-8) - (-R)$

27. $d - (-y) - y - 10 - (-d)$

28. $x - (-7) + x - (-y) + 7 + y$

Guess the solution to the following equations. It will help to change the differences to sums before you begin to guess the solutions.

29. $x - 7 = -9$

30. $a - 9 = -3$

31. $m - (-3) = -7$

32. $16 - (-m) = 8$

33. $8 - a = 9$

34. $5 - n = 9$

The amount of the change in temperature between two extremes can be found by using $|T_1 - T_2|$. The name of the temperature unit is degrees Fahrenheit (°F) or degrees Celsius (°C). For each of the following, find the *magnitude* of the temperature change.

35. $T_1 = 35°F, T_2 = -10°F$ **36.** $T_1 = -80°F, T_2 = -200°F$

37. $T_1 = -15°C, T_2 = 90°C$ **38.** $T_1 = 5°C, T_2 = 86°C$

39. $T_1 = a°F, T_2 = -a°F$

1.6 MULTIPLYING INTEGERS AND RAISING INTEGERS TO POWERS

Objectives

- ⊡ To multiply a positive and a negative integer.
- ⊡ To multiply two negative integers.
- ⊡ To write exponential expressions as products and then to evaluate them.
- ⊡ To simplify an algebraic expression containing repeated addition of negatives.
- ⊡ To simplify an algebraic expression containing multiplication and exponents.

⊡ Multiplying a Positive and a Negative Number

You already know how to multiply two positive integers. For example, $5(4) = 20$, $5(3) = 15$, etc. How might we multiply a positive and a negative number? Look at the following multiplications. Do you see a pattern?

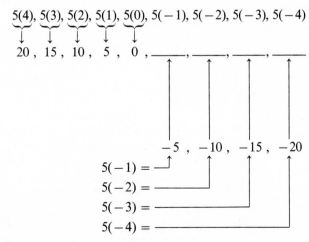

If you wished the pattern to continue, what values would you assign to $5(-1)$, $5(-2)$, $5(-3)$, and $5(-4)$?

Look at the following multiplications. Do you see the pattern again?

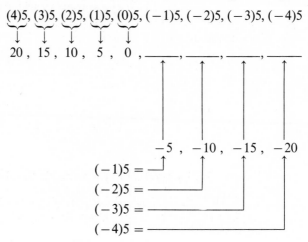

$$(4)5, (3)5, (2)5, (1)5, (0)5, (-1)5, (-2)5, (-3)5, (-4)5$$

$$20, \quad 15, \quad 10, \quad 5, \quad 0, \underline{\quad}, \underline{\quad}, \underline{\quad}, \underline{\quad}$$

$$-5, \quad -10, \quad -15, \quad -20$$

$$(-1)5 =$$
$$(-2)5 =$$
$$(-3)5 =$$
$$(-4)5 =$$

If you wished this pattern to continue, what values would you assign to $(-1)5$, $(-2)5$, $(-3)5$, and $(-4)5$?

As you can see, when multiplying numbers with different signs, the result is always negative.

MULTIPLICATION RULE 1: Multiplying a Positive and a Negative Number

Multiply the magnitudes of the numbers and retain the negative $-$ sign.

EXAMPLE 1

Find the product $2(-3)$.

Solution:

$$|2| = 2$$
$$|-3| = 3$$
$$\overline{6} \text{ multiplying their magnitudes}$$

Therefore, $2(-3) = -6$. ∎

Note: In Section 1.1 you learned that $2a = a + a$. If $a = -3$, then

$$2(-3) = -3 + (-3).$$

Note: We can now apply addition rule 1.

$$= -6 \quad \text{agrees with the above solution}$$

EXAMPLE 2

Find the product $-12(6)$.

Solution:

$$|-12| = 12$$
$$|6| = \underline{6}$$
$$72 \quad \text{multiplying their magnitudes}$$

Therefore, $\qquad -12(6) = -72.$ ∎

Multiplying Two Negative Numbers

Now that we know that a negative number times a positive number is negative, let's look at the following multiplications.

$$\underbrace{-5(4),}_{\downarrow}\ \underbrace{-5(3),}_{\downarrow}\ \underbrace{-5(2),}_{\downarrow}\ \underbrace{-5(1),}_{\downarrow}\ \underbrace{-5(0),}_{\downarrow}\ -5(-1),\ -5(-2),\ -5(-3),\ -5(-4)$$

$$-20,\ \ -15,\ \ -10,\ \ -5,\ \ \ 0\ \ ,\ \underline{\hspace{1cm}},\ \underline{\hspace{1cm}},\ \underline{\hspace{1cm}},\ \underline{\hspace{1cm}}$$

$$5\ \ ,\ \ 10\ \ ,\ \ 15\ \ ,\ \ 20$$

$$-5(-1) = $$
$$-5(-2) = $$
$$-5(-3) = $$
$$-5(-4) = $$

If you wished the pattern to continue, what values would you assign to $-5(-1)$, $-5(-2)$, $-5(-3)$, and $-5(-4)$?

As you can see, when multiplying two negative numbers, the result is always positive.

MULTIPLICATION RULE 2: Multiplying Two Negative Numbers
Multiply the magnitudes of the numbers. The result is always positive.

EXAMPLE 3

Find the product $(-15)(-3)$.

Solution:

$$|-15| = 15$$
$$|-3| = \underline{3}$$
$$45 \quad \text{multiplying their magnitudes}$$

Therefore, $(-15)(-3) = 45.$ ∎

⠒ Exponents

A number called an *exponent* is used to describe the multiplication of a number by itself. For example, $(6)(6)$ can also be written as 6^2 (read "six squared" or "six to the second power"). Notice that the 2 is raised. The 6 is called the **base** and the 2 is called the **exponent** or **power**. Of course, $6^2 = (6)(6) = 36$.

For any number a, $(a)(a) = a^2$.

EXAMPLE 4

Write $(-6)^2$ as a product; then evaluate.

Solution:

$$(-6)^2 = (-6)(-6)$$
$$= 36 \quad ∎ \quad \text{multiplication rule 2}$$

EXAMPLE 5

Write $-(6)^2$ as a product; then evaluate.

Solution: $-(6)^2$ is read "the opposite of six squared," and

$$-(6)^2 = -(6)(6)$$
$$= -36 \quad ∎$$

Note: Be sure you can distinguish the difference between Example 4 and Example 5. You may also see $-(6)^2$ written without parentheses as -6^2. $-(6)^2 = -6^2 = -36$, while $(-6)^2 = 36$.

Numbers other than 2 can be used for exponents as well. 6^3 (read "six cubed" or "six to the third power") means to multiply *three* sixes, or to use 3 *factors* of 6.

$$6^3 = \underbrace{(6)(6)(6)}_{3 \text{ factors}}$$

6^4 (read "six to the fourth power") means to multiply *four* sixes.

$$6^4 = \underbrace{(6)(6)(6)(6)}_{4 \text{ factors}}$$

If n is a counting number $(1, 2, 3, 4, \ldots)$, then

$$a^n = \underbrace{(a)(a)(a) \ldots (a)}_{n \text{ factors}}.$$

 Do not get this rule confused with

$$\underbrace{a + a + a + \cdots + a}_{n \text{ times}} = na.$$

For example, $(x)(x)(x) = x^3$, while $x + x + x = 3x$.

EXAMPLE 6

Write $(-3)^4$ as a product; then evaluate.

Solution:

$$(-3)^4 = \underbrace{(-3)(-3)}\underbrace{(-3)(-3)}$$

$$= \qquad (9)(9) \qquad \text{multiplication rule 2}$$

$$= \qquad 81 \quad \blacksquare$$

Note: An *even* number $(2, 4, 6, \ldots)$ of negatives being multiplied will always have a positive result. Therefore, a negative number raised to an *even* power will always be positive.

EXAMPLE 7

Write $(-1)^5$ as a product; then evaluate.

Solution:

$$(-1)^5 = \underbrace{(-1)(-1)}\underbrace{(-1)(-1)}(-1)$$

$$= \qquad \underbrace{(1)(1)}(-1) \qquad \text{multiplication rule 2}$$

$$= \qquad (1)(-1)$$

$$= \qquad -1 \quad \blacksquare \qquad \text{multiplication rule 1}$$

Note: An *odd* number (1, 3, 5, . . .) of negatives being multiplied will always have a negative result. Therefore, a negative number raised to an *odd* power will always be negative.

⚃ Repeated Addition of Negatives

To simplify an expression like $-3 - 3 - 3 - 3$, read "negative three minus three minus three minus three," we have two methods of reasoning.

Method 1

$$-3 - 3 - 3 - 3 = \underbrace{-3 + (-3)} + \underbrace{(-3) + (-3)} \quad \text{subtraction rule}$$

$$= \qquad (-6) + (-6)$$

$$= \qquad\qquad -12 \qquad\qquad \text{addition rule 1}$$

Method 2

$$-3 - 3 - 3 - 3 = \underbrace{-3 + (-3) + (-3) + (-3)}_{4 \text{ times}} \quad \text{subtraction rule}$$

$$= 4(-3) \qquad\qquad \underbrace{a + a + a + \cdots + a}_{n \text{ times}} = na$$

$$= -12 \qquad\qquad \text{multiplication rule 1}$$

EXAMPLE 8

Simplify $-6 - 6 - t - t - t$.

Solution:

$$-6 - 6 - t - t - t = \underbrace{-6 + (-6)}_{2 \text{ times}} + \underbrace{(-t) + (-t) + (-t)}_{3 \text{ times}} \quad \text{subtraction rule}$$

$$= \underbrace{2(-6)} \quad + \quad \underbrace{3(-t)}$$

$$= \qquad -12 \quad + (-3t) \qquad \text{multiplication rule 1}$$

$$= \qquad -12 \quad - 3t \quad \blacksquare \qquad \begin{array}{l}\text{subtraction rule} \\ \text{in reverse}\end{array}$$

⚃ Some Properties of Multiplication

Any number times one will remain the same.

$$\boxed{\text{For any number } a,\ 1(a) = a(1) = a.}$$

However, any number times negative one will become its opposite.

$$\text{For any number } a, -1(a) = a(-1) = -a.$$

EXAMPLE 9

Simplify $(-x)(1)(y)(-1)$.

Solution:

$$\underbrace{(-x)(1)(y)(-1)} = (-x)\underbrace{(y)(-1)} \qquad a(1) = a$$

$$= (-x)(-y) \qquad a(-1) = -a$$

$$= xy \quad \blacksquare \qquad \text{multiplication rule 2}$$

Any number multiplied by zero equals zero.

$$\text{For any number } a, 0(a) = a(0) = 0.$$

EXAMPLE 10

Simplify $5(x)(0)(-y)$.

Solution:

$$5(x)\overbrace{(0)(-y)} = 5(x)\underbrace{(0)} \qquad 0(a) = 0$$

$$= 5(0) \qquad a(0) = 0$$

$$= 0 \quad \blacksquare \qquad a(0) = 0$$

Note: If one or more zeros are contained in a product, then the whole product becomes zero.

Does the order in which you multiply two numbers together matter?

$$(-5)(-2) = 10, \text{ but also } (-2)(-5) = 10.$$

Similarly,

$$(-5)(2) = -10, \text{ but also } (2)(-5) = -10.$$

As you can see, the order in which you multiply is immaterial.

> **COMMUTATIVE LAW OF MULTIPLICATION**
> For any numbers a and b, $ab = ba$.

EXAMPLE 11

Simplify $(-4)(n)(-n)(8)(n)$.

Solution: In the product $(-4)(n)(-n)(8)(n)$ there are two negative factors. Since two is *even*, the product must be positive. Therefore,

$$(-4)(n)(-n)(8)(n) = +\underbrace{(4)(8)}\underbrace{(n)(n)(n)} \qquad \text{commutative law of multiplication}$$
$$= + \quad (32) \quad (n^3)$$
$$= 32n^3 \quad \blacksquare$$

EXAMPLE 12

Simplify $(-2)^3(-t)^2(5)$.

Solution: In the product, $(-2)^3(-t)^2(5) = (-2)(-2)(-2)(-t)(-t)(5)$, there are five negative factors. Since five is *odd*, the product must be negative. Therefore,

$$(-2)^3(-t)^2(5) = -\underbrace{(2)(2)(2)(5)}\underbrace{(t)(t)} \qquad \text{commutative law of multiplication}$$
$$= - \quad (40) \quad (t^2)$$
$$= -40t^2 \quad \blacksquare$$

EXERCISES 1.6

Find the following products.

1. $16(-3)$
2. $-5(6)$
3. $-12(8)$
4. $-17(-3)$
5. $-15(4)$
6. $22(-4)$
7. $-17(0)$
8. $0(-98)$
9. $-25(-1)$
10. $-1(16)$
11. $16(-2)(-3)$
12. $5(-1)(-3)$
13. $-5(-3)(6)$
14. $-6(-2)(-8)$
15. $-8(-2)(-3)(-1)$
16. $-13(-3)(1)(-1)$
17. $5(-6)(2)(-13)(0)$
18. $-13(-1)(-3)(-1)$
19. $93(-1)(-1)(-1)(1)$
20. $87(-1)(-1)(-1)(-1)$

Write each of the following as a product; then evaluate.

21. $(-3)^2$
22. 7^2
23. $(-4)^3$
24. $(-2)^5$

25. $-(-2)^4$

26. $-(-5)^3$

27. $(-1)^{12}$

28. $-(-1)^{19}$

29. $(-5)^2(2)^2$

30. $-3^2(-2)^3$

Simplify the following algebraic expressions.

31. $3 - y - y - y$

32. $-x - x + y - y$

33. $-t + 3 - t + 3 - t$

34. $-4 - w - w - w + 4 - w$

35. $2(x)(-9)$

36. $(-3)(y)(-4)$

37. $(-1)(n)(1)(n)(-1)$

38. $1(m)(-m)(-1)$

39. $(-3)(t)(0)(-t)$

40. $t(-3)(0)(t)(8)$

41. $(-5)(-1)(n)(-d)(-n)$

42. $(-3)(-p)(-p)(-1)(k)$

43. $x(y)(-x)(-x)(-y)$

44. $Z(-Z)(-Y)(-Z)(-Y)$

45. $(-3)^2(-x)^3$

46. $(-4)^2(-y)^4$

47. $-5^2(-x)^4(y)^2$

48. $(-2)^4(-w)^3(-w^3)$

Guess the solution to each of the following equations. Use the integers as possible replacements for the variables.

49. $2x = 32$

50. $-32d = 0$

51. $-3x = 48$

52. $7k = -35$

53. $-5n = -125$

54. $-8v = -96$

55. $-17y = 17$

56. $-51H = -51$

57. $x^2 = 9$

58. $y^3 = 8$

59. $x^2 = -9$

60. $y^3 = -8$

1.7 DIVIDING INTEGERS

Objectives

⊡ To find the quotients $\dfrac{a}{0}$ and $\dfrac{0}{a}$.

⊡ To divide a positive and a negative integer.

⊡ To divide two negative integers.

⊡ To simplify algebraic expressions containing division.

⊡ Divisions Containing a Zero

You already know how to divide two positive integers. For example, $\frac{15}{3} = 5$. Division problems can be *checked* by multiplying as follows:

$$\frac{15}{3} = 5 \quad \text{the } check: \quad 3(5) = 15$$

Suppose we didn't know that $\frac{15}{3} = 5$. We could say that $\frac{15}{3} = n$, where n is some unknown number. We could then write the check as follows:

$$\frac{15}{3} = n \quad \text{the } \textit{check}: \quad 3n = 15$$

Can you guess the solution to the equation $3n = 15$? Of course, $n = 5$, and therefore $\frac{15}{3} = 5$.

EXAMPLE 1

Find the quotient of $\frac{15}{0}$ using the check.

Solution: $\frac{15}{0} = n$, where n is some unknown number. Therefore,

$$\frac{15}{0} = n \quad \text{the } \textit{check}: \quad 0n = 15.$$

What number times 0 will equal 15? We already have a rule which states that zero times any number is zero: $a(0) = 0(a) = 0$. So, to find a number n that multiplies by zero to equal 15 is impossible. Therefore, we say that division by zero is *undefined*. Thus, $\frac{15}{0}$ is *undefined*. ■

For any number a, $\dfrac{a}{0}$ is undefined.

EXAMPLE 2

Find the quotient of $\frac{0}{15}$ using the check.

Solution: $\frac{0}{15} = n$, where n is some unknown number. Therefore,

$$\frac{0}{15} = n \quad \text{the } \textit{check}: \quad 15n = 0$$

Of course, the solution to the equation $15n = 0$ is $n = 0$ (since $a(0) = 0(a) = 0$). Thus, $\frac{0}{15} = 0$. ■

Note: Zero divided by any number (except 0) is always zero.

For any number a (except 0), $\dfrac{0}{a} = 0$.

Dividing a Positive and a Negative Integer

We can use the *check* to help us develop a rule for the division of a positive and negative integer.

What is the quotient of $\dfrac{-15}{3}$? Let $\dfrac{-15}{3} = n$, where n is some unknown number. Therefore,

$$\dfrac{-15}{3} = n \quad \text{the } \textit{check}\text{:} \quad 3n = -15$$

Of course, the solution to the equation, $3n = -15$, is $n = \boxed{-5}$, since $3(\boxed{-5}) = -15$ (multiplication rule 1). Thus, $\dfrac{-15}{3} = \boxed{-5}$.

Any negative number divided by a positive number is always negative.

What is the quotient of $\dfrac{15}{-3}$? Let $\dfrac{15}{-3} = n$, where n is some unknown number. Therefore,

$$\dfrac{15}{-3} = n \quad \text{the check:} \quad -3n = 15.$$

Of course, the solution to the equation, $-3n = 15$, is $n = \boxed{-5}$, since $-3(\boxed{-5}) = 15$ (multiplication rule 2). Thus, $\dfrac{15}{-3} = \boxed{-5}$.

Any positive number divided by a negative number is always negative. Comparing these quotients should convince you that for any numbers a and b,

$$\dfrac{-a}{b} = \dfrac{a}{-b} = -\dfrac{a}{b} \, (b \neq 0).$$

DIVISION RULE 1: Dividing a Positive and a Negative Number
Divide the magnitudes of the numbers and retain the negative $-$ sign.

EXAMPLE 3

Find the quotient of $\dfrac{-42}{6}$.

Solution:

$$\frac{|-42|}{|6|} = \frac{42}{6} = 7 \qquad \text{dividing their magnitudes}$$

Therefore, $\dfrac{-42}{6} = -7.$ ■

Note: Also, $\dfrac{42}{-6} = -7.$

⋰ Dividing Two Negative Integers

Again, we can use the *check* to help us develop a rule for the division of two negative integers.

What is the quotient of $\dfrac{-15}{-3}$? Let $\dfrac{-15}{-3} = n$, where n is some unknown number. Therefore,

$$\frac{-15}{-3} = n \quad \text{the } check: \quad -3n = -15.$$

Of course, the solution to this equation, $-3n = -15$, is $n = \boxed{5}$, since $-3\,\boxed{(5)} = -15$ (multiplication rule 1). Thus, $\dfrac{-15}{-3} = \boxed{5}$.

Any negative number divided by a negative number is always positive.

DIVISION RULE 2:   Dividing Two Negative Numbers
Divide the magnitudes of the numbers. The result is always positive.

EXAMPLE 4

Find the quotient of $\dfrac{-44}{-4}$.

Solution: $\dfrac{|-44|}{|-4|} = \dfrac{44}{4} = 11 \qquad \text{dividing their magnitudes}$

Therefore, $\dfrac{-44}{-4} = 11.$ ■

⚃ Some Properties of Division

Any number divided by one will remain the same.

$$\text{For any number } a, \frac{a}{1} = a.$$

However, any number divided by negative one will become its opposite.

$$\text{For any number } a, \frac{a}{-1} = -a.$$

EXAMPLE 5

Simplify **a)** $\dfrac{5xyz}{1}$ **b)** $\dfrac{5xyz}{-1}$

Solution:

a) $\dfrac{5xyz}{1} = 5xyz$ ■ $\dfrac{a}{1} = a$

b) $\dfrac{5xyz}{-1} = -5xyz$ ■ $\dfrac{a}{-1} = -a$

Any number (except 0) divided by itself equals one.

$$\text{For any number } a \text{ (except 0)}, \frac{a}{a} = 1.$$

However, any number (except 0) divided by its opposite equals negative one.

$$\text{For any number } a \text{ (except 0)}, \frac{a}{-a} = \frac{-a}{a} = -1.$$

EXAMPLE 6

Simplify **a)** $\dfrac{3y}{3y}, y \neq 0$ **b)** $\dfrac{3y}{-3y}, y \neq 0$

Solution:

a) $\dfrac{3y}{3y} = 1$, provided $y \neq 0$. ■ $\qquad \dfrac{a}{a} = 1$

b) $\dfrac{3y}{-3y} = \dfrac{-3y}{3y} = -1$, provided $y \neq 0$. ■ $\qquad \dfrac{a}{-a} = \dfrac{-a}{a} = -1$

Application

AVERAGE VALUE

The *average value* or *mean* of several measurements can be found by adding the measurements and then dividing this sum by the number of measurements in the sample.

FORMULA: **Average Value**

$\bar{x} = \dfrac{x_1 + x_2 + x_3 + \cdots + x_n}{n}$, where $\bar{x}$ (read "x bar") is the average value, x_1, x_2, x_3, and so on, are the measurements, and n is the number of measurements.

EXAMPLE 7

In an industrial experiment, the temperature of a coolant is recorded as it passes a particular point in a steel pipe. The readings are recorded in Table 1.3. Find the average temperature of the coolant as it passes this point.

TABLE 1.3

Reading number	Temperature
1	$-8°C$
2	$-15°C$
3	$-6°C$
4	$0°C$
5	$2°C$
6	$3°C$

Solution:　There are 6 temperature readings, so $n = 6$. Therefore,

$$\bar{x} = \frac{-8 + (-15) + (-6) + 0 + 2 + 3}{6}$$

$$\bar{x} = \frac{-24}{6} \qquad \text{adding the measurements}$$

$$\bar{x} = -4°C. \quad \blacksquare \qquad \text{dividing (division rule 1)}$$

EXERCISES 1.7

Find the following quotients.

1. $\dfrac{32}{4}$　　　　**2.** $\dfrac{72}{8}$　　　　**3.** $\dfrac{-16}{2}$　　　　**4.** $\dfrac{-24}{3}$

5. $\dfrac{-42}{-7}$　　　**6.** $\dfrac{-81}{-9}$　　　**7.** $\dfrac{125}{-5}$　　　**8.** $\dfrac{200}{-10}$

9. $\dfrac{-81}{-81}$　　　**10.** $\dfrac{-97}{-97}$　　　**11.** $\dfrac{13}{0}$　　　**12.** $\dfrac{73}{0}$

13. $\dfrac{79}{-1}$　　　**14.** $\dfrac{-13}{1}$　　　**15.** $\dfrac{0}{-83}$　　　**16.** $\dfrac{0}{-105}$

Simplify the following algebraic expressions.

17. $\dfrac{-w}{w}, w \neq 0$　　　　　　　　**18.** $\dfrac{-7t}{-7t}, t \neq 0$

19. $\dfrac{tv}{1}$　　　　　　　　　　　**20.** $\dfrac{-xyz^2}{1}$

21. $\dfrac{3p}{-1}$　　　　　　　　　　　**22.** $\dfrac{-6xy}{-1}$

23. $\dfrac{x-5}{x-5}, x \neq 5$　　　　　　　**24.** $\dfrac{t+7}{t+7}, t \neq -7$

Guess the solution to the following equations. Use the integers as possible replacements for the variables.

25. $\dfrac{n}{5} = -2$　　　　　　　　　**26.** $\dfrac{p}{3} = -9$

27. $\dfrac{36}{a} = -6$　　　　　　　　**28.** $\dfrac{54}{h} = -9$

29. $\dfrac{x}{-8} = -6$　　　　　　　　**30.** $\dfrac{y}{-12} = -4$

31. $\dfrac{-99}{f} = 11$　　　　　　　　**32.** $\dfrac{-144}{n} = 12$

33. $\dfrac{-64}{-t} = 2$　　　　　　　　**34.** $\dfrac{-81}{-R} = -1$

Find the average temperature.

35. Reading number	Temperature
1	$-10°F$
2	$-21°F$
3	$3°F$
4	$-6°F$
5	$9°F$

36. Reading number	Temperature
1	$-18°C$
2	$-16°C$
3	$-3°C$
4	$-9°C$
5	$-14°C$
6	$-6°C$
7	$-11°C$

1.8 PARENTHESES AND THE ORDER OF OPERATIONS

Objectives

⊡ To evaluate an algebraic expression given certain values for the variables.

⊡ To find the opposite of the sum of two integers.

⊡ The Order of Operations

In expressions like $5(-2)$ or $5 + (-2)$, the parentheses are used for the purpose of clarity. It may be confusing to omit the parentheses and write $5 \cdot -2$ or $5 + -2$. It would be hard to distinguish between subtraction symbols and negative signs, or addition symbols and plus signs, etc. The *parentheses* help organize our thinking.

Look at this problem: $3 + 2(7)$. Does this mean to add 3 to the product of 2 and 7?

$$3 + 2(7) = 3 + 14 = 17?$$

Or does it mean to multiply the sum of 3 and 2 by 7?

$$3 + 2(7) = 5(7) = 35?$$

To eliminate the confusion, mathematicians have agreed to always perform multiplications before additions. Thus, $3 + 2(7) = 17$, *not* 35. Suppose we had wanted to add $3 + 2$ first, and then multiply by 7. How might we show these operations? Use parentheses.

If you saw $(3 + 2)7$, it would mean to do what's inside the parentheses first *before* you multiply. So,

$$(3 + 2)7 = (5)7 = 35, \text{ whereas}$$

$$3 + 2(7) = 3 + 14 = 17.$$

Look at this problem: $5 \cdot 2^2$. Does this mean to multiply 5 by 2^2?

$$5 \cdot 2^2 = 5(4) = 20?$$

Or does it mean to multiply 5 times 2, and then square?

$$5 \cdot 2^2 = 10^2 = 100?$$

Again, notice the confusion. Mathematicians have agreed to always do powers first before multiplication. Thus, $5 \cdot 2^2 = 20$, *not* 100. Suppose we had wanted to multiply $5 \cdot 2$ first, and then square that result. How might we show this? Again, use parentheses.

If you saw $(5 \cdot 2)^2$, it would mean that you do what's inside the parentheses first, before you square. So,

$$(5 \cdot 2)^2 = (10)^2 = 100, \text{ whereas}$$

$$5 \cdot 2^2 = 5(4) = 20.$$

To eliminate the confusion that can occur, mathematicians have agreed upon the following **order of operations**.

ORDER OF OPERATIONS

1. Always perform the operations inside parentheses. If there is more than one set of parentheses, do the innermost parentheses first.

2. Take care of any powers (exponents).

3. Do the multiplications and divisions in order from left to right.

4. Do the additions and subtractions in order from left to right.

To evaluate an algebraic expression, simply replace the variables with the given values and follow the *order of operations*.

EXAMPLE 1

Evaluate $x - xy^3$, when $x = -2$ and $y = 3$.

Solution:

$$x - xy^3 = -2 - (-2)3^3 \quad\quad \text{replacing } x \text{ with } -2 \text{ and } y \text{ with } 3$$

$$= -2 - (-2)27 \quad\quad \text{powers first}$$

$$= -2 - (-54) \quad\quad \text{multiplication next}$$

$$= 52 \quad\blacksquare \quad\quad\quad\quad \text{subtraction last}$$

EXAMPLE 2

Evaluate $x(y - y(x + z))$ when $x = -1$, $y = 2$, and $z = -3$.

Solution:

$$x(y - y(x + z)) = -1(2 - 2[-1 + (-3)]) \quad\quad \text{replacing } x \text{ with } -1, y \text{ with } 2, \text{ and } z \text{ with } -3$$

$$= -1(2 - 2(-4)) \quad\quad \text{inner parentheses first, addition}$$

$$= -1(2 - (-8)) \quad\quad \text{now working inside the next set of parentheses, multiplication before subtraction}$$

$$= -1(10)$$

$$= -10 \quad\blacksquare \quad\quad \text{finally, multiplication}$$

EXAMPLE 3

Evaluate $\dfrac{xy + x}{y + x}$ when $x = -2$ and $y = 4$.

Solution: The division bar takes the place of parentheses. Another way to write the problem would be as follows:

$$(xy + x) \div (y + x)$$

The order of operations tells us that the operations inside the parentheses must be simplified before we can divide. Therefore, the numbers above and below the division bar must be simplified to one number before the division operation can take place.

$$\frac{xy + x}{y + x} = \frac{-2(4) + (-2)}{4 + (-2)} \quad \text{replacing } x \text{ with } -2 \text{ and } y \text{ with } 4$$

$$= \frac{-8 + (-2)}{2} \quad \text{for the top part, multiplication before addition}$$

$$= \frac{-10}{2}$$

$$= -5 \quad \blacksquare \quad \text{finally, division}$$

⊡ The Opposite of the Sum of Two Numbers

Does $-(x + y) = -x + (-y)$ for all values of x and y? Often, in technical mathematics, you must decide whether two algebraic expressions are equivalent. To do this, assign the variables some numbers and then evaluate the two expressions to see if they yield the same numerical value. When evaluating the expressions, be sure to follow the order of operations. For $-(x + y)$, you must add first and then take the opposite of the sum. For $-x + (-y)$, you must find the opposite of each number first and then add (see Table 1.4).

As you can see, $-(x + y)$ and $-x + (-y)$ yield the same numerical values when x and y are replaced with the arbitrary choices listed in Table 1.4. Therefore, it appears that the opposite of the sum of two numbers equals the sum of their opposites.

> For any numbers a and b, $-(a + b) = -a + (-b)$.

TABLE 1.4

Arbitrary choices for x and y	$-(x + y)$	$-x + (-y)$
$x = 2, y = 3$	$-(2 + 3) = -(5) = -5$	$-2 + (-3) = -5$
$x = -2, y = 3$	$-(-2 + 3) = -(1) = -1$	$-(-2) + (-3) = 2 + (-3) = -1$
$x = 2, y = -3$	$-(2 + (-3)) = -(-1) = 1$	$-2 + (-(-3)) = -2 + 3 = 1$
$x = -2, y = -3$	$-(-2 + (-3)) = -(-5) = 5$	$-(-2) + (-(-3)) = 2 + 3 = 5$

EXAMPLE 4

Find the opposite of $-x + 5$.

Solution: The opposite of $-x + 5$ is written $-(-x + 5)$, and

$$-(-x + 5) = -(-x) + (-5) \qquad -(a + b) = -a + (-b)$$
$$= x - 5 \quad \blacksquare \qquad \text{subtraction rule in reverse}$$

Note: Each term inside the parentheses changes to its opposite sign.

EXAMPLE 5

Find the opposite of $8 - y$.

Solution: The opposite of $8 - y$ is written $-(8 - y)$.

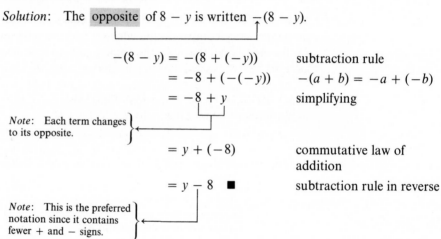

$$-(8 - y) = -(8 + (-y)) \qquad \text{subtraction rule}$$
$$= -8 + (-(-y)) \qquad -(a + b) = -a + (-b)$$
$$= -8 + y \qquad \text{simplifying}$$

Note: Each term changes to its opposite.

$$= y + (-8) \qquad \text{commutative law of addition}$$
$$= y - 8 \quad \blacksquare \qquad \text{subtraction rule in reverse}$$

Note: This is the preferred notation since it contains fewer + and − signs.

Application

TEMPERATURE CONVERSION

There are two temperature scales that are used in most technical work. One is the Fahrenheit scale and the other is the Celsius scale. On the Fahrenheit scale, water boils at 212°F and freezes at 32°F. On the Celsius scale, water boils at 100°C and freezes at 0°C (see Fig. 1.19). In many technical applications it is necessary to convert a Fahrenheit temperature to a Celsius temperature or vice versa. If you know the Fahrenheit temperature and wish to find the equivalent Celsius temperature, use this formula.

FORMULA: **Temperature Conversion—Fahrenheit to Celsius**

$$T_c = \frac{5(T_f - 32)}{9},$$

where T_c is the Celsius temperature and T_f is the Fahrenheit temperature.

If you know the Celsius temperature and wish to find the equivalent Fahrenheit temperature, use this formula.

FORMULA: **Temperature Conversion—Celsius to Fahrenheit**

$$T_f = \frac{9T_c}{5} + 32,$$

where T_f is the Fahrenheit temperature and T_c is the Celsius temperature.

EXAMPLE 6

The boiling point of a certain type of alcohol was found to be 176°F. What is the equivalent Celsius temperature?

Solution: Replace T_f with 176 and follow the order of operations.

$$T_c = \frac{5(T_f - 32)}{9} = \frac{5(176 - 32)}{9}$$

$$= \frac{5(144)}{9} \qquad \text{parentheses first, subtract}$$

$$= \frac{720}{9} \qquad \text{multiply next}$$

$$= 80°C \quad \blacksquare \qquad \text{finally, divide}$$

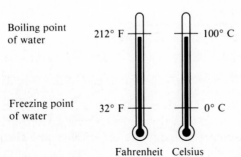

Boiling point
of water   212° F   100° C

Freezing point
of water   32° F   0° C

Fahrenheit Celsius   **Figure 1.19**

EXAMPLE 7

The freezing point of a certain type of alcohol was found to be $-120°C$. What is the equivalent Fahrenheit temperature?

Solution: Replace T_c with -120 and follow the order of operations.

$$T_f = \frac{9T_c}{5} + 32 = \frac{9(-120)}{5} + 32$$

$$= \frac{-1080}{5} + 32 \qquad \text{multiply first}$$

$$= -216 + 32 \qquad \text{divide next}$$

$$= -184°F \quad \blacksquare \qquad \text{finally, add}$$

EXERCISES 1.8

PEMDAS

Evaluate the following using the order of operations.

1. $(2)5 - 3$

2. $8(2) + 6$

3. $(3)4^2$

4. $(2 \cdot 6)^2$

5. $3(6)^2 + 8$

6. $4(5)^2 - (-4)$

7. $4(2)^3 - 8(7)$

8. $4^3 - 3(-9)$

9. $(13 - 16)^2$

10. $-(2 + (-5))^2$

11. $\dfrac{2(5)^2 - 5}{3 - 8}$

12. $\dfrac{2(8 - 5)^3}{6 - 3^2}$

Evaluate each of the following algebraic expressions by replacing x with 2, y with -1, and z with -3.

13. $xyz - z$

14. $xy(z - z)$

15. $4x^3$

16. $(4x)^3$

17. $3x^2 - 7y$

18. $3(x^2 - 7)y$

19. $4(x - y)^2$

20. $(4(x - y))^2$

21. $\dfrac{(x - y) + z}{y}$

22. $\dfrac{x - (y + z)}{y}$

23. $\dfrac{z^4}{(x - (x + z))^3}$

24. $\dfrac{z^4}{x - (x + z)^3}$

Rewrite each of the following without using parentheses. Simplify.

25. $-(10 + t)$

26. $-(y + 9)$

27. $-(-t + 7)$

28. $-(-3 + w)$

29. $-(p - q)$

30. $-(x + (-y))$

31. $-(-19 - h)$

32. $-(-y - 13)$

33. Does $2x + 3x = 5x$? Fill in the table below and state your conclusion.

	$2x + 3x$	$5x$
$x = 3$		
$x = 5$		
$x = -3$		
$x = -5$		

34. Does $(-x)^2 = -x^2$? Fill in the table below and state your conclusion. (Hint: $-x$ means $-1 \cdot x$).

	$(-x)^2$	$-x^2$
$x = 3$		
$x = 5$		
$x = -3$		
$x = -5$		

35. Does $(x + y)^2 = x^2 + y^2$? Fill in the table below and state your conclusion.

	$(x^2 + y)^2$	$x^2 + y^2$
$x = 2$ $y = 3$		
$x = -2$ $y = -3$		
$x = -2$ $y = 3$		
$x = 2$ $y = -3$		

36. Does $(xy)^2 = x^2 y^2$? Fill in the table below and state your conclusion.

	$(xy)^2$	$x^2 y^2$
$x = 2$ $y = 3$		
$x = -2$ $y = -3$		
$x = -2$ $y = 3$		
$x = 2$ $x = -3$		

Fill in the table below.

	Element	Melting Point °C	Melting Point °F
37.	Aluminum (Al)	660°C	
38.	Copper (Cu)	1080°C	
39.	Iron (Fe)	1540°C	
40.	Lead (Pb)		617°F
41.	Silver (Ag)		1760°F
42.	Zinc (Zn)		788°F

Fill in the table below.

Liquid	Boiling Point °C	Boiling Point °F	Freezing Point °C	Freezing Point °F
43. Mercury		671°F	−40°C	
44. Alcohol	75°C			−175°F

1.9   SOME PROPERTIES OF THE INTEGERS

Objectives

⊡ To simplify an algebraic expression using the associative law of addition.

⊡ To simplify an algebraic expression using the associative law of multiplication.

⊡ To simplify an algebraic expression using the distributive law.

⊡ To simplify an algebraic expression by combining similar terms.

⊡ The Associative Law of Addition

To add three numbers, a, b, and c, you could add a and b first, and then add c.

$$(a + b) + c$$

Or you could add a to the sum of b and c.

$$a + (b + c)$$

It doesn't matter which numbers are grouped together. The result will be the same. For example,

$$(-3 + 4) + 7 = 1 + 7 = 8, \text{ while}$$

$$-3 + (4 + 7) = -3 + 11 = 8.$$

The order of operations states that we must perform the operations inside the parentheses first. However, in some algebraic expressions, this is impossible. The commutative and associative laws of addition allow us to rearrange numbers and parentheses so that we may simplify an algebraic expression.

ASSOCIATIVE LAW OF ADDITION
For any numbers a, b, and c, $(a + b) + c = a + (b + c)$.

EXAMPLE 1

Simplify $5 + (x - 8)$.

Solution: Parentheses first! However, if we don't know what value x represents, how can we subtract 8 from it? To simplify this expression, we must proceed as follows:

$$5 + (x - 8) = (x - 8) + 5 \qquad \text{commutative law of addition}$$
$$= x + (-8 + 5) \qquad \text{associative law of addition}$$
$$= x - 3 \quad \blacksquare$$

EXAMPLE 2

Simplify $(t + 3) + (9 - t)$.

Solution: $(t + 3) + (9 - t) = (3 + 9) + (t + (-t))$ commutative and associative laws of addition

$$= 12 + 0$$
$$= 12 \quad \blacksquare$$

 There is no associative law of subtraction. $(a - b) - c$ and $a - (b - c)$ are not equivalent. For example,

$$\underbrace{(5 - 3)} - 2 = 2 - 2 = 0, \text{ while}$$

$$5 - \underbrace{(3 - 2)} = 5 - 1 = 4.$$

EXAMPLE 3

Simplify $6 - (y - 3)$.

Solution: Since there is no associative law of subtraction, we cannot yet rearrange the numbers and the parentheses. To simplify this expression, we must proceed as follows:

$$6 - (y - 3) = 6 + \underbrace{(-(y - 3))} \qquad \text{subtraction rule}$$

Note: Each term changes to its opposite.

$$= 6 + (-y + 3) \qquad -(a + b) = -a + (-b)$$

$$= (6 + 3) + (-y) \qquad \text{commutative and associative laws of addition}$$

$$= 9 - y \quad \blacksquare$$

Note: An addition sign + must precede a set of parentheses before the associative law of addition can be applied.

⊡ The Associative Law of Multiplication

To multiply three numbers, a, b, and c, you could multiply a and b first, and then multiply by c.

$$(ab)c$$

Or you could multiply a by the product of b and c.

$$a(bc)$$

For example,

$$(-3 \cdot 2)4 = (-6)4 = -24, \text{ while}$$

$$-3(2 \cdot 4) = -3(8) = -24.$$

The commutative and associative laws of multiplication allow us to rearrange numbers and parentheses so that we may simplify an algebraic expression.

ASSOCIATIVE LAW OF MULTIPLICATION
For any numbers a, b, and c, $(ab)c = a(bc)$.

EXAMPLE 4

Simplify $(8x)3$.

Solution: Parentheses first! However, if we don't know what number x represents, how can we multiply it by 8? To simplify this expression, we must proceed as follows:

$$
\begin{aligned}
(8x)3 &= 3(8x) && \text{commutative law of multiplication} \\
&= (3 \cdot 8)x && \text{associative law of multiplication} \\
&= 24x \quad \blacksquare
\end{aligned}
$$

EXAMPLE 5

Simplify $(3a)(-4ab)(2b)$.

Solution:

$$
\begin{aligned}
(3a)(-4ab)(2b) &= (-4 \cdot 3 \cdot 2)(a \cdot a)(b \cdot b) && \text{commutative and associative} \\
& && \text{laws of multiplication} \\
&= -24a^2b^2 \quad \blacksquare
\end{aligned}
$$

⠒ The Distributive Law

Look at the array of dots in Fig. 1.20. We could find the total number of dots without actually counting them by simply multiplying 4(5). There are 20 dots. Now suppose we took this array of dots and separated them into two groups, as shown in Fig. 1.21.

There are now 4(3) dots on the left and 4(2) dots on the right, or 4(3) + 4(2) dots altogether. Of course, there are still 4(5) dots. The 5 has simply been renamed as 3 + 2. Therefore,

$$4(5) = 4(3 + 2) = 4(3) + 4(2) = 20.$$

The expression 4(3 + 2) = 4(3) + 4(2) is an example of the distributive law of multiplication over addition. Notice how the 4 gets distributed to each of the numbers inside the parentheses.

$$4\,(3 + 2) = 4\,(3) + 4\,(2)$$

> **DISTRIBUTIVE LAW**
> For any numbers a, b, and c, $a(b + c) = ab + ac$.

EXAMPLE 6

Simplify $5(-4 + 9)$.

Solution: There are two approaches.

Method 1: $5(-4 + 9) = 5(5)$ parentheses first

$$= 25 \quad \blacksquare$$

Method 2: $5\,(-4 + 9) = 5\,(-4) + 5\,(9)$ distributive law

$$= -20 + 45 \qquad \text{multiply first, then add}$$

$$= 25 \quad \blacksquare$$

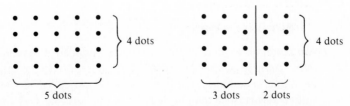

Figure 1.20   **Figure 1.21**

EXAMPLE 7

Simplify $3(x + 8)$.

Solution: $3(x + 8) = 3(x) + 3(8)$ distributive law

$\qquad\qquad\qquad = 3x + 24$ ■

EXAMPLE 8

Simplify $(5 - 7)(3y - 8)$.

Solution:

$(5 - 7)(3y - 8) = -2(3y - 8)$ parentheses first, subtract

$\qquad\qquad = -2(3y + (-8))$ subtraction rule

$\qquad\qquad = -2(3y) + (-2)(-8)$ distributive law

$\qquad\qquad = -6y + 16 \text{ or } 16 - 6y$ ■

EXAMPLE 9

Simplify $5 - 7(3y - 8)$.

Solution:

$5 - 7(3y - 8) = 5 + (-7(3y - 8))$ subtraction rule

$\qquad\qquad = 5 + (\ominus 21y \oplus 56)$ distributive law

Note: Be sure to change
signs when distributing
a negative.

$\qquad\qquad = 61 - 21y$ ■   associative and commutative
laws of addition

Note: Be sure you understand the difference between Example 8 and Example 9. For $5 - 7(3y - 8)$, you must *multiply first before you subtract* 7 from 5. If you don't, you violate the order of operations.

⸬ Similar Terms

An algebraic expression like $3x + 2x$ is called a **binomial** since it contains *two terms*, namely $3x$ and $2x$. Terms that have exactly the same variables and exactly the same exponents for these variables are called **similar terms**. $3x$ and $2x$ are similar terms since they both contain the variable x. However, $3x^2$ and $2x$ are *not* similar terms, since one x is squared and the other is not. Similar terms can always be combined into one term by using the *distributive law in reverse*.

DISTRIBUTIVE LAW IN REVERSE
For any numbers a, b, and c, $ab + ac = a(b + c)$.

EXAMPLE 10

Simplify $3x + 2x$.

Solution:

$$3x + 2x = x(3 + 2) \qquad \text{distributive law in reverse}$$

$$= x(5) \qquad \text{simplifying}$$

$$= 5x \quad \blacksquare \qquad \text{commutative law of multiplication}$$

Note: Also, $\underbrace{a + a + a + \cdots + a}_{n \text{ times}} = na$.

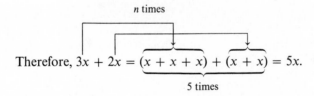

Therefore, $3x + 2x = \underbrace{(x + x + x) + (x + x)}_{5 \text{ times}} = 5x$.

EXAMPLE 11

Simplify $15y^2 - y^2$.

Solution:

$$15y^2 - y^2 = 15y^2 - 1y^2 \qquad 1(a) = a$$

$$= y^2(15 - 1) \qquad \text{distributive law in reverse}$$

$$= 14y^2 \quad \blacksquare \qquad \text{simplifying}$$

Note: To add similar terms simply add their numerical coefficients, and to subtract similar terms simply subtract their numerical coefficients.

EXAMPLE 12

Simplify $(12t^2 - t) - (8t - 2t^2)$.

Solution:

$$(12t^2 - t) \ominus (8t - 2t^2) = (12t^2 - t) + (\ominus 8t \oplus 2t^2) \qquad -(a + b) = -a + (-b)$$

Note: The negative distributes over the parentheses and changes each sign.

$$= (12t^2 + 2t^2) + (-8t - t) \qquad \text{commutative and associative laws of addition}$$

$$= 14t^2 - 9t \quad \blacksquare \qquad \text{combining similar terms}$$

EXAMPLE 13

Simplify $5(4 - 2w) + 3(15 - w)$.

Solution:

$$5(4 - 2w) + 3(15 - w) = (20 - 10w) + (45 - 3w) \qquad \text{distributive law}$$

$$= (20 + 45) + (-10w - 3w) \qquad \text{commutative and associative laws of addition}$$

$$= 65 - 13w \quad \blacksquare \qquad \text{combining similar terms}$$

EXERCISES 1.9

Use the commutative and associative laws of addition and multiplication to simplify the following algebraic expressions.

1. $(x + 5) + 8$ **2.** $(y - 8) - 8$

3. $-6 + (y - 9)$ **4.** $-8 + (3y + 2)$

5. $3 - (5 + 4x)$ **6.** $-9 - (2t - 9)$

7. $(15 - t) - (t - 11)$ **8.** $-(y + 8) + (-7 + y)$

9. $2(5t)(3t)$ **10.** $6t^2(-8)$

11. $9a(-3a^2)$ **12.** $-3t^3(-8t)$

13. $2x(-3y)(-xy)$ **14.** $5x(-2y)(-3xy)$

15. $(4mn)^2$ **16.** $3xy(-3y)^3$

Use the distributive law to eliminate the parentheses and simplify.

17. $-5(x + 4)$ **18.** $-4(d - 4)$

19. $3p(p - 2)$ **20.** $-3t^2(1 - t)$

21. $(5 - v)v$ **22.** $(7 - 3h)2h$

23. $(3 + 2x)(-6x^2)$ **24.** $(9 - t)(-t)$

25. $(3 - 9)(2 + 3n)$ **26.** $(4 + 3)(5y - 4)$

27. $-6 + 5(4 - w)$ **28.** $-11 - 4(3w + 1)$

29. $-5(3y)(4 - 3y)$ **30.** $7(-y - 3)(-5y)$

Simplify by combining similar terms.

31. $2y + 7y$ **32.** $12f - 6f$

33. $2n - 8n$ **34.** $3a - (-12a)$

35. $17a^3 - a^3$ **36.** $t^3 - 13t^3$

37. $x - (3x - 4)$ **38.** $p - (p - 5)$

39. $2y + 2y(5 - y)$ **40.** $y - 6(3y - 4)$

41. $(6x - 4) + 5(1 - 3x)$ **42.** $3(5 - x) - (4x - 1)$

43. $(14n^2 - 3n^2)(4n - 3)$ **44.** $(z - 2z)(6z - 2)$

45. $3(2x - y) + 2(3x + y)$ **46.** $4(a - 2b) - 5(b + 4a)$

47. $5w(3 - f) - (w - f)$ **48.** $3t(t - u) - (t - u)t$

For exercises 49–52, refer to Fig. 1.22.

49. Write a formula that describes the distance y.

50. Using the formula from exercise 49, find y when $x = 3$.

51. Write a formula that describes the distance w.

52. Using the formula from exercise 51, find z when $w = 17$.

For exercises 53–56, refer to Fig. 1.23.

53. Write a formula that describes the distance z.

54. Using the formula from exercise 53, find z when $y = -1$.

55. Write a formula that describes the distance x.

56. Using the formula from exercise 55, find y when $x = 16$.

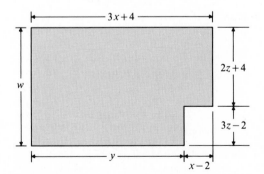

Figure 1.22

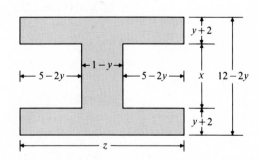

Figure 1.23

Chapter Review

The following review exercises are grouped according to the objectives that should have been mastered in Chapter 1. Work each problem carefully. If any weaknesses appear, immediately refer to and read the subsection that matches that objective.

1.1 SIMPLE ALGEBRAIC EXPRESSIONS

Objectives

⊡ To write a simple algebraic expression that contains one operation.

Write as an algebraic expression.

1. five plus some number

2. some number minus some different number

3. some number times the same number

④. the product of some number and nine

5. eight divided by some number

⊡ To write an algebraic expression that contains more than one operation.

Write as an algebraic expression.

6. four times a number decreased by one

7. five less twice some number

⑧. five less than twice some number

⊡ To evaluate an algebraic expression that contains one operation.

Evaluate.

9. $24n$, if $n = 12$

10. $\dfrac{196}{x}$, if $x = 7$

⊡ To simplify an algebraic expression containing repeated addition.

Simplify.

11. $y + y + y + y + y + y$

⑫. $x + x + x + x + y + y + y$

13. $t + t + t + 3 + 9$

1.2 A FIRST LOOK AT EQUATIONS AND FORMULAS

Objectives

⊡ To write a simple equation from words.

Write as an equation.

14. The product of three and some voltage V equals eighteen.

15. Twice some force F plus eight equals twenty.

⊡ To guess the solution to a simple equation.

Guess the solution.

⑯ $t + 3 = 14$

17. $9x = 72$

18. $4n - 2 = 18$

⊡ To write a formula that describes a special relationship between the dimensions of a technical drawing.

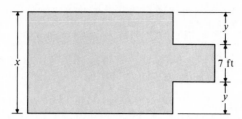

Figure 1.24

19. Given Fig. 1.24, write a formula that describes the distance x.

20. Using the formula from exercise 19, solve for y when $x = 19$ ft.

1.3 THE NUMBER LINE AND THE INTEGERS

Objectives

⊡ To represent certain words with integers.

21. If an *upward* force is designated by a positive integer, then what integer would designate a *downward* force of 600 lb?

⊡ To find the opposite of an integer.

Simplify.

22. $-(18)$ **23.** $-(-8)$

⊡ To find the magnitude of an integer.

Simplify.

24. $|-9|$ **25.** $|14|$

⊡ To place $>$, $<$, or $=$ between a pair of integers.

26. Place $>$, $<$, or $=$ between -8 ___ -2.

27. Place $>$, $<$, or $=$ between -2 ___ -8.

28. Place $>$, $<$, or $=$ between $|-5|$ ___ $-(-5)$.

1.4 ADDING INTEGERS

Objectives

⊡ To add two negative integers.

Find the sum.

29. $(-13) + (-24)$

⊡ To add a negative and a positive integer.

Find the sum.

30. $14 + (-18)$

31. $-14 + 18$

⊡ To simplify an algebraic expression containing addition.

Simplify.

32. $y + 7 + y + (-2)$

33. $-x + 6 + (-3) + x + (-4)$

1.5 SUBTRACTING INTEGERS

Objectives

⊡ To subtract two integers.

Find the difference.

34. $-6 - 12$ **35.** $-6 - (-12)$

⊡ To simplify an algebraic expression containing addition and subtraction.

Simplify.

36. $x + (-y) - (-4)$

37. $6 - a - 4 - (-a) - 3$

38. $-y - (-7) + y - 8 - (-y)$

1.6 MULTIPLYING INTEGERS AND RAISING INTEGERS TO POWERS

Objectives

⊡ To multiply a positive and a negative integer.

Find the product.

39. $8(-3)$ **40.** $-15(4)$

⊡ To multiply two negative integers.

Find the product.

41. $(-12)(-6)$

⊡ To write exponential expressions as products and then to evaluate them.

Evaluate.

42. $(-5)^2$ **43.** -5^2 **44.** $(-2)^6$ **45.** $(-2)^5$

⊡ To simplify an algebraic expression containing repeated addition of negatives.

Simplify.

46. $-t - t - x - x - x$

⊡ To simplify an algebraic expression containing multiplication and exponents.

Simplify.

47. $x(-1)(-x)(1)$

48. $3(x)(y)(x)(0)$

49. $(-2)(n)(8)(n)(-n)$

50. $(-3)^2(-t)^5(2)$

1.7 DIVIDING INTEGERS

Objectives

⊡ To find the quotients $\dfrac{a}{0}$ and $\dfrac{0}{a}$.

Find the quotient.

51. $\dfrac{10}{0}$ **52.** $\dfrac{0}{10}$

⊡ To divide a positive and a negative integer.

Find the quotient.

53. $\dfrac{-63}{9}$ **54.** $\dfrac{63}{-9}$

⊡ To divide two negative integers.

Find the quotient.   **55.** $\dfrac{-63}{-9}$

⊡ To simplify algebraic expressions containing division.

Simplify.   **56a.** $\dfrac{-2xy}{1}$ **56b.** $\dfrac{-2xy}{-1}$

57a. $\dfrac{-2xy}{-2xy}, x \neq 0, y \neq 0$ **57b.** $\dfrac{2xy}{-2xy}, x \neq 0, y \neq 0$

1.8 PARENTHESES AND THE ORDER OF OPERATIONS

Objectives

⊡ To evaluate an algebraic expression given certain values for the variables.

Evaluate.   **58.** $x + xy^4$ if $x = -3, y = -2$

59. $x(y + (x - z)^3)$ if $x = -1, y = 2, z = 3$

60. $\dfrac{(x - y)^3}{(x + y)^3}$ if $x = 2, y = -3$

⊡ To find the opposite of the sum of two integers.

Simplify.   **61.** $-(x + 3)$ **62.** $-(3 - x)$

1.9 SOME PROPERTIES OF THE INTEGERS

Objectives

⊡ To simplify an algebraic expression using the associative law of addition.

Simplify.   **63.** $6 + (x - 2)$

64. $(y + 8) + (5 - y)$

65. $9 - (t - 4)$

⊡ To simplify an algebraic expression using the associative law of multiplication.

Simplify.   **66.** $(-9x)3$ **67.** $(2a)(-6ab)(-ab)$

⊡ To simplify an algebraic expression using the distributive law.

Simplify.   **68.** $6(-2 + 5)$ **69.** $5(x + 3)$

70. $(8 - 10)(4y - 3)$ **71.** $8 - 10(4y - 3)$

⊡ To simplify an algebraic expression by combining similar terms.

Simplify.   **72.** $9y + 3y$

73. $7x^2 - x^2$

74. $(10p^2 - 3p) - (5p - 3p^2)$

75. $6(2 - 3z) + 9(2z - 1)$

If you have worked through the Review Exercises and corrected any weaknesses, then you are ready to take the following Chapter Test.

1. Write an algebraic expression for "twice some number less seven."

2. Write an equation for "three less than twice some pressure P_1 equals twenty-one," and then guess the solution to the equation.

Perform the indicated operations.

3. $15 + (-12)$ 4. $-19 + 3$ 5. $-13 + (-7)$

6. $32 - 49$ 7. $-8 - 7$ 8. $-12(3)$

9. $15 - (-3)$ 10. -8^2 11. $-6(-14)$

12. $(-3)^4$ 13. $(-8)^2$ 14. $-(-6) + 3$

15. $(-3)^3$ 16. $\dfrac{-72}{9}$ 17. $\dfrac{-18}{-6}$

18. $\dfrac{-21}{0}$ 19. $\dfrac{0}{-15}$ 20. $\dfrac{56}{-8}$

Evaluate the given algebraic expressions replacing x with -2 and y with 1.

21. $3x^2$ 22. $(3x)^2$ 23. $x^3 + y^3$ 24. $(x + y)^3$

Simplify.

25. $T + T + T$ 26. $-x + 17 + (-12) + x$

27. $-a - (-b)$ 28. $x - 9 - (-x) + (-4)$

29. $-p - (-4) - p - 4 - p$ 30. $-5(-a)(-a)(-4)(a)$

31. $(-2)^2(-x)^3$ 32. $-5^2(-x)^3(-y)^3$

33. $\dfrac{-3st}{-1}$ 34. $\dfrac{-3st}{3st}, s \neq 0, t \neq 0$

35. $-5 + (x + 4)$ 36. $6 - (-2 \mp 5y)$

37. $-6x(2 - x)$ 38. $-(4 + 3x)$

39. $12x - 16x$ 40. $13a - (-a)$

41. $(6s^2 - 3s^2)(2s - 2)$ 42. $2h - 3h(2 - 3h)$

43. $6f(2 - f) - f(f - 6)$ 44. $3x(x - y) + y(x - y)$

45. Given Fig. 1.25, write a formula that describes the distance x.

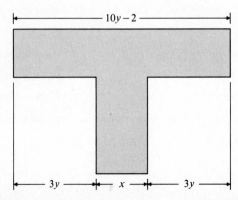

Figure 1.25

$$2h - 6h + 9h$$
$$-4n + 9h$$
$$5h$$

YOU KNOW YOU WANT TO GO OUT WITH ME!

2

WORKING WITH RATIONAL NUMBERS

2.1 BUILDING UP AND BREAKING DOWN RATIONAL NUMBERS

Objectives

⊡ To locate a rational number on a number line.

⊡ To find an equivalent fraction whose denominator is known.

⊡ To reduce a fraction to lowest terms.

⊡ Rational Numbers

In arithmetic, you learned about numbers called **fractions**. For example, if a piece of lumber is cut (divided) into three equal parts, each part would be one-third of the whole piece. The fraction one-third is represented by the quotient $\frac{1}{3}$ (see Fig. 2.1). Two-thirds ($\frac{2}{3}$) represents *two* of the three equal parts (see Fig. 2.2). In technical mathematics, we often refer to fractions like $\frac{1}{3}$ and $\frac{2}{3}$ as **rational numbers**.

DEFINITION

Any number in the form $\overset{\text{numerator}}{\underset{\text{denominator}}{\dfrac{a}{b}}}$ is called a **rational number**, provided a and b are integers and $b \neq 0$.

All integers are rational numbers since they can be written in the form $\frac{a}{b}$ by using $b = 1$. For example, $-8 = \dfrac{-8}{1}$, $16 = \dfrac{16}{1}$, and so on.

EXAMPLE 1

Locate the following rational numbers on a number line.

a) $\dfrac{11}{8}$ b) $\dfrac{-11}{8}$ c) $\dfrac{-1}{-2}$

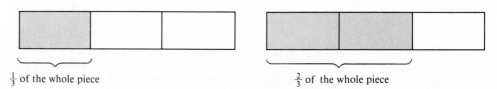

$\frac{1}{3}$ of the whole piece $\frac{2}{3}$ of the whole piece

Figure 2.1   **Figure 2.2**

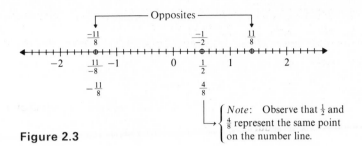

Figure 2.3

Note: Observe that $\frac{1}{2}$ and $\frac{4}{8}$ represent the same point on the number line.

Solution:

a) For $\dfrac{-11}{8}$, divide each unit space to the right of zero into *eight* equal parts. The *eleventh* mark to the right of zero is $\dfrac{11}{8}$ (see Fig. 2.3).

b) Recall from Chapter 1 that $\dfrac{-a}{b} = \dfrac{a}{-b} = -\dfrac{a}{b}$. Thus, $\dfrac{-11}{8} = \dfrac{11}{-8} = -\dfrac{11}{8}$ all represent the same *negative* number. So, divide each unit space to the *left* of zero into eight equal parts. The eleventh mark to the *left* of zero is $\dfrac{-11}{8}$ (see Fig. 2.3).

c) Recall from Chapter 1 that $\dfrac{-a}{-b} = \dfrac{a}{b}$. Thus, $\dfrac{-1}{-2}$ and $\dfrac{1}{2}$ both represent the same *positive* number. Of course, halfway between 0 and 1 is the correct location for $\dfrac{-1}{-2}$, or $\dfrac{1}{2}$ (see Fig. 2.3). ■

Note: When working with rational numbers, it is always important to keep in mind the following:

1. $\dfrac{-a}{b} = \dfrac{a}{-b} = -\dfrac{a}{b}$

2. $\dfrac{-a}{-b} = \dfrac{a}{b}$

⊡ Fundamental Property of Fractions

In Fig. 2.3, you should have observed that $\frac{1}{2}$ and $\frac{4}{8}$ have the same location on the number line. Rational numbers that represent the same quantity are called **equivalent fractions**. The rational numbers $\frac{1}{2}$ and $\frac{4}{8}$ are examples of equivalent

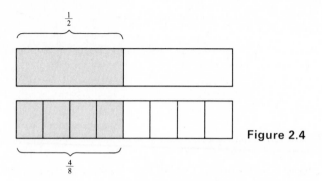

Figure 2.4

fractions (see Fig. 2.4). Note that if we multiply both the numerator and the denominator of $\frac{1}{2}$ by 4, we obtain $\frac{4}{8}$.

$$\frac{1}{2} = \frac{4\,(1)}{4\,(2)} = \frac{4}{8}$$

equivalent
fractions

We can always "build up" a rational number to higher terms by multiplying its numerator and denominator by the same quantity. We refer to this property of rational numbers as the **fundamental property of fractions**.

FUNDAMENTAL PROPERTY OF FRACTIONS

$$\frac{a}{b} = \frac{ka}{kb},$$

provided $b \neq 0, k \neq 0$.

In Section 2.4, you will discover that the fundamental property of fractions plays a critical part in the addition and subtraction of rational numbers.

EXAMPLE 2

Change $\dfrac{3}{4}$ to an equivalent fraction whose denominator is 20.

Solution:

$$\frac{3}{4} = \frac{?}{20}$$

We must multiply
4 by 5 to obtain 20.

Thus,

$$\frac{3}{4} = \frac{5\,(3)}{5\,(4)} = \frac{15}{20}. \quad \blacksquare \qquad \text{fundamental property of fractions}$$

equivalent
fractions

EXAMPLE 3

Change $\dfrac{-6}{7}$ to an equivalent fraction whose denominator is 42.

Solution:

$$\frac{-6}{7} = \frac{?}{42}$$

We must multiply 7
by 6 to obtain 42.

Thus,

$$\frac{-6}{7} = \frac{6\,(-6)}{6\,(7)} = \frac{-36}{42}. \quad \blacksquare \qquad \text{fundamental property of fractions}$$

EXAMPLE 4

Change $\dfrac{2x}{-5}$ to an equivalent fraction whose denominator is $15x$.

Solution:

$$\frac{2x}{-5} = \frac{?}{15x}$$

We must multiply
-5 by $-3x$ to obtain $15x$.

Thus,

$$\frac{2x}{-5} = \frac{-3x\,(2x)}{-3x\,(-5)} = \frac{-6x^2}{15x}. \quad \blacksquare \qquad \text{fundamental property of fractions}$$

EXAMPLE 5

Change $\dfrac{9y}{4x}$ to an equivalent fraction whose denominator is $12x^2y$.

Solution:

$$\frac{9y}{4x} = \frac{?}{12x^2y}$$

We must multiply $4x$
by $3xy$ to obtain $12\,x^2y$.

Thus,

$$\frac{9y}{4x} = \frac{3xy\,(9y)}{3xy\,(4x)} = \frac{27xy^2}{12x^2y}. \quad \blacksquare \qquad \text{fundamental property of fractions}$$

∴ Cancellation Law

By applying the fundamental property of fractions in reverse, we can "break down" a rational number to its lowest terms by *cancelling* common factors from the numerator and denominator. In doing so, we obtain a *reduced fraction*. We will refer to the fundamental property of fractions, when used in reverse, as the **cancellation law**.

CANCELLATION LAW

$$\frac{ka}{kb} = \frac{\overset{1}{\cancel{k}a}}{\underset{1}{\cancel{k}b}} = \frac{a}{b},$$

provided $b \neq 0, k \neq 0$.

For example, to show that $\frac{4}{8}$ reduces to $\frac{1}{2}$, we factor 4 and 8 and then cancel any common factors that appear in the numerator and denominator. To **factor** a number simply means to write it as a product. Thus,

$$\frac{4}{8} = \frac{2 \cdot 2}{2 \cdot 4} = \frac{\overset{1}{\cancel{2}} \cdot 2}{\underset{1}{\cancel{2}} \cdot 4} = \frac{2}{4}.$$

factoring ⟶ ⟵ cancelling common factors

However,

$$\frac{2}{4} = \frac{2 \cdot 1}{2 \cdot 2} = \frac{\overset{1}{\cancel{2}} \cdot 1}{\underset{1}{\cancel{2}} \cdot 2} = \frac{1}{2}.$$

factoring again ⟶ ⟵ cancelling common factors again

Or, more quickly,

$$\frac{4}{8} = \frac{4 \cdot 1}{4 \cdot 2} = \frac{\overset{1}{\cancel{4} \cdot 1}}{\underset{1}{\cancel{4} \cdot 2}} = \frac{1}{2}.$$

factoring ———   ——— cancelling common factors

As you can see, it is always advantageous to work with the largest common factor of the numerator and denominator.

 The numerator and denominator of a fraction must be entirely written as *products* if you are to apply the cancellation law and reduce a fraction correctly. For example,

$$\frac{5 \cdot 10}{5} = \frac{\overset{1}{\cancel{5}} \cdot 10}{\underset{1}{\cancel{5}}} \text{ is } correct, \text{ while}$$

$$\frac{5 \oplus 10}{5} = \frac{\overset{1}{\cancel{5}} \oplus 10}{\underset{1}{\cancel{5}}} \text{ is } wrong.$$

EXAMPLE 6

Reduce $\dfrac{15}{20}$ to lowest terms.

Solution:

$$\frac{15}{20} = \frac{5 \cdot 3}{5 \cdot 4} \qquad \text{factoring}$$

$$= \frac{\overset{1}{\cancel{5}} \cdot 3}{\underset{1}{\cancel{5}} \cdot 4} \qquad \text{cancellation law}$$

$$= \frac{3}{4} \quad \blacksquare \qquad \text{simplifying}$$

EXAMPLE 7

Reduce $\dfrac{-48}{36}$ to lowest terms.

Solution:

$$\frac{-48}{36} = -\frac{48}{36} \qquad \frac{-a}{b} = -\frac{a}{b}$$

$$= -\frac{12 \cdot 4}{12 \cdot 3} \qquad \text{factoring}$$

$$= -\frac{\overset{1}{\cancel{12}} \cdot 4}{\underset{1}{\cancel{12}} \cdot 3} \qquad \text{cancellation law}$$

$$= -\frac{4}{3} \quad \blacksquare \qquad \text{simplifying}$$

Note: In arithmetic, you may have written $\frac{4}{3}$ as the mixed number $1\frac{1}{3}$. However, in technical mathematics, we generally leave our answer as a single fraction in the form $\frac{a}{b}$. Thus, we prefer $-\frac{4}{3}$ rather than $-1\frac{1}{3}$. We will discuss mixed numbers further in Section 2.5.

EXAMPLE 8

Reduce $\dfrac{-6x}{-2x}$ to lowest terms.

Solution:

$$\frac{-6x}{-2x} = \frac{6x}{2x} \qquad \frac{-a}{-b} = \frac{a}{b}$$

$$= \frac{2 \cdot 3 \cdot x}{2 \cdot x} \qquad \text{factoring}$$

$$= \frac{\overset{1}{\cancel{2}} \cdot 3 \cdot \overset{1}{\cancel{x}}}{\underset{1}{\cancel{2}} \cdot \underset{1}{\cancel{x}}} \qquad \text{cancellation law}$$

$$= \frac{3}{1} \qquad \text{simplifying}$$

$$= 3 \quad \blacksquare \qquad \frac{a}{1} = a$$

Note: We cancel the xs under the assumption that $x \neq 0$. For if $x = 0$, $\dfrac{x}{x} = \dfrac{0}{0}$, and *we can never cancel zeros*. Thus, $\dfrac{-6x}{-2x} = 3$, provided $x \neq 0$. In this chapter, we will assume that any variable in the denominator of a fraction does not equal zero.

EXAMPLE 9

Reduce $\dfrac{8x^2y}{12xy^2}$ to lowest terms.

Solution:

$$\frac{8x^2y}{12xy^2} = \frac{4\cdot 2\cdot x\cdot x\cdot y}{4\cdot 3\cdot x\cdot y\cdot y} \qquad \text{factoring}$$

$$= \frac{\overset{1}{\cancel{4}}\cdot 2\cdot \overset{1}{\cancel{x}}\cdot x\cdot \overset{1}{\cancel{y}}}{\underset{1}{\cancel{4}}\cdot 3\cdot \underset{1}{\cancel{x}}\cdot y\cdot \underset{1}{\cancel{y}}} \qquad \text{cancellation law}$$

$$= \frac{2x}{3y} \quad\blacksquare \qquad \text{simplifying}$$

EXERCISES 2.1

State the rational number associated with each letter on the following number lines.

1.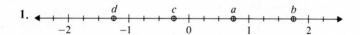

2.

Find the missing numerator so that the fractions are equivalent.

3. $\dfrac{5}{8} = \dfrac{?}{24}$

4. $\dfrac{3}{7} = \dfrac{?}{35}$

5. $\dfrac{-11}{9} = \dfrac{?}{36}$

6. $\dfrac{-5}{3} = \dfrac{?}{18}$

7. $\dfrac{3}{-4} = \dfrac{?}{32}$

8. $\dfrac{3}{-5} = \dfrac{?}{-25}$

9. $\dfrac{1}{4} = \dfrac{?}{12y}$

10. $\dfrac{1}{3} = \dfrac{?}{15t}$

11. $\dfrac{-5y}{4x} = \dfrac{?}{16x^2}$

12. $\dfrac{-5z}{6y} = \dfrac{?}{24y^2}$

13. $\dfrac{7}{xy} = \dfrac{?}{3x^2y}$

14. $\dfrac{3}{-xy} = \dfrac{?}{5xy^2}$

15. $\dfrac{2u}{-7t} = \dfrac{?}{28t^3u}$

16. $\dfrac{-x}{15z} = \dfrac{?}{45xz^3}$

17. $10 = \dfrac{?}{3x}$

18. $7 = \dfrac{?}{9w^2}$

19. $2x = \dfrac{?}{5xy^2}$

20. $3p = \dfrac{?}{8p^2q^2}$

Reduce the following fractions to lowest terms.

21. $\dfrac{16}{18}$

22. $\dfrac{24}{28}$

23. $\dfrac{-36}{20}$

24. $\dfrac{40}{-24}$

25. $\dfrac{-8x}{-12x}$

26. $\dfrac{-21y}{-49y}$

27. $\dfrac{45t^2}{60t}$

28. $\dfrac{45w}{25w^2}$

29. $\dfrac{22x^2y}{-33y}$

30. $\dfrac{-27d^2}{45d^2g}$

31. $\dfrac{ab}{2a^2b^2}$

32. $\dfrac{4ab}{28a^2b}$

33. $\dfrac{-16p^2q}{2p^2q}$

34. $\dfrac{-64m^2n^2}{4m^2n^2}$

35. $\dfrac{-3xy^2z}{-21xyz^2}$

36. $\dfrac{-42x^2yz^2}{-30xyz^2}$

2.2 MULTIPLYING AND DIVIDING RATIONAL NUMBERS

Objectives

⊡ To multiply fractions.

⊡ To divide fractions.

⊡ Multiplying Fractions

As can be seen in Fig. 2.5, half *of* two-thirds is one-third. The word *of*, when used in this context, means *times*. Thus, *half of two-thirds* can be represented by the *product*

$$\frac{1}{2} \cdot \frac{2}{3}.$$

Now, what procedure can we use to multiply $\frac{1}{2} \cdot \frac{2}{3}$ so that we will obtain the answer $\frac{1}{3}$? Suppose we just multiply the numbers in the numerators and then multiply the

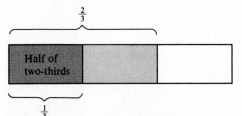

Figure 2.5

numbers in the denominators as follows:

$$\frac{1}{2} \cdot \frac{2}{3} = \frac{1 \cdot 2}{2 \cdot 3} \qquad \text{multiplying numerators and multiplying denominators}$$

$$= \frac{1 \cdot \overset{1}{\cancel{2}}}{\underset{1}{\cancel{2}} \cdot 3} \qquad \text{cancellation law}$$

$$= \frac{1}{3} \qquad \text{simplifying}$$

To multiply two fractions, simply multiply their numerators and then multiply their denominators. Remember to cancel first (if possible) before you actually find any products. This way, the answer will always be in reduced form.

MULTIPLICATION RULE FOR FRACTIONS

$$\frac{a}{b} \cdot \frac{c}{d} = \frac{ac}{bd},$$

provided $b \neq 0$, $d \neq 0$.

EXAMPLE 1

Find the product $\dfrac{6}{7} \cdot \dfrac{3}{10}$.

Solution:

$$\frac{6}{7} \cdot \frac{3}{10} = \frac{6 \cdot 3}{7 \cdot 10} \qquad \text{multiplication rule for fractions}$$

$$= \frac{\overset{3}{\cancel{6}} \cdot 3}{7 \cdot \underset{5}{\cancel{10}}} \qquad \text{cancellation law} \left(\frac{6}{10} = \frac{2 \cdot 3}{2 \cdot 5} = \frac{3}{5} \right)$$

$$= \frac{9}{35} \quad \blacksquare \qquad \text{simplifying}$$

EXAMPLE 2

Find the product $\dfrac{-2}{3} \cdot \dfrac{9}{10}$.

Solution: This product contains *one* negative. Therefore, the result must be *negative*.

$$\frac{-2}{3} \cdot \frac{9}{10} = -\frac{2}{3} \cdot \frac{9}{10}$$

$$= -\frac{2 \cdot 9}{3 \cdot 10} \qquad \text{multiplication rule for fractions}$$

$$= -\frac{\overset{1}{\cancel{2}} \cdot \overset{3}{\cancel{9}}}{\underset{1}{\cancel{3}} \cdot \underset{5}{\cancel{10}}} \qquad \text{cancellation law} \left(\frac{2}{10} = \frac{\cancel{2} \cdot 1}{\cancel{2} \cdot 5} = \frac{1}{5} \text{ and } \frac{9}{3} = \frac{\cancel{3} \cdot 3}{\cancel{3} \cdot 1} = \frac{3}{1} \right)$$

$$= -\frac{3}{5} \quad \blacksquare \qquad \text{simplifying}$$

EXAMPLE 3

Find the product $\left(\dfrac{3x}{-5} \right)(-10x)$.

Solution: In this product, there are *two* negatives. Thus, the product must be *positive*.

$$\left(\frac{3x}{-5} \right)(-10x) = + \left(\frac{3x}{5} \right)(10x)$$

$$= \frac{3x}{5} \cdot \frac{10x}{1} \qquad a = \frac{a}{1}$$

$$= \frac{3x \cdot 10x}{5 \cdot 1} \qquad \text{multiplication rule for fractions}$$

$$= \frac{3x \cdot \overset{2x}{\cancel{10x}}}{\underset{1}{\cancel{5}} \cdot 1} \qquad \text{cancellation law} \left(\frac{10x}{5} = \frac{\cancel{5} \cdot 2x}{\cancel{5} \cdot 1} = \frac{2x}{1} \right)$$

$$= \frac{6x^2}{1} \qquad \text{simplifying}$$

$$= 6x^2 \quad \blacksquare \qquad \frac{a}{1} = a$$

EXAMPLE 4

Find the product $\dfrac{12x}{7y} \cdot \dfrac{14y^2}{3x^2}$.

Solution:

$$\frac{12x}{7y} \cdot \frac{14y^2}{3x^2} = \frac{12x \cdot 14y^2}{7y \cdot 3x^2} \qquad \text{multiplication rule for fractions}$$

$$= \frac{\overset{4}{\cancel{12x}} \cdot \overset{2y}{\cancel{14y^2}}}{\underset{1}{\cancel{7y}} \cdot \underset{x}{\cancel{3x^2}}} \qquad \text{cancellation law}$$

$$\left(\frac{12x}{3x^2} = \frac{\cancel{3x} \cdot 4}{\cancel{3x} \cdot x} = \frac{4}{x} \text{ and } \frac{14y^2}{7y} = \frac{\cancel{7y} \cdot 2y}{\cancel{7y} \cdot 1} = \frac{2y}{1} \right)$$

$$= \frac{8y}{x} \quad \blacksquare \qquad \text{simplifying}$$

EXAMPLE 5

Simplify $\left(-\dfrac{3n}{2} \right)^3$.

Solution: $\left(-\dfrac{3n}{2} \right)^3 = \underbrace{\left(-\dfrac{3n}{2} \right)\left(-\dfrac{3n}{2} \right)\left(-\dfrac{3n}{2} \right)}_{\text{3 factors}} \qquad a^3 = a \cdot a \cdot a$

Note: 3 negative factors will yield a negative result.

$$= -\left(\frac{3n}{2} \right)\left(\frac{3n}{2} \right)\left(\frac{3n}{2} \right)$$

$$= -\frac{3n \cdot 3n \cdot 3n}{2 \cdot 2 \cdot 2} \qquad \begin{array}{l}\text{multiplication rule} \\ \text{for fractions}\end{array}$$

$$= -\frac{27n^3}{8} \quad \blacksquare \qquad \text{simplifying}$$

⊡ Dividing Fractions

Two numbers whose product equals 1 are said to be **reciprocals** of each other. For example, the reciprocal of $\frac{5}{6}$ is $\frac{6}{5}$ since

$$\frac{5}{6} \cdot \frac{6}{5} = \frac{5 \cdot 6}{6 \cdot 5} \qquad \text{multiplication rule for fractions}$$

$$= \frac{\overset{1}{\cancel{5}} \cdot \overset{1}{\cancel{6}}}{\underset{1}{\cancel{6}} \cdot \underset{1}{\cancel{5}}} \qquad \text{cancellation law}$$

$$= 1 \qquad \text{simplifying}$$

DEFINITION

The **reciprocal** of $\dfrac{a}{b}$ is $\dfrac{b}{a}$ and $\dfrac{a}{b} \cdot \dfrac{b}{a} = 1$, provided $a \neq 0, b \neq 0$.

Now, suppose we wished to divide $\frac{3}{8}$ by $\frac{5}{6}$. There are two ways to symbolize the division:

$$\frac{3}{8} \div \frac{5}{6} \quad \text{or} \quad \frac{\dfrac{3}{8}}{\dfrac{5}{6}}$$

We could proceed with the division as follows:

$$\frac{3}{8} \div \frac{5}{6} = \frac{\dfrac{3}{8}}{\dfrac{5}{6}} = \frac{\dfrac{3}{8} \cdot \dfrac{6}{5}}{\dfrac{5}{6} \cdot \dfrac{6}{5}}$$

fundamental property of fractions

Note: If we multiply both numerator and denominator by the **reciprocal of $\frac{5}{6}$**, we will force a division by 1, and anything divided by 1 is itself.

$$= \frac{\dfrac{3}{8} \cdot \dfrac{6}{5}}{1}$$

$\dfrac{a}{b} \cdot \dfrac{b}{a} = 1$

$$= \frac{3}{8} \Big| \frac{6}{5}$$

$\dfrac{a}{1} = a$

Note: We have now changed the division problem into a multiplication problem.

Observe that

change the divisor to its reciprocal

$$\frac{3}{8} \div \frac{5}{6} = \frac{3}{8} \cdot \frac{6}{5}$$

change division to multiplication

Thus, to divide two fractions, change the division symbol to a multiplication symbol and multiply the dividend by the reciprocal of the divisor.

DIVISION RULE FOR FRACTIONS

$$\frac{a}{b} \div \frac{c}{d} = \frac{\dfrac{a}{b}}{\dfrac{c}{d}} = \frac{a}{b} \cdot \frac{d}{c}, \text{ provided } b \neq 0,$$
$$c \neq 0,$$
$$d \neq 0.$$

dividend — $\dfrac{a}{b}$

divisor — $\dfrac{c}{d}$

reciprocal of divisor — $\dfrac{d}{c}$

EXAMPLE 6

Find the quotient $\dfrac{3}{8} \div \dfrac{5}{6}$.

Solution: $\dfrac{3}{8} \div \dfrac{5}{6} = \dfrac{3}{8} \cdot \dfrac{6}{5}$ division rule for fractions

$= \dfrac{3 \cdot 6}{8 \cdot 5}$ multiplication rule for fractions

$= \dfrac{3 \cdot \overset{3}{\cancel{6}}}{\underset{4}{\cancel{8}} \cdot 5}$ cancellation law $\left(\dfrac{6}{8} = \dfrac{2 \cdot 3}{2 \cdot 4} = \dfrac{3}{4}\right)$

$= \dfrac{9}{20}$ ■   simplifying

EXAMPLE 7

Find the quotient $\dfrac{5}{-6} \div \dfrac{5}{12}$.

Solution: $\dfrac{5}{-6} \div \dfrac{5}{12} = \dfrac{5}{-6} \cdot \dfrac{12}{5}$ division rule for fractions

Note: This product contains *one* negative. Thus, the result must be negative.

$= -\dfrac{5}{6} \cdot \dfrac{12}{5}$

$= -\dfrac{5 \cdot 12}{6 \cdot 5}$ multiplication rule for fractions

$= -\dfrac{\overset{1}{\cancel{5}} \cdot \overset{2}{\cancel{12}}}{\underset{1}{\cancel{6}} \cdot \underset{1}{\cancel{5}}}$ cancellation law $\left(\dfrac{5}{5} = 1 \text{ and } \dfrac{12}{6} = 2\right)$

$= -2$ ■   simplifying

EXAMPLE 8

Find the quotient $\dfrac{\dfrac{t}{-6}}{\dfrac{-3t}{16}}$.

Solution: $\quad \dfrac{\dfrac{t}{-6}}{\dfrac{-3t}{16}} = \dfrac{t}{-6} \cdot \dfrac{16}{-3t}$ division rule for fractions

Note: This product contains *two* negatives. Thus, the result must be positive.

$$= + \frac{t}{6} \cdot \frac{16}{3t}$$

$$= \frac{t \cdot 16}{6 \cdot 3t} \qquad \text{multiplication rule for fractions}$$

$$= \frac{\overset{1}{\cancel{t}} \cdot \overset{8}{\cancel{16}}}{\underset{3}{\cancel{6}} \cdot \underset{3}{\cancel{3t}}} \qquad \text{cancellation law}$$

$$\left(\frac{t}{3t} = \frac{t \cdot 1}{t \cdot 3} = \frac{1}{3} \text{ and } \frac{16}{6} = \frac{2 \cdot 8}{2 \cdot 3} = \frac{8}{3} \right)$$

$$= \frac{8}{9} \quad \blacksquare \qquad \text{simplifying}$$

EXAMPLE 9

Find the quotient $\left(\dfrac{2m}{9} \right) \div (6m^2)$.

Solution: $\quad \left(\dfrac{2m}{9} \right) \div (6m^2) = \dfrac{2m}{9} \div \dfrac{6m^2}{1} \qquad a = \dfrac{a}{1}$

$$= \frac{2m}{9} \cdot \frac{1}{6m^2} \qquad \text{division rule for fractions}$$

$$= \frac{2m \cdot 1}{9 \cdot 6m^2} \qquad \text{multiplication rule for fractions}$$

$$= \frac{\overset{1}{\cancel{2m}} \cdot 1}{9 \cdot \underset{3m}{\cancel{6m^2}}} \qquad \text{cancellation law}$$

$$\left(\frac{2m}{6m^2} = \frac{2m \cdot 1}{2m \cdot 3m} = \frac{1}{3m} \right)$$

$$= \frac{1}{27m} \quad \blacksquare \qquad \text{simplifying}$$

Application

WOOD SCREWS

The common distance between the peaks of the threads in a wood screw is called its **pitch**. If a wood screw has 4 threads per inch, then it must have a $\frac{1}{4}$ inch pitch (see Fig. 2.6). Notice that the *number of threads per inch* and the *pitch* of the wood screw are reciprocals of each other.

> **FORMULA: Pitch of a wood screw**
>
> $$P = \frac{1}{N},$$
>
> where P is the pitch in inches, and N is the number of threads per inch.

The **lead** of a wood screw is the distance the screw advances into the wood in one complete turn of the screw. In a *single threaded screw* the lead equals the pitch (see Fig. 2.7), while in a *double threaded screw* the lead equals *twice* the pitch (see Fig. 2.8).

Figure 2.6

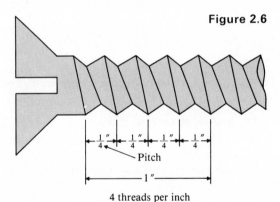

4 threads per inch

Figure 2.7

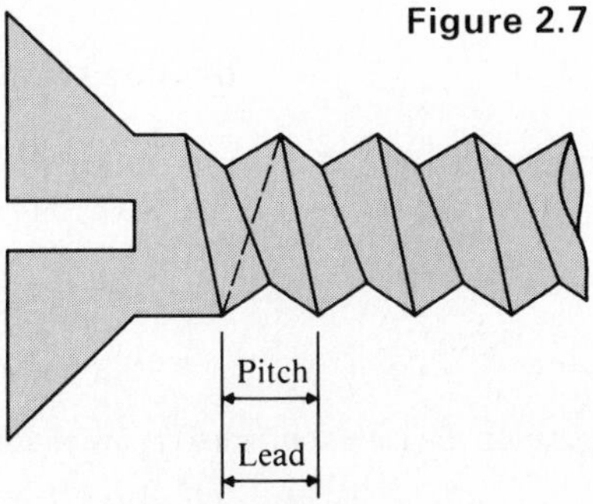

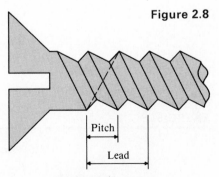

Single-threaded screw
Lead = pitch: $L = P$

Figure 2.8

Double-threaded screw
Lead = twice pitch: $L = 2P$

EXAMPLE 10

What is the pitch and lead of a double threaded screw with 16 threads per inch?

Solution:

$$\text{pitch} \quad P = \frac{1}{N} = \frac{1}{16} \text{ in.}$$

$$\text{lead} \quad L = 2P = 2\left(\frac{1}{16}\right)$$

$$= \frac{2}{1} \cdot \frac{1}{16} \qquad a = \frac{a}{1}$$

$$= \frac{2 \cdot 1}{1 \cdot \overset{1}{\cancel{16}}_{8}} \qquad \text{multiplication rule for fractions and cancellation law}$$

$$= \frac{1}{8} \text{ in.} \quad \blacksquare \qquad \text{simplifying}$$

EXAMPLE 11

Referring to the wood screw described in Example 10, how many turns are necessary to fully tighten the screw if its overall length is 4 in.?

Solution: The *lead* represents the distance the screw advances into the wood in *one* complete turn of the screw. From Example 10, *one* complete turn advances this screw $\frac{1}{8}$ in. To determine how many turns are necessary to fully tighten this 4-inch screw, we must find how many one-eighths are contained in four. Thus, we divide as follows:

$$4 \div \frac{1}{8} = \frac{4}{1} \div \frac{1}{8} \qquad a = \frac{a}{1}$$

$$= \frac{4}{1} \cdot \frac{8}{1} \qquad \text{division rule for fractions}$$

$$= 32 \text{ complete turns} \quad \blacksquare \qquad \text{multiplication rule for fractions}$$

EXERCISES 2.2

Perform the indicated operation. Be sure your answers are reduced to lowest terms.

1. $\dfrac{-2}{3} \cdot \dfrac{1}{5}$

2. $\dfrac{1}{-6} \cdot \dfrac{5}{7}$

3. $\dfrac{-2}{3} \div \dfrac{1}{5}$

4. $\dfrac{1}{-6} \div \dfrac{5}{7}$

5. $(-5)\left(\dfrac{13}{-4}\right)$

6. $(-3)\left(\dfrac{-13}{2}\right)$

7. $\dfrac{7}{10} \cdot \dfrac{5}{7}$

8. $\dfrac{3}{4} \cdot \dfrac{14}{9}$

9. $\dfrac{18}{5} \cdot \dfrac{10}{-9}$

10. $\dfrac{-6}{15} \cdot \dfrac{5}{-4}$

11. $\dfrac{4}{-9} \div \dfrac{-10}{21}$

12. $\dfrac{-8}{35} \div \dfrac{6}{5}$

13. $\dfrac{-13}{4} \div 13$

14. $-6 \div \dfrac{14}{-3}$

15. $\dfrac{\frac{5}{8}}{\frac{25}{12}}$

16. $\dfrac{\frac{-9}{10}}{\frac{-21}{20}}$

17. $\left(-\dfrac{2}{3}\right)^2$

18. $\left(-\dfrac{5}{8}\right)^2$

19. $\left(\dfrac{-4}{3}\right)^3$

20. $-\left(\dfrac{1}{-2}\right)^3$

21. $\dfrac{3t}{4} \cdot \dfrac{-5t}{8}$

22. $\dfrac{-v}{3} \cdot \dfrac{-5v}{7}$

23. $\dfrac{3p}{4q} \cdot \dfrac{8p}{9q}$

24. $\dfrac{-3x}{2y} \cdot \dfrac{-8x}{-15y}$

25. $\left(\dfrac{zy}{-3}\right)\left(\dfrac{-8}{5y}\right)$

26. $\left(\dfrac{xy}{2}\right)\left(\dfrac{-1}{4x}\right)$

27. $\dfrac{3}{-5m} \div \dfrac{9}{10m}$

28. $\dfrac{-5p^2}{11} \div \dfrac{15p^2}{22}$

29. $\dfrac{\frac{-xy}{9}}{\frac{-y^2}{6}}$

30. $\dfrac{\frac{ef}{-18}}{\frac{-e^2}{24}}$

31. $\left(\dfrac{12m^3}{5}\right) \div (-8m)$

32. $(6m^2n^2) \div \left(\dfrac{9mn}{10}\right)$

33. $\left(\dfrac{-3x}{5}\right)^2$

34. $\left(\dfrac{xy}{-4}\right)^2$

35. $\left(-\dfrac{x}{2y}\right)^3$

36. $\left(\dfrac{-2}{-5m}\right)^3$

37. $-10\left(\dfrac{3x}{4} \div \dfrac{15xy}{-8}\right)$

38. $3y\left(\dfrac{5y^2}{6} \div \dfrac{15y}{16}\right)$

39. $\left(\dfrac{-10}{7xyz} \cdot \dfrac{21x^2y}{2} \right) \div (5xy)$　　　　**40.** $(-6x^2y) \div \left(\dfrac{3x^2}{7y} \cdot \dfrac{14y^2}{15x} \right)$

41. What is the pitch and lead of a double threaded screw with 24 threads per inch?

42. What is the pitch and lead of a single threaded screw with 36 threads per inch?

43. Referring to the wood screw described in exercise 41, how many turns are necessary to fully tighten the screw if its overall length is 3 in.?

44. Referring to the wood screw described in exercise 42, how many turns are necessary to fully tighten the screw if its overall length is x in.?

2.3　ADDING AND SUBTRACTING RATIONAL NUMBERS WITH THE SAME DENOMINATOR

Objectives

⊡ To add fractions that have the same denominator.

⊡ To subtract fractions that have the same denominator.

⊡ To add or subtract fractions whose denominators are opposites.

⊡ Adding Fractions with the Same Denominator

Do you remember how to add fractions that have the same denominator? To add $\frac{3}{6} + \frac{2}{6}$, observe Fig. 2.9.

$$\overset{\displaystyle\longrightarrow \text{add the numerators} \longrightarrow}{\dfrac{3}{6} + \dfrac{2}{6} = \dfrac{3 + 2}{6} = \dfrac{5}{6}}$$

$\longrightarrow$ keep the same denominator $\longrightarrow$

To add fractions having the same denominator, simply add their numerators and retain the same denominator.

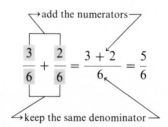

Figure 2.9

ADDITION RULE FOR FRACTIONS

$$\frac{a}{c} + \frac{b}{c} = \frac{a+b}{c}, \text{ provided } c \neq 0.$$

EXAMPLE 1

Find the sum $\dfrac{3}{16} + \dfrac{5}{16}$.

Solution:

$$\frac{3}{16} + \frac{5}{16} = \frac{3+5}{16} \qquad \text{addition rule for fractions}$$

$$= \frac{8}{16} \qquad \text{simplifying}$$

Note: Always write the answer in reduced form.

$$= \frac{\overset{1}{\cancel{8}} \cdot 1}{\underset{1}{\cancel{8}} \cdot 2} \qquad \text{cancellation law}$$

$$= \frac{1}{2} \;\blacksquare \qquad \text{reduced form}$$

EXAMPLE 2

Find the sum $\dfrac{13}{8} + \dfrac{-1}{8}$.

Solution:

$$\frac{13}{8} + \frac{-1}{8} = \frac{13 + (-1)}{8} \qquad \text{addition rule for fractions}$$

$$= \frac{12}{8} \qquad \text{simplifying}$$

$$= \frac{3 \cdot \overset{1}{\cancel{4}}}{2 \cdot \underset{1}{\cancel{4}}} \qquad \text{cancellation law}$$

$$= \frac{3}{2} \;\blacksquare \qquad \text{reduced form}$$

EXAMPLE 3

Find the sum $\dfrac{-2y}{5} + \dfrac{3y}{5}$.

Solution: $\quad \dfrac{-2y}{5} + \dfrac{3y}{5} = \dfrac{-2y + 3y}{5}$ $\qquad$ addition rule for fractions

$\qquad\qquad\qquad = \dfrac{y}{5}$ ■ $\qquad$ combining similar terms
$(-2y + 3y = 1y = y)$

EXAMPLE 4

Find the sum $\dfrac{5}{9x^2} + \dfrac{4}{9x^2}$.

Solution: $\quad \dfrac{5}{9x^2} + \dfrac{4}{9x^2} = \dfrac{5 + 4}{9x^2}$ $\qquad$ addition rule for fractions

$\qquad\qquad\qquad = \dfrac{9}{9x^2}$ $\qquad$ simplifying

$\qquad\qquad\qquad = \dfrac{\overset{1}{\cancel{9}} \cdot 1}{\underset{1}{\cancel{9}} \cdot x^2}$ $\qquad$ cancellation law

$\qquad\qquad\qquad = \dfrac{1}{x^2}$ ■ $\qquad$ reduced form

⚃ Subtracting Fractions with the Same Denominator

Do you remember how to subtract fractions that have the same denominator? To subtract $\frac{3}{6} - \frac{2}{6}$, observe Fig. 2.10.

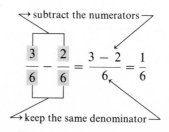

To subtract fractions having the same denominator, simply subtract their numerators and retain the same denominator.

SUBTRACTION RULE FOR FRACTIONS

$$\frac{a}{c} - \frac{b}{c} = \frac{a - b}{c}, \text{ provided } c \neq 0.$$

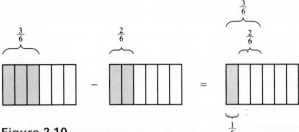

Figure 2.10

EXAMPLE 5

Find the difference $\dfrac{5}{16} - \dfrac{3}{16}$.

Solution:

$$\dfrac{5}{16} - \dfrac{3}{16} = \dfrac{5-3}{16} \qquad \text{subtraction rule for fractions}$$

$$= \dfrac{2}{16} \qquad \text{simplifying}$$

Note: Always write an answer in reduced form.

$$= \dfrac{\overset{1}{\cancel{2}} \cdot 1}{\underset{1}{\cancel{2}} \cdot 8} \qquad \text{cancellation law}$$

$$= \dfrac{1}{8} \quad \blacksquare \qquad \text{reduced form}$$

EXAMPLE 6

Find the difference $\dfrac{5}{12} - \dfrac{-11}{12}$.

Solution:

$$\dfrac{5}{12} - \dfrac{-11}{12} = \dfrac{5 - (-11)}{12} \qquad \text{subtraction rule for fractions}$$

$$= \dfrac{16}{12} \qquad \text{simplifying}$$

$$= \dfrac{\overset{1}{\cancel{4}} \cdot 4}{\underset{1}{\cancel{4}} \cdot 3} \qquad \text{cancellation law}$$

$$= \dfrac{4}{3} \quad \blacksquare \qquad \text{reduced form}$$

EXAMPLE 7

Find the difference $\dfrac{2x}{3} - \dfrac{8x}{3}$.

Solution: $\dfrac{2x}{3} - \dfrac{8x}{3} = \dfrac{2x - 8x}{3}$ subtraction rule for fractions

$\qquad\qquad\quad = \dfrac{-6x}{3}$ combining similar terms in the numerator

$\qquad\qquad\quad = -\dfrac{\overset{1}{\cancel{3}} \cdot 2x}{\underset{1}{\cancel{3}} \cdot 1}$ cancellation law

$\qquad\qquad\quad = -2x$ ■   reduced form

EXAMPLE 8

Find the difference $\dfrac{4y}{4y} - \dfrac{3}{4y}$.

Solution: $\dfrac{4y}{4y} - \dfrac{3}{4y} = \dfrac{4y - 3}{4y}$ ■   subtraction rule for fractions

Note: To cancel as follows: $\dfrac{\overset{1}{\cancel{4y}} - 3}{\underset{1}{\cancel{4y}}}$ is *wrong*. Since the numerator is not written

entirely as a product, the cancellation law does not apply. The expression $\dfrac{4y - 3}{4y}$
is in reduced form.

⬚ Adding and Subtracting Fractions whose Denominators Are Opposites

How might we find the sum $\dfrac{3}{5} + \dfrac{1}{-5}$ or the difference $\dfrac{3}{5} - \dfrac{1}{-5}$? You should first note that the denominators in each problem are *opposites*. We could proceed with the addition problem as follows:

$$\frac{3}{5} + \frac{1}{-5} = \frac{3}{5} + \left(-\frac{1}{5}\right) \qquad \frac{a}{-b} = -\frac{a}{b}$$

$$= \frac{3}{5} \ominus \frac{1}{5} \qquad\qquad \text{subtraction rule in reverse,}$$
$$a + (-b) = a - b$$

Note: We have now changed the addition problem to a subtraction problem containing the *same* denominators.

Observe that

$$\overset{\displaystyle \longrightarrow \text{change addition to subtraction} \longrightarrow}{\frac{3}{5} \oplus \frac{1}{\boxed{-5}}} = \frac{3}{5} \ominus \frac{1}{\boxed{5}} = \frac{3-1}{5} = \frac{2}{5}.$$

$\longrightarrow$ change this denominator to its opposite $\Longrightarrow$

Thus,

$$\frac{a}{c} \oplus \frac{b}{\ominus c} = \frac{a}{c} + \left(-\frac{b}{c}\right) = \frac{a}{c} \ominus \frac{b}{c} = \frac{a-b}{c}.$$

$\longrightarrow$ To add fractions whose denominators are opposites, subtract the second numerator from the first and retain the denominator of the first fraction. $\longrightarrow$

To find the difference $\dfrac{3}{5} - \dfrac{1}{-5}$, we could proceed as follows:

$$\frac{3}{5} - \frac{1}{-5} = \frac{3}{5} - \left(-\frac{1}{5}\right) \qquad \frac{a}{-b} = -\frac{a}{b}$$

$$= \frac{3}{5} \oplus \frac{1}{5} \qquad \begin{array}{l} \text{subtraction rule,} \\ a - (-b) = a + b \end{array}$$

Note: We have now changed the subtraction problem to an addition problem containing the *same* denominators.

Observe that

$$\overset{\displaystyle \longrightarrow \text{change subtraction to addition} \longrightarrow}{\frac{3}{5} \ominus \frac{1}{\boxed{-5}}} = \frac{3}{5} \oplus \frac{1}{\boxed{5}} = \frac{3+1}{5} = \frac{4}{5}.$$

$\longrightarrow$ change this denominator to its opposite $\Longrightarrow$

Thus,

$$\frac{a}{c} \ominus \frac{b}{\ominus c} = \frac{a}{c} - \left(-\frac{b}{c}\right) = \frac{a}{c} \oplus \frac{b}{c} = \frac{a+b}{c}.$$

$\longrightarrow$ To subtract fractions whose denominators are opposites, add the numerators and retain the denominator of the first fraction. $\longrightarrow$

EXAMPLE 9

Find the sum $\dfrac{2w^2}{3} + \dfrac{7w^2}{-3}$.

Solution:

$$\dfrac{2w^2}{3} \oplus \dfrac{7w^2}{-3} = \dfrac{2w^2}{3} \ominus \dfrac{7w^2}{3}$$

$$= \dfrac{2w^2 - 7w^2}{3} \qquad \text{subtraction rule for fractions}$$

$$= \dfrac{-5w^2}{3} \text{ or } -\dfrac{5w^2}{3} \quad \blacksquare \qquad \text{combining similar terms in the numerator}$$

EXAMPLE 10

Find the difference $\dfrac{5xy}{6z} - \dfrac{xy}{-6z}$.

Solution:

$$\dfrac{5xy}{6z} \ominus \dfrac{xy}{-6z} = \dfrac{5xy}{6z} \oplus \dfrac{xy}{6z}$$

$$= \dfrac{5xy + xy}{6z} \qquad \text{addition rule for fractions}$$

$$= \dfrac{6xy}{6z} \qquad \text{combining similar terms in the numerator}$$

$$= \dfrac{\overset{1}{\cancel{6}} \cdot xy}{\underset{1}{\cancel{6}} \cdot z} \qquad \text{cancellation law}$$

$$= \dfrac{xy}{z} \quad \blacksquare \qquad \text{reduced form}$$

EXERCISES 2.3

Perform the indicated operations. Be sure your answer is in reduced form.

1. $\dfrac{7}{8} + \dfrac{5}{8}$

2. $\dfrac{5}{9} + \dfrac{7}{9}$

3. $\dfrac{-9}{10} + \dfrac{3}{10}$

4. $\dfrac{3}{8} + \dfrac{-5}{8}$

5. $\dfrac{22}{55} - \dfrac{7}{55}$

6. $\dfrac{-4}{15} - \dfrac{1}{15}$

7. $\dfrac{1}{6} - \dfrac{-5}{6}$

8. $\dfrac{-1}{25} - \dfrac{-6}{25}$

9. $\dfrac{4}{15} - \dfrac{8}{-15}$

10. $\dfrac{2}{14} - \dfrac{5}{-14}$

11. $\dfrac{7}{20} + \dfrac{3}{-20}$

12. $\dfrac{-3}{11} + \dfrac{8}{-11}$

13. $\dfrac{5x}{6} + \dfrac{x}{6}$

14. $\dfrac{5t}{12} + \dfrac{7t}{12}$

15. $\dfrac{y}{8} - \dfrac{3y}{8}$

16. $\dfrac{20h^2}{21} - \dfrac{2h^2}{21}$

17. $\dfrac{-7}{15n} + \dfrac{1}{15n}$

18. $\dfrac{-5}{3n} + \dfrac{-4}{3n}$

19. $\dfrac{x}{5y} + \dfrac{-5y}{5y}$

20. $\dfrac{y}{6y} + \dfrac{-5}{6y}$

21. $\dfrac{3m}{2n} - \dfrac{n}{2n}$

22. $\dfrac{-3v}{4t} - \dfrac{-6}{4t}$

23. $\dfrac{2w^2}{xy} - \dfrac{3w^2}{xy}$

24. $\dfrac{-4t^2}{pq} - \dfrac{-5t^2}{pq}$

25. $\dfrac{5a}{12x} - \dfrac{a}{-12x}$

26. $\dfrac{-z^3}{6} + \dfrac{5z^3}{-6}$

27. $\left(\dfrac{2x}{3y} + \dfrac{8x}{3y}\right) + \dfrac{4x}{-3y}$

28. $\left(\dfrac{a}{8y^2} - \dfrac{3a}{-8y^2}\right) - \dfrac{7a}{8y^2}$

29. $\dfrac{2x}{-10ab} - \left(\dfrac{3x}{10ab} + \dfrac{5x}{10ab}\right)$

30. $\dfrac{9k}{5} - \left(\dfrac{2k}{5} - \dfrac{7k}{-5}\right)$

31. $\left(\dfrac{5}{6m^2} - \dfrac{1}{6m^2}\right) - \dfrac{2}{3m^2}$

32. $\dfrac{5y^2}{4a^2} - \left(\dfrac{5y^2}{8a^2} + \dfrac{y^2}{8a^2}\right)$

33. $3m\left(\dfrac{5}{9n^2} + \dfrac{7}{9n^2}\right)$

34. $\dfrac{8x^2}{9}\left(\dfrac{-3}{4x} - \dfrac{6}{4x}\right)$

35. $\left(\dfrac{5x^2}{4y} - \dfrac{x^2}{-4y}\right) \div \dfrac{3x}{2y}$

36. $\left(\dfrac{-2x}{5y} + \dfrac{4x}{-5y}\right) \div (3xy)$

2.4 ADDING AND SUBTRACTING RATIONAL NUMBERS WITH DIFFERENT DENOMINATORS

Objectives

▣ To completely factor a given number.

▣ Given fractions with different denominators, to find the least common denominator (LCD).

▣ To add or subtract fractions that have different denominators.

▣ Defining the Least Common Denominator (LCD)

The addition and subtraction rules for fractions allow us to add or subtract fractions that have the *same* denominator. How might we find the sum $\frac{1}{2} + \frac{1}{3}$?

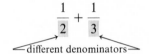

If we thought of $\frac{1}{2}$ as $\frac{3}{6}$ and $\frac{1}{3}$ as $\frac{2}{6}$, we could then find the sum as we did in Section 2.3 (see Fig. 2.11).

To add or subtract fractions with different denominators, we must first change them to equivalent fractions that share a *common denominator*. For the

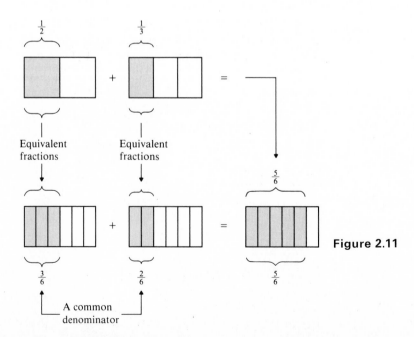

Figure 2.11

fractions $\frac{1}{2}$ and $\frac{1}{3}$, we used a common denominator of 6. However, any number exactly divisible by both 2 and 3 could serve as a common denominator.

$$6, 12, 18, 24, 30, 36, 42, 48, \ldots$$
possible common denominators for $\frac{1}{2}$ and $\frac{1}{3}$

If we use 12 as a common denominator, we can also find the sum as follows:

equivalent fractions

$$\frac{1}{2} + \frac{1}{3} = \frac{6}{12} + \frac{4}{12} = \frac{10}{12} = \frac{5 \cdot 2}{6 \cdot 2} = \frac{5}{6}$$

equivalent fractions   cancellation law

Six is called the **least common denominator (LCD)** for the fractions $\frac{1}{2}$ and $\frac{1}{3}$ since it is the *smallest* number exactly divisible by both 2 and 3. When adding or subtracting fractions with different denominators, it is always best to work with the LCD.

DEFINITION
The **least common denominator (LCD)** is the smallest number exactly divisible by each denominator.

For some addition and subtraction problems, the LCD can be found by inspection. However, if the denominators are fairly large numbers, you will need to **completely factor** each denominator in order to determine the LCD.

To completely factor a number simply means to write it as the product of **prime numbers**. Recall from arithmetic that a prime number is any counting number (excluding 1) that is divisible only by 1 and itself.

$$2, 3, 5, 7, 11, 13, 17, 19, 23, 29, \ldots$$
the first ten prime numbers

EXAMPLE 1

Completely factor 42.

Solution:

$$42 = 6 \cdot 7$$

Note: Keep going until you get all prime factors.

$$= 2 \cdot 3 \cdot 7 \quad \blacksquare \quad \text{factored completely}$$

EXAMPLE 2

Completely factor 24.

Solution:

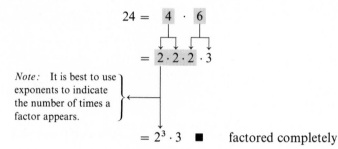

$$= 2^3 \cdot 3 \quad \blacksquare \qquad \text{factored completely}$$

Note: It is best to use exponents to indicate the number of times a factor appears.

Note: Alternately, we could have written

$$24 = 8 \cdot 3 = 2 \cdot 2 \cdot 2 \cdot 3 = 2^3 \cdot 3.$$

It doesn't matter with what two factors you start. The completely factored forms will always be exactly the same.

⊡ Determining the Least Common Denominator (LCD)

If you cannot determine the LCD by inspection, then use the following 2-step procedure.

TWO-STEP PROCEDURE FOR DETERMINING THE LCD

Step 1: Completely factor each denominator using exponents to indicate the number of times a factor appears.

Step 2: Multiply together each different factor to the highest power that it appears in any one of the individual denominators.

EXAMPLE 3

Find the LCD for $\dfrac{1}{24} + \dfrac{5}{42}$.

Solution:

Step 1:
$$\frac{1}{24} + \frac{5}{42} = \frac{1}{2^3 \cdot 3} + \frac{5}{2 \cdot 3 \cdot 7}$$

Note: The different factors are 2, 3, and 7. The highest powers to which they appear are 2^3, 3, and 7.

Step 2: Thus,   LCD $=$ $2^3 \cdot 3 \cdot 7$ or 168. ∎

EXAMPLE 4

Find the LCD for $\dfrac{-2}{25} - \dfrac{1}{18}$.

Solution:

Step 1:
$$\frac{-2}{25} - \frac{1}{18} = \frac{-2}{5^2} - \frac{1}{3^2 \cdot 2}$$

Note: The different factors are 5, 3, and 2. The highest powers to which they appear are 5^2, 3^2, and 2.

Step 2: Thus,   LCD $= 5^2 \cdot 3^2 \cdot 2$ or 450. ∎

Note: Notice that if the denominators do not contain any common factors, the LCD is simply the product of the denominators. LCD $= (25)(18) = 450$.

EXAMPLE 5

Find the LCD for $\dfrac{7x}{12} - \dfrac{5x}{8}$.

Solution:

Step 1:
$$\frac{7x}{12} - \frac{5x}{8} = \frac{7x}{2^2 \cdot 3} - \frac{5x}{2^3}$$

Note: The different factors are 2 and 3. The highest powers to which they appear are 2^3 and 3.

Step 2: Thus,   LCD $=$ $2^3 \cdot 3$ or 24. ∎

EXAMPLE 6

Find the LCD for $\dfrac{1}{3x^2y} + \dfrac{-5}{9xy^2}$.

Solution:

Step 1:

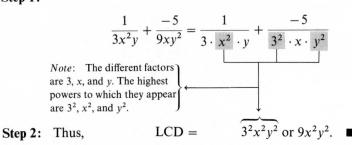

$$\frac{1}{3x^2y} + \frac{-5}{9xy^2} = \frac{1}{3 \cdot x^2 \cdot y} + \frac{-5}{3^2 \cdot x \cdot y^2}$$

Note: The different factors are 3, x, and y. The highest powers to which they appear are 3^2, x^2, and y^2.

Step 2: Thus, $\qquad\qquad$ LCD $= \qquad 3^2x^2y^2$ or $9x^2y^2$. ∎

When the denominators are different prime numbers, the LCD is simply the product of the denominators. For example, the LCD for

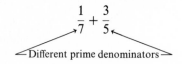

$$\frac{1}{7} + \frac{3}{5}$$

Different prime denominators

is simply $(7)(5) = 35$.

EXAMPLE 7

Find the LCD for $\dfrac{4m}{x} - \dfrac{3m}{y}$.

Solution:

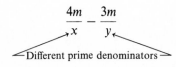

$$\frac{4m}{x} - \frac{3m}{y}$$

Different prime denominators

The LCD is simply xy. ∎

⬚ Adding and Subtracting Fractions that Have Different Denominators

Once the LCD is chosen, we use the fundamental property of fractions to change each fraction to an equivalent fraction whose denominator is the LCD. Only then can we apply the addition or subtraction rule for fractions to find the sum or difference.

EXAMPLE 8

Find the sum $\dfrac{1}{24} + \dfrac{5}{42}$.

Solution:

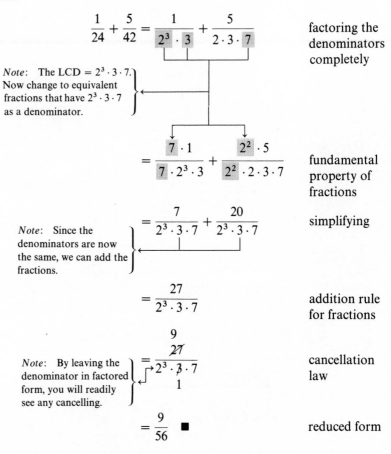

$$\frac{1}{24} + \frac{5}{42} = \frac{1}{2^3 \cdot 3} + \frac{5}{2 \cdot 3 \cdot 7}$$

factoring the
denominators
completely

Note: The LCD $= 2^3 \cdot 3 \cdot 7$.
Now change to equivalent
fractions that have $2^3 \cdot 3 \cdot 7$
as a denominator.

$$= \frac{7 \cdot 1}{7 \cdot 2^3 \cdot 3} + \frac{2^2 \cdot 5}{2^2 \cdot 2 \cdot 3 \cdot 7}$$

fundamental
property of
fractions

Note: Since the
denominators are now
the same, we can add the
fractions.

$$= \frac{7}{2^3 \cdot 3 \cdot 7} + \frac{20}{2^3 \cdot 3 \cdot 7}$$

simplifying

$$= \frac{27}{2^3 \cdot 3 \cdot 7}$$

addition rule
for fractions

Note: By leaving the
denominator in factored
form, you will readily
see any cancelling.

$$= \frac{\overset{9}{\cancel{27}}}{2^3 \cdot \underset{1}{\cancel{3}} \cdot 7}$$

cancellation
law

$$= \frac{9}{56} \quad \blacksquare$$

reduced form

EXAMPLE 9

Find the difference $\dfrac{-2}{25} - \dfrac{1}{18}$.

Solution:

$$\frac{-2}{25} - \frac{1}{18} = \frac{-2}{5^2} - \frac{1}{3^2 \cdot 2}$$

factoring the denominators completely

Note: The LCD $= 5^2 \cdot 3^2 \cdot 2$. Now change to equivalent fractions that have $5^2 \cdot 3^2 \cdot 2$ as a denominator.

$$= \frac{-2 \cdot 3^2 \cdot 2}{5^2 \cdot 3^2 \cdot 2} - \frac{5^2 \cdot 1}{5^2 \cdot 3^2 \cdot 2}$$

fundamental property of fractions

Note: Since the denominators are now the same, we can subtract the fractions.

$$= \frac{-36}{5^2 \cdot 3^2 \cdot 2} - \frac{25}{5^2 \cdot 3^2 \cdot 2}$$

simplifying

Note: Since 61 is prime and not a factor of the denominator, we cannot reduce further.

$$= \frac{-61}{5^2 \cdot 3^2 \cdot 2}$$

subtraction rule for fractions

$$= \frac{-61}{450} \text{ or } -\frac{61}{450} \quad \blacksquare$$

reduced form

EXAMPLE 10

Find the difference $\dfrac{7x}{12} - \dfrac{5x}{8}$.

Solution:

$$\frac{7x}{12} - \frac{5x}{8} = \frac{7x}{2^2 \cdot 3} - \frac{5x}{2^3}$$

factoring the denominators completely

Note: The LCD $= 2^3 \cdot 3$. Now change to equivalent fractions that have $2^3 \cdot 3$ as a denominator.

$$= \frac{2 \cdot 7x}{2 \cdot 2^2 \cdot 3} - \frac{3 \cdot 5x}{3 \cdot 2^3}$$

fundamental property of fractions

$$= \frac{14x}{2^3 \cdot 3} - \frac{15x}{2^3 \cdot 3} \qquad \text{simplifying}$$

$$= \frac{-x}{2^3 \cdot 3} \qquad \begin{array}{l}\text{subtraction rule} \\ \text{for fractions}\end{array}$$

$$= \frac{-x}{24} \text{ or } -\frac{x}{24} \quad \blacksquare \qquad \text{reduced form}$$

EXAMPLE 11

Find the sum $\dfrac{1}{3x^2 y} + \dfrac{-5}{9xy^2}$.

Solution:

$$\frac{1}{3x^2 y} + \frac{-5}{9xy^2} = \frac{1}{3 \cdot \boxed{x^2} \cdot y} + \frac{-5}{\boxed{3^2} \cdot x \cdot \boxed{y^2}} \qquad \begin{array}{l}\text{factoring the} \\ \text{denominators} \\ \text{completely}\end{array}$$

Note: The LCD $= 3^2 x^2 y^2$. Now change to equivalent fractions having $3^2 x^2 y^2$ as a denominator.

$$= \frac{\boxed{3y} \cdot 1}{\boxed{3y} \cdot 3x^2 y} + \frac{\boxed{x} \cdot (-5)}{\boxed{x} \cdot 3^2 xy^2} \qquad \begin{array}{l}\text{fundamental} \\ \text{property of} \\ \text{fractions}\end{array}$$

$$= \frac{3y}{3^2 x^2 y^2} + \frac{-5x}{3^2 x^2 y^2} \qquad \text{simplifying}$$

$$= \frac{3y + (-5x)}{3^2 x^2 y^2} \qquad \begin{array}{l}\text{addition rule} \\ \text{for fractions}\end{array}$$

$$= \frac{3y - 5x}{9x^2 y^2} \quad \blacksquare \qquad \text{reduced form}$$

Note: Just a reminder that you can never cancel as follows:

$$\frac{\dfrac{1}{\cancel{3y} \ominus 5x}}{\cancel{9x^2 y^2}} \qquad \textit{wrong}$$
$$3x^2 y$$

EXAMPLE 12

Find the difference $\dfrac{4m}{x} - \dfrac{3m}{y}$.

Solution: The LCD = xy. Now change to equivalent fractions that have xy as a denominator.

$$\frac{4m}{x} - \frac{3m}{y} = \frac{y \cdot 4m}{y \cdot x} - \frac{x \cdot 3m}{x \cdot y} \qquad \text{fundamental property of fractions}$$

$$= \frac{4my}{xy} - \frac{3mx}{xy} \qquad \text{simplifying}$$

$$= \frac{4my - 3mx}{xy} \qquad \blacksquare \qquad \text{subtraction rule for fractions}$$

Application

NUTS AND BOLTS

Referring to Fig. 2.12, observe that the *major diameter*, *minor diameter*, and *depth* are related by the following formula.

FORMULA: Minor Diameter of a Bolt

$$D_m = D_M - 2d,$$

where D_m is the minor diameter, D_M is the major diameter, and d is the depth.

In the manufacturing of nuts and bolts, it is necessary to drill a hole of the proper size in a *nut blank* so that the nut will fit onto the bolt's threads properly. The drill used to make this hole in the nut is referred to as a *tap drill*.

If the hole in the nut blank were drilled to exactly the same dimension as the minor diameter of the bolt, it would be extremely difficult to screw the nut onto the threads of the bolt. For this reason, the hole to be drilled into the nut blank, called the *tap drill diameter*, is usually increased by $\frac{1}{4}$ of the *depth*. This extra *clearance* will facilitate the screwing of the nut onto the bolt's threads (see Fig. 2.12).

FORMULA: Tap Drill Diameter

$$D_t = D_m + \frac{1}{4}d,$$

where D_t is the tap drill diameter, D_m is the minor diameter, and d is the depth.

EXAMPLE 13

What size tap drill should be used to drill a hole in a nut blank that is to be used with a bolt with a major diameter of $\frac{1}{2}$ in. and a depth of $\frac{1}{16}$ in.?

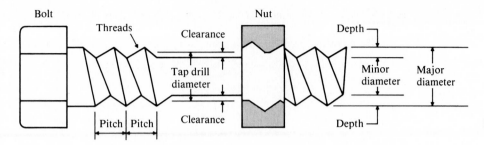

Figure 2.12

Solution: First, find the minor diameter.

$$D_m = D_M - 2d = \frac{1}{2} - 2\left(\frac{1}{16}\right)$$

$$= \frac{1}{2} - \frac{\overset{1}{\cancel{2}} \cdot 1}{1 \cdot \underset{8}{\cancel{16}}} \qquad \text{multiplication rule for fractions}$$

Note: Remember the order of operations: multiply before you add or subtract.

$$= \frac{1}{2} - \frac{1}{8} \qquad \text{simplifying}$$

$$= \frac{4}{8} - \frac{1}{8} \qquad \text{fundamental property of fractions}$$

$$D_m = \frac{3}{8} \text{ in.} \qquad \text{subtraction rule for fractions}$$

Since we now know the minor diameter, we can find the tap drill size.

$$D_t = D_m + \frac{1}{4}d = \frac{3}{8} + \frac{1}{4}\left(\frac{1}{16}\right)$$

$$= \frac{3}{8} + \frac{1 \cdot 1}{4 \cdot 16} \qquad \text{multiplication rule for fractions}$$

$$= \frac{3}{8} + \frac{1}{64} \qquad \text{simplifying}$$

$$= \frac{24}{64} + \frac{1}{64} \qquad \text{fundamental property of fractions}$$

$$D_t = \frac{25}{64} \text{ in.} \quad \blacksquare \qquad \text{addition property of fractions}$$

EXERCISES 2.4

Completely factor the following numbers. Use exponents to indicate the number of times a factor appears.

1. 28   **2.** 45

3. 54   **4.** 40

5. 100   **6.** 147

7. 144   **8.** 162

9. 132   **10.** 312

Find the LCD; then perform the indicated operation.

11. $\dfrac{3}{8} + \dfrac{1}{6}$ **12.** $\dfrac{5}{12} + \dfrac{5}{9}$

13. $\dfrac{9}{4} - \dfrac{7}{10}$ **14.** $\dfrac{7}{15} - \dfrac{1}{9}$

15. $\dfrac{1}{25} + \dfrac{-7}{20}$ **16.** $\dfrac{-11}{12} + \dfrac{-3}{16}$

17. $\dfrac{-1}{63} - \dfrac{5}{27}$ **18.** $\dfrac{-7}{108} - \dfrac{-5}{72}$

19. $\dfrac{5}{21} + \dfrac{1}{-14}$ **20.** $\dfrac{2}{15} - \dfrac{3}{-10}$

21. $\dfrac{3x}{5} + \dfrac{x}{6}$ **22.** $\dfrac{y}{9} + \dfrac{y}{4}$

23. $\dfrac{3a^2}{8} - \dfrac{7a^2}{10}$ **24.** $\dfrac{5h^2}{9} - \dfrac{20h^2}{21}$

25. $\dfrac{5y}{18x} - \dfrac{-y}{12x}$ **26.** $\dfrac{11y}{24t} + \dfrac{-3y}{16t}$

27. $\dfrac{3}{m} + \dfrac{4}{n}$ **28.** $\dfrac{6}{y} - \dfrac{5}{z}$

29. $\dfrac{1}{4x^2} - \dfrac{3}{2x}$ **30.** $\dfrac{5}{2y} - \dfrac{7}{16y^2}$

31. $\dfrac{4}{m^2n} + \dfrac{-2}{n^2}$ **32.** $\dfrac{1}{3p} - \dfrac{-3}{p^2q^2}$

33. $\dfrac{a}{b} + \dfrac{b}{a}$ **34.** $\dfrac{2p}{q} - \dfrac{3q}{p}$

35. $\dfrac{4}{63x^2y} - \dfrac{5}{18xy}$ **36.** $\dfrac{7}{36t^2w^2} - \dfrac{2}{45t}$

37. $\dfrac{b^2}{4a^2} - \dfrac{c}{a}$ **38.** $\dfrac{b^2}{12a^2} + \dfrac{2}{3a}$

39. $\dfrac{7}{12x} - \dfrac{5}{18x^2} + \dfrac{4}{3x}$ **40.** $\dfrac{2}{15p^2} + \dfrac{2}{9pq} - \dfrac{3}{10p^2}$

41. The major diameter of a bolt is $\frac{3}{4}$ in. and the thread depth is $\frac{1}{8}$ in. What is the minor diameter of the bolt?

42. The major diameter of a bolt is $\frac{15}{16}$ in. and the thread depth is $\frac{3}{16}$ in. What is the minor diameter of the bolt?

43. What size tap drill should be used to drill a hole in a nut blank that is to be used with the bolt described in exercise 41?

44. What size tap drill should be used to drill a hole in a nut blank that is to be used with the bolt described in exercise 42?

2.5 MIXED EXPRESSIONS

Objectives

- [·] To change a mixed expression to a single fraction.
- [··] To change a single fraction to a mixed expression.
- [··] To write a rational expression from an English phrase.

[·] Changing Mixed Expressions to Single Fractions

Recall from arithmetic that a **mixed number** represents the sum of a positive integer and a fraction. For example,

$$2\frac{1}{4} = 2 + \frac{1}{4}.$$

mixed number

In technical mathematics, we often refer to the sum or difference of an integer and a rational number as a **mixed expression**. To change a mixed expression to a single fraction, place the integer over 1 and add the fractions. To change $2\frac{1}{4}$ to a single fraction, we can proceed as follows:

$$2\frac{1}{4} = 2 + \frac{1}{4} \qquad \text{definition of a mixed number}$$

$$= \frac{2}{1} + \frac{1}{4} \qquad a = \frac{a}{1}$$

$$= \frac{4 \cdot 2}{4 \cdot 1} + \frac{1}{4} \qquad \text{fundamental property of fractions}$$

$$= \frac{8}{4} + \frac{1}{4} \qquad \text{simplifying}$$

$$= \frac{9}{4} \qquad \text{addition property of fractions}$$

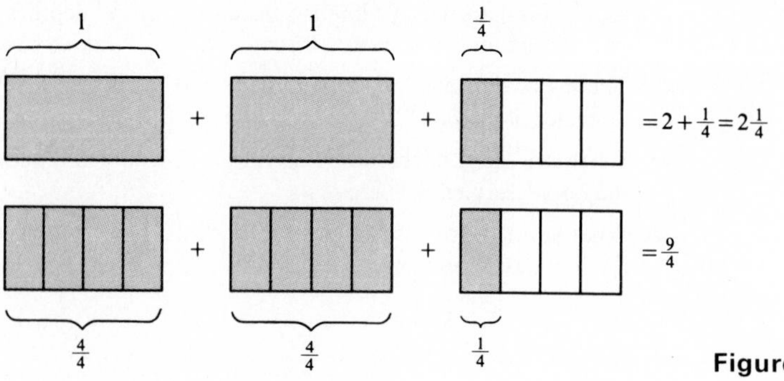

Figure 2.13

Thus,

$$2\frac{1}{4} = 2 + \frac{1}{4} = \frac{9}{4} \text{ (see Fig. 2.13)}.$$

EXAMPLE 1

Change $\dfrac{3}{10} - 2$ to a single fraction.

Solution:

$$\frac{3}{10} - 2 = \frac{3}{10} - \frac{2}{1} \qquad\qquad a = \frac{a}{1}$$

$$= \frac{3}{10} - \frac{2 \cdot \boxed{10}}{1 \cdot \boxed{10}} \qquad\qquad \text{fundamental property of fractions}$$

$$= \frac{3}{10} - \frac{20}{10} \qquad\qquad \text{simplifying}$$

$$= \frac{-17}{10} \text{ or } -\frac{17}{10} \quad\blacksquare \qquad \text{subtraction rule for fractions}$$

EXAMPLE 2

Change $1 + \dfrac{3}{2y}$ to a single fraction.

Solution:

$$1 + \frac{3}{2y} = \frac{1}{1} + \frac{3}{2y} \qquad a = \frac{a}{1}$$

$$= \frac{2y \cdot 1}{2y \cdot 1} + \frac{3}{2y} \qquad \text{fundamental property of fractions}$$

$$= \frac{2y}{2y} + \frac{3}{2y} \qquad \text{simplifying}$$

$$= \frac{2y + 3}{2y} \quad \blacksquare \qquad \text{addition rule for fractions}$$

EXAMPLE 3

Change $2x - \dfrac{3x}{5}$ to a single fraction.

Solution:

$$2x - \frac{3x}{5} = \frac{2x}{1} - \frac{3x}{5} \qquad a = \frac{a}{1}$$

$$= \frac{5 \cdot 2x}{5 \cdot 1} - \frac{3x}{5} \qquad \text{fundamental property of fractions}$$

$$= \frac{10x}{5} - \frac{3x}{5} \qquad \text{simplifying}$$

$$= \frac{7x}{5} \quad \blacksquare \qquad \text{subtraction rule for fractions}$$

⬚ Changing Single Fractions to Mixed Expressions

In Example 2, you found that

$$1 + \frac{3}{2y} = \frac{2y + 3}{2y}.$$

mixed expression —⟍   single fraction —⟍

Now, suppose you were given $\dfrac{2y + 3}{2y}$ and asked to write it as a mixed expression.

Of course, you can't cancel as follows:

$$\frac{\overset{1}{\cancel{2y}} \oplus 3}{\underset{1}{\cancel{2y}}} \quad \textit{wrong}$$

However, you could apply the addition rule for fractions in reverse

$$\frac{a + b}{c} = \frac{a}{c} + \frac{b}{c}$$

and proceed as follows:

$$\frac{2y + 3}{2y} = \frac{2y}{\boxed{2y}} + \frac{3}{\boxed{2y}} \qquad \text{addition rule for fractions in reverse}$$

Note: The denominator divides into each term of the numerator $\Bigg\}$

$$= \frac{\overset{1}{\cancel{2y}}}{\underset{1}{\cancel{2y}}} + \frac{3}{2y} \qquad \text{cancellation law}$$

$$= 1 + \frac{3}{2y} \qquad \text{simplifying}$$

EXAMPLE 4

Change $\dfrac{x^2 + 5}{x}$ to a mixed expression.

Solution: $\dfrac{x^2 + 5}{x} = \dfrac{x^2}{x} + \dfrac{5}{x}$ addition rule for fractions in reverse

$$= \frac{\overset{x}{\cancel{x^2}}}{\underset{1}{\cancel{x}}} + \frac{5}{x} \qquad \text{cancellation law}$$

$$= x + \frac{5}{x} \quad \blacksquare \qquad \text{simplifying}$$

EXAMPLE 5

Change $\dfrac{14 - 8n^2}{2n}$ to a mixed expression.

Solution:

$$\frac{14 - 8n^2}{2n} = \frac{14}{2n} - \frac{8n^2}{2n} \qquad \text{subtraction rule for fractions in reverse}$$

$$= \frac{\overset{7}{\cancel{14}}}{\underset{n}{\cancel{2n}}} - \frac{\overset{4n}{\cancel{8n^2}}}{\underset{1}{\cancel{2n}}} \qquad \text{cancellation law}$$

$$= \frac{7}{n} - 4n \quad \blacksquare \qquad \text{simplifying}$$

⋰ Rational Expressions

In Section 2.2, you learned that the word "of" means "times." For example, half *of* two thirds can be represented by the product

$$\frac{1}{2} \cdot \frac{2}{3}.$$

If you were asked to write an algebraic expression for

half of some number, you could write

$$\frac{1}{2} \cdot x = \frac{1}{2} \cdot \frac{x}{1} = \frac{x}{2}.$$

a single fraction

The single fraction $\frac{x}{2}$ is referred to as a **rational expression**.

EXAMPLE 6

Write a rational expression for "two-thirds of some number."

Solution:

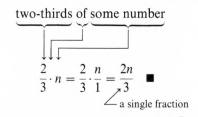

two-thirds of some number

$$\frac{2}{3} \cdot n = \frac{2}{3} \cdot \frac{n}{1} = \frac{2n}{3} \quad \blacksquare$$

a single fraction

EXAMPLE 7

Write a rational expression for "one-fourth of some number increased by five."

Solution:

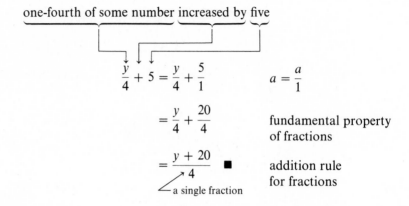

one-fourth of some number increased by five

$$\frac{y}{4} + 5 = \frac{y}{4} + \frac{5}{1} \qquad a = \frac{a}{1}$$

$$= \frac{y}{4} + \frac{20}{4} \qquad \text{fundamental property of fractions}$$

$$= \frac{y + 20}{4} \quad\blacksquare \qquad \text{addition rule for fractions}$$

a single fraction

EXAMPLE 8

Write a rational expression for "six less three-eighths of some number."

Solution:

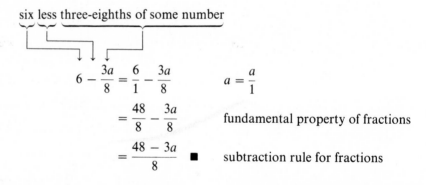

six less three-eighths of some number

$$6 - \frac{3a}{8} = \frac{6}{1} - \frac{3a}{8} \qquad a = \frac{a}{1}$$

$$= \frac{48}{8} - \frac{3a}{8} \qquad \text{fundamental property of fractions}$$

$$= \frac{48 - 3a}{8} \quad\blacksquare \qquad \text{subtraction rule for fractions}$$

EXERCISES 2.5

Change each of the following to a single fraction.

1. $3\frac{5}{8}$ **2.** $5\frac{4}{5}$ **3.** $-4\frac{1}{2}$ **4.** $-8\frac{2}{3}$

5. $\frac{3}{7} - 4$ **6.** $6 - \frac{7}{8}$ **7.** $4 + \frac{4}{y}$ **8.** $p - \frac{3}{8}$

9. $\frac{3}{5m} - m$ **10.** $\frac{5}{2y} + 3y^2$ **11.** $x + \frac{2x}{5}$ **12.** $\frac{mn}{4} + 6mn$

13. $t^2 - \frac{t^2}{-4}$ **14.** $\frac{x^2y^2}{-5} + 2x^2y^2$ **15.** $x + x + \frac{x}{2}$ **16.** $3y - y - \frac{y}{3}$

Change each of the following to a mixed expression.

17. $\dfrac{a + 1}{a}$

18. $\dfrac{6 - x}{6}$

19. $\dfrac{x^2 - 5}{x}$

20. $\dfrac{x^2 y - 3}{xy}$

21. $\dfrac{4y^2 + 6xy}{3}$

22. $\dfrac{7 + 8t^3}{8t^2}$

23. $\dfrac{12 - 6y^2}{-4}$

24. $\dfrac{18x - 2y^2}{-6}$

25. $\dfrac{24 + 12xy^2}{4xy^2}$

26. $\dfrac{36xy - 24x^2 y^2}{24x^2 y}$

Write a rational expression for each of the following.

27. One-tenth of some voltage v.

28. Three-fourths of some current I.

29. One-tenth more than some voltage v.

30. Three-fourths less than some current I.

31. Three less than two-thirds of some force F.

32. Five less one-sixth of some temperature T.

33. Half of some pressure P_1 increased by some pressure P_2.

34. Twice some area A_1 decreased by nine-tenths of some area A_2.

Refer to Fig. 2.14.

35. If d_2 is one-twelfth of d_1, find a rational expression that describes the distance L.

36. If d_1 is seven-eighths of L, find a rational expression that describes the distance d_2.

Refer to Figure 2.15.

37. If d_1 is three-sixteenths of d_2, find a rational expression that describes the distance L.

38. If d_2 is five-eighths of L, find a rational expression that describes the distance d_1.

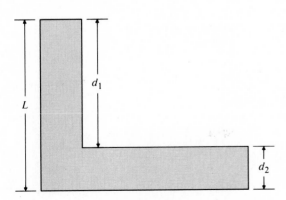

Figure 2.14

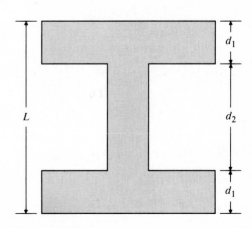

Figure 2.15

Chapter Review

REVIEW EXERCISES

The following review exercises are grouped according to the objectives that should have been mastered in Chapter 2. Work each problem carefully. If any weaknesses appear, immediately refer to and read the subsection that matches that objective.

2.1 BUILDING UP AND BREAKING DOWN RATIONAL NUMBERS

Objectives

⊡ To locate a rational number on a number line.

Find the position on the number line.

1. a) $\dfrac{7}{6}$ **b)** $\dfrac{-7}{6}$ **c)** $\dfrac{-2}{-3}$

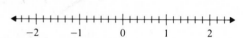

⊡ To find an equivalent fraction whose denominator is known.

Find the missing numerator.

2. $\dfrac{5}{8} = \dfrac{?}{32}$

3. $\dfrac{-3}{5} = \dfrac{?}{30}$

4. $\dfrac{3y}{-4} = \dfrac{?}{12y}$

5. $\dfrac{11x}{6y} = \dfrac{?}{18xy^2}$

⊡ To reduce a fraction to lowest terms.

Reduce.

6. $\dfrac{12}{28}$ **7.** $\dfrac{-36}{20}$ **8.** $\dfrac{-9z}{-3z}$ **9.** $\dfrac{16xy^2}{12x^2y}$

2.2 MULTIPLYING AND DIVIDING RATIONAL NUMBERS

Objectives

⊡ To multiply fractions.

Find the product.

10. $\dfrac{8}{9} \cdot \dfrac{5}{12}$

11. $\dfrac{-3}{4} \cdot \dfrac{14}{15}$

12. $\left(\dfrac{2y}{-7}\right)(-21y)$

13. $\dfrac{18t^2}{5w^2} \cdot \dfrac{15w}{2t}$

14. $\left(\dfrac{-2y}{5}\right)^3$

⊡ To divide fractions.

Find the quotient.

15. $\dfrac{7}{8} \div \dfrac{11}{12}$

16. $\dfrac{3}{-10} \div \dfrac{9}{5}$

17. $\dfrac{\dfrac{2x}{-3}}{\dfrac{-4x}{15}}$

18. $\dfrac{11t}{6} \div (22t^2)$

2.3 ADDING AND SUBTRACTING RATIONAL NUMBERS WITH THE SAME DENOMINATOR

Objectives

⊡ To add fractions that have the same denominator.

Find the sum.

19. $\dfrac{11}{12} + \dfrac{5}{12}$

20. $\dfrac{7}{6} + \dfrac{-5}{6}$

21. $\dfrac{-5t}{4} + \dfrac{3t}{4}$

22. $\dfrac{9}{32p^2} + \dfrac{7}{32p^2}$

⊡ To subtract fractions that have the same denominator.

Find the difference.

23. $\dfrac{11}{12} - \dfrac{5}{12}$

24. $\dfrac{9}{20} - \dfrac{-7}{20}$

25. $\dfrac{5x}{8} - \dfrac{21x}{8}$

26. $\dfrac{6y}{6y} - \dfrac{5}{6y}$

⊡ To add or subtract fractions whose denominators are opposites.

Find the sum or difference.

27. $\dfrac{5a^2}{7} + \dfrac{2a^2}{-7}$

28. $\dfrac{4mn}{3p} - \dfrac{2mn}{-3p}$

2.4 ADDING AND SUBTRACTING RATIONAL NUMBERS WITH DIFFERENT DENOMINATORS

Objectives

⊡ To completely factor a given number.

Factor completely.

29. 30

30. 72

⊡ Given fractions with different denominators, to find the least common denominator (LCD).

Find the LCD.

31. $\dfrac{1}{18} + \dfrac{5}{66}$

32. $\dfrac{-3}{40} - \dfrac{1}{9}$

33. $\dfrac{5x}{12} - \dfrac{7x}{18}$

34. $\dfrac{1}{2x^2y^2} + \dfrac{-3}{8x^3y}$

35. $\dfrac{9}{m} - \dfrac{-3}{n}$

⊡ To add or subtract fractions that have different denominators.

Find the sum or difference.

36. $\dfrac{1}{18} + \dfrac{5}{66}$

37. $\dfrac{-3}{40} - \dfrac{1}{9}$

38. $\dfrac{5x}{12} - \dfrac{7x}{18}$

39. $\dfrac{1}{2x^2y^2} + \dfrac{-3}{8x^3y}$

40. $\dfrac{9}{m} - \dfrac{-3}{n}$

2.5 MIXED EXPRESSIONS

Objectives

⬚ To change a mixed expression to a single fraction.

Write as a single
fraction.

41. $\dfrac{5}{8} - 3$

42. $1 + \dfrac{5}{4x}$

 43. $7y - \dfrac{2}{3y}$

⬚ To change a single fraction to a mixed expression.

Write as a mixed
expression.

44. $\dfrac{3 - x^2}{x}$

45. $\dfrac{20 - 6n^2}{2n}$

⬚ To write a rational expression given an English phrase.

Write as a rational
expression.

46. "three-fifths of some number"

47. "one-ninth of some number decreased by three"

48. "four more than five-sixths some number"

CHAPTER TEST
Time
50 minutes

Score
30–27 excellent
26–24 good
23–18 fair
below 17 poor

If you have worked through the Review Exercises and corrected any weaknesses, then you are ready to take the following Chapter Test.

Find the missing numerator.

1. $\dfrac{3}{5} = \dfrac{?}{30}$

2. $\dfrac{-5t}{6} = \dfrac{?}{18t}$

Reduce to lowest terms.

3. $\dfrac{-18}{42}$

4. $\dfrac{-6x}{-24x^2}$

Perform the indicated operations. Be sure your answer is in reduced form.

5. $\dfrac{7}{10} \cdot \dfrac{-5}{7}$

6. $\dfrac{8}{9} \div \dfrac{4}{21}$

7. $\dfrac{7}{8} + \dfrac{5}{8}$

8. $\dfrac{-7}{15} - \dfrac{14}{15}$

9. $\dfrac{5}{16} + \dfrac{1}{12}$

10. $\dfrac{3}{28} - \dfrac{5}{98}$

11. $\dfrac{3}{26} + \dfrac{1}{9}$

12. $\dfrac{-1}{18} - \dfrac{2}{35}$

13. $\left(\dfrac{4}{9x}\right)(6x)$

14. $\dfrac{\dfrac{3x}{-5}}{\dfrac{6x^2}{25}}$

15. $\dfrac{-t}{6} + \dfrac{5t}{6}$ **16.** $\dfrac{3x}{7y} - \dfrac{4x}{7y}$

17. $\dfrac{3x}{8} + \dfrac{7x}{10}$ **18.** $\dfrac{3}{4xy^2} - \dfrac{-5}{6x^2}$

19. $1 + \dfrac{6w}{5}$ **20.** $\left(\dfrac{-x}{3}\right)^3$

21. $\left(\dfrac{4y}{-5}\right) \div (-6y)$ **22.** $\dfrac{3a^2}{8} + \dfrac{7a^2}{-8}$

23. $\dfrac{3}{5t} - \dfrac{7}{-5t}$ **24.** $\dfrac{2}{x} + \dfrac{-3}{y}$

25. $\dfrac{t^2}{3} - \dfrac{t^2}{15}$ **26.** $\dfrac{3y}{4} - 8y$

Change to a mixed expression.

27. $\dfrac{4n^2 - 5}{4n}$ **28.** $\dfrac{36y - 20xy^2}{4xy}$

Write a rational expression for the following English phrases.

29. Six less than two-thirds of some voltage V

30. Twice some force F_1 increased by half of some force F_2

3

MENSURATION

3.1 DECIMAL FRACTIONS AND ROUNDING OFF

Objectives

⊡ To write a decimal fraction in decimal form.

⊡ To round off a decimal to a given position.

⊡ Decimal Fractions

Recall from arithmetic, a *decimal fraction* is a rational number having as its denominator some power of ten (10, 100, 1000, and so on). However, unlike a common fraction, the denominator of a decimal fraction is not usually written, but is instead expressed by **place value** (see Fig. 3.1).

Observe in the place value chart of Fig. 3.1 that

1. Places to the left of the decimal point end with *s* while places to the right end with *ths*.

2. The decimal point separates the whole number part from the fractional part of a decimal number.

3. Any position in the place value chart is one-tenth the value of the position to its immediate left.

Although the decimal fraction "thirty seven thousand*ths*" can be written in **fractional form** as $\frac{37}{1000}$, it is more conveniently written in **decimal form** as .037.

> *Note:* We need this place-holding zero to force the last digit (the 7) into the thousand*ths* position. $\left.\right\} \rightarrow .037$

Place value chart

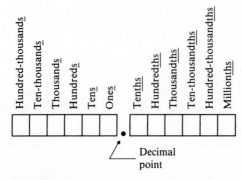

Decimal point

Figure 3.1

EXAMPLE 1

Write in decimal form.
  a) four hundred ten-thousandths
  b) four hundred ten thousandths
  c) four hundred and ten thousandths

Solution:

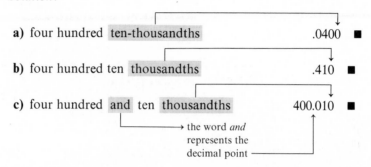

a) four hundred ten-thousandths   .0400 ■

b) four hundred ten thousandths   .410 ■

c) four hundred and ten thousandths   400.010 ■

→ the word *and*
represents the
decimal point

Note: Four hundred and ten thousandths (400.010) is called a *mixed decimal fraction*, since it contains a whole number and a decimal fraction.

⸬ Rounding Off

Often, when working with decimals, it is necessary to **round off** an answer to the nearest whole number, nearest tenth, or some other position. The rounded result represents an approximation of the given decimal number. We will use the symbol ≈ to mean "is approximately equal to."

 To round off a decimal number, use the following rule.

RULE FOR ROUNDING OFF DECIMALS

 1. Circle the digit in the place you wish to round off.

 2. If the first digit to the right of the circled digit is

   a) *less than 5*, drop all digits to the right of the circled digit, or

   b) *greater than or equal to 5*, increase the circled digit by 1; then drop all digits to its right.

 3. Any digit to the left of the decimal point that is dropped must be replaced by a place-holding zero.

EXAMPLE 2

Round off 319.6359 to

  a) tenths   **b)** hundredths   **c)** tens   **d)** ones

Solution:

a)

319.⟨6⟩359 (to tenths)

tenths ⟶↑ L→ Since the first digit to the right
place   of ⟨6⟩ is *less than* 5, drop all digits
    to the right of ⟨6⟩.

319.6359 ≈ 319.6   (to tenths) ■

b)

319.6⟨3⟩59   (to hundredths)

hundredths ⟶↑ L→ Since the first digit to the right
place   of ⟨3⟩ is *equal to* 5, increase ⟨3⟩ to ⟨4⟩
    and drop all digits to its right.

319.6359 ≈ 319.64   (to hundredths)   ■

c)

3⟨1⟩9.6359   (to tens) ■

tens ⟶↑ L→ Since the first digit to the right
place   of ⟨1⟩ is *greater than* 5, increase ⟨1⟩ to ⟨2⟩
    and drop all digits to its right.

319.6359 ≈ 320   (to tens)   ■

L→ { *Note*: When dropping digits to
the left of the decimal point,
be sure to fill them in with
place-holding zeros. }

d)

31⟨9⟩.6359   (to ones) ■

ones ⟶↑ L→ Since the first digit to the right
place   of ⟨9⟩ is *greater than* 5, increase ⟨9⟩ to ⟨10⟩
    and drop all digits to its right.

319.6359 ≈ 32$\overline{0}$ (to ones)   ■

L→ { *Note*: A bar is placed above
the zero to indicate we have
rounded to *ones* and not to *tens*
as in Example 2c. }

EXERCISES 3.1

Write each of the following in decimal form.

1. five hundredths
2. ninety hundredths
3. eighteen thousandths
4. two thousandths
5. six hundred sixteen thousandths
6. three thousand three ten-thousandths
7. six hundred and sixteen thousandths

8. three thousand and three ten-thousandths

9. nine hundred-thousandths

10. one thousand one hundred-thousandths

11. nine and one hundred thousandths

12. one thousand and one hundred thousandths

Round off each number to the position indicated.

13. 57.65 (ones)

14. 57.45 (ones)

15. 2.743 (tenths)

16. 2.793 (tenths)

17. 187.2 (tens)

18. 184.8 (tens)

19. .003445 (ten-thousandths)

20. .003451 (ten-thousandths)

21. 27.6954 (hundredths)

22. 27.6946 (hundredths)

23. 8849.9 (hundreds)

24. 7236.9 (thousands)

25. .0003996 (millionths)

26. .0069995 (millionths)

27. 999.573 (ones)

28. 999.573 (tens)

3.2 SIGNIFICANT DIGITS

Objectives

⊡ To state if a given number is approximate or exact.

⊡ To state the accuracy and precision of a given measurement.

⊡ To determine the upper and lower limits of a given measurement.

⊡ Approximate and Exact Numbers

Consider measuring the block of wood with the ruler pictured in Fig. 3.2.

The ruler is marked in .1-in. intervals. The length of the block of wood, to the nearest tenth of an inch, is 2.3 in. You may be tempted to say that the measure is *exactly* 2.3 in. However, this would not be correct. Does the end of the block of wood fall exactly in the middle of the marking for 2.3 in. or slightly to the left or right of the mark? If you used a powerful magnifying glass, you might attempt to answer this question, but you still couldn't determine the exact length of the block of wood. Every measurement involves some error. Thus, any number found by a

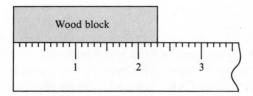

Figure 3.2

measuring process is an **approximate number**. The length 2.3 in. is an approximate length.

If there are 16 blocks of wood in a plastic container, each having a length of 2.3 in., the number 16 is *exact* since it would be possible to actually *count* the 16 blocks. Thus, any number found by a counting process is ar **exact number**.

EXAMPLE 1

In the following statement, identify each number as approximate or exact.

"A 6.25 -acre piece of land is to be subdivided into 5 house lots. Each lot is to have 175 ft of frontage and sell for $ 25,000 ."

Solution:

6.35	is approximate	(found by measuring)
5	is exact	(found by counting)
175	is approximate	(found by measuring)
25,000	is exact	(found by counting) ∎

Accuracy and Precision

With the ruler in Fig. 3.2, we found the length of the block of wood to be 2.3 in. The *3* in the tenths position of 2.3 in. is not exactly known. However, it does have some *significance* since the length seems to be closer to 2.3 in. than either 2.2 in. or 2.4 in. If a number contributes to our knowledge of how good an approximation is, then it is called a **significant digit**.

The significant digits in a measurement can be summarized as follows:

SIGNIFICANT DIGITS

1. All nonzero digits are significant.

2. Zeros *between* nonzero digits are significant.

3. Zeros appearing at the *end of a decimal fraction or a mixed decimal fraction* are significant.

4. Zeros at the *beginning of a decimal fraction* are *not* significant and serve only to locate the decimal point correctly.

5. Zeros at the *end of an integer* are *not* significant unless a bar ¯ is placed above one of the zeros. The bar is placed over the last zero that is significant.

The **accuracy** of a measurement refers to the *number of significant digits* in a measurement, while the **precision** of a measurement refers to the *place value of the last significant digit* in a measurement. For example, the measurement 2.3 in. is *accurate* to two significant digits while being *precise* to tenths.

EXAMPLE 2

State the accuracy and precision of the following measurements.

a) 132.7°F   c) 100.00 ft   e) 2200 lb

b) 6.002 in.   d) .0006 V   f) 22$\overline{0}$0 lb

Solution:

a) The measurement 132.7°F is accurate to four significant digits and precise to tenths.

b) The measurement 6.002 in. is accurate to four significant digits and precise to thousandths.

c) The measurement 100.00 ft is accurate to five significant digits and precise to hundredths.

d) The measurement .0006 V is accurate to only one significant digit and precise to ten-thousandths.

e) The measurement 2200 lb is accurate to two significant digits and precise to hundreds.

f) The measurement 22$\overline{0}$0 pounds is accurate to three significant digits and precise to tens. ∎

Note: In the above measurements, .0006 V is the *least* accurate measurement since it contains only one significant digit. However, it is also the *most* precise measurement, being measured to the nearest ten-thousandth of a volt.

⁚ Greatest Possible Error (g.p.e.)

Unless stated otherwise, the **greatest possible error** in a measurement is *half of the precision* of the measurement. The precision of the measurement 2.3 in. is *one tenth* of an inch. Thus,

$$\text{g.p.e.} = \frac{1}{2} \, of \; .1 \; = \frac{1}{2}(.1) = \frac{.1}{2} \qquad = .05 \text{ in.}$$

means "times"

$$\begin{array}{r} .05 \\ 2\overline{).10} \end{array}$$

Therefore, the *upper limit* of the measurement is

$$2.3 \text{ in.} \; + .05 \text{ in.} = 2.35 \text{ in.}$$

$$\begin{array}{r} 2.30 \\ + \; .05 \\ \hline 2.35 \end{array}$$

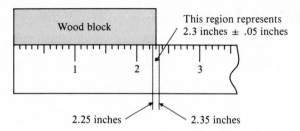

Figure 3.3

2.25 inches   2.35 inches

while the *lower limit* of the measurement is

$$2.3 \text{ in.} \; - \; .05 \text{ in.} = 2.25 \text{ in.}$$

$$\begin{array}{r} 2.30 \\ - \;\; .05 \\ \hline 2.25 \end{array}$$

Although we don't know the exact length of the block of wood, we do know that the length is definitely between 2.25 in. and 2.35 in., and that we can consistently measure the length to be between these two values. Notice that any measurement between 2.25 in. and 2.35 in., when rounded to tenths, would be 2.3 in. (see Fig. 3.3).

EXAMPLE 3

Determine the upper and lower limits of the measurements.

a) .0006 V   **b)** 2200 lb

Solution:

a) The precision of the measurement .0006 V is *one ten-thousandth* of a volt. Thus,

$$\text{g.p.e.} = \frac{1}{2} \text{ of } \underbrace{.0001}_{\text{precision}} = \frac{1}{2}(.0001) = \frac{.0001}{2} = .00005 \text{ V.}$$

$$2\overline{)\,.00010} \;\; .00005$$

$$\text{upper limit} = .0006 \text{ V} \; + \; .00005 \text{ V} = .00065 \text{ V}$$

$$\begin{array}{r} .00060 \\ + .00005 \\ \hline .00065 \end{array}$$

$$\text{lower limit} = .0006 \text{ V} \; - \; .00005 \text{ V} = .00055 \text{ V}$$

$$\begin{array}{r} .00060 \\ - .00005 \\ \hline .00055 \end{array}$$

Thus, .0006 V represents a measurement between .00055 V and .00065 V. ■

b) The precision of the measurement 2200 lb is *one hundred* pounds. Thus,

$$\text{g.p.e.} = \frac{1}{2} \text{ of } \underset{\text{precision}}{\underline{100}} = \frac{1}{2}(100) = \frac{100}{2} = 50 \text{ lb.}$$

upper limit = 2200 lb $+$ 50 lb = 2250 lb

lower limit = 2200 lb $-$ 50 lb = 2150 lb

Thus, 2200 lb represents a measurement between 2150 lb and 2250 lb.   ■

EXERCISES 3.2

In the following statements, identify each number as approximate or exact.

1. A 4 -cylinder automobile equipped with a 5 -speed transmission traveled 500 mi on 12.2 gal of gasoline.

2. A certain roof deck needs 6 steel beams spaced 8 ft apart in order to carry a load of 22,000 lb.

Fill in the table.

(handwritten margin notes:)
Ones = .5
Tenths = .05
Hundredths = .005
Thou. = .0005
Hun. Thou. = .000005

Measurement	Accuracy	Precision	g.p.e.	Lower limit	Upper limit
3. 481 ft	3	Ones	.5	480.5	481.5
4. 9267 ft	4	Ones	.5	9266.5	9267.5
5. 10.0 gal	3	Tenths	.05	9.95	10.05
6. 8.07 gal	3	Hundredths	.005	8.065	8.075
7. 20.00°F	4	Hundredths	.005	19.995	20.005
8. 190.0°C	4	Tenths	.05	189.95	190.05
9. .06 V	1	Hundredths	.005	.0055	.065
10. .00042 V	2	Hundred-Thousandths	.000005	.00415	.00425
11. .060 V	2	Thousandths	.0005	.0595	.0605
12. .000402 V	3	Millionths			
13. 4500 lb	2	Hundreds	.005		
14. 16000 lb	2	Thousandths			
15. 4500 lb	4	Ones	.5		
16. 16000 lb	4	Tens	.05		

State which of the given two measurements is the more precise and which is the more accurate. If they have the same precision or accuracy, write "same."

17. 1.8 mi or 1.80 mi

18. 16.3 m or .750 m

19. 1500 Ω or 1600 Ω

20. .012 A or .162 A

21. .35 V or 15$\overline{0}$ V

22. 15 ft or 15.00 ft

23. 2560 yd or 256$\overline{0}$ yd

24. .05 g or .050 g

3.3 CALCULATING WITH APPROXIMATE NUMBERS

Objectives

☐ To add and subtract approximate numbers using a calculator.

☐ To multiply and divide approximate numbers using a calculator.

☐ To raise an approximate number to a power using a calculator.

☐ To perform a basic arithmetic operation that involves an exact number.

☐ Adding and Subtracting Approximate Numbers

Suppose we wished to *add* the approximate numbers, 13.602 and 2.6. The *maximum possible sum* would be the sum of the **upper limits**.

$$\begin{array}{ll} 13.6025 & \text{upper limit of } 13.602 \\ +\ \ 2.65 & \text{upper limit of } 2.6 \\ \hline 16.2525 & \text{maximum possible sum} \end{array}$$

The *minimum possible sum* would be the sum of the **lower limits**.

$$\begin{array}{ll} 13.6015 & \text{lower limit of } 13.602 \\ +\ \ 2.55 & \text{lower limit of } 2.6 \\ \hline 16.1515 & \text{minimum possible sum} \end{array}$$

Therefore, the *actual* sum is somewhere between a minimum value of 16.1515 and a maximum value of 16.2525. We cannot approximate the sum any more closely than this.

Notice that the minimum and maximum values of the sum begin with the number 16, but start to differ when we look at the *tenths* position in each sum. Because the first uncertain digit appears in the *tenths* position, all digits after the *tenths* position can be considered superfluous. If the sum is to be rounded, then it should be rounded to *tenths*. Thus,

$$13.602 + 2.6 = 16.2,$$

since 16.2 is about in the middle of the range of possible values. Keep in mind that the **rounded sum**, 16.2, is also an approximate number, and the last significant digit (2) is not really known. However, 16.2 is an approximate number that best describes the actual sum.

Now, suppose we simply added the approximate numbers as they were given We would obtain the following sum:

$$\begin{array}{r} 13.602 \\ 2.6 \\ \hline 16.202 \end{array}$$

13.602 and 2.6 → addends

16.202   sum of the approximate numbers

Notice that the *last significant digit* in the *least precise* addend (the 6 in 2.6) is in the *tenths* position. Keeping this in mind, we could simply round 16.202 to *tenths* and obtain 16.2 (the same result as was obtained by the longer maximum–minimum procedure outlined above).

This example allows us to state the following rule of thumb for adding and subtracting approximate numbers.

RULE FOR ADDING OR SUBTRACTING APPROXIMATE NUMBERS

Add or subtract the approximate numbers that are given. Then round off this result to the **precision** of the *least precise* approximate number.

When working with approximate numbers, a calculator is usually used to perform the basic operations of addition $\boxed{+}$, and subtraction $\boxed{-}$.

calculator keys

EXAMPLE 1

Using a calculator, find the sum of the approximate numbers.

$$13.75 + 1.098 + .0082$$

Solution: Enter

$\boxed{13.75}$ $\boxed{+}$ $\boxed{1.098}$ $\boxed{+}$ $\boxed{.0082}$ $\boxed{=}$.   14.8562

least precise (hundredths)   Display

The sum, 14.8562, should be rounded to *hundredths*, since the least precise number (13.75) is precise to *hundredths*.

Thus, $13.75 + 1.098 + .0082 = 14.86$.   ■

EXAMPLE 2

Using a calculator, find the difference of the approximate numbers.

$$7.23 - 17.2$$

Solution: Enter

$$\boxed{7.23} \boxed{-} \boxed{17.2} \boxed{=} \quad \underbrace{-9.97}$$
least precise (tenths)⟋   Display

The difference, -9.97, should be rounded to *tenths*, since the least precise number (17.2) is precise to *tenths*.

Thus, $7.23 - 17.2 = -10.0$.   ∎

⸿ Multiplying and Dividing Approximate Numbers

Suppose we wished to multiply the approximate numbers 13.602 and 2.6. The *maximum possible product* of the approximate numbers would be the product of the **upper limits**.

$$
\begin{array}{r}
13.6025 \\
\times \quad 2.65 \\
\hline
680125 \\
816150 \\
272050 \\
\hline
36.046625
\end{array}
$$

13.6025 upper limit of 13.602
2.65 upper limit of 2.6
36.046625 maximum possible product

The *minimum possible product* would be the product of the **lower limits**.

$$
\begin{array}{r}
13.6015 \\
\times \quad 2.55 \\
\hline
680075 \\
680075 \\
272030 \\
\hline
34.683825
\end{array}
$$

13.6015 lower limit of 13.602
2.55 lower limit of 2.6
34.683825 minimum possible product

Therefore, the actual product is somewhere between a minimum value of 34.683825 and a maximum value of 36.046625. We cannot approximate the product any more closely than this.

Notice that the minimum and maximum values of the product start to differ when we look at the *ones* position in each product. Because the first uncertain digit appears in the *ones* position, all digits after the *ones* position can be considered superfluous. If the product is to be rounded, then it should be rounded to a whole number (ones position). Thus,

$$13.602 \times 2.6 = 35,$$

since 35 is about in the middle of the range of possible values. The *rounded product*, 35, is also an approximate number, and the last significant digit (the 5) is not exactly known but does contribute to our knowledge of just how good the approximation truly is.

Now, suppose we simply multiplied the approximate numbers as they were given. We would obtain the following product:

$$
\begin{array}{r}
13.602 \\
\times \quad 2.6 \\
\hline
81612 \\
27204 \\
\hline
35.3652
\end{array}
$$

factors

35.3652   product of the approximate numbers

Notice that the *least accurate factor* (the 2.6) has an *accuracy* of *two significant digits*. Keeping this in mind, we could round 35.3652 to two significant digits and obtain 35 (the same result obtained by the longer maximum–minimum procedure outlined above).

This example allows us to state the following rule of thumb for multiplying and dividing approximate numbers.

RULE FOR MULTIPLYING OR DIVIDING APPROXIMATE NUMBERS

Multiply or divide the approximate numbers that are given. Then round off this result to the **accuracy** of the *least accurate* approximate number.

When working with approximate numbers, a calculator is usually used to perform the basic operations of multiplication $\boxed{\times}$ and division $\boxed{\div}$.

calculator keys

EXAMPLE 3

Using a calculator, find the product of the approximate numbers.

$$(.0024)(5.61)$$

Solution: Enter

$$\boxed{.0024} \ \boxed{\times} \ \boxed{5.61} \ \boxed{=} \qquad .013464$$

least accurate (two significant digits)   Display

The product, .013464, should be rounded to an accuracy of two significant digits, since the least accurate number (.0024) contains two significant digits.

Thus, $(.0024)(5.61) = .013$. ∎

To display a negative number on a calculator, enter the number and then press the $\boxed{+/-}$ key.

EXAMPLE 4

Using a calculator, find the quotient of the approximate numbers.

$$\frac{-16.757}{.24}$$

Solution: Enter

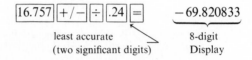

      least accurate   8-digit
   (two significant digits) Display

The quotient, -69.820833, should be rounded to an accuracy of two significant digits, since the least accurate number (.24) contains two significant digits.

Thus, $\dfrac{-16.757}{.24} = -7\bar{0}.$ ∎

        Note: The bar above the zero
        indicates that this zero is a
        significant digit.

⬚ Raising Approximate Numbers to Powers

Recall from Chapter 1 that raising a number to a power represents repeated multiplication.

$$a^n = \underbrace{(a)(a)(a)\ldots(a)}_{n \text{ factors}}$$

Thus, when raising an approximate number to a power, you should use the rule for *multiplying approximate numbers.*

To *square* a number using a calculator, enter the number and then press the $\boxed{x^2}$ key.

EXAMPLE 5

Using a calculator, evaluate $(-3.74)^2$.

Solution: Enter

$$\boxed{3.74}\ \boxed{+/-}\ \boxed{x^2} \quad \underbrace{13.9876}_{\text{Display}}$$

Since -3.74 is accurate to three significant digits, 13.9876 should also be rounded to three significant digits.

Thus, $(-3.74)^2 = 14.0.$ ∎

To raise an approximate number to a higher power, use the $\boxed{y^x}$ key on a calculator.

EXAMPLE 6

Using a calculator, evaluate $(2.593)^4$.

Solution: Enter

$$\boxed{2.593}\ \boxed{y^x}\ \boxed{4}\ \boxed{=} \qquad \underbrace{45.207456}_{\substack{\text{8-digit}\\ \text{Display}}}$$

Since 2.593 is accurate to four significant digits, 45.207456 should also be rounded to four significant digits.

Thus, $(2.593)^4 = 45.21$.   ∎

EXAMPLE 7

Using a calculator, evaluate $(-1.8)^3$.

Solution: Enter

$$\boxed{1.8}\ \boxed{+/-}\ \boxed{y^x}\ \boxed{3}\ \boxed{=} \qquad \underbrace{\overset{\textit{blinking}}{5.832}\ \ \textit{or error}}_{\text{Display}}$$

The *blinking* 5.832 represents the magnitude of the result. It is up to you to attach the correct sign to this magnitude. Recall from Chapter 1 that *a negative number to an even power is always positive, while a negative number to an odd power is always negative*. Thus,

$$(-1.8)^3 = -5.8$$

(odd) → two significant digits

If your calculator displays *error*, you must clear the calculator and proceed as follows:

$$\text{Enter}\quad \boxed{1.8}\ \boxed{y^x}\ \boxed{3}\ \boxed{=} \qquad \underbrace{5.832}_{\text{Display}}$$

Note: No $\boxed{+/-}$ is entered.

Again, it is up to you to attach the correct sign. Thus,

$$(-1.8)^3 = -5.8 \qquad ∎$$

(odd) → two significant digits

Note: You must be careful when using the $\boxed{y^x}$ key to raise a negative number to a power. It is always up to you to attach the correct sign.

⬚ Operations with Exact Numbers

An *exact* number has no limitation to the number of significant digits that it contains. For example, if the number 2 is known to be exact, then it can be thought of as 2.000 ... (the zeros continue forever). If exact numbers are mixed with approximate numbers in a calculation, then the number of significant digits retained in the answer is governed only by the approximate numbers involved.

EXAMPLE 8

Each length of copper pipe in a stock pile measures 11.75 ft. What is the *total* length when 17 pieces of this pipe are laid end to end?

Solution: Enter

$$\overset{\text{exact number}}{\boxed{11.75}\times\boxed{17}\boxed{=}} \quad \underset{\text{Display}}{199.75}$$

approximate number

The number of significant digits to be retained depends entirely upon the approximate number 11.75 ft. Thus, the *total* length should be recorded as 199.8 ft (retain four significant digits).   ∎

Application

ELECTRICAL RESISTANCES IN PARALLEL

In electronics when two resistors are connected in *parallel* (see Fig. 3.4), the total resistance R_t between points A and B can be found by dividing the sum of the resistances into the product of the resistances.

FORMULA: Total Parallel Resistance

$$R_t = \frac{R_1 R_2}{R_1 + R_2}$$

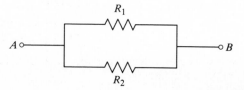

Figure 3.4

EXAMPLE 9

Two resistors are measured on an ohmmeter and found to have values of 68Ω and 330Ω. What is the total resistance R_t when these resistors are connected in parallel?

Solution:

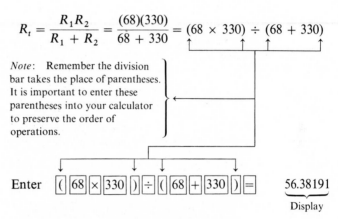

$$R_t = \frac{R_1 R_2}{R_1 + R_2} = \frac{(68)(330)}{68 + 330} = (68 \times 330) \div (68 + 330)$$

Note: Remember the division bar takes the place of parentheses. It is important to enter these parentheses into your calculator to preserve the order of operations.

Enter (68 × 330) ÷ (68 + 330) =   56.38191

Display

Thus, $R_t = 56\,\Omega$ (retain two significant digits). ∎

EXERCISES 3.3

Using the rules for approximate numbers and a calculator, evaluate each of the following.

1. $2.68 + 13.5$
2. $21.84 + 540$
3. $1.090 + .03 + .1534$
4. $59.002 + .0075 + .93$
5. $6.32 - 50.7$
6. $.0079 - 5.310$
7. $.006 + 5.34 - 8.002$
8. $18\overline{0}0 - 160 - 58.7$
9. $(321)(.6)$
10. $(3.04)(9.325)$
11. $(810\overline{0})(.542)(-12.2)$
12. $(-2100)(6.30)(-.032)$
13. $\dfrac{1.843}{9.0}$
14. $\dfrac{930}{-.83}$
15. $(.042)^2$
16. $(12.95)^2$
17. $(15.0)^3$
18. $(18.0)^4$
19. $(-1.063)^5$
20. $(-1.05)^6$
21. $\dfrac{-54.5}{(320)(16.0)}$
22. $\dfrac{.92}{64.1 + 30.0}$
23. $\dfrac{(-3.25)^2}{4.23 - 1.09}$
24. $\dfrac{1015.8 - 18.75}{298 + 5.645}$

25. The thickness of a microprocessor chip is .0524 in. What is the thickness of 67 such chips when piled one on top of the other?

26. One piece of copper pipe weighs 4.94 lb. What is the total weight of 16 of these pipes?

27. An electronic card sorter sorts 196 cards in 15 sec. How many cards does it sort per second?

28. An 18-ft board is cut into 7 equal parts. How long is each piece?

Find the total resistance R_t between points A and B.

29.

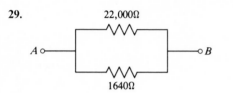

30.

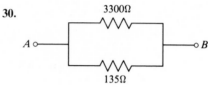

31.

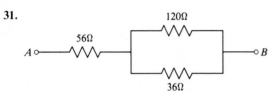

32.

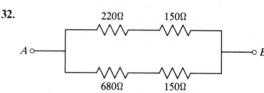

3.4 **LINEAR MEASURE**

Objectives

⊡ To convert an English unit of linear measure to another English unit of linear measure.

⊡ To convert a metric unit of linear measure to another metric unit of linear measure.

⊡ To convert an English unit of linear measure to a metric unit of linear measure or vice versa.

⊡ **The English System**

When measuring a length, width, height, depth, or any distance, we use **linear measure**. The basic unit of linear measure in the English system is called the *foot*. For smaller measurements we use the *inch* and for larger measurements we use the *yard* or *mile*.

When converting from one unit of measure to another, we multiply by a **conversion factor**. A conversion factor is a fraction that is equal to 1. It is formed by taking equal measures and placing one of the equal measures in the numerator and the other measure in the denominator of the fraction. Multiplying by a conversation factor is like multiplying by *exactly 1*. Therefore, conversion factors can be considered *exact* numbers and should be treated as such when converting from one measure to another. Listed in Table 3.1 are the common relationships and conversion factors for the units of linear measure in the English System.

TABLE 3.1

Equal Measures	Conversion Factors		
1 foot = 12 inches (1 ft = 12 in.)	$\dfrac{1 \text{ ft}}{12 \text{ in.}}$	or	$\dfrac{12 \text{ in.}}{1 \text{ ft}}$
1 yard = 3 feet (1 yd = 3 ft)	$\dfrac{1 \text{ yd}}{3 \text{ ft}}$	or	$\dfrac{3 \text{ ft}}{1 \text{ yd}}$
1 mile = 5280 feet (1 mi = 5280 ft)	$\dfrac{1 \text{ mi}}{5280 \text{ ft}}$	or	$\dfrac{5280 \text{ ft}}{1 \text{ mi}}$

EXAMPLE 1

The height of a building is 47.6 ft. How many inches is this?

Solution:

$$47.6 \text{ ft} = \frac{47.6 \text{ ft}}{1} \left(\frac{12 \text{ in.}}{1 \text{ ft}} \right) \qquad \text{multiplying by a} \atop \text{conversion factor}$$

Note: Always choose a conversion factor that allows you to *cancel* the unwanted units.

$$= (47.6)(12) \text{ in.} \qquad \text{simplifying}$$
$$= 571 \text{ in.} \quad \blacksquare \qquad \text{retain three significant digits}$$

EXAMPLE 2

The length of a highway is 3.907 mi. How many yards is this?

Solution:

$$3.907 \text{ mi} = \frac{3.907 \text{ mi}}{1} \left(\frac{5280 \text{ ft}}{1 \text{ mi}} \right) \left(\frac{1 \text{ yd}}{3 \text{ ft}} \right) \qquad \text{multiplying by} \atop \text{conversion factors}$$

Note: To obtain the desired units, more than one conversion factor may have to be applied.

$$= \frac{(3.907)(5280)}{3} \text{ yd} \qquad \text{simplifying}$$

$$= 6876 \text{ yd} \quad \blacksquare \qquad \text{retain four significant digits}$$

⊡ The Metric System

The basic unit of length in the metric system is called the *meter*. The meter is also the accepted unit of length of the Système International d'Unités (abbreviated SI).

TABLE 3.2

Prefix	Symbol	Meaning
mega	M	1,000,000
kilo	k	1000
centi	c	.01
milli	m	.001
micro	μ	.000001

SI is a uniform system of measurements used by most industrialized nations. The meter is slightly longer (about three inches longer) than a yard. Prefixes are attached to the word "meter" in order to represent distances that are greater than or less than a meter. The most commonly used prefixes are listed in Table 3.2. Table 3.3 lists some of the common relationships and conversion factors for the units of linear measure in the metric system. The great advantage of the metric system is that it is based on multiples of ten. Changing from one unit of measure to another unit of measure in the system simply means changing the location of the decimal point.

EXAMPLE 3

The length of a ski trail is 2.64 km. How many meters is this?

Solution: $2.64 \text{ km} = \dfrac{2.64 \text{ km}}{1}\left(\dfrac{1000 \text{ m}}{1 \text{ km}}\right)$ multiplying by a conversion factor

$= (2.64)(1000) \text{ m}$ simplifying

Note: To *multiply* by 1000, simply move the decimal point *three* places to the *right.*

$= 2640 \text{ m}$ ■   retain three significant digits

TABLE 3.3

Equal Measures	Conversion Factors	
1 meter = 100 centimeters (1 m = 100 cm)	$\dfrac{1 \text{ m}}{100 \text{ cm}}$ or	$\dfrac{100 \text{ cm}}{1 \text{ m}}$
1 meter = 1000 millimeters (1 m = 1000 mm)	$\dfrac{1 \text{ m}}{1000 \text{ mm}}$ or	$\dfrac{1000 \text{ mm}}{1 \text{ m}}$
1 kilometer = 1000 meters (1 km = 1000 m)	$\dfrac{1 \text{ km}}{1000 \text{ m}}$ or	$\dfrac{1000 \text{ m}}{1 \text{ km}}$

EXAMPLE 4

The major diameter of a bolt is 19.0 mm. How many centimeters is this?

$$\textit{Solution:} \quad 19.0 \text{ mm} = \frac{19.0 \cancel{\text{ mm}}}{1}\left(\frac{1 \cancel{\text{ m}}}{1000 \cancel{\text{ mm}}}\right)\left(\frac{100 \text{ cm}}{1 \cancel{\text{ m}}}\right) \quad \begin{array}{l}\text{multiplying by}\\ \text{conversion factors}\end{array}$$

$$= \frac{\overset{1}{(19.0)(\cancel{100})}}{\underset{10}{\cancel{1000}}} \text{ cm} \qquad \text{cancellation law}$$

$$= \frac{19.0}{10} \text{ cm} \qquad \text{simplifying}$$

Note: To *divide* by 10, simply move the decimal point *one* place to the *left.*

$$= 1.90 \text{ cm} \quad \blacksquare \qquad \begin{array}{l}\text{retain three}\\ \text{significant digits}\end{array}$$

∴ Conversions Between the Systems

In order to convert from the English system of measurement to the metric system, or vice versa, it is necessary to compare the basic units of length in each system and formulate a conversion factor. We will use the fact that 1 ft and .3048 m are exactly equal measures. For smaller units of measure, we will use the fact that 1 in. and 2.54 cm are exactly equal measures (see Table 3.4).

EXAMPLE 5

An artesian well is drilled to a depth of 102 m. How many feet is this?

$$\textit{Solution:} \quad 102 \text{ m} = \frac{102 \cancel{\text{ m}}}{1}\left(\frac{1 \text{ ft}}{.3048 \cancel{\text{ m}}}\right) \quad \begin{array}{l}\text{multiplying by a}\\ \text{conversion factor}\end{array}$$

$$= \frac{102}{.3048} \text{ ft} \qquad \text{simplifying}$$

$$= 335 \text{ ft} \quad \blacksquare \qquad \text{retain three significant digits}$$

TABLE 3.4

Equal Measures	Conversion Factors	
1 foot = .3048 meters (1 ft = .3048 m)	$\dfrac{1 \text{ ft}}{.3048 \text{ m}}$ or	$\dfrac{.3048 \text{ m}}{1 \text{ ft}}$
1 inch = 2.54 centimeters (1 in. = 2.54 cm)	$\dfrac{1 \text{ in.}}{2.54 \text{ cm}}$ or	$\dfrac{2.54 \text{ cm}}{1 \text{ in.}}$

TABLE 3.5 Decimal Equivalents: English-Metric

Fraction	$\frac{1}{64}$ths	Decimal Equivalent	Millimeters	Fraction	$\frac{1}{64}$ths	Decimal Equivalent	Millimeters
. .	1	.015635	0.39688	. .	33	.515625	13.09688
$\frac{1}{32}$	2	.03125	0.79375	$\frac{17}{32}$	34	.53125	13.49375
. .	3	.046875	1.19063	. .	35	.546875	13.89063
$\frac{1}{16}$	4	.0625	1.58750	$\frac{9}{16}$	36	.5625	14.28750
. .	5	.078125	1.98438	. .	37	.578125	14.68438
$\frac{3}{32}$	6	.09375	2.38125	$\frac{19}{32}$	38	.59375	15.08125
. .	7	.109375	2.77813	. .	39	.609375	15.47813
$\frac{1}{8}$	8	.125	3.17500	$\frac{5}{8}$	40	.625	15.87500
. .	9	.140625	3.57188	. .	41	.640625	16.27188
$\frac{5}{32}$	10	.15625	3.96875	$\frac{21}{32}$	42	.65625	16.66875
. .	11	.171875	4.36563	. .	43	.671875	17.06563
$\frac{3}{16}$	12	.1875	4.76250	$\frac{11}{16}$	44	.6875	17.46250
. .	13	.203125	5.15938	. .	45	.703125	17.85938
$\frac{7}{32}$	14	.21875	5.55625	$\frac{23}{32}$	46	.71875	18.25625
. .	15	.234375	5.95313	. .	47	.734375	18.65313
$\frac{1}{4}$	16	.25	6.35000	$\frac{3}{4}$	48	.75	19.05000
. .	17	.265625	6.74688	. .	49	.765625	19.44688
$\frac{9}{32}$	18	.28125	7.14375	$\frac{25}{32}$	50	.78125	19.84375
. .	19	.296875	7.54063	. .	51	.796875	20.24063
$\frac{5}{16}$	20	.3125	7.93750	$\frac{13}{16}$	52	.8125	20.63750
. .	21	.328125	8.33438	. .	53	.828125	21.03438
$\frac{11}{32}$	22	.34375	8.73125	$\frac{27}{32}$	54	.84375	21.43125
. .	23	.359375	9.12813	. .	55	.859375	21.82813
$\frac{3}{8}$	24	.375	9.52500	$\frac{7}{8}$	56	.875	22.22500
. .	25	.390625	9.92188	. .	57	.890625	22.62188
$\frac{13}{32}$	26	.40625	10.31875	$\frac{29}{32}$	58	.90625	23.01875
. .	27	.421875	10.71563	. .	59	.921875	23.41563
$\frac{7}{16}$	28	.4375	11.11250	$\frac{15}{16}$	60	.9375	23.81250
. .	29	.453125	11.50938	. .	61	.953125	24.20938
$\frac{15}{32}$	30	.46875	11.90625	$\frac{31}{32}$	62	.96875	24.60625
. .	31	.484375	12.30313	. .	63	.984375	25.00313
$\frac{1}{2}$	32	.5	12.70000	1	64	1.	25.40000

EXAMPLE 6

The pitch of a wood screw is $\frac{3}{32}$ in. How many millimeters is this? (Retain four significant digits.)

Solution:

$$\frac{3}{32}\text{ in.} = \frac{3\text{ in.}}{32}\left(\frac{2.54\text{ cm}}{1\text{ in.}}\right)\left(\frac{1\text{ m}}{100\text{ cm}}\right)\left(\frac{1000\text{ mm}}{1\text{ m}}\right) \qquad \text{multiplying by conversion factors}$$

$$= \frac{(3)(2.54)(\overset{10}{\cancel{1000}})}{(32)(\underset{1}{\cancel{100}})}\text{ mm} \qquad \text{cancellation law}$$

$$= \frac{3(25.4)}{32}\text{ mm} \qquad \text{simplifying}$$

$$= 2.381\text{ mm} \quad \blacksquare \qquad \text{retain four significant digits}$$

Note: See Table 3.5 for a list of decimal equivalents of common fractions.

EXERCISES 3.4

Convert the following measurements to the indicated units. *2, 6, 8, 12, 14*

1. 192 in. to feet
2. 3.75 ft to inches
3. 92.5 in. to yards
4. 75 yd to feet
5. 32.6 mi to feet
6. 11,576 ft to miles
7. 37 mm to centimeters
8. 587.3 cm to meters
9. 75.0 km to meter
10. 6752 m to kilometers
11. 154 m to feet
12. 17.35 ft to meters
13. 1.000 mi to kilometers
14. 1572 km to miles
15. 16.7 cm to inches
16. 1500 m to yards
17. 19.4 mm to inches
18. 18.0 mm to inches
19. $\frac{5}{8}$ in. to millimeters
20. $\frac{13}{64}$ in. to millimeters
21. $1\frac{1}{32}$ in. to millimeters
22. $2\frac{1}{2}$ in. to millimeters

Metric wrenches are usually produced and sold in 1-mm intervals while standard wrenches are usually produced and sold in $\frac{1}{64}$-in. intervals. Fill in the table below for the equivalent sized wrench. Use Table 3.5 to check your work.

	Metric Wrench Size	Standard Wrench Size
23.	8 mm	?
24.	11 mm	?
25.	?	$\frac{19}{32}$ in.
26.	?	$\frac{29}{32}$ in.
27.	21 mm	?
28.	?	$\frac{63}{64}$ in.

29. If you have only a metric drill bit set that is based on 1-mm intervals ranging from 7 mm to 19 mm, what drill bit would you select to drill a $\frac{3}{4}$-in. hole?

30. You must tap a hole in metal for a metric screw that has 11 threads per centimeter and a major diameter of 11 mm. Would a tap drill of $\frac{7}{16}$ in., 28 threads per inch suffice?

3.5 PERIMETER AND CIRCUMFERENCE

Objectives

⊡ To find the perimeter of a simple polygon.

⊡ To find the circumference of a circle.

⊡ To find the perimeter of a composite geometric figure.

⊡ Perimeter of a Simple Polygon

If three or more line segments lie in the same plane (flat surface) and are joined at their endpoints to form a closed figure, then the figure is called a **polygon**. The line segments are called the **sides** of the polygon. Shown in Fig. 3.5 are two very common polygons, the *triangle* and the *quadrilateral*.

The distance around a polygon is called its *perimeter. The perimeter of a polygon is the sum of the lengths of its sides.* Listed in Table 3.6 are the perimeters of some triangles and quadrilaterals.

EXAMPLE 1

Find the perimeter of the triangle in Fig. 3.6.

Solution:
$$P = a + b + c$$
$$= 52 \text{ ft } 9 \text{ in.} + 41 \text{ ft } 8 \text{ in.} + 30 \text{ ft } 6 \text{ in.}$$

Note:
$$\begin{array}{r} 52 \text{ ft } 9 \text{ in.} \\ 41 \text{ ft } 8 \text{ in.} \\ +30 \text{ ft } 6 \text{ in.} \\ \hline 123 \text{ ft } 23 \text{ in.} \end{array}$$

$$= 123 \text{ ft } 23 \text{ in.}$$

Note: However,
23 in. = 1 ft 11 in.

$$P = 124 \text{ ft } 11 \text{ in.} \quad \blacksquare$$

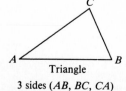
Triangle
3 sides (*AB*, *BC*, *CA*)

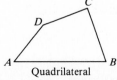
Quadrilateral
4 sides (*AB*, *BC*, *CD*, *DA*)   **Figure 3.5**

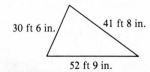
30 ft 6 in.   41 ft 8 in.
52 ft 9 in.

Figure 3.6

TABLE 3.6

Name	Figure	Perimeter (P)
Triangle: a three-sided polygon		$P = a + b + c$
Isosceles triangle: a triangle with two equal sides		$P = 2a + b$
Equilateral triangle: a triangle with all sides equal		$P = 3a$
Quadrilateral: a four-sided polygon		$P = a + b + c + d$
Parallelogram: a quadrilateral with opposite sides parallel and equal		$P = 2a + 2b$
Rhombus: a parallelogram with all sides equal		$P = 4a$
Rectangle: a parallelogram with adjacent sides perpendicular		$P = 2l + 2w$
Square: a rectangle with all sides equal		$P = 4s$

EXAMPLE 2

Find the perimeter in *inches* of the rectangle in Fig. 3.7.

13.0 cm

24.3 cm   **Figure 3.7**

Solution: $P = 2l + 2w$

$$= 2(24.3 \text{ cm}) + 2(13.0 \text{ cm})$$

$$P = 74.6 \text{ cm}$$

$$= \frac{74.6 \, \cancel{cm}}{1} \left(\frac{1 \text{ in.}}{2.54 \, \cancel{cm}} \right)$$

$$\swarrow \text{conversion factor}$$

$$P = 29.4 \text{ in.} \quad \blacksquare$$

EXAMPLE 3

Find a simplified rational expression that describes the perimeter of the isosceles triangle in Fig. 3.8.

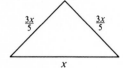

Figure 3.8

Solution: $P = 2a + b$

$$= 2\left(\frac{3x}{5}\right) + x \qquad \text{substituting}$$

$$= \frac{6x}{5} + x \qquad \begin{array}{l}\text{multiplication rule} \\ \text{for fractions}\end{array}$$

$$= \frac{6x}{5} + \frac{5\,x}{5} \qquad \begin{array}{l}\text{fundamental property} \\ \text{of fractions}\end{array}$$

$$= \frac{11x}{5} \quad \blacksquare \qquad \begin{array}{l}\text{addition rule} \\ \text{for fractions}\end{array}$$

⬚ Circumference of a Circle

A **circle** is a closed curve that lies on a plane (a flat surface) such that all points on the curve are exactly the same distance from a fixed point. The fixed point is called the **center** of the circle. In Fig. 3.9, the point O is the center of the circle. The **radius** of a circle is the length of a line segment that has one endpoint at the center of the circle and the other endpoint on the circle. In Fig. 3.9, $\overline{OA}$, $\overline{OB}$, and $\overline{OC}$ are radii of the circle. The **diameter** of a circle is the length of a line segment that passes

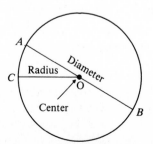

Figure 3.9

through the center of the circle and has both endpoints on the circle. In Fig. 3.9, $\overline{AB}$ is a diameter of the circle. Note that it takes two radii to form one diameter.

In any circle,

$$d = 2r$$

or $\quad r = \dfrac{d}{2}$, where d is the diameter and r is the radius of the circle.

The **circumference** of a circle is the distance around the circle. Circumference is to a circle what perimeter is to a polygon. The circumference of any circle divided by its diameter is always the same constant value. This constant value is assigned the Greek letter π (pi).

In any circle,

$$\dfrac{C}{d} = 3.14159265359\ldots = \pi,$$ where C is the circumference and d the diameter of the circle.

We can obtain a formula for the circumference of a circle by remembering that division can be *checked* by multiplication.

$$\dfrac{C}{d} = \pi, \qquad \text{the } \textit{check}: d(\pi) = C$$

Thus,

$$C = \pi d.$$

Since a diameter equals two radii ($d = 2r$), we could also write

$$C = \pi d$$
$$C = \pi(2r) \qquad \text{replacing } d \text{ with } 2r$$
$$C = 2\pi r \qquad \text{simplifying}$$

FORMULA: Circumference of a Circle

$$C = \pi d$$

or $\quad C = 2\pi r$, where C is the circumference,
d is the diameter,
r is the radius of the circle.

Pi is different from the numbers you have studied so far. It is called an **irrational number** since its decimal value never terminates or repeats. Most likely your calculator has a π button. If you push this button, you will see that π, displayed to eight digits, is

$$\pi = \underbrace{3.1415927}_{\text{8-digit display}}$$

EXAMPLE 4

Find the circumference of the circle in Fig. 3.10.

Solution:

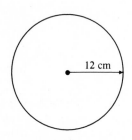

12 cm

Figure 3.10

$$
\begin{aligned}
C &= 2\pi r \\
&= 2\pi(12 \text{ cm}) \\
&= 24\pi \text{ cm} \qquad \text{(answer in terms of } \pi\text{)}
\end{aligned}
$$

Note: Using a calculator, enter $\boxed{24}\ \boxed{\times}\ \boxed{\pi}\ \boxed{=}$.

$$= 75 \text{ cm} \quad \blacksquare \qquad \text{(answer in decimal form to two significant digits)}$$

⠌ Perimeter of a Composite Geometric Figure

A composite geometric figure is one made up of simpler geometric figures such as rectangles, triangles, circles, and so on. The perimeter of a composite geometric figure that contains straight sides and circular sides is simply the *sum* of the straight parts and the circular parts.

EXAMPLE 5

Find the perimeter of the composite figure in Fig. 3.11a (a rectangle surmounted by a semicircle).

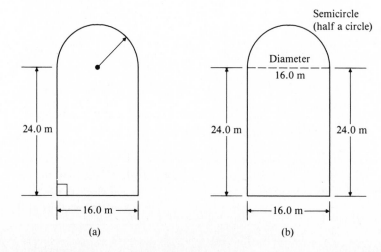

Semicircle (half a circle)

Diameter 16.0 m

24.0 m

24.0 m

24.0 m

24.0 m

|←— 16.0 m —→|

|←— 16.0 m —→|

(a)

(b)

Figure 3.11

Solution: First, find any missing dimensions (see Fig. 3.11b).

$$\text{perimeter of the straight parts} = 24.0 \text{ m} + 16.0 \text{ m} + 24.0 \text{ m} = 64.0 \text{ m}$$

$$\text{circumference of the } \textit{semicircle} = \frac{C}{2} = \frac{\pi d}{2} = \frac{\pi(16.0 \text{ m})}{2} = 25.1 \text{ m}$$

$$\underbrace{\text{total perimeter}} = \underbrace{\text{perimeter of the straight parts}} + \underbrace{\text{circumference of the semicircle}}$$

$$\longrightarrow P = \qquad 64.0 \text{ m} \qquad + \qquad 25.1 \text{ m}$$

$$P = 89.1 \text{ m} \quad \blacksquare$$

EXAMPLE 6

Find a simplified algebraic expression that describes the perimeter of the composite figure in Fig. 3.12a (a square surmounted by a quarter circle). Use 3.142 as a decimal approximation of π.

Solution: First, find any missing dimensions (see Fig. 3.12b).

$$\text{perimeter of the straight parts} = x + x + 2x = 4x$$

$$\text{circumference of the } \textit{quarter} \text{ circle} = \frac{C}{4} = \frac{\overset{1}{\cancel{2}}\pi r}{\underset{2}{\cancel{4}}} = \frac{\pi r}{2} = \frac{3.142(x)}{2} = 1.571x$$

$$\underbrace{\text{total perimeter}} = \underbrace{\text{perimeter of the straight parts}} + \underbrace{\text{circumference of the quarter circle}}$$

$$\longrightarrow P = \qquad 4x \qquad + \qquad 1.571x$$

$$P = 5.571x \quad \blacksquare$$

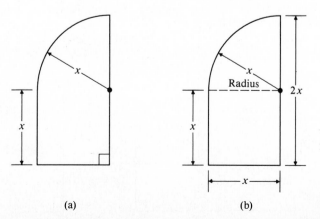

(a)　　　　　　　　　　(b)　　　　　**Figure 3.12**

EXERCISES 3.5

1. Find the perimeter of a quadrilateral whose sides measure 4 ft 3 in., 12 ft 4 in., 10 ft 8 in., and 6 ft 9 in.

2. Find the perimeter of a triangle whose sides measure 19 ft 8 in., 40 ft 7 in., and 45 ft 6 in.

3. Find the perimeter of a parallelogram whose sides measure 6 ft $8\frac{1}{4}$ in. and 14 ft $9\frac{1}{2}$ in.

4. Find the perimeter of a rectangle whose length measures 1 ft $3\frac{3}{8}$ in. and whose width measures $9\frac{1}{4}$ in.

5. Find the perimeter of an equilateral triangle if one side measures 18.6 cm.

6. Find the perimeter of an isosceles triangle whose equal sides each measure 32.25 m and whose other side measures 27.83 m.

7. Find the perimeter in *meters* of a square if one side measures 16.68 ft.

8. Find the perimeter in *feet* of a rhombus if one side measures 2.18 m.

9. Find the circumference of a circle whose diameter measures 192 mm (leave your answer in terms of π).

10. Find the circumference of a circle whose radius measures 32 in. (leave your answer in terms of π).

Find a simplified expression that describes the perimeter of the following polygons.

11.

12.
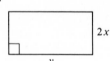

13.

14.

15.
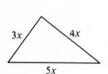

16.
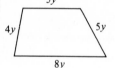

17.

18.

Find the perimeter of the following geometric figures. In exercises 23–26, use 3.142 as a decimal approximation of π.

19.
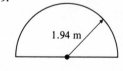

20.
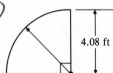

21.
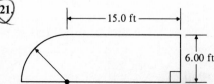

22.
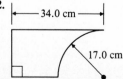

CIRCLE $\pi = 3.142$
PROD

23.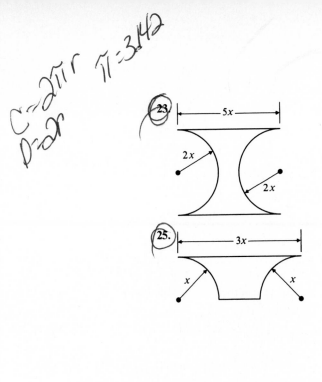

24.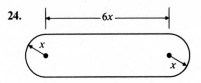

25.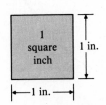

26.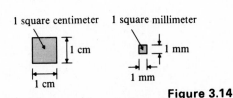

3.6 SQUARE MEASURE AND AREA

Objectives

- ☐ To convert a unit of square measure to another unit of square measure.
- ☐ To find the area of a simple geometric figure.
- ☐ To find the area of a composite geometric figure.

☐ Square Measure and Conversion

The region enclosed by a geometric figure is referred to as its **area**. The unit of measure used to find an area is the **square unit**. Each side of the square unit is exactly 1 unit in length (see Fig. 3.13). The area of a geometric figure is the number of these square units contained within the figure.

If the area is fairly small, then we use a square inch (sq in or in^2), square centimeter (sq cm or cm^2), or a square millimeter (sq mm or mm^2) as the unit of measure to find the area (see Fig. 3.14).

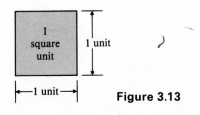

Figure 3.13

Figure 3.14

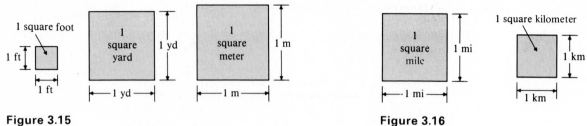

Figure 3.15

Figure 3.16

For larger areas, we would use the square foot (sq ft or ft^2), square yard (sq yd or yd^2), or square meter (sq m or m^2) (see Fig. 3.15).

For still larger areas, we would use the square mile (sq mi or mi^2) or square kilometer (sq km or km^2) (see Fig. 3.16).

In technical work it is often necessary to change a measurement of area from one unit of measure to another.

EXAMPLE 1

The area of a floor in a room is 144 sq ft (144 ft^2). How many square yards (yd^2) is this?

Solution:

$$144 \text{ ft}^2 = \frac{144 \text{ ft}^2}{1}\left(\frac{1 \text{ yd}}{3 \text{ ft}}\right)\left(\frac{1 \text{ yd}}{3 \text{ ft}}\right) \qquad \text{multiplying by a conversion factor}$$

Note: The conversion factor is applied *twice* to cancel ft^2.

$$= \frac{144}{9} \text{ yd}^2 \qquad \text{simplifying}$$

$$= 16.0 \text{ yd}^2 \quad \blacksquare \qquad \text{retain three significant digits}$$

EXAMPLE 2

A piece of land has an area of 6100 sq m (6100 m^2). How many square feet (ft^2) is this?

Solution:

$$6100 \text{ m}^2 = \frac{6100 \text{ m}^2}{1}\left(\frac{1 \text{ ft}}{.3048 \text{ m}}\right)\left(\frac{1 \text{ ft}}{.3048 \text{ m}}\right) \qquad \text{multiplying by a conversion factor}$$

$$= \frac{6100}{(.3048)^2} \text{ ft}^2 \qquad \text{simplifying}$$

$$= 66{,}000 \text{ ft}^2 \quad \blacksquare \qquad \text{retain two significant digits}$$

TABLE 3.7

Equal Measures	Conversion Factor
1 acre = 43,560 square feet (1 Ac = 43,560 ft²)	$\dfrac{1 \text{ Ac}}{43{,}560 \text{ ft}^2}$ or $\dfrac{43{,}560 \text{ ft}^2}{1 \text{ Ac}}$

The area of a large piece of land is often written in terms of *acres*. One acre is equivalent to 43,560 sq ft (see Table 3.7).

EXAMPLE 3

A piece of land has an area of 97,460 sq ft. How many acres is this?

Solution:

$$97{,}460 \text{ ft}^2 = \frac{97{,}460 \ \cancel{\text{ft}^2}}{1} \left(\frac{1 \text{ Ac}}{43{,}560 \ \cancel{\text{ft}^2}} \right) \qquad \text{multiplying by a conversion factor}$$

$$= \frac{97{,}460}{43{,}560} \text{ Ac} \qquad \text{simplifying}$$

$$= 2.237 \text{ Ac} \quad \blacksquare \qquad \text{retain four significant digits}$$

⊡ Area of a Simple Geometric Figure

Suppose we wished to find the area of a rectangle whose length is 3 in. and whose width is 2 in. The *number* of square inches contained within the rectangle is its area (see Fig. 3.17). By counting the squares in Fig. 3.17, we find that there are six square inches (6 in²) enclosed within this rectangle. Therefore, the area A of the rectangle is 6 sq in. ($A = 6 \text{ in}^2$). We didn't actually have to count the number of squares. We could have found the same result by multiplying the length (3 in.) by the width (2 in.) as follows:

$$A = (3 \text{ in.})(2 \text{ in.}) = 6 \text{ in}^2$$

The area of a rectangle can always be indirectly found by multiplying its length (l) by its width (w), provided the length and width have the same unit of measure.

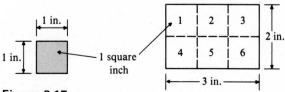

Figure 3.17

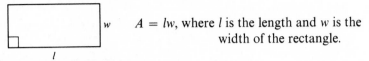

FORMULA: **Area of a Rectangle**

$A = lw$, where l is the length and w is the width of the rectangle.

A square is a rectangle whose length equals its width. Thus, if s is the length of a side of a square, then $A = lw$ becomes $A = (s)(s) = s^2$.

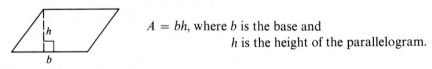

FORMULA: **Area of a Square**

$A = s^2$, where s is the side of the square.

If we cut the triangle off the end of the parallelogram as shown in Fig. 3.18, and place this triangle on the other end of the parallelogram, we would obtain a *rectangle* of length b and width h. Thus, the area of a parallelogram can be found by multiplying its base (b) by its height (h).

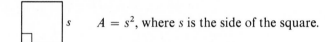

FORMULA: **Area of a Parallelogram**

$A = bh$, where b is the base and h is the height of the parallelogram.

Notice that in a parallelogram, the height (h) is the *perpendicular distance* to the base (b). If we cut the parallelogram as shown in Fig. 3.19, two identical triangles would be formed. The area of each triangle would be one-half that of the entire parallelogram's area. Thus the area of each triangle is $\dfrac{bh}{2}$.

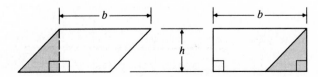

Figure 3.18

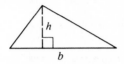

Figure 3.19

FORMULA: **Area of a Triangle**

$A = \dfrac{bh}{2}$, where b is the base and

h is the height of the triangle.

Notice that in a triangle, the height (h) is the *perpendicular distance* to the base (b).

Suppose we cut the area of a circle into several pie-shaped pieces as shown in Fig. 3.20a, and then rearranged these pie-shaped pieces as shown in Fig. 3.20b. As you can see, the area that is formed in Fig. 3.20b is very close to the area of a parallelogram.

The *base* of the parallelogram is approximately *half the circumference* $\left(\dfrac{C}{2}\right)$ of the circle, and the *height* is approximately the *radius* of the circle. The greater the number of pie-shaped pieces, the better the area of a parallelogram approximates the area in Fig. 3.20b. Recall that the area of a parallelogram is the product of its base (b) and height (h). Thus,

$$A = bh \qquad \text{area of a parallelogram}$$

$$A = \frac{C}{2}\,(r) \qquad \text{replacing } b \text{ with } \frac{C}{2} \text{ and } h \text{ with } r$$

$$A = \frac{2\pi r}{2}\,(r) \qquad \text{replacing } C \text{ with } 2\pi r$$

$$A = \frac{\overset{1}{\cancel{2}} \pi r}{\underset{1}{\cancel{2}}}\,(r) \qquad \text{cancellation law}$$

$$A = \pi r^2 \qquad \text{simplifying}$$

(a)

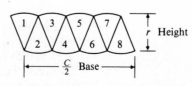

(b)

Figure 3.20

FORMULA: Area of a Circle

 $A = \pi r^2$, where r is the radius of the circle.

EXAMPLE 4

Find the area of the rectangle in Fig. 3.21.

Solution:

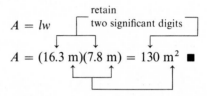

$$A = lw$$

retain two significant digits

$$A = (16.3 \text{ m})(7.8 \text{ m}) = 130 \text{ m}^2 \quad \blacksquare$$

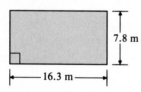

Figure 3.21

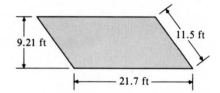

Figure 3.22

EXAMPLE 5

Find the area in *square meters* for the parallelogram in Fig. 3.22.

Solution:

$$A = bh$$

$$A = (21.7 \text{ ft})(9.21 \text{ ft}) = 20\bar{0} \text{ ft}^2$$

$$= \frac{20\bar{0} \text{ ft}^2}{1} \cdot \frac{.3048 \text{ m}}{1 \text{ ft}} \cdot \frac{.3048 \text{ m}}{1 \text{ ft}}$$

Note: The conversion factor is applied twice to cancel ft².

$$= (20\bar{0})(.3048)^2 \text{ m}^2$$

$$A = 18.6 \text{ m}^2 \quad \blacksquare$$

EXAMPLE 6

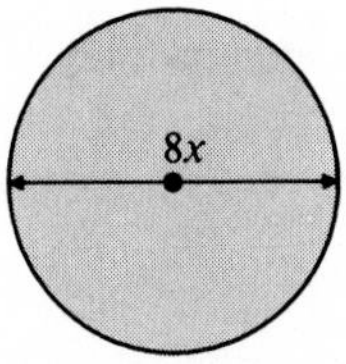

Find a simplified algebraic expression that describes the area of the circle shown in Fig. 3.23. Leave your answer in terms of π.

Solution:

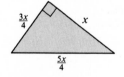

Figure 3.23

$$r = \frac{d}{2} = \frac{8x}{2} = \frac{\overset{4x}{\cancel{8}x}}{\underset{1}{\cancel{2}}} = 4x$$

$$A = \pi r^2 = \pi(4x)^2 \qquad \text{substituting}$$

Note: $(4x)^2$ means $(4x)(4x) = 16x^2$.

$$= \pi(16x^2) \qquad \text{squaring}$$

$$A = 16\pi x^2 \quad \blacksquare \qquad \text{preferred notation}$$

If two sides of a triangle are perpendicular to each other, the triangle is called a **right triangle**. In a right triangle, the perpendicular sides form the *base* and *height* of the triangle.

EXAMPLE 7

Find a simplified rational expression that describes the area of the right triangle in Fig. 3.24.

Solution:

$$A = \frac{bh}{2}$$

$$= \frac{(x)\left(\dfrac{3x}{4}\right)}{2} \qquad \text{substituting}$$

$$= \frac{\dfrac{3x^2}{4}}{2} \qquad \text{multiplication rule for fractions}$$

$$= \frac{3x^2}{4} \cdot \frac{1}{2} \qquad \text{division rule for fractions}$$

$$A = \frac{3x^2}{8} \quad \blacksquare \qquad \text{multiplication rule for fractions}$$

Figure 3.24

⠻ Area of a Composite Geometric Figure

If we split a composite geometric figure into simpler geometric shapes such as rectangles, triangles, circles, and so on, then its area can be found by *summing* the areas of the simpler geometric shapes.

EXAMPLE 8

Find the area of the composite figure in Fig. 3.25a.

Solution: First split the composite figure into simpler shapes and find any missing dimensions (see Fig. 3.25b).

$$\text{area of the semicircle} = \frac{\pi r^2}{2} = \frac{\pi (3.34 \text{ m})^2}{2} = 17.5 \text{ m}^2$$

$$\text{area of the square} = s^2 = (3.34 \text{ m})^2 = 11.2 \text{ m}^2$$

$$\underbrace{\text{total area}} = \underbrace{\begin{array}{c}\text{area of the}\\\text{semicircle}\end{array}} + \underbrace{\begin{array}{c}\text{area of the}\\\text{square}\end{array}}$$

$$A = 17.5 \text{ m}^2 \quad + \quad 11.2 \text{ m}^2$$

$$A = 28.7 \text{ m}^2 \quad \blacksquare$$

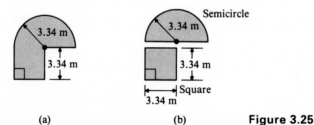

(a)   (b)   **Figure 3.25**

EXAMPLE 9

Find a simplified algebraic expression that describes the area of the composite figure in Fig. 3.26a.

Solution: First, split the composite figure into simpler shapes and find any missing dimensions (see Fig. 3.26b).

$$\text{area of the rectangle } A_1 = lw = (6x)(x) = 6x^2$$

$$\text{area of the rectangle } A_2 = lw = (5x)(x) = 5x^2$$

$$\text{area of the rectangle } A_3 = lw = (6x)(x) = 6x^2$$

$$\text{total area} = A_1 + A_2 + A_3$$

$$A = 6x^2 + 5x^2 + 6x^2 = 17x^2 \quad \blacksquare$$

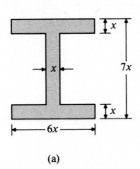
(a)

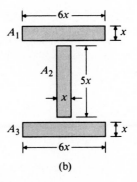
(b)

Figure 3.26

Application

AIRCRAFT WINGS

In the design of an aircraft wing, the *aspect ratio* has a critical effect on the flying characteristics of the airplane. The aspect ratio is defined as the *square of the span* of the wing divided by the total *area* of the wing. The span is the length of the wing from wingtip to wingtip.

> **FORMULA: Aspect Ratio**
>
> $$\text{a.r.} = \frac{S^2}{A},$$ where S is the span and A is the area of the wing.

The greater the aspect ratio, the greater the lateral stability of the aircraft, which means the aircraft will not roll (see Fig. 3.27).

EXAMPLE 10

Find the aspect ratio of the wing shown in Fig. 3.28.

Solution: The span of the wing is the total distance from wingtip to wingtip.

$$S = 3.00 \text{ ft} + 18.0 \text{ ft} + 3.00 \text{ ft} = 24.0 \text{ ft}$$

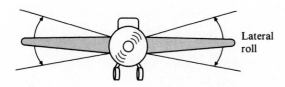

Lateral roll

Figure 3.27

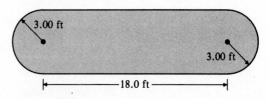

3.00 ft

3.00 ft

18.0 ft

Figure 3.28

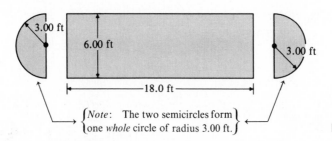

Figure 3.29

To find the area of the wing, split the wing into simpler shapes and find any missing dimensions as shown in Fig. 3.29.

$$\text{area of the } two \text{ semicircles} = \pi r^2 = \pi(3.00 \text{ ft})^2 = 28.3 \text{ ft}^2$$
$$\text{area of the rectangle} = lw = (18.0 \text{ ft})(6.00 \text{ ft}) = 108 \text{ ft}^2$$
$$\text{total area} = 28.3 \text{ ft}^2 + 108 \text{ ft}^2 = 136.3 \text{ ft}^2$$

Thus,

$$\text{a.r.} = \frac{S^2}{A} = \frac{(24.0 \text{ ft})^2}{136.3 \text{ ft}^2} = \frac{576 \text{ ft}^2}{136.3 \text{ ft}^2} = 4.23 \quad \blacksquare$$

Note: Any aspect ratio between the values of 4.00 and 6.00 would signify a wing design with *average* lateral stability.

EXERCISES 3.6

Convert the following measurements to the indicated units.

1. 1 sq yd to square feet
2. 918 sq ft to square yards
3. 170,000 sq ft to acres
4. 3.97 Ac to square feet
5. 32.6 m² to square feet
6. 574.2 ft² to square meters
7. 57,500 cm² to square meters
8. 1.895 cm² to square millimeters
9. 68,000 ft² to square kilometers
10. 79.06 cm² to square inches
11. $6\frac{3}{4}$ sq ft to square yards
12. $28\frac{4}{5}$ in² to square feet
13. 1.00 sq mi to acres
14. 957.2 Ac to square kilometers
15. Find the area of a rectangle whose length measures 6.28 ft and whose width measures 4.02 ft.
16. Find the area of a square if one of its sides measures 16.68 km.
17. Find the area of a triangle whose base measures 28.0 m and whose height measures 8.6 m.
18. Find the area of a parallelogram whose base measures 12.1 cm and whose height measures 7.9 cm.
19. Find the area of a circle whose radius measures 22.0 in.
20. Find the area of a circle whose diameter measures 8.6 mm.
21. Find the area in *square feet* of a square whose side measures 3 ft 4 in.

22. Find the area in *square feet* of a triangle whose base is 28 ft 6 in. and whose height is 6 ft 8 in.

23. Find the area in *square meters* of a circle whose diameter measures 8.8 ft.

24. Find the area in *acres* of a rectangle whose length measures 1195 ft and whose width measures 425 ft.

Find a simplified expression that describes the area of the following simple geometric figures.

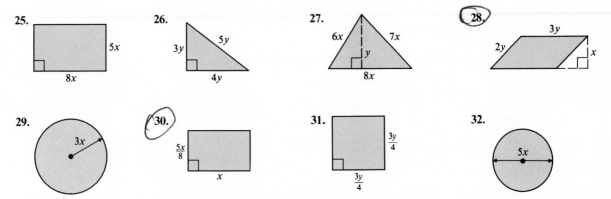

25.   5x   8x

26.   3y   5y   4y

27.   6x   7x   y   8x

28.   3y   2y   x

29.   3x

30. $\frac{5x}{8}$ x

31. $\frac{3y}{4}$ $\frac{3y}{4}$

32.   5x

Find the area of the following composite figures. In exercises 39 and 40, use 3.142 as a decimal approximation for π.

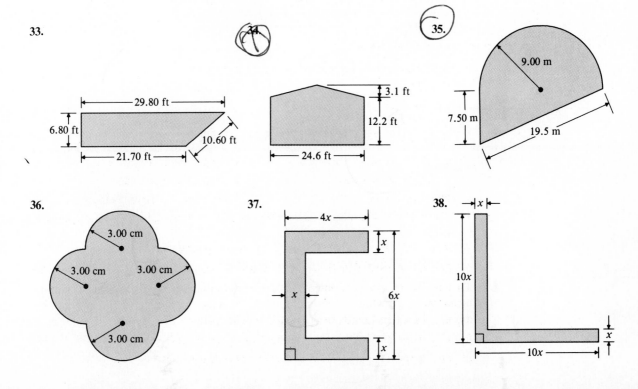

33.   29.80 ft   6.80 ft   10.60 ft   21.70 ft

34.   3.1 ft   12.2 ft   24.6 ft

35.   9.00 m   7.50 m   19.5 m

36.   3.00 cm   3.00 cm   3.00 cm   3.00 cm

37.   4x   x   x   6x

38.   x   10x   10x   x

39.

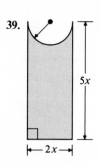

40.

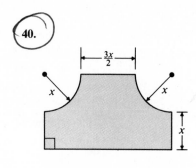

Find the aspect ratio for the following aircraft wings.

41.

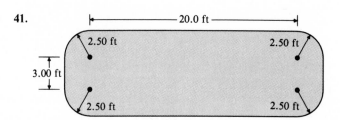

42.

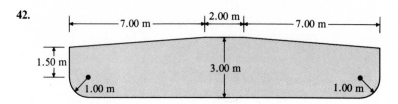

3.7 CUBIC MEASURE AND VOLUME

Objectives

☐ To convert a unit of cubic measure to another unit of cubic measure.

☐ To find the volume of a right prism.

☐ To find the volume of a cylinder.

☐ Cubic Measure and Conversion

The amount of space occupied by a geometric solid is called its **volume**. The unit of measure used to find a volume is the   **unit**. Each edge of the cubic unit is exactly 1 unit in length (see Fig. 3.30). The volume of a geometric solid is the number of these cubic units contained within the solid. If the volume is fairly small, then we use a cubic inch (cu in. or in^3), cubic centimeter (cu cm or cm^3) or

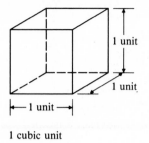

1 cubic unit

Figure 3.30

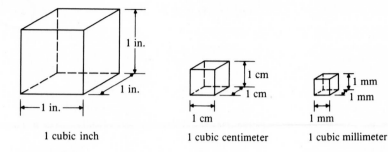

1 cubic inch

Figure 3.31

1 cubic centimeter

1 cubic millimeter

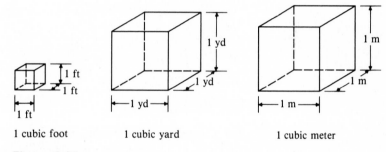

1 cubic foot   1 cubic yard   1 cubic meter

Figure 3.32

a cubic millimeter (cu mm or mm³) as the unit of measure to find the volume (see Fig. 3.31). For larger volumes, we would use the cubic foot (cu ft or ft³), cubic yard (cu yd or yd³), or the cubic meter (cu m or m³) (see Fig. 3.32).

In the English system, the **gallon** is generally used to express the volume occupied by a liquid. One gallon of a liquid occupies a space equivalent to 231 cu in. (231 in³). In the metric system, the **liter** is generally used to express the volume occupied by a liquid or gas. One liter of a liquid or gas occupies a space equivalent to 1000 cu cm (1000 cm³) (see Table 3.8). In technical work, it is often necessary to change a measurement of volume from one unit of measure to another.

TABLE 3.8

Equal Measures	Conversion Factors
1 gallon = 231 cubic inches (1 gal = 231 in³)	$\dfrac{1\ \text{gal}}{231\ \text{in}^3}$ or $\dfrac{231\ \text{in}^3}{1\ \text{gal}}$
1 liter = 1000 cubic centimeters (1 L = 1000 cm³)	$\dfrac{1\ \text{L}}{1000\ \text{cm}^3}$ or $\dfrac{1000\ \text{cm}^3}{1\ \text{L}}$

EXAMPLE 1

A foundation requires 1700 cu ft (1700 ft^3) of concrete. How many cubic yards (yd^3) is this?

Solution:

$$1700 \text{ ft}^3 = \frac{1700 \cancel{\text{ft}^3}}{1}\left(\frac{1 \text{ yd}}{3 \cancel{\text{ft}}}\right)\left(\frac{1 \text{ yd}}{3 \cancel{\text{ft}}}\right)\left(\frac{1 \text{ yd}}{3 \cancel{\text{ft}}}\right) \qquad \text{multiplying by a conversion factor}$$

Note: The conversion factor is applied three times to cancel ft^3.

$$= \frac{1700}{27} \text{ yd}^3 \qquad \text{simplifying}$$

$$= 63 \text{ yd}^3 \quad \blacksquare \qquad \text{retain two significant digits}$$

EXAMPLE 2

The displacement of a four-cylinder automobile engine is rated as 1.8 liters (1.8 L). How many cubic inches (in^3) is this?

Solution:

$$1.8 \text{ L} = \frac{1.8 \cancel{\text{L}}}{1}\left(\frac{1000 \cancel{\text{cm}^3}}{1 \cancel{\text{L}}}\right)\left(\frac{1 \text{ in.}}{2.54 \cancel{\text{cm}}}\right)\left(\frac{1 \text{ in.}}{2.54 \cancel{\text{cm}}}\right)\left(\frac{1 \text{ in.}}{2.54 \cancel{\text{cm}}}\right) \qquad \text{multiplying by conversion factors}$$

$$= \frac{1800}{(2.54)^3} \text{ in}^3 \qquad \text{simplifying}$$

$$= 110 \text{ in}^3 \quad \blacksquare \qquad \text{retain two significant digits}$$

EXAMPLE 3

A wheelbarrow is rated as having a 4.0 cu ft (4.0 ft^3) capacity. How many gallons of water will it hold?

Solution:

$$4.0 \text{ ft}^3 = \frac{4.0 \cancel{\text{ft}^3}}{1}\left(\frac{12 \cancel{\text{in.}}}{1 \cancel{\text{ft}}}\right)\left(\frac{12 \cancel{\text{in.}}}{1 \cancel{\text{ft}}}\right)\left(\frac{12 \cancel{\text{in.}}}{1 \cancel{\text{ft}}}\right)\left(\frac{1 \text{ gal}}{231 \cancel{\text{in}^3}}\right) \qquad \text{multiplying by conversion factors}$$

$$= \frac{(4.0)(12)^3}{231} \text{ gal} \qquad \text{simplifying}$$

$$= 3\bar{0} \text{ gal} \quad \blacksquare \qquad \text{retain two significant digits}$$

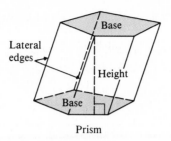

Figure 3.33

Figure 3.34

⠫ Volume of a Prism

If two identical polygons lie in different but parallel planes (flat surfaces) and their corresponding vertices are joined with parallel line segments, then a **geometric solid** is formed that is called a **prism**. The identical and parallel polygonal figures are called the **bases** of the prism, and the perpendicular distance between the bases is called the **height** of the prism (see Fig. 3.33). If the **lateral edges** of a prism are perpendicular to the bases, then the prism is referred to as a **right prism** (see Fig. 3.34). In this section, we will work only with right prisms.

Suppose we wish to find the volume of the right prism shown in Fig. 3.35. The number of cubic inches contained within the prism is its volume. By counting the cubes in Fig. 3.35, we find that there are 9 cubes in row 1, 9 cubes in row 2, and 9 cubes in row 3. Therefore, there are 27 cubes contained within the prism. Its volume (V) is thus 27 cu in., which is abbreviated $V = 27\ \text{in}^3$.

We didn't actually have to count the number of cubes. Notice that there are 9 *square inches* ($9\ \text{in}^2$) contained in the *base* of the prism in Fig. 3.35. If we multiply the *area of the base* ($9\ \text{in}^2$) by the *height* of the prism (3 in.), we would obtain the volume of $27\ \text{in}^3$ as follows:

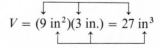

$$V = (9\ \text{in}^2)(3\ \text{in.}) = 27\ \text{in}^3$$

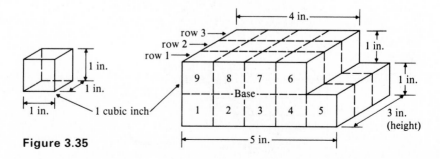

Figure 3.35

The volume of any prism can be indirectly found by multiplying the *area of its base* times its *height*.

FORMULA: **Volume of a Prism**

$V = BH$, where B is the area of the base and
H is the height of the prism.

EXAMPLE 4

Find the volume of the prism shown in Fig. 3.36.

Solution: The bases of the prism are the right triangles. Thus,

$$V = BH = \left(\frac{bh}{2}\right)H = \frac{(15.4 \text{ ft})(12.8 \text{ ft})}{2} \cdot (25.0 \text{ ft})$$

area of a
triangle

$$= \frac{(15.4)(12.8)(25.0)}{2} \text{ ft}^3$$

$$V = 2460 \text{ ft}^3 \quad \blacksquare$$

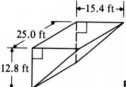

Figure 3.36

If all the faces of a prism are rectangles, the prism is called a *rectangular box*. In a rectangular box, any opposite pair of faces can be chosen as the bases, since all opposite faces are identical and parallel. In the rectangular box shown in Fig. 3.37, let us choose the top and bottom as the bases. Thus,

$$V = BH \text{ becomes } V = (lw)H \text{ or } V = lwh.$$

area of a
rectangle   height

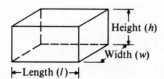

Rectangular box   **Figure 3.37**

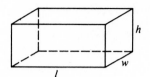

FORMULA: **Volume of a Rectangular Box**

$V = lwh$, where l is the length,
$\quad\quad\quad w$ is the width, and
$\quad\quad\quad h$ is the height of the box.

A rectangular box having all its edges the same length is called a **cube** (see Fig. 3.38). For a cube with edge s,

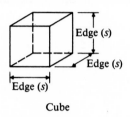

Edge (s)
Edge (s)
Edge (s)

Cube

Figure 3.38

$$V = BH \text{ becomes } V = (s^2)s \text{ or } V = s^3.$$

with "area of a square" labeling s^2 and "height" labeling s.

FORMULA: **Volume of a Cube**

$V = s^3$, where s is the length of an edge of the cube.

EXAMPLE 5

Find the volume in *cubic centimeters* for the cube in Fig. 3.39.

Solution:

3.75 in.

Figure 3.39

$$V = s^3 = (3.75 \text{ in.})^3 = 52.7 \text{ in}^3$$

$$= \frac{52.7 \text{ in}^3}{1} \left(\frac{2.54 \text{ cm}}{1 \text{ in.}}\right)\left(\frac{2.54 \text{ cm}}{1 \text{ in.}}\right)\left(\frac{2.54 \text{ cm}}{1 \text{ in.}}\right)$$

Note: The conversion factor
is applied three times to
cancel in³.

$$= (52.7)(2.54)^3 \text{ cm}^3$$
$$V = 864 \text{ cm}^3 \quad\blacksquare$$

EXAMPLE 6

Find a simplified algebraic expression that describes the volume of the rectangular box in Fig. 3.40.

Solution:

$$V = lwh$$

$$= (3x)(2x)\left(\frac{3x}{2}\right) \qquad \text{substituting}$$

$$= \frac{(3x)(2x)(3x)}{2} \qquad \text{multiplication rule for fractions}$$

$$= \frac{(3x)(\overset{x}{\cancel{2x}})(3x)}{\underset{1}{\cancel{2}}} \qquad \text{cancellation law}$$

$$V = 9x^3 \quad \blacksquare \qquad \text{simplifying}$$

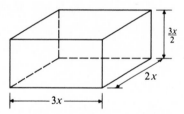

Figure 3.40

⠇ Volume of a Cylinder

A geometric solid with two identical and parallel circular bases is called a *cylinder*. The perpendicular distance between the bases is the height of the cylinder (see Fig. 3.41).

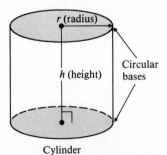

Cylinder   **Figure 3.41**

Since a cylinder can be considered a **circular prism**, the same basic formula, $V = BH$, applies for the volume of a cylinder. Thus,

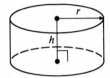

$$V = BH \text{ becomes } V = (\pi r^2)h \text{ or } V = \pi r^2 h.$$

FORMULA: Volume of a Cylinder

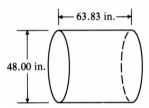

$V = \pi r^2 h$, where r is the radius and
h is the height of the cylinder.

EXAMPLE 7

How many *gallons* of oil can be stored in the oil drum shown in Fig. 3.42?

← 63.83 in. →
48.00 in.

Figure 3.42

Solution:

$$r = \frac{d}{2} = \frac{48.00 \text{ in.}}{2} = 24.00 \text{ in.}$$

$$V = \pi r^2 h = \pi(24.00 \text{ in})^2(63.83 \text{ in}) = 115{,}500 \text{ in}^3$$

$$= \frac{115{,}500 \text{ in}^3}{1}\left(\frac{1 \text{ gal}}{231 \text{ in}^3}\right)$$

$$V = 500.0 \text{ gal} \quad \blacksquare$$

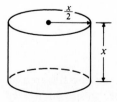

Figure 3.43

EXAMPLE 8

Find a simplified rational expression that describes the volume of the cylinder in Fig. 3.43. Leave your answer in terms of π.

Solution:

$$V = \pi r^2 h$$

$$= \pi \left(\frac{x}{2}\right)^2 (x) \qquad \text{substituting}$$

Note: $\left(\frac{x}{2}\right)^2$ means $\left(\frac{x}{2}\right)\left(\frac{x}{2}\right) = \frac{x^2}{4}.$

$$= \pi \left(\frac{x^2}{4}\right)(x) \qquad \text{squaring}$$

$$= \frac{\pi(x^2)(x)}{4} \qquad \text{multiplication rule for fractions}$$

$$= \frac{\pi x^3}{4} \quad \blacksquare \qquad \text{simplifying}$$

Application

SWIMMING POOLS

Most inground swimming pools contain a shallow end that gradually slopes to a deep end as shown in Fig. 3.44. The swimming pool can be considered a prism with a *side wall* as the *base* and the distance between the side walls (20.0 ft) as the *height*.

EXAMPLE 9

How many gallons of water will it take to fill the swimming pool in Fig. 3.44?

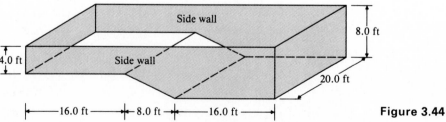

Figure 3.44

Solution: We must first find its volume in terms of cubic feet. To find the area of the base (side wall) split the base into simpler geometric shapes as shown in Fig. 3.45.

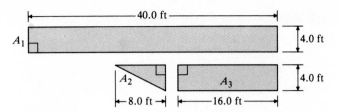

Figure 3.45

area of the rectangle $A_1 = lw = (40.0 \text{ ft})(4.0 \text{ ft}) = 160 \text{ ft}^2$

area of the triangle $A_2 = \dfrac{bh}{2} = \dfrac{(8.0 \text{ ft})(4.0 \text{ ft})}{2} = 16 \text{ ft}^2$

area of the rectangle $A_3 = lw = (16.0 \text{ ft})(4.0 \text{ ft}) = 64 \text{ ft}^2$

total base area $= A_1 + A_2 + A_3 = 160 \text{ ft}^2 + 16 \text{ ft}^2 + 64 \text{ ft}^2 = 240 \text{ ft}^2$

Thus,

$$V = BH = (240 \text{ ft}^2)(20.0 \text{ ft})$$
$$= 4800 \text{ ft}^3$$
$$= \frac{4800 \text{ ft}^3}{1}\left(\frac{12 \text{ in}}{1 \text{ ft}}\right)\left(\frac{12 \text{ in}}{1 \text{ ft}}\right)\left(\frac{12 \text{ in}}{1 \text{ ft}}\right)\left(\frac{1 \text{ gal}}{231 \text{ in}^3}\right)$$
$$= \frac{(4800)(12)^3}{231} \text{ gal}$$
$$V = 36{,}000 \text{ gal} \quad \blacksquare$$

EXERCISES 3.7

Convert the following measurements to the indicated units.

1. 1.0 cu yd to cubic feet
2. 513 cu ft to cubic yards
3. .053 m³ to cubic centimeters
4. 189,000 cm³ to cubic meters
5. 173.9 cm³ to cubic inches
6. .0075 in³ to cubic mm
7. 175,000 in³ to cubic feet
8. 207,000 in³ to cubic yards
9. 5.00 L to cubic inches
10. 427 cu in. to liters
11. 7.5 gal to cubic feet
12. $\frac{1}{7}$ gal to cubic inches
13. 1.00 gal to liters
14. 1.00 L to quarts (Hint: 4 qt = 1 gal)
15. Find the volume of a rectangular box whose length measures 13.2 m, width measures 10.6 m, and height measures 11.8 m.
16. Find the volume of a cube if one of its edges measures 23.4 in.
17. Find the volume in *cubic meters* of a cube whose edge measures 67.9 ft.
18. Find the volume in *cubic yards* of a rectangular box whose length and width each measure 24.8 ft and whose height measures 5.7 ft.

19. Find the volume of a cylinder whose diameter measures 7.5 in. and whose height measures 4.5 in.

20. Find the volume in *liters* of a cylinder whose radius measures 56.6 cm and whose height measures 34.9 cm.

Find the volume of the following geometric solids.

21. BcH

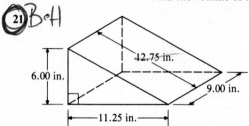

6.00 in. 12.75 in. 9.00 in. 11.25 in.

22.

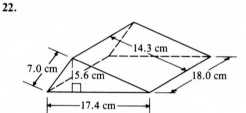

14.3 cm 7.0 cm 5.6 cm 18.0 cm 17.4 cm

23.

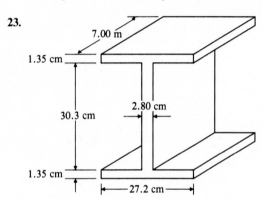

7.00 m 1.35 cm 2.80 cm 30.3 cm 1.35 cm 27.2 cm

24.

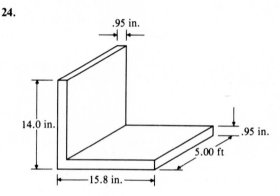

.95 in. 14.0 in. .95 in. 5.00 ft 15.8 in.

25.

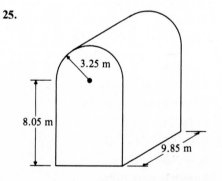

3.25 m 8.05 m 9.85 m

26.

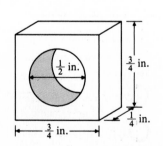

$\frac{1}{2}$ in. $\frac{3}{4}$ in. $\frac{1}{4}$ in. $\frac{3}{4}$ in.

Find a simplified expression that describes the volume of the following geometric solids.

27.

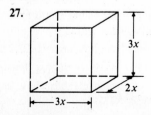

$3x$ $2x$ $3x$

28.

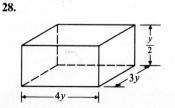

$\frac{y}{2}$ $3y$ $4y$

29.

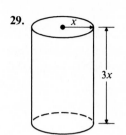

30.

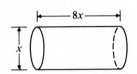

31.

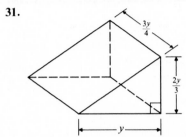

32.

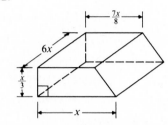

33.

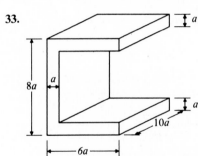

34.

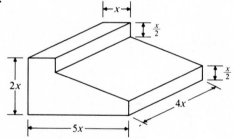

35. How many gallons of oil can be stored in the oil tank shown in Fig. 3.46?

36. How many liters of water will it take to fill the swimming pool in Fig. 3.47 level to the top?

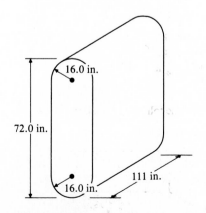

Figure 3.46

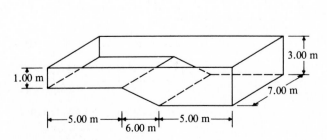

Figure 3.47

Chapter Review

**REVIEW
EXERCISES**

The following review exercises are grouped according to the objectives that should have been mastered in Chapter 3. Work each problem carefully. If any weaknesses appear, immediately refer to and read the subsection that matches that objective.

3.1 DECIMAL FRACTIONS AND ROUNDING OFF

Objectives

⊡ To write a decimal fraction in decimal form.

Write in decimal form.

1. **a)** one hundred ten-thousandths
   b) one hundred ten thousandths
   c) one hundred and ten thousandths

⊡ To round off a decimal to a given position.

Round off to the position indicated.

2. 479.8453
   a) to tenths **b)** to hundredths **c)** to tens **d)** to ones

3.2 SIGNIFICANT DIGITS

Objectives

⊡ To state if a given number is approximate or exact.

Identify each number as approximate or exact.

3. A 10,000 -square-foot office building is divided into 10 individual office spaces. Each office is to be 35 feet by 25 feet and rent for $750 per month.

⊡ To state the accuracy and precision of a given number.

State the accuracy and precision.

4. **a)** 12.8°C **b)** 10.02 L **c)** 64.00 ft
   d) .003 A **e)** 28,000 gal **f)** 28,0$\overline{0}$0 gal

⊡ To determine the upper and lower limits of a given measurement.

Find the upper and lower limits.

5. **a)** .003 A **b)** 28,000 gal

3.3 CALCULATING WITH APPROXIMATE NUMBERS

Objectives

⊡ To add and subtract approximate numbers using a calculator.

Evaluate.

6. $128.9 + 16.95 + .0835$ 148.9 7. $19.74 - 28.7$

⊡ To multiply and divide approximate numbers using a calculator.

Evaluate.

8. $(18.74)(.0063)$

9. $\dfrac{-.832}{128.7} =$ $-.006464 =$ $-.006466$

☐ To raise an approximate number to a power using a calculator.

Evaluate.

10. $(-18.5)^2$ =342.25 **11.** $(6.003)^4$ **12.** $(-.81)^5$

☐ To perform a basic arithmetic operation that involves an exact number.

Evaluate.

13. The thickness of a piece of metal measures .173 cm. What is the thickness of 39 such pieces when piled one on top of the other?

3.4 LINEAR MEASURE

Objectives

☐ To convert an English unit of linear measure to another English unit of linear measure.

Convert to the indicated units.

14. 187.9 ft to inches

15. 8.6 mi to yards

☐ To convert a metric unit of linear measure to another metric unit of linear measure.

Convert to the indicated units.

16. 8.73 km to meters

17. 257 mm to centimeters

☐ To convert an English unit of linear measure to a metric unit of linear measure or vice versa.

Convert to the indicated units.

18. 1750 m to feet

19. $\frac{3}{64}$ in. to millimeters (retain four significant digits)

3.5 PERIMETER AND CIRCUMFERENCE

Objectives

☐ To find the perimeter of a simple polygon.

Find the perimeter.

20.

15 ft 5 in.

13 ft 7 in.

18 ft 11 in.

25 ft 9 in.

Find the perimeter in *inches*.

21.

12.7 cm

16.9 cm

Find the perimeter.

$P = 2a + b$

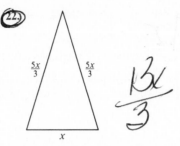

22.

$\dfrac{13x}{3}$

⊡ To find the circumference of a circle.

Find the circumference.

23.

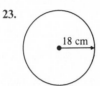

18 cm

⊡ To find the perimeter of a composite geometric figure.

Find the perimeter.

$L = 2\pi r$ or πd

24.

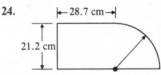

|← 28.7 cm →|

21.2 cm

$21.2 + 28.7 + 28.7 + 21.2 =$
$98.8 + 33.3$
$133 / cm$

Using 3.142 for π, find the perimeter.

25.

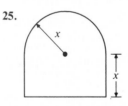

x

x

3.6 SQUARE MEASURE AND AREA

Objectives

⊡ To convert a unit of square measure to another unit of square measure.

Convert to the indicated units.

26. 558 sq ft to square yards

27. 7230 m^2 to square feet

28. 203,000 ft^2 to acres

⊡ To find the area of a simple geometric figure.

Find the area.

29.

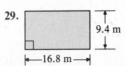

9.4 m

|← 16.8 m →|

Find the area in *square meters*.

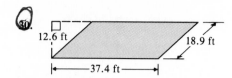

30.

12.6 ft   18.9 ft

37.4 ft

Find the area, leaving your answer in terms of π.

31.

12x

Find the area.

32.

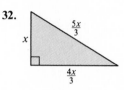

$\frac{5x}{3}$

x

$\frac{4x}{3}$

⊡ To find the area of a composite geometric figure.

Find the area.

33.

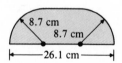

8.7 cm

8.7 cm

26.1 cm

34.

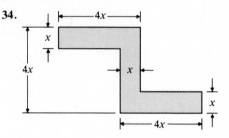

4x

x

4x

x

x

4x

3.7 CUBIC MEASURE AND VOLUME

Objectives

⊡ To convert a unit of cubic measure to another unit of cubic measure.

Convert to the indicated units.

35. 920 cu ft to cubic yards

36. 2.0 L to cubic inches

37. 18.3 ft^3 to gallons

⊡ To find the volume of a right prism.

Find the volume.

38.

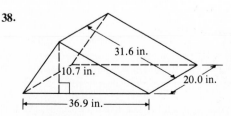

31.6 in.

10.7 in.

20.0 in.

36.9 in.

Find the volume of the cube in *cubic feet*.

39.

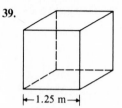

|←— 1.25 m —→|

Find the volume.

40.

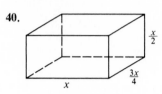

$\frac{x}{2}$

$\frac{3x}{4}$

x

⊡ To find the volume of a cylinder.

Find the volume in *gallons*.

41.

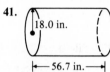

18.0 in.

|←— 56.7 in. —→|

Find the volume and leave your answer in terms of π.

42.

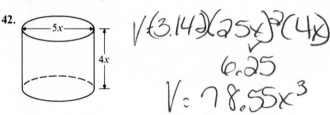

5x

4x

$V (3.14)(25x)^2(4x)$

6.25

$V = 78.55x^3$

If you have worked through the Review Exercises and corrected any weaknesses, then you are ready to take the following Chapter Test.

Write in decimal form.

1. five hundred thousandths

2. five and one hundred-thousandths

Round off to the position indicated.

3. 80.2954 (hundredths)

4. 5744.96 (tens)

Fill in the table.

Measurement	Accuracy	Precision	g.p.e.	Lower limit	Upper limit
5. .0040 A					
6. 82$\overline{0}$0 lb					

Using a calculator, evaluate each of the following. Assume that all numbers are approximate and retain the correct number of significant digits.

7. $-18.95 + 46.7 - 927.6$

8. $(.0724)(-124.6)^2$

9. $(-1.030)^6$

10. $\dfrac{(37.9)(-18.5)^3}{2700}$

Convert to the indicated units.

11. $\frac{9}{32}$ in. to millimeters (retain four significant digits)

12. 267,000 ft² to acres

13. 750 sq ft to square meters

14. 7173 in³ to cubic feet

15. Find the perimeter of a parallelogram whose sides measure 54 ft 9 in. and 62 ft 8 in.

16. Find the perimeter in *meters* of a square whose side measures 83.5 cm.

17. Find the area of a triangle whose base measures 19.3 cm and whose height measures 20.7 cm.

18. Find the area in *square yards* of a circle if its diameter measures 14.9 ft.

19. Find the volume of a cube if one of its edges measures 13.26 m.

20. Find the volume in *gallons* of a cylinder with radius 12.0 ft and height 3.0 ft.

21. Find the perimeter of the geometric figure in Fig. 3.48.

22. Find the area of the geometric figure in Fig. 3.48.

23. Find the volume of the geometric solid in Fig. 3.49.

24. Find the area of the circle in Fig. 3.50. Leave your answer in terms of π.

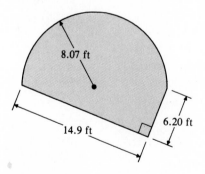

Figure 3.48

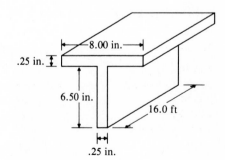

Figure 3.49

Figure 3.50

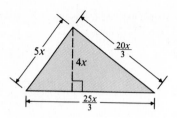

Figure 3.51

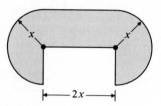

Figure 3.52

25. Find the perimeter of the triangle in Fig. 3.51.

26. Find the area of the triangle in Fig. 3.51.

27. Find the perimeter of the geometric figure in Fig. 3.52.

28. Find the area of the geometric figure in Fig. 3.52.

29. Find the volume of the rectangular box in Fig. 3.53.

30. Find the volume of the cylinder in Fig. 3.54.

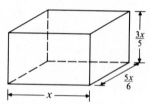

Figure 3.53

Figure 3.54

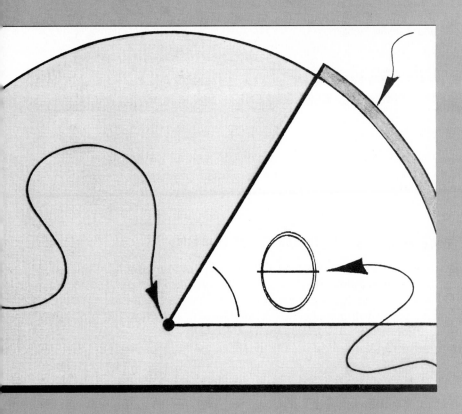

4

SIMPLE
EQUATIONS
AND
VERBAL
PROBLEMS

4.1 THE MULTIPLICATION AND ADDITION PROPERTIES OF EQUALITY

Objectives

☐ To use the Multiplication Property of Equality to solve a simple equation.

☐ To use the Addition Property of Equality to solve a simple equation.

☐ To apply either property to solve a formula for a given variable.

☐ The Multiplication Property of Equality

The equations that were presented in Chapter 1 were so simple that you could find their solutions by guessing. However, if you were asked to solve the equation $3(x - 2) = 12 - 2(x + 5)$, you would not have much luck in guessing its solution. In this section, you will discover two properties that will allow you to solve an equation without having to guess its solution.

Consider the equation $6x = 18$. On the *left side* of this equation is the algebraic expression $6x$. On the *right side* is the number 18. The *two sides* of an equation are always separated by the equal sign $=$. By guessing the solution to this equation, we know that $x = 3$, since this equation becomes true when x is replaced by 3.

Suppose we *multiplied each side* of the equation $6x = 18$ by 2. We would form a new equation as follows:

$$\text{left side} \leftarrow 6x = 18 \rightarrow \text{right side}$$

$$2 \cdot 6x = 18 \cdot 2 \qquad \text{multiplying each side by 2}$$

$$12x = 36 \qquad \text{simplifying}$$

Notice the solution to this new equation, $12x = 36$, is still $x = 3$, since this new equation becomes true when x is replaced by 3.

Now, suppose we *multiplied each side* of the equation $6x = 18$ by $\frac{1}{2}$. Of course, multiplying each side by $\frac{1}{2}$ is the same as dividing each side by 2. We would form a new equation as follows:

$$\text{left side} \leftarrow 6x = 18 \rightarrow \text{right side}$$

$$\frac{1}{2} \cdot 6x = 18 \cdot \frac{1}{2} \qquad \text{multiplying each side by } \frac{1}{2}$$

$$\frac{6x}{2} = \frac{18}{2} \qquad \text{multiplication rule for fractions}$$

$$3x = 9 \qquad \text{simplifying}$$

Again, notice the solution to this new equation, $3x = 9$, is also $x = 3$, since this new equation becomes true when x is replaced by 3.

The equations $6x = 18$, $12x = 36$, and $3x = 9$, are called **equivalent equations**, since each has the same solution.

MULTIPLICATION PROPERTY OF EQUALITY
If each side of an equation is *multiplied by* (or *divided by*) the same number (other than 0), the new equation that is formed is *equivalent* to the original equation.

The Multiplication Property of Equality allows you to alter an equation without changing the value of its solution.

The goal to keep in mind when solving an equation by using the Multiplication Property of Equality is to *isolate* the variable on one side of the equation. When the variable is isolated, the solution to the equation will appear on the other side.

If we multiplied each side of the equation $6x = 18$ by $\frac{1}{6}$, which is the **reciprocal** of 6, we would isolate the variable x and immediately obtain the equivalent equation $x = 3$ as follows:

$$6x = 18$$

$$\frac{1}{6} \cdot 6x = 18 \cdot \frac{1}{6} \qquad \text{multiplying each side by } \frac{1}{6}$$

Note: Any number times its reciprocal equals 1.

$$1x = 3 \qquad \text{simplifying}$$

$$x = 3 \qquad 1(a) = a$$

the isolated variable —↑   ↑— the solution

EXAMPLE 1

Solve $4x = -30$.

Solution: Isolate the x by multiplying each side by $\frac{1}{4}$ (the reciprocal of 4).

$$4x = -30$$

$$\frac{1}{4} \cdot 4x = -30 \cdot \frac{1}{4} \qquad \text{multiplying each side by } \frac{1}{4}$$

$$1x = \frac{\overset{-15}{(-30)}(1)}{\underset{2}{\cancel{4}}} \qquad \text{multiplication rule for fractions}$$

$$x = \frac{-15}{2} \text{ or } -7\frac{1}{2} \quad \blacksquare \qquad \text{simplifying}$$

Note: Does $4x = -30$, when $x = \dfrac{-15}{2}$? Replace the x with $\dfrac{-15}{2}$ to see if the equation becomes true. This is called *checking* the solution.

$$4x = -30$$

$$4\left(\frac{-15}{2}\right) = -30?$$

$$\frac{\overset{2}{(\cancel{4})}(-15)}{\underset{1}{\cancel{2}}} = -30?$$

$$-30 = -30 \qquad \text{true}$$

EXAMPLE 2

Solve $-\frac{2}{3}y = 10$.

Solution: Isolate the y by multiplying each side by $-\frac{3}{2}$ (the reciprocal of $-\frac{2}{3}$).

$$-\frac{2}{3}y = 10$$

$$\left(-\frac{3}{2}\right)\left(-\frac{2}{3}y\right) = (10)\left(-\frac{3}{2}\right) \qquad \text{multiplying each side by } -\frac{3}{2}$$

$$1y = \frac{(\overset{5}{\cancel{10}})(-3)}{\underset{1}{\cancel{2}}} \qquad \text{multiplication rule for fractions}$$

$$y = -15 \quad \blacksquare \qquad \text{simplifying}$$

Note: *Check* by replacing y with -15 to see if the equation becomes true.

$$-\frac{2}{3}y = 10$$

$$-\frac{2}{3}(-15) = 10$$

$$\frac{(-2)(\overset{-5}{\cancel{-15}})}{\underset{1}{\cancel{3}}} = 10$$

$$10 = 10 \qquad \text{true}$$

EXAMPLE 3

Solve $\dfrac{n}{3} = -2.6$.

Solution: $\frac{n}{3} = \frac{1}{3}n$. Thus, isolate the n by multiplying each side by 3 (the reciprocal of $\frac{1}{3}$).

$$\frac{n}{3} = -2.6$$

$$3 \cdot \frac{n}{3} = (-2.6)(3) \qquad \text{multiplying each side by 3}$$

$$n = -7.8 \quad \blacksquare \qquad \text{simplifying}$$

Note: Try checking the equation by replacing n with -7.8.

EXAMPLE 4

Solve $\dfrac{5}{t} = -15$

Solution:

$$\frac{5}{t} = -15$$

$$t\left(\frac{5}{t}\right) = (-15)t \qquad \text{multiplying each side by } t \ (t \neq 0)$$

$$5 = -15t \qquad \text{simplifying}$$

$$-\frac{1}{15}(5) = -\frac{1}{15}(-15t) \qquad \text{multiplying each side by } -\frac{1}{15}$$

$$-\frac{1}{3} = t \quad \blacksquare \qquad \text{simplifying}$$

Note: Be sure to avoid this common mistake:

$$\frac{5}{t} = -15$$

$$\frac{1}{5}\left(\frac{5}{t}\right) = (-15)\frac{1}{5} \qquad \text{multiplying each side by } \frac{1}{5}$$

$$t = -3 \quad \textit{Wrong}$$

Of course, $\dfrac{1}{5}\left(\dfrac{5}{t}\right) = \dfrac{1}{t}$ not t. Thus, $\dfrac{1}{t} = -3$ so $t = -\dfrac{1}{3}$.

Note: $\dfrac{1}{t} = -3$ is read "the reciprocal of t equals -3". If the reciprocal of t is -3, then $t = -\frac{1}{3}$.

⊡ The Addition Property of Equality

Consider the equation $x + 3 = 8$. By guessing its solution, we know that $x = 5$, since the equation becomes true when x is replaced by 5.

Suppose we *added* 4 to each side of the equation $x + 3 = 8$. We would form a new equation as follows:

$$\text{left side} \leftarrow x + 3 = 8 \rightarrow \text{right side}$$

$$x + 3 + \boxed{4} = 8 + \boxed{4} \qquad \text{adding 4 to each side}$$

$$x + 7 = 12 \qquad \text{simplifying}$$

Notice the solution to this new equation, $x + 7 = 12$, is still $x = 5$, since this new equation becomes true when x is replaced by 5.

Now, suppose we *added* -4 to each side of the equation $x + 3 = 8$. Of course, adding -4 to each side is the same as subtracting 4 from each side. We would form a new equation as follows:

$$x + 3 = 8$$

$$x + 3 + \boxed{(-4)} = 8 + \boxed{(-4)} \qquad \text{adding } -4 \text{ to each side}$$

$$x - 1 = 4 \qquad \text{simplifying}$$

Again, notice that the solution to this new equation, $x - 1 = 4$, is also $x = 5$, since this new equation becomes true when x is replaced by 5.

The equations $x + 3 = 8$, $x + 7 = 12$, and $x - 1 = 4$, are called equivalent equations, since each has the same solution.

ADDITION PROPERTY OF EQUALITY
If the same number is *added to* (or *subtracted from*) each side of an equation, then the new equation that is formed is *equivalent* to the original equation.

The Addition Property of Equality allows you to alter an equation without changing the value of its solution.

The goal to keep in mind when solving an equation by using the Addition Property of Equality is to *isolate* the variable on one side of the equation. When the variable is isolated, the solution to the equation will appear on the other side.

If we added -3, which is the *opposite* of 3, to each side of the equation $x + 3 = 8$, we would isolate the variable x and immediately obtain the equivalent equation $x = 5$ as follows:

$$x + 3 = 8$$

$$x + 3 + \boxed{(-3)} = 8 + \boxed{(-3)} \qquad \text{adding } -3 \text{ to each side}$$

Note: Any number added to its opposite is 0.

$$x + 0 = 5 \qquad \text{simplifying}$$

$$x = 5 \qquad a + 0 = a$$

the isolated variable ⟶ ⟵ the solution

EXAMPLE 5

Solve $x + 61 = -25$.

Solution: Isolate the x by adding -61 (the opposite of 61) to each side.

$$x + 61 = -25$$

$$x + 61 + (\,\boxed{-61}\,) = -25 + (\,\boxed{-61}\,) \qquad \text{adding } -61 \text{ to each side}$$

$$x + 0 = -86$$

$$x = -86 \quad \blacksquare \qquad \text{simplifying}$$

Note: *Check* by replacing x with -86 to see if the equation becomes true.

$$x + 61 = -25$$

$$(-86) + 61 = -25?$$

$$-25 = -25 \qquad \text{true}$$

EXAMPLE 6

Solve $\frac{7}{16} = m - \frac{3}{4}$.

Solution: Isolate the m by adding $\frac{3}{4}$ (the opposite of $-\frac{3}{4}$) to each side.

$$\frac{7}{16} = m - \frac{3}{4}$$

$$\frac{7}{16} + \boxed{\frac{3}{4}} = m - \frac{3}{4} + \boxed{\frac{3}{4}} \qquad \text{adding } \frac{3}{4} \text{ to each side}$$

$$\frac{7}{16} + \boxed{\frac{12}{16}} = m + 0 \qquad \text{fundamental property of fractions}$$

$$\frac{19}{16} = m \quad \blacksquare \qquad \text{addition rule for fractions}$$

Note: Try checking the equation by replacing m with $\frac{19}{16}$.

EXAMPLE 7

Solve $.75 - p = 1.61$.

Solution:

$$.75 - p = 1.61$$

$$.75 + (\boxed{-.75}) - p = 1.61 + (\boxed{-.75}) \qquad \text{adding } -.75 \text{ to each side}$$

$$0 \ominus p = .86 \qquad \text{simplifying}$$

Note: Do not lose this negative sign.

$$-p = .86$$

Note: We must remove the negative sign from in front of the p. If the opposite of p is $.86$, then $p = -.86$.

$$p = -.86 \quad \blacksquare \qquad \text{taking the opposite of each side}$$

Note: To avoid the negative variable term, you may wish to proceed as follows:

$$.75 - p = 1.61$$

$$.75 - p + \boxed{p} = 1.61 + \boxed{p} \qquad \text{adding } p \text{ to each side}$$

$$.75 = 1.61 + p \qquad \text{simplifying}$$

$$.75 + (\boxed{-1.61}) = 1.61 + (\boxed{-1.61}) + p \qquad \text{adding } -1.61 \text{ to each side}$$

$$-.86 = p \qquad \text{simplifying}$$

⋮ Formulas

There are times when it is necessary to solve a *formula* for one of its variables. The Multiplication and Addition Properties of Equality can be used to isolate the variable for which we wish to solve.

EXAMPLE 8

Solve $V = lwh$, for l (volume of a rectangular box).

Solution:

$$V = lwh$$

$$V\left(\frac{1}{w}\right) = lw\left(\frac{1}{w}\right)h \qquad \text{multiplying each side by } \frac{1}{w}$$

$$\frac{V}{w} = lh \qquad \text{simplifying}$$

$$\frac{V}{w}\left(\frac{1}{h}\right) = lh\left(\frac{1}{h}\right) \qquad \text{multiplying each side by } \frac{1}{h}$$

$$\frac{V}{wh} = l \text{ or } l = \frac{V}{wh} \quad \blacksquare \qquad \text{simplifying}$$

EXAMPLE 9

Solve $P = a + b + c$, for a (perimeter of a triangle).

Solution:

$$P = a + b + c$$
$$P + (\,-b\,) = a + b + (\,-b\,) + c \qquad \text{adding } -b \text{ to each side}$$
$$P - b = a + c \qquad \text{simplifying}$$
$$P - b + (\,-c\,) = a + c + (\,-c\,) \qquad \text{adding } -c \text{ to each side}$$
$$P - b - c = a \text{ or } a = P - b - c \quad \blacksquare \qquad \text{simplifying}$$

EXERCISES 4.1

Solve and check.

1. $x + 73 = 120$

2. $32 + y = 17$

3. $z - 15 = 19$

4. $19 - y = -14$

5. $5x = 75$

6. $96 = 6x$

7. $-20 = 8x$

8. $-24t = 16$

9. $\dfrac{y}{5} = 10$

10. $\dfrac{y}{6} = -12$

11. $12 = \dfrac{6}{y}$

12. $-25 = \dfrac{15}{-p}$

13. $\frac{5}{8} = w + \frac{3}{4}$

14. $12\frac{1}{2} = 3\frac{1}{3} + j$

15. $\frac{2}{3} = h - \frac{7}{8}$

16. $7\frac{1}{2} = \frac{5}{8} - v$

17. $\frac{5}{6}y = -15$

18. $-\frac{2}{3}n = 18$ $-\frac{3}{2} \cdot -\frac{2}{3}n = 18 \cdot -\frac{3}{2} = -27$

19. $3 = \dfrac{-6.45}{m}$

20. $\dfrac{-8}{t} = -.25$

21. $x + 1.8 = -9.6$

22. $18.3 = 5.2 + y$

23. $3 = f - 3\frac{1}{2}$

24. $-14.6 = 7.8 - p$

25. $-.80 = 2.5z$

26. $-.04p = -96$

27. $\dfrac{3g}{5} = -9$

28. $-5 = \dfrac{10z}{3}$

29. $8 = z + \frac{7}{8}$

30. $.75 + x = 15$

31. $t - .73 = -.82$

32. $-\frac{7}{16} = m + \frac{1}{2}$

33. $-.8 - p = 4.2$

34. $.15 - g = 10$

35. $-.5 - z = -5.3$

36. $-\frac{8}{9} = -\frac{3}{4} - m$

37. $\dfrac{7y}{18} = \dfrac{14}{15}$

38. $\dfrac{5y}{12} = \dfrac{-5}{18}$

39. $\dfrac{n}{9} = 0$

40. $\dfrac{6y}{7} = 0$

Solve each formula for the indicated quantity.

41. Distance, $d = rt$. Solve for time, t.

42. Interest, $I = prt$. Solve for rate, r.

43. Total series resistance, $R_t = R_1 + R_2 + R_3$. Solve for resistance R_2.

44. Current, $I = \dfrac{E}{R}$. Solve for voltage, E.

45. Supplementary angle, $A = 180 - B$. Solve for angle B.

46. Bolt size, $D_M = D_m + 2d$. Solve for minor diameter, D_m.

47. Velocity, $v = v_0 + at$. Solve for initial velocity, v_0.

48. Force exerted to a spring, $F = -ks$. Solve for the constant of proportionality, k.

49. Speed of the driving pulley, $S = \dfrac{ds}{D}$. Solve for speed of the driven pulley, s.

50. S.A.E. horsepower, $H.P. = \dfrac{D^2 N}{2.5}$. Solve for the number of cylinders, N.

51. Rear axle ratio, $r = \dfrac{T}{t}$. Solve for the number of teeth in the pinion gear, t.

52. Wavelength, $\lambda = \dfrac{h}{mv}$. Solve for velocity, v. (λ is the Greek letter lambda.)

53. Referring to exercise 43, find the value of R_2 if $R_t = 66.0\,\Omega$, $R_1 = 14.8\,\Omega$ and $R_3 = 29.6\,\Omega$.

54. Referring to exercise 46, find the value of D_m if $D_M = \frac{1}{2}''$ and $d = \frac{3}{63}''$.

55. Referring to exercise 51, find the value of t if $r = 4.2$ and $T = 63$ teeth.

4.2 USING THE PROPERTIES TOGETHER TO SOLVE SIMPLE EQUATIONS

Objectives

⊡ To solve an equation of the form $ax + b = c$, for x.

⊡ To solve a simple equation in which the variable appears on each side of the equation.

⊡ To apply the properties together to solve a formula for a given variable.

⊡ Equations of the Form $ax + b = c$

In more complicated equations, the Multiplication and Addition Properties of Equality can be used *together* to help solve the equation.

In order to solve the equation $3x + 5 = 17$, we must isolate the x by

1. removing the *addition* of 5 and

2. removing the *multiplication* by 3

from the left side of the equation.

If we want to remove the multiplication by 3 first, we would proceed as follows:

Method 1

$$3x + 5 = 17$$

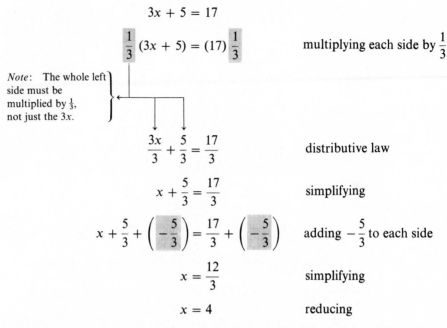

$$\frac{1}{3}(3x + 5) = (17)\frac{1}{3}$$ multiplying each side by $\frac{1}{3}$

Note: The whole left side must be multiplied by $\frac{1}{3}$, not just the $3x$.

$$\frac{3x}{3} + \frac{5}{3} = \frac{17}{3}$$ distributive law

$$x + \frac{5}{3} = \frac{17}{3}$$ simplifying

$$x + \frac{5}{3} + \left(-\frac{5}{3}\right) = \frac{17}{3} + \left(-\frac{5}{3}\right)$$ adding $-\frac{5}{3}$ to each side

$$x = \frac{12}{3}$$ simplifying

$$x = 4$$ reducing

However, if we remove the addition of 5 first, we would proceed as follows:

Method 2

$$3x + 5 = 17$$

$$3x + 5 + (-5) = 17 + (-5)$$ adding -5 to each side

$$3x = 12$$ simplifying

$$\frac{1}{3}(3x) = (12)\frac{1}{3}$$ multiplying each side by $\frac{1}{3}$

$$x = 4$$ simplifying

I'm sure you would agree that Method Two was easier, since it avoided using the distributive law and fractional computations. *In general, when solving an equation of the form ax + b = c for x, remove the addition of b before you remove the multiplication by a.*

EXAMPLE 1

Solve $\dfrac{5x}{8} - 3 = 7$.

Solution:

$$\frac{5x}{8} - 3 = 7$$

$$\frac{5x}{8} - 3 + \boxed{3} = 7 + \boxed{3} \qquad \text{adding 3 to each side}$$

$$\frac{5x}{8} = 10 \qquad \text{simplifying}$$

$$\frac{8}{5}\left(\frac{5x}{8}\right) = (10)\frac{8}{5} \qquad \text{multiplying each side by } \frac{8}{5}$$

$$x = 16 \quad \blacksquare \qquad \text{simplifying}$$

Note: To *check*, replace *x* with 16 to see if the equation becomes true.

$$\frac{5x}{8} - 3 = 7$$

$$\frac{5(16)}{8} - 3 = 7?$$

$$10 - 3 = 7?$$

$$7 = 7 \quad \text{true}$$

⠘ Variables on Each Side of an Equation

In some equations, the variable appears on each side of the equation. The first step in solving an equation of this type is to remove the variable term from either the left or right side of the equation.

EXAMPLE 2

Solve $9w - 25 = 4w - 10$.

Solution: Let's remove the variable term $4w$ from the right side.

$$9w - 25 = 4w - 10$$

$$9w + (\boxed{-4w}) - 25 = 4w + (\boxed{-4w}) - 10 \qquad \text{adding } -4w \text{ to each side}$$

$$5w - 25 = -10 \qquad \text{simplifying}$$

Note: This is an equation of the form $ax + b = c$.

$$5w - 25 + \boxed{25} = -10 + \boxed{25} \qquad \text{adding 25 to each side}$$

$$5w = 15 \qquad \text{simplifying}$$

$$\frac{1}{5}(5w) = (15)\frac{1}{5} \qquad \text{multiplying each side by } \frac{1}{5}$$

$$w = 3 \quad \blacksquare \qquad \text{simplifying}$$

Note: If the variable term $9w$ is removed from the left side first, we could find the solution as follows:

$$9w - 25 = 4w - 10$$

$$9w + (\boxed{-9w}) - 25 = 4w + (\boxed{-9w}) - 10 \qquad \text{adding } -9w \text{ to each side}$$

$$-25 = -5w - 10 \qquad \text{simplifying}$$

Note: This method develops a negative variable term, which you may want to avoid.

$$-25 + \boxed{10} = -5w - 10 + \boxed{10} \qquad \text{adding 10 to each side}$$

$$-15 = -5w \qquad \text{simplifying}$$

$$\left(\boxed{-\frac{1}{5}}\right)(-15) = (-5w)\left(\boxed{-\frac{1}{5}}\right) \qquad \text{multiplying each side by } -\frac{1}{5}$$

$$3 = w \qquad \text{simplifying}$$

If a variable term appears inside a set of parentheses, we must first remove the parentheses and combine any similar terms that are on the *same side* of the equation.

EXAMPLE 3

Solve $5x + 7 - 2(3 - 4x) = 2(5x + 8)$.

Solution:

$$5x + 7 - 2(3 - 4x) = 2(5x + 8)$$

$$5x + 7 - 6 \boxed{+} 8x = 10x + 16 \qquad \text{distributive law}$$

$$13x + 1 = 10x + 16 \qquad \text{combining similar terms on the } \textit{same side}$$

$$13x + (\boxed{-10x}) + 1 = 10x + (\boxed{-10x}) + 16 \qquad \text{adding } -10x \text{ to each side}$$

$$3x + 1 = 16 \qquad \text{simplifying}$$

$$3x + 1 + (\boxed{-1}) = 16 + (\boxed{-1}) \qquad \text{adding } -1 \text{ to each side}$$

$$3x = 15 \qquad \text{simplifying}$$

$$\boxed{\frac{1}{3}}(3x) = (15)\boxed{\frac{1}{3}} \qquad \text{multiplying each side by } \frac{1}{3}$$

$$x = 5 \quad \blacksquare \qquad \text{simplifying}$$

Note: To *check*, replace *x* with 5 and follow the order of operations you learned in Chapter 1.

$$5x + 7 - 2(3 - 4x) = 2(5x + 8)$$
$$5 \cdot 5 + 7 - 2(\underbrace{3 - 4 \cdot 5}) = 2(\underbrace{5 \cdot 5 + 8})?$$

$5 \cdot 5 + 7 - 2(-17) = 2(33)?$	parentheses first
$25 + 7 + 34 = 66?$	multiplication next
$66 = 66$ true	addition last

When the division bar extends all the way across one side of the equation, remove this division immediately, by multiplying each side of the equation by the denominator of the algebraic fraction.

EXAMPLE 4

Solve $\dfrac{5t - 7}{2} = 7 - t$.

Solution:

$$\frac{5t - 7}{2} = 7 - t$$

$2 \cdot \dfrac{5t - 7}{2} = 2\,(7 - t)$	multiplying each side by 2
$5t - 7 = 14 - 2t$	simplifying and distributing
$5t + 2t - 7 = 14 - 2t + 2t$	adding $2t$ to each side
$7t - 7 = 14$	simplifying
$7t - 7 + 7 = 14 + 7$	adding 7 to each side
$7t = 21$	simplifying
$\dfrac{1}{7}\,(7t) = (21)\,\dfrac{1}{7}$	multiplying each side by $\dfrac{1}{7}$
$t = 3$ ∎	simplifying

Note: Try checking the equation by replacing *t* with 3.

⬚ Formulas

The Multiplication and Addition Properties of Equality can be used *together* to solve a formula for one of its variables.

EXAMPLE 5

Solve $P = 2l + 2w$, for l (perimeter of a rectangle).

Solution:

$$P = 2l + 2w$$

$$P + (\boxed{-2w}) = 2l + 2w + (\boxed{-2w})$$ adding $-2w$ to each side

$$P - 2w = 2l$$ simplifying

$$\boxed{\frac{1}{2}}(P - 2w) = \boxed{\frac{1}{2}}(2l)$$ multiplying each side by $\frac{1}{2}$

$$\frac{P - 2w}{2} = l \text{ or } l = \frac{P - 2w}{2} \quad \blacksquare$$ simplifying

Note: You may also wish to write your answer as a *mixed expression*.

$$l = \frac{P - 2w}{2} = \frac{P}{2} - \frac{2w}{2} = \frac{P}{2} - w$$

single fraction ———⟍   ⟍— mixed expression

EXAMPLE 6

Solve $T_c = \dfrac{5(T_f - 32)}{9}$, for T_f (temperature conversion).

Solution:

$$T_c = \frac{5(T_f - 32)}{9}$$

$$\boxed{9} \cdot T_c = \frac{5(T_f - 32)}{9} \cdot \boxed{9}$$ multiplying each side by 9

$$9T_c = 5(T_f - 32)$$ simplifying

$$9T_c = 5T_f - 160$$ distributive law

$$9T_c + \boxed{160} = 5T_f - 160 + \boxed{160}$$ adding 160 to each side

$$9T_c + 160 = 5T_f$$ simplifying

$$\boxed{\frac{1}{5}}(9T_c + 160) = \boxed{\frac{1}{5}} 5T_f$$ multiplying each side by $\frac{1}{5}$

$$\frac{9T_c + 160}{5} = T_f \text{ or } T_f = \frac{9T_c + 160}{5} \quad \blacksquare$$ simplifying

Note:

$$T_f = \frac{9T_c + 160}{5} = \frac{9T_c}{5} + \frac{160}{5} = \frac{9T_c}{5} + 32$$

single fraction ⎯⎯⎯⎯⎯⎯⎯⎯⎯⎯⎯ mixed expression

EXERCISES 4.2

Solve and check.

1. $3x - 7 = 5$

2. $18 = 7x + 4$

3. $5 = 3x + 8$

4. $6m - 15 = -5$

5. $3 + \dfrac{3p}{5} = -42$

6. $16 = -5 - \dfrac{3k}{4}$

7. $36 - .60v = 51$

8. $-52 = -16 - 1.8t$

9. $4x + 15 = 7x$

10. $-3x = 12 - 6x$

11. $11t - 18 = 38 - 3t$

12. $6h + 7 = 3h + 28$

13. $3 + 5n = 3n - 13$

14. $10h - 12 = 8h - 11$

15. $20 - 4g = 6 - 7g$

16. $7y - 15 = 27 - y$

17. $w + w = -18$

18. $2w - 7w = 35$

19. $p + \dfrac{3p}{4} = 28$

20. $\dfrac{4n}{25} - n = 42$

21. $6y + 10 - y = 3y$

22. $5k + 2k - 6 = -16 - k - 6$

23. $7 - 4x + 2 = 7x - 3 - 5x$

24. $20 + 3t = 12 + 3t - 8t + 8$

25. $(x + 4) - (x - 5) + x = 10$

26. $3(y - 2) = 12 - 2(y + 4)$

27. $7x - 2(3x - 6) = 3(5x - 10)$

28. $-10z + 2(z - 4) = -2(3z - 2) + 8$

29. $\dfrac{x - 3}{5} = 7$

30. $18 = \dfrac{2y - 4}{3}$

31. $\dfrac{3x - 5}{9} = 1 - x$

32. $2x - 7 = \dfrac{8 - 7x}{-9}$

33. $5(y - 4) = \dfrac{8 - 2y}{6}$

34. $\dfrac{6 - 3n}{5} = 0$

35. $\dfrac{42}{x} - 3 = 4$

36. $48 = \dfrac{28}{y} - 8$

37. $\dfrac{8k}{9} = k - 9$

38. $7 - x = \dfrac{-x}{5}$

39. $\dfrac{5z}{3} - 8 = 2$

40. $6 - \dfrac{y}{8} = 0$

Solve each formula for the indicated quantity.

41. Velocity, $v = v_0 + at$. Solve for time, t.

42. Pressure, $P = P_a + hdg$. Solve for the density of the liquid, d.

43. First moment, $M = F_1 d_1 + F_2 d_2 + F_3 d_3$. Solve for distance, d_2.

44. Moment of inertia, $I = F_1 d_1{}^2 + F_2 d_2{}^2$. Solve for force, F_1.

45. Area of a trapezoid, $A = \dfrac{h}{2}(b_1 + b_2)$. Solve for height, h.

46. Last term of a progression, $L = a + (n - 1)d$. Solve for the common difference, d.

47. Resistance, $R = R_0(1 + \alpha t)$. Solve for the temperature coefficient, α. (α is the Greek letter alpha).

48. Shear of a beam, $V = w\left(-x + \dfrac{l}{2}\right)$. Solve for the length of the beam, l.

49. Internal resistance of a battery, $R_i = \dfrac{E - IR}{I}$. Solve for the battery voltage, E.

50. Coefficient of volume expansion, $\beta = \dfrac{V - V_i}{V_i t}$. Solve for the volume, V. (β is the Greek letter beta).

51. Deflection of a beam, $\delta = \dfrac{Px^2(3l - x)}{6EI}$. Solve for the length of the beam, l. (δ is the Greek letter delta).

52. Heat, $H = \dfrac{KAt(T_1 - T_2)}{d}$. Solve for temperature, T_2.

53. Maximum deflection of a simply supported beam, $\delta_{max} = \dfrac{5wl^4}{384EI}$. Solve for the moment of inertia, I.

54. Referring to exercise 45, find the value of h if $A = 16.0 \text{ ft}^2$, $b_1 = 11.3 \text{ ft}$, and $b_2 = 8.7 \text{ ft}$.

55. Referring to exercise 49, find the value of E, If $I = .25 \text{ A}$, $R_i = 1.6 \,\Omega$, and $R = 33 \,\Omega$.

4.3 AN INTRODUCTION TO SOLVING VERBAL PROBLEMS

Objectives

⊡ To solve a verbal problem that contains one unknown.

⊡ To solve a verbal problem that contains more than one unknown.

⊡ One Unknown

Being able to understand and solve verbal problems is a critical part of technical mathematics. It is extremely important to become skillful in reading words and changing these words into algebraic expressions and equations. You may recall from Chapter 1 that a table was given that listed many words that are used to suggest the operations of addition, subtraction, multiplication, and division. For your convenience, this table is given again.

The expression	The operation
plus, sum, increased by, more than	addition
minus, difference, less, less than, take away, decreased by	subtraction
times, product, twice, doubled, tripled	multiplication
divided by, quotient	division

Students often have difficulty with verbal problems. One of the reasons that they do have difficulty is that they do not have an orderly procedure to follow. Organization of your thoughts is extremely important if you are to be successful in solving verbal problems. For this reason, this text has adopted a **five-step procedure** that will assist you in solving verbal problems.

FIVE-STEP PROCEDURE FOR SOLVING A VERBAL PROBLEM

1. Formula
   a) Always carefully read the problem at least twice: the first time for a general overview, and the second (and possibly third) time to allow you to write a **general formula** that describes and relates all the quantities that are given and asked for in the problem.
   b) If appropriate, draw a sketch to help you write this formula.

2. Setup
   a) Record the information that is **known** (given).
   b) Identify the **unknown** and assign to it some variable.
   c) Apply conversion factors if necessary.

3. Equation
   a) Develop an **equation** by replacing all the letters in the FORMULA (Step 1) with the appropriate information that was gathered in the SETUP (Step 2).
   b) Solve this equation for the unknown.

4. Conclusion
   Carefully list the **values** of all the known and unknown quantities described in the SETUP (Step 2). Survey this list of values and be certain that your **conclusion** seems reasonable and logically satisfies the conditions of the problem.

5. Check
   Check by substituting the values listed in the CONCLUSION (Step 4) for the variables in the FORMULA (Step 1).

EXAMPLE 1

Two hundred twenty-five feet of fencing are needed to fence all sides of a rectangular vegetable garden. The width of the garden is 42.7 ft. What is the length?

Solution:

1. Formula

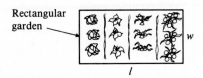

Rectangular garden

$P = 2l + 2w$, where P is the perimeter
l is the length
w is the width

2. Setup

$$P = 225 \text{ ft}$$
$$w = 42.7 \text{ ft}$$
$$l = ? \quad \text{unknown}$$

3. Equation

$$P = 2l + 2w$$
$$225 = 2l + 2(42.7) \qquad \text{substituting}$$
$$225 = 2l + 85.4 \qquad \text{simplifying}$$
$$225 + (\boxed{-85.4}) = 2l + 85.4 + (\boxed{-85.4}) \qquad \text{adding } -85.4 \text{ to each side}$$
$$139.6 = 2l \qquad \text{simplifying}$$
$$\boxed{\frac{1}{2}}(139.6) = \boxed{\frac{1}{2}}(2l) \qquad \text{multiplying each side by } \frac{1}{2}$$
$$69.8 = l \qquad \text{simplifying}$$

4. Conclusion

$$P = 225 \text{ ft}$$
$$w = 42.7 \text{ ft}$$
$$l = 69.8 \text{ ft} \quad \text{now known} \quad \blacksquare$$

5. Check

$$P = 2l + 2w$$
$$225 \text{ ft} = 2(69.8 \text{ ft}) + 2(42.7 \text{ ft})?$$
$$225 \text{ ft} = 139.6 \text{ ft} + 85.4 \text{ ft}?$$
$$225 \text{ ft} = 225 \text{ ft} \quad \text{true}$$

EXAMPLE 2

In an industrial experiment, three temperature readings of a coolant were taken. The average temperature was found to be 4.8°C. The first two readings were 5.3°C and 4.7°C respectively. What was the third reading?

Solution:

1. Formula

$$\bar{t} = \frac{t_1 + t_2 + t_3}{3}, \text{ where } \bar{t} \text{ is the average temperature}$$
$$t_1 \text{ is the first reading}$$
$$t_2 \text{ is the second reading}$$
$$t_3 \text{ is the third reading}$$

2. Setup

$$\bar{t} = 4.8°C$$
$$t_1 = 5.3°C$$
$$t_2 = 4.7°C$$
$$t_3 = ? \quad \text{unknown}$$

3. Equation

$$\bar{t} = \frac{t_1 + t_2 + t_3}{3}$$

$$4.8 = \frac{5.3 + 4.7 + t_3}{3} \qquad \text{substituting}$$

$$4.8 = \frac{10.0 + t_3}{3} \qquad \text{simplifying}$$

$$3\,(4.8) = \frac{10.0 + t_3}{3} \cdot 3 \qquad \text{multiplying by 3}$$

$$14.4 = 10.0 + t_3 \qquad \text{simplifying}$$

$$14.4 + (-10.0) = 10.0 + (-10.0) + t_3 \qquad \text{adding } -10.0 \text{ to each side}$$

$$4.4 = t_3 \qquad \text{simplifying}$$

4. Conclusion

$$\bar{t} = 4.8°C$$
$$t_1 = 5.3°C$$
$$t_2 = 4.7°C$$
$$t_3 = 4.4°C \qquad \text{now known} \quad \blacksquare$$

5. Check: The check is left for you.

⚃ More Than One Unknown

If a verbal problem contains *more than one unknown*, assign one of the unknowns the variable x and record each of the other unknowns in terms of x. Usually it is best to let x represent the unknown that all other unknowns are compared to. The quantity named directly after a word of comparison (*than* or *as*) should be assigned the variable x.

EXAMPLE 3

Two hundred twenty-five feet of fencing are needed to fence all sides of a rectangular vegetable garden. The width of the garden is 42.7 ft *less than* the length. What are the dimensions of the garden?

Solution:

1. Formula

Rectangular garden ⟶ 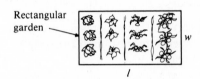 w

$P = 2l + 2w$, where P is the perimeter
l is the length
w is the width

l

2. Setup

$$P = 225 \text{ ft}$$

$$l = x$$

$$w = x - 42.7 \text{ ft}$$

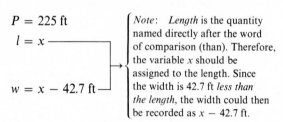

Note: *Length* is the quantity named directly after the word of comparison (than). Therefore, the variable x should be assigned to the length. Since the width is 42.7 ft *less than the length*, the width could then be recorded as $x - 42.7$ ft.

3. Equation

$$P = 2l + 2w$$

$225 = 2x + 2(x - 42.7)$	substituting
$225 = 2x + 2x - 85.4$	distributive law
$225 = 4x - 85.4$	simplifying
$225 + \boxed{85.4} = 4x - 85.4 + \boxed{85.4}$	adding 85.4 to each side
$310.4 = 4x$	simplifying
$\dfrac{1}{4}(310.4) = \dfrac{1}{4}(4x)$	multiplying each side by $\dfrac{1}{4}$
$77.6 = x$	simplifying

4. Conclusion

$$P = 225 \text{ ft}$$
$$l = x = 77.6 \text{ ft}$$
$$w = x - 42.7 \text{ ft} = 34.9 \text{ ft} \quad \blacksquare$$

5. Check

$$P = 2l + 2w$$
$$225 \text{ ft} = 2(77.6 \text{ ft}) + 2(34.9 \text{ ft})?$$
$$225 \text{ ft} = 155.2 \text{ ft} + 69.8 \text{ ft}?$$
$$225 \text{ ft} = 225 \text{ ft} \qquad \text{true}$$

EXAMPLE 4

In an industrial experiment, three temperature readings of a coolant were taken and the average temperature was found to be 4.8°C. The first two readings were the same. The third reading was 3.2°C *more than half* the first reading. What were the three temperature readings?

Solution:

1. Formula

$$\bar{t} = \frac{t_1 + t_2 + t_3}{3}, \text{ where } \bar{t} \text{ is the average temperature}$$
$$t_1 \text{ is the first reading}$$
$$t_2 \text{ is the second reading}$$
$$t_3 \text{ is the third reading}$$

2. Setup

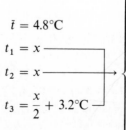

$$\bar{t} = 4.8°C$$
$$t_1 = x$$
$$t_2 = x$$
$$t_3 = \frac{x}{2} + 3.2°C$$

Note: The *first reading* is the quantity named directly after the word of comparison (than). Therefore, the variable x should be assigned to the first reading. Since the first two readings are the *same*, the second reading can also be assigned the variable x. *Half* the first reading can be written $\frac{x}{2}$. Thus, if the third reading is 3.2°C *more than half the first*, then the third reading can be recorded as $\frac{x}{2} + 3.2°C$.

3. Equation

$$\bar{t} = \frac{t_1 + t_2 + t_3}{3}$$

$$4.8 = \frac{x + x + \left(\dfrac{x}{2} + 3.2\right)}{3} \qquad \text{substituting}$$

$$4.8 = \frac{\dfrac{5x}{2} + 3.2}{3} \qquad \text{simplifying}$$

$$3\,(4.8) = \frac{\dfrac{5x}{2} + 3.2}{3} \cdot 3 \qquad \text{multiplying each side by 3}$$

$$14.4 = \frac{5x}{2} + 3.2 \qquad \text{simplifying}$$

$$14.4 + (\;-3.2\;) = \frac{5x}{2} + 3.2 + (\;-3.2\;) \qquad \text{adding } -3.2 \text{ to each side}$$

$$11.2 = \frac{5x}{2} \qquad \text{simplifying}$$

$$\frac{2}{5}\,(11.2) = \frac{5x}{2} \cdot \frac{2}{5} \qquad \text{multiplying each side by } \frac{2}{5}$$

$$4.48 = x \qquad \text{simplifying}$$

4. Conclusion

$$\bar{t} = 4.8°\text{C}$$
$$t_1 = 4.5°\text{C} \qquad \text{(retain two significant digits)}$$
$$t_2 = 4.5°\text{C} \qquad \text{(retain two significant digits)}$$
$$t_3 = 5.4°\text{C} \qquad \text{(retain two significant digits)} \quad \blacksquare$$

5. Check: The check is left for you.

EXERCISES 4.3

Using the five-step procedure outlined in this section, solve the following verbal problems.

1. The perimeter of an equilateral triangle is 98.4 m. What is the length of a side?
2. The perimeter of a square is 168 cm. What is the length of a side?
3. The circumference of a truck tire is 10.3 ft. What is its diameter?
4. The area of a rectangular piece of land is 209,000 sq ft. The length is 602 ft. What is the width?
5. A swimming pool in the shape of a cylinder holds 2500 gal of water. Its diameter is 12 ft. What is its height?

6. The area of a right triangular piece of land is 1.15 km². Of the two sides that are per-pendicular, one measures 827 m and the length of the other side is unknown. Find the length of the unknown side.

7. An office building in the shape of a rectangular box occupies 12,000 cu ft of space. The area of the flat roof is 710 sq ft. What is the height of the building?

8. When two resistors are connected in parallel, their total resistance is 75 Ω. If one of the resistances is 100 Ω, then what is the value of the other resistor?

9. The perimeter of an isosceles triangle is 59.6 cm. The nonequal side measures 15.4 cm. What is the length of one of the equal sides?

10. The perimeter of a rectangle is 9820 km. The length is 2950 km. What is the width?

11. The perimeter of an isosceles triangle is 59.6 cm. The nonequal side measures 15.4 cm less than one of the equal sides. What are the dimensions of the triangle?

12. The perimeter of a rectangle is 9820 km. The length is 2950 km more than the width. What are the dimensions of the rectangle?

13. A student's average score on two exams is 64 points. The score on the first exam is 5 points more than twice the score on the second exam. What are the two exam scores?

14. The average weight of four wood screws is .115 lb. Three of the four weigh the same, and one weighs .100 lb less than twice one of the others. What is the weight of each wood screw?

15. A nine-foot board is cut into two pieces. The larger piece is 8 in. longer than four times the shorter piece. What are the lengths, in inches, of each piece?

16. When two resistors are connected in series, their total resistance is 68.0 Ω. The smaller resistance is 6.0 Ω less than one-third of the larger resistance. What is the value of each resistor?

17. A nine-foot board is cut into three pieces. The first piece is six times as long as the second piece. The third piece is 8 in. longer than half the first piece. What is the length, in inches, of each piece?

18. When three resistors are connected in series, their total resistance is 68.0 Ω. The first resistance is 18.0 Ω less than the second resistance. The third resistance is the sum of the first two resistances. What is the value of each resistor?

19. The perimeter of a rectangle is four times the length, decreased by 7.0 cm. The length is twice the width, increased by 1.0 cm. Find the dimensions of the rectangle.

20. The average diameter of three ball bearings is twice the diameter of the first, decreased by .132 cm. The diameters of the second and third ball bearings are each .006 cm less than the diameter of the first ball bearing. What is the diameter of each ball bearing?

4.4 VERBAL PROBLEMS INVOLVING CURRENTS AND VOLTAGES (OPTIONAL)

Objectives

☐ To determine the currents in a parallel circuit such that the circuit conforms to Kirchhoff's current law.

☐ To determine the voltage drops in a series circuit such that the circuit conforms to Kirchhoff's voltage law.

⊡ Kirchhoff's Current Law

In Chapter 1 we used Kirchhoff's current law as an application for the addition of integers. Kirchhoff's current law states that the algebraic *sum* of all the currents at a junction point P in a circuit must be zero. Currents flowing toward point P were considered positive and currents flowing away from point P were considered negative. In mathematics, the symbol for summing is the Greek letter sigma Σ. Current is measured in units of *amperes*, which is simply abbreviated A.

FORMULA: Kirchhoff's Current Law

$\Sigma I = 0$, that is, the sum of all currents (I) equals zero.

If two lamps are connected in parallel to a voltage source (battery) as shown in Fig. 4.1, a current I_1 will start to flow in the circuit. When current I_1 reaches the junction point P, it will branch into currents I_2 and I_3 (see Fig. 4.1).

Notice current I_1 is flowing *toward* point P and currents I_2 and I_3 are flowing *away* from point P. According to Kirchhoff's current law, $\Sigma I = 0$ at point P. Thus,

$$I_1 + (-I_2) + (-I_3) = 0$$

or

$$I_1 = I_2 + I_3.$$

Battery (E)

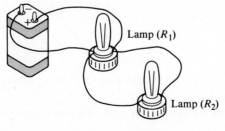

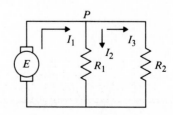

(a) Pictorial diagram of a parallel circuit

(b) Schematic diagram of a parallel circuit

Figure 4.1

EXAMPLE 1

Given the circuit in Fig. 4.1, determine the currents I_2 and I_3 if I_3 is four times as large as I_2, and $I_1 = .135$ A.

Solution:

1. Formula

$$\Sigma I = 0$$

or

$$I_1 = I_2 + I_3 \text{ for the circuit in Fig. 4.1}$$

2. Setup

$$I_1 = .135 \text{ A}$$
$$I_2 = x$$
$$I_3 = 4x$$

3. Equation

$$I_1 = I_2 + I_3$$

$.135 = x + 4x$ substituting

$.135 = 5x$ simplifying

$\dfrac{1}{5}(.135) = \dfrac{1}{5}(5x)$ multiplying each side by $\dfrac{1}{5}$

$.027 = x$ simplifying

4. Conclusion

$$I_1 = .135 \text{ A}$$
$$I_2 = x = .027 \text{ A}$$
$$I_3 = 4x = .108 \text{ A} \quad \blacksquare$$

5. Check

$$I_1 = I_2 + I_3$$
$$.135 \text{ A} = .027 \text{ A} + .108 \text{ A}?$$
$$.135 \text{ A} = .135 \text{ A} \quad \text{true}$$

∴ Kirchhoff's Voltage Law

In electronics, there is also a rule called Kirchhoff's voltage law. It states that in a closed loop the *sum* of the voltage drops across each electrical component must equal the voltage supply in that loop. Voltage is measured in *volts* (V).

FORMULA: Kirchhoff's Voltage Law

$\Sigma V = E$, that is, the sum of the voltage drops (V)
equals the voltage supply (E).

If two lamps are connected in *series* as shown in Fig. 4.2, a voltage drop (V_1) will occur across one lamp, and another voltage drop V_2 will occur across the other lamp.

According to Kirchhoff's voltage law, the *sum* of the voltage drops will equal the voltage supplied by the battery, $\Sigma V = E$. Thus,

$$V_1 + V_2 = E$$

EXAMPLE 2

Given the circuit in Fig. 4.2, determine the voltage drops V_1 and V_2, if V_2 is 2.00 V more than one-third V_1 and $E = 8.24$ V.

Solution:

1. Formula

$$\Sigma V = E$$

or $\qquad V_1 + V_2 = E$ for the circuit in Fig. 4.2

2. Setup

$$E = 8.24 \text{ V}$$

$$V_1 = x$$

$$V_2 = \frac{x}{3} + 2.00 \text{ V}$$

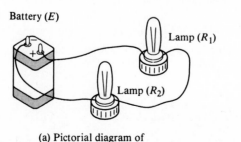

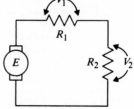

Battery (E)

Lamp (R_1)

Lamp (R_2)

(a) Pictorial diagram of
a series circuit

(b) Schematic diagram of
a series circuit

Figure 4.2

3. Equation

$$V_1 + V_2 = E$$

$$x + \left(\frac{x}{3} + 2.00\right) = 8.24 \qquad \text{substituting}$$

$$\frac{4x}{3} + 2.00 = 8.24 \qquad \text{simplifying}$$

$$\frac{4x}{3} + 2.00 + (\,-2.00\,) = 8.24 + (\,-2.00\,) \qquad \text{adding } -2.00 \text{ to each side}$$

$$\frac{4x}{3} = 6.24 \qquad \text{simplifying}$$

$$\frac{3}{4} \cdot \frac{4x}{3} = (6.24)\frac{3}{4} \qquad \text{multiplying each side by } \frac{3}{4}$$

$$x = 4.68 \qquad \text{simplifying}$$

4. Conclusion

$$E = 8.24 \text{ V}$$
$$V_1 = 4.68 \text{ V}$$
$$V_2 = 3.56 \text{ V} \quad \blacksquare$$

5. Check: The check is left for you.

EXERCISES 4.4

For Exercises 1–6, refer to Fig. 4.3.

1. If $I_1 = .37$ A and $I_3 = .14$ A, what is the value of current I_2?

2. If $I_1 = .083$ A and $I_2 = .008$ A, what is the value of current I_3?

3. If I_2 is .012 A and I_1 is five times larger than I_3, what are the values of the currents I_1 and I_3?

4. If I_1 is .083 A and I_2 is .008 A more than half I_3, what are the values of the currents I_2 and I_3?

5. If I_1 is .37 A and I_3 is .14 A less than twice I_2, what are the values of the currents I_2 and I_3?

6. If I_1 is 1 A more than I_2, and I_2 is 3 A more than twice I_3, what are the values of the currents I_1, I_2, and I_3?

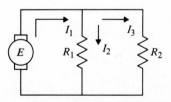

Figure 4.3

For Exercises 7–12, refer to Fig. 4.4.

7. If $E = 16.0$ V and $V_2 = 6.2$ V, what is the value of the voltage V_1?

8. If $E = 65.6$ V and $V_1 = 12.8$ V, what is the value of the voltage V_2?

9. If E is 16.0 V and V_2 is 6.2 V more than V_1, what are the values of the voltages V_1 and V_2?

10. If E is 65.6 V and V_1 is 12.8 V less than one-third of V_2, what are the values of the voltages V_1 and V_2?

11. If V_1 is 122.2 V and E is 31.6 V more than twice V_2, what are the values of the voltages E and V_2?

12. If E is 16 V more than twice V_1 and V_1 is 2 V more than half of V_2, what are the values of the voltages V_1, V_2, and E?

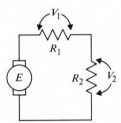

Figure 4.4

For Exercise 13, refer to Fig. 4.5.

13. I_1 is 9 A more than I_4, and I_2 is twice as large as I_3. I_4 is the sum of I_2 and I_3. Find the currents I_1, I_2, I_3, and I_4.

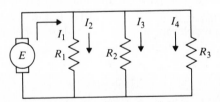

Figure 4.5

For Exercise 14, refer to Fig. 4.6.

14. E is 4 V more than V_4, and V_1 and V_2 are both half as large as V_3. V_4 is the sum of V_1, V_2, and V_3. Find the voltages V_1, V_2, V_3, V_4, and E.

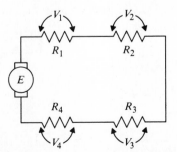

Figure 4.6

4.5 VERBAL PROBLEMS INVOLVING PARALLEL FORCES (OPTIONAL)

Objectives

⊡ To determine the forces acting on an object such that the object conforms to the force law of equilibrium.

⊡ To determine the distances to the forces acting on an object such that the object conforms to the moment law of equilibrium.

⊡ Force Law of Equilibrium

In Chapter 1 we used the *force law of equilibrium* as an application for the addition of integers. The force law of equilibrium states that if an object is at rest (in equilibrium), then the algebraic *sum* of all the forces acting on that object must be zero. Upward forces were considered positive and downward forces negative. In mathematics, the symbol for summing is the Greek letter sigma Σ. Forces are measured in pounds (lb) or Newtons (N). One Newton is approximately .225 lb.

FORMULA: Force Law of Equilibrium

$\Sigma F = 0$, that is, the sum of all forces (F) equals zero.

The beam shown in Fig. 4.7 is also shown in a force diagram. The load on the beam represents a downward force (F_3), and upward forces (F_1 and F_2) are exerted by each column in order to hold up this load. (We will assume that the beams in this section are of negligible weight.)

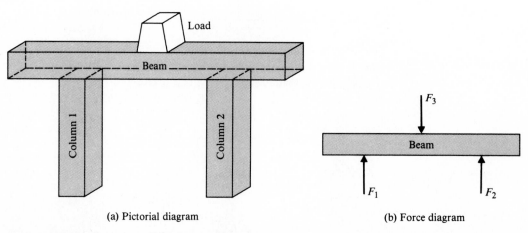

(a) Pictorial diagram (b) Force diagram

Figure 4.7

If the beam shown in Fig. 4.7 is at rest, then according to the force law of equilibrium, $\Sigma F = 0$. Thus,

$$F_1 + F_2 + (-F_3) = 0$$

or

$$F_1 + F_2 = F_3.$$

EXAMPLE 1

Given the beam in Fig. 4.7, determine the forces F_1 and F_3 if F_3 is four times as large as F_1, and $F_2 = 975$ lb.

Solution:

1. Formula

$$\Sigma F = 0$$

or $F_1 + F_2 = F_3$ for the beam in Fig. 4.7

2. Setup

$$F_1 = x$$
$$F_2 = 975 \text{ lb}$$
$$F_3 = 4x$$

3. Equation

$$F_1 + F_2 = F_3$$

$x + 975 = 4x$	substituting
$x + (-x) + 975 = 4x + (-x)$	adding $-x$ to each side
$975 = 3x$	simplifying
$\frac{1}{3}(975) = \frac{1}{3}(3x)$	multiplying each side by $\frac{1}{3}$
$325 = x$	simplifying

4. Conclusion

$$F_1 = x = 325 \text{ lb}$$
$$F_2 = 975 \text{ lb}$$
$$F_3 = 4x = 1300 \text{ lb} \quad \blacksquare$$

5. Check

$$F_1 + F_2 = F_3$$
$$325 \text{ lb} + 975 \text{ lb} = 1300 \text{ lb}?$$
$$1300 \text{ lb} = 1300 \text{ lb} \qquad \text{true}$$

⦂ Moment Law of Equilibrium

What would happen to the beam in Fig. 4.7 if column two were removed? The load (F_3) would cause the beam to rotate about column one (see Fig. 4.8). This tendency of force to cause rotation about some point is called the **moment** of force. The moment of force is the product of the force times the distance from the force to a **datum line**. A datum line is a line from which we measure the distances to all the forces. For consistency, let's always measure the distances to the forces from a datum line drawn vertically through the *left* end of the object. A moment is usually expressed in pound·feet (lb·ft), pound·inches (lb·in), or Newton·meters (N·m).

FORMULA: Moment of Force

$M = Fd$, where M is the moment, F is the force, and
d is the distance to the force.

In mechanics, there is a rule called the **moment law of equilibrium**. It states that the *sum* of all moments acting on an object at rest (in equilibrium) must also be equal to zero. If it is not, the object will start to rotate like the beam in Fig. 4.8.

FORMULA: Moment Law of Equilibrium

$\Sigma M = 0$, that is, the sum of all moments ($M = Fd$) equals zero.

Thus, if the beam in Fig. 4.9 is in equilibrium, two conditions must hold:

$$1. \ \Sigma F = 0 \text{ or } F_1 + F_2 = F_3$$
$$\text{and} \quad 2. \ \Sigma M = 0$$
$$F_1 d_1 + F_2 d_2 + (-F_3 d_3) = 0$$
$$\text{or} \quad F_1 d_1 + F_2 d_2 = F_3 d_3$$

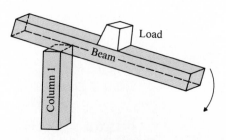

Figure 4.8

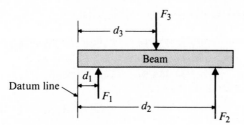

Figure 4.9

EXAMPLE 2

If the beam in Fig. 4.9 is to conform to the moment law of equilibrium, determine the distances d_1 and d_3 if d_3 is 5 ft more than d_1 and d_2 is 20 ft. Assume $F_1 = 325$ lb, $F_2 = 975$ lb, and $F_3 = 1300$ lb (from Example 1).

Solution:

1. Formula

$$\Sigma M = 0$$

or $F_1 d_1 + F_2 d_2 = F_3 d_3$ for the beam in Fig. 4.9

2. Setup

$$F_1 = 325 \text{ lb} \qquad d_1 = x$$
$$F_2 = 975 \text{ lb} \qquad d_2 = 20 \text{ ft}$$
$$F_3 = 1300 \text{ lb} \qquad d_3 = x + 5 \text{ ft}$$

3. Equation

$$F_1 d_1 + F_2 d_2 = F_3 d_3$$

$325x + (975)(20) = 1300(x + 5)$	substituting
$325x + 19500 = 1300x + 6500$	simplifying and distributing
$325x + (\boxed{-325x}) + 19500 = 1300x + (\boxed{-325x}) + 6500$	adding $-325x$ to each side
$19500 = 975x + 6500$	simplifying
$19500 + (\boxed{-6500}) = 975x + 6500 + (\boxed{-6500})$	adding -6500 to each side
$13000 = 975x$	simplifying
$\dfrac{1}{\boxed{975}}(13000) = \dfrac{1}{\boxed{975}}(975x)$	multiplying each side by $\dfrac{1}{975}$
$13\dfrac{1}{3} = x$	simplifying

4. Conclusion

$$F_1 = 325 \text{ lb} \qquad d_1 = 13\frac{1}{3} \text{ ft or 13 ft 4 in.}$$

$$F_2 = 975 \text{ lb} \qquad d_2 = 20 \text{ ft}$$

$$F_3 = 1300 \text{ lb} \qquad d_3 = 18\frac{1}{3} \text{ ft or 18 ft 4 in.} \quad \blacksquare$$

5. Check: The check is left for you.

Application

LOADING A SMALL AIRCRAFT

If the total weight of a loaded aircraft is above the maximum allowable limit, or if its center of gravity is located outside the allowable limit, the stability of the aircraft will be seriously affected. Before takeoff, the pilot must check these conditions to assure a safe flight.

EXAMPLE 3

Suppose a single-engine airplane is being prepared for flight. The load to be carried consists of

a) three passengers weighing 60 lb, 120 lb, and 210 lb

b) the pilot weighing 170 lb

c) baggage weighing 180 lb

d) 90 gal of fuel weighing 6 lb per gal (full tank)

The loads are distributed as shown in Fig. 4.10.

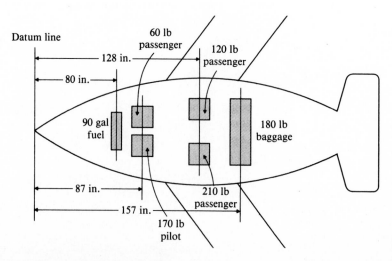

Figure 4.10

Suppose the owner's manual for the aircraft states the following:

a) Empty weight of the aircraft—3950 lb

b) Empty weight center of gravity—86 in. from the datum line

c) Maximum allowable total weight—5350 lb

d) Safe limits for the center of gravity—83.5–89.5 in.

Is the airplane safe to fly?

Solution: First, check to see if the total weight of the loaded aircraft is beyond its maximum allowable limit of 5350 lb.

$$\Sigma \text{ Weights} < 5350 \text{ lb}?$$

aircraft   passengers   pilot   baggage   fuel

$$3950 \text{ lb} + 390 \text{ lb} + 170 \text{ lb} + 180 \text{ lb} + 90 \text{ gal}\left(6\,\frac{\text{lb}}{\text{gal}}\right) < 5350 \text{ lb}?$$

$$4690 \text{ lb} + 540 \text{ lb} < 5350 \text{ lb}?$$

$$\text{total weight} \rightarrow \boxed{5230 \text{ lb}} \; < 5350 \text{ lb}$$

Therefore, the aircraft's total weight is within the maximum allowable limits.

Second, determine whether the center of gravity is within the limits set forth in the owner's manual.

1. Formula: $\Sigma M = 0$

$$F_1 d_1 + F_2 d_2 + F_3 d_3 + F_4 d_4 + F_5 d_5 = F_t d_t$$

where F_1 is empty weight of aircraft, d_1 is empty weight center of gravity
F_2 is weight in front seats, d_2 is distance to front seats
F_3 is weight in rear seats, d_3 is distance to rear seats
F_4 is weight of baggage, d_4 is distance to baggage compartment
F_5 is weight of fuel, d_5 is distance to fuel
F_t is total weight, d_t is center of gravity of loaded aircraft

2. Setup: $F_1 = 3950$ lb, $d_1 = 86$ in.
$F_2 = 230$ lb, $d_2 = 87$ in.
$F_3 = 330$ lb, $d_3 = 128$ in.
$F_4 = 180$ lb, $d_4 = 157$ in.
$F_5 = 540$ lb, $d_5 = 80$ in.
$F_t = 5230$ lb, $d_t = ?$ the unknown

3. Equation

$$F_1 d_1 + F_2 d_2 + F_3 d_3 + F_4 d_4 + F_5 d_5 = F_t d_t$$
$$(3950)(86) + (230)(87) + (330)(128) + (180)(157) + (540)(80) = 5230 d_t$$
$$473{,}410 = 5230 d_t$$
$$90.5 = d_t$$

4. Conclusion: The center of gravity of the loaded aircraft is located *outside* the allowable limit, since 90.5 in. is *not* between 83.5 in. and 89.5 in. Therefore, *the aircraft is not yet safe to fly.*   ■

5. Check: The check is left for you.

Note: The center of gravity of the loaded aircraft must be corrected before a safe flight can be assured. The center of gravity can be brought within the allowable limit by doing one or more of the following:

a) Limiting baggage (see Exercise 9)

b) Limiting the amount of fuel to that required for flight

c) Rearranging the position of the passengers and cargo (see Exercise 10)

EXERCISES 4.5

For Exercises 1–6, refer to Fig. 4.11.

1. Given that $F_1 = 174$ lb, $d_1 = 6.00$ ft, $F_2 = 348$ lb, and $d_2 = 14.0$ ft, determine
   a) Force F_3
   b) Distance d_3

2. Given that $F_3 = 95.0$ N, $d_3 = 2.20$ m, $F_2 = 55.0$ N, and $d_2 = 3.00$ m, determine
   a) Force F_1
   b) Distance d_1

3. Given that F_1 is 174 lb and d_1 is 6.00 ft, determine
   a) Forces F_2 and F_3, if F_3 is seven times larger than F_2.
   b) Distances d_2 and d_3, if d_2 is 4.00 ft more than d_3.

4. Given that F_3 is 95.0 N and d_3 is 2.20 m, determine
   a) Forces F_1 and F_2, if F_1 is 20.0 N more than twice F_2.
   b) Distances d_1 and d_2, if d_2 is 1.00 m more than twice d_1.

5. Given that F_3 is four times as large as F_1, and d_3 is 3 m less than d_2, determine
   a) Forces F_1, F_2 and F_3, if F_2 is 400 N more than F_1.
   b) Distances d_1, d_2, and d_3, if d_2 is 4 m less than twice d_1.

6. Given that F_1 is 75 lb less than one-fourth F_3 and that d_1 is one-third d_2, determine
   a) Forces F_1, F_2, and F_3, if F_2 is seven-eighths of F_3.
   b) Distances d_1, d_2, and d_3, if d_3 is 1 ft less than d_2.

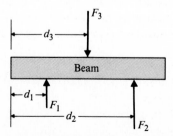

Figure 4.11

For Exercises 7 and 8, refer to Fig. 4.12.

7. Given that F_3 is 180 lb more than F_4, and that d_2 is 9 in. more than d_3, determine
    a) Forces F_1, F_2, F_3, and F_4, if F_2 is twice F_4, and F_1 is 200 lb less than F_3.
    b) Distances d_1, d_2, d_3, and d_4, if d_3 is twice d_4 and d_1 is one-third d_4.

8. Given that F_3 is twice F_4, and that d_3 is 250 cm less than d_2, determine
    a) Forces F_1, F_2, F_3, and F_4, if both F_1 and F_2 are 20 N less than F_3.
    b) Distances d_1, d_2, d_3, and d_4, if d_4 is 1 m more than d_1, and d_2 is five times d_1.

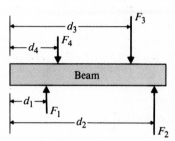

Figure 4.12

For Exercises 9 and 10, refer to Example 3 and Fig. 4.10.

9. How many pounds of baggage must be off-loaded to bring the center of gravity within the allowable limits set forth in the owner's manual?

10. If all the baggage must go on the flight and the 210-lb passenger and the 60-lb passenger switched seats, would the center of gravity of the loaded aircraft fall within the allowable limits as set forth in the owner's manual?

4.6 RATIO

Objectives

- ⊡ To write the ratio of one measurement to another using various forms.
- ⊡ To solve a verbal problem that involves ratios.

⊡ Various Forms of a Ratio

When a quotient is used to compare two quantities, the comparison is called a **ratio**. For example, if a certain type of solder is made up of eight parts tin to five parts lead, then we can describe this situation by writing a quotient as follows:

$$\frac{\text{parts of tin}}{\text{parts of lead}} = \frac{8}{5} \qquad (\text{read "the ratio of 8 to 5"})$$

This is the **fractional form** of a ratio.

This same ratio can also be written in **colon form** as follows:

parts of tin: parts of lead = 8:5 (read "the ratio of 8 to 5")

As you can see, a ratio consists of *two* terms.

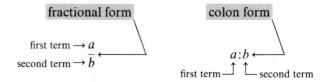

If the terms are inverted, we obtain the *inverse ratio*. Thus,

$$\frac{\text{parts of lead}}{\text{parts of tin}} = \frac{5}{8} = 5:8$$

the inverse ratio of 8 to 5

It is often desirable to express a ratio as a **decimal number compared to one**. To do this, multiply both terms of the ratio by the reciprocal of the second term. Thus,

$$\frac{8}{5} = \frac{8\left(\frac{1}{5}\right)}{5\left(\frac{1}{5}\right)} = \frac{\frac{8}{5}}{1} = \frac{1.6}{1} \text{ or } 1.6:1, \text{ and}$$

decimal-to-one form

$$\frac{5}{8} = \frac{5\left(\frac{1}{8}\right)}{8\left(\frac{1}{8}\right)} = \frac{\frac{5}{8}}{1} = \frac{.625}{1} \text{ or } .625:1.$$

A ratio is usually made between quantities having the *same* units of measure. Therefore, conversion factors may have to be used before a ratio between two quantities can be described. *Ratios should always be reduced to lowest terms.*

EXAMPLE 1

Write a ratio for 12 ft to 45 in. using

a) fractional form
b) colon form
c) decimal-to-one form

Solution: Either change 12 ft to inches or 45 in. to feet. It is usually easier to change the larger unit of measure to the smaller unit of measure. Therefore, convert 12 ft to inches.

$$12 \text{ ft} = \frac{12 \cancel{\text{ft}}}{1} \cdot \left(\frac{12 \text{ in.}}{1 \cancel{\text{ft}}}\right) = 144 \text{ in.}$$

a)   fractional form $\dfrac{144 \text{ in.}}{45 \text{ in.}} = \dfrac{144}{45} = \dfrac{\cancel{9} \cdot 16}{\cancel{9} \cdot 5} = \dfrac{16}{5}$ ∎

Note: The common unit of measure (inches) cancels. $\Bigg\}$

b)   colon form   16:5   ∎

c)   decimal-to-one form $\dfrac{16}{5} = \dfrac{16\left(\dfrac{1}{5}\right)}{5\left(\dfrac{1}{5}\right)} = \dfrac{\dfrac{16}{5}}{1} = \dfrac{3.2}{1}$ or 3.2:1   ∎

Note: The inverse ratio of 12 ft to 45 in. would be

$$\frac{5}{16} = 5\!:\!16 = \frac{.3125}{1} \text{ or } .3125\!:\!1.$$

⸫ Verbal Problems

The fundamental property of fractions allows us to say that the ratio of 8 to 5 not only compares the numbers 8 and 5, but also compares 16 and 10, 24 and 15, -80 and -50, and in general $8x$ and $5x$ ($x \neq 0$).

$$\frac{8}{5} = \frac{16}{10} = \frac{24}{15} = \frac{-80}{-50} = \frac{8x}{5x} \, (x \neq 0)$$

reduced form of the ratio $\longrightarrow$ $\llcorner$ general form of the ratio

When working with verbal problems, we use the *general form of a ratio* to allow for all possible comparisons.

EXAMPLE 2

The ratio of the number of pounds of tin to the number of pounds of lead in a certain type of solder is 8:5. How many pounds of each metal are contained in 52 lb of this solder?

Solution:

1. Formula

$$W_t = W_1 + W_2, \text{ where } W_t \text{ is the total weight}$$
$$W_1 \text{ is the weight of the tin}$$
$$W_2 \text{ is the weight of the lead}$$

2. Setup

$$W_t = 52 \text{ lb}$$
$$W_1 = 8x$$
$$W_2 = 5x$$

Note: $\dfrac{8x}{5x} = \dfrac{8}{5} = 8:5$ when reduced.

3. Equation

$$W_t = W_1 + W_2$$

$$52 = 8x + 5x \qquad \text{substituting}$$

$$52 = 13x \qquad \text{simplifying}$$

$$\frac{1}{13}(52) = \frac{1}{13}(13x) \qquad \text{multiplying each side by } \frac{1}{13}$$

$$4 = x \qquad \text{simplifying}$$

4. Conclusion

$$W_t = 52 \text{ lb}$$
$$W_1 = 8x = 32 \text{ lb}$$
$$W_2 = 5x = 20 \text{ lb} \quad \blacksquare$$

Note: $\dfrac{32}{20} = \dfrac{8}{5} = 8:5$ when reduced.

5. Check: The check is left for you.

The idea of a ratio can easily be extended to compare three or more quantities. For example, if the quantities a, b, and c are in the ratio $3:4:5$, then

$$a:b = 3:4,$$
$$b:c = 4:5, \text{ and}$$
$$a:c = 3:5.$$

EXAMPLE 3

The perimeter of a triangle is 84 in. The lengths of the sides are in the ratio $3:4:5$. What are the lengths of the sides?

Solution:

1. Formula

$$P = a + b + c, \text{ where } P \text{ is the perimeter}$$

a is the length of one side
b is the length of the second side
c is the length of the third side

2. Setup

$$P = 84 \text{ in.}$$
$$a = 3x$$
$$b = 4x$$
$$c = 5x$$

3. Equation

$$P = a + b + c$$
$$84 = 3x + 4x + 5x \qquad \text{substituting}$$
$$84 = 12x \qquad \text{simplifying}$$
$$\frac{1}{12}(84) = \frac{1}{12}(12x) \qquad \text{multiplying each side by } \frac{1}{12}$$
$$7 = x \qquad \text{simplifying}$$

4. Conclusion

$$P = 84 \text{ in.}$$
$$a = 3x = 21 \text{ in.} \leftarrow$$
$$b = 4x = 28 \text{ in.} \leftarrow$$
$$c = 5x = 35 \text{ in.} \leftarrow \blacksquare$$

$$\begin{cases} Note: \\ \dfrac{a}{b} = \dfrac{21}{28} = \dfrac{3}{4} = 3:4 \\ \\ \dfrac{b}{c} = \dfrac{28}{35} = \dfrac{4}{5} = 4:5 \\ \\ \dfrac{a}{c} = \dfrac{21}{35} = \dfrac{3}{5} = 3:5 \end{cases}$$

5. Check

$$P = a + b + c$$
$$84 \text{ in.} = 21 \text{ in.} + 28 \text{ in.} + 35 \text{ in.}?$$
$$84 \text{ in.} = 84 \text{ in.} \qquad \text{true}$$

Application

OHM'S LAW

Over a hundred years ago, Georg Simon Ohm discovered that when a lamp was connected directly across the terminals of a voltage supply, the current through the lamp always had the same constant value. He also discovered that if he doubled the voltage in the circuit, the current through the same lamp doubled. Similarly if he tripled the voltage, the current tripled, and so on. Therefore, he concluded that the *ratio* of the voltage (*E*) to the current (*I*) was a constant (*k*).

$$\frac{E}{I} = k$$

With different sized lamps, different constants occurred. The constant appeared to be a property of the individual circuit. Finally, he discovered that for a given voltage supply, the value of the constant k increased as the current decreased. He thought of this constant as representing an opposition to the flow of current. Therefore, he called this constant k the **resistance** of the circuit.

By replacing the constant k with the resistance symbol R, we have Ohm's law.

FORMULA: Ohm's Law

$$\frac{E}{I} = R, \text{ where } E \text{ is the voltage in volts,}$$
$$I \text{ is the current in amps, and}$$
$$R \text{ is the resistance in ohms.}$$

The unit of measurement for the resistance R could have been labeled volts per amperes $\left(\dfrac{\text{volts}}{\text{amperes}}\right)$. However, it was decided to honor Ohm's discovery by calling the unit of measure for electrical resistance the **ohm** (Ω).

EXAMPLE 4

A television set draws a current of 2.25 A. If its resistance is 52.0 Ω, to what voltage supply is it connected (see Fig. 4.13)?

Solution:

1. Formula

$$\frac{E}{I} = R, \text{ where } E \text{ is the voltage}$$
$$I \text{ is the current}$$
$$R \text{ is the resistance}$$

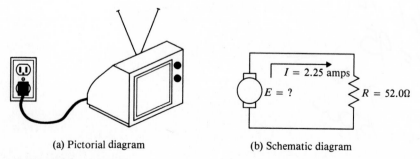

(a) Pictorial diagram   (b) Schematic diagram

Figure 4.13

2. Setup

$$E = ? \qquad \text{unknown}$$
$$I = 2.25 \text{ A}$$
$$R = 52.0 \ \Omega$$

3. Equation

$$\frac{E}{I} = R$$

$$\frac{E}{2.25} = 52.0 \qquad\qquad \text{substituting}$$

$$(\boxed{2.25}) \frac{E}{2.25} = (52.0)(\boxed{2.25}) \qquad \text{multiplying each side by 2.25}$$

$$E = 117 \text{ V} \qquad\qquad \text{simplifying}$$

4. Conclusion: The television set is connected to a 117 V source. ∎

5. Check: The check is left for you.

EXERCISES 4.6

Write the *inverse* ratio of each of the following ratios.

1. $\frac{9}{5}$ $\frac{5}{9}$ **2.** $\frac{3}{10}$ $\frac{10}{3}$ **3.** 6:19   **4.** .25:1

Express each of the following as a ratio using
  a) fractional form
  b) colon form
  c) decimal-to-one form

5. 4 in. to 16 in.   **6.** 9 cm to 12 cm

7. $\frac{5}{16}$ sq in. to $\frac{1}{2}$ sq in.   **8.** 4 cu yd to $6\frac{2}{3}$ cu yd

9. 550 m to 2 km   **10.** $66\overline{0}$ ft to 3.00 mi

11. 45 in² to 2.0 ft²   **12.** 800 mm³ to 2 cm³

13. The fuel mixture for a certain two-cycle chain saw engine requires that the ratio of gasoline to oil be 15 to 1. How many liters of gas and oil are there in a 4 L mixture?

14. In a step-up transformer, the ratio of the number of turns of wire in the primary coil to the number of turns of wire in the secondary coil is 1:160. If there are 15,295 turns of wire altogether, then how many turns of wire are there in each coil?

15. The perimeter of a rectangle is 560 in. The ratio of the width to the length is 3:4. Find the length and width of the rectangle.

16. The perimeter of a quadrilateral is 68 cm. The sides are in the ratio 5:8:10:11. What are the lengths of each side of the quadrilateral?

17. A dry mixture of concrete is made up of cement, sand, and gravel in the ratio 2:3:3 by volume. How many cubic yards of each must be used to make 42 cu yd of this mixture?

18. When three resistors are connected in series, they have a total resistance of $70\,\Omega$. The resistances are in the ratio of $2:3:5$. Find the values of each resistor.

19. One square has a side twice as large as another square. What is the ratio of their areas?

20. One cube has an edge three times as large as another cube. What is the ratio of their volumes?

21. Two cylinders have the same height, yet one has a radius four times as large as the other. What is the ratio of their volumes?

22. Two cylinders have the same diameter, yet one has a height five times as large as the other. What is the ratio of their volumes?

23. A toaster is connected to a 117 V outlet and has a resistance of $12.0\,\Omega$. What current does the toaster draw?

24. An electric dryer draws a current of 11.0 A. If its resistance is $20.0\,\Omega$, to what voltage supply is it connected?

4.7 PROPORTIONS

Objectives

☐ To use the cross product property to solve a proportion.

☐ To solve a verbal problem that involves a direct proportion.

☐ To solve a verbal problem that involves an inverse proportion.

☐ Cross Product Property

If we state the fact that one ratio is equal to another ratio, we form what is called a **proportion**. For example, the ratio of 8 to 5 equals the ratio of 16 to 10, can be written as the proportion

$$\frac{8}{5} = \frac{16}{10}.$$

The proportion $\dfrac{a}{b} = \dfrac{c}{d}$ contains four terms, which can be numbered as follows:

$$\begin{array}{c} \text{first} \to a \\ \text{second} \to b \end{array} = \begin{array}{c} c \leftarrow \text{third} \\ d \leftarrow \text{fourth} \end{array}$$

Mathematicians refer to the first and fourth terms as the **extremes**, while the second and third terms are called the **means**. Notice that in the proportion $\frac{8}{5} = \frac{16}{10}$, the

product of the extremes	equals	the product of the means.
(8)(10)	=	(5)(16)
80	=	80

We can show that this is true of every proportion as follows:

$$\frac{a}{b} = \frac{c}{d}$$

$$b\left(\frac{a}{b}\right) = b\left(\frac{c}{d}\right) \qquad \text{multiplying each side by } b$$

$$a = \frac{bc}{d} \qquad \text{simplifying}$$

$$(a)\, d = \left(\frac{bc}{d}\right) d \qquad \text{multiplying each side by } d$$

$$ad = bc \qquad \text{simplifying}$$

product of the extremes ⌐   ⌐product of the means

CROSS PRODUCT PROPERTY
In any proportion

$$\frac{a}{b} = \frac{c}{d}, \; ad = bc. \qquad \begin{array}{l} b \neq 0 \\ d \neq 0 \end{array}$$

When one or more terms of a proportion are unknown, the *cross product property* can be used to help solve the proportion.

EXAMPLE 1

Solve $\dfrac{5}{2} = \dfrac{3}{x}$.

Solution:

$$\frac{5}{2} = \frac{3}{x}$$

$$(5)(x) = (2)(3) \qquad \text{cross product property}$$

$$5x = 6 \qquad \text{simplifying}$$

$$\frac{1}{5}(5x) = \frac{1}{5}(6) \qquad \text{multiplying each side by } \frac{1}{5}$$

$$x = \frac{6}{5} \; \blacksquare \qquad \text{simplifying}$$

Note: Does $\dfrac{5}{2} = \dfrac{3}{x}$, when $x = \dfrac{6}{5}$?

$$\frac{5}{2} = \frac{3}{\frac{6}{5}}?$$

$$\frac{5}{2} = \frac{3}{1} \cdot \frac{5}{6}?$$

$$\frac{5}{2} = \frac{5}{2} \qquad \text{true}$$

EXAMPLE 2

Solve $\dfrac{5y}{8} = \dfrac{3y + 5}{4}$.

Solution:

$$\frac{5y}{8} = \frac{3y + 5}{4}$$

$(5y)(4) = (8)(3y + 5)$ cross product property

$20y = 24y + 40$ simplifying and distributing

$20y + (-24y) = 24y + (-24y) + 40$ adding $-24y$ to each side

$-4y = 40$ simplifying

$-\dfrac{1}{4}(-4y) = -\dfrac{1}{4}(40)$ multiplying each side by $-\dfrac{1}{4}$

$y = -10$ ■   simplifying

Note: Try checking by replacing y with -10 in the original proportion.

⬚ Direct Proportions

We can say that the weight of a uniform steel beam is *directly proportional* to its length, since an *increase* in its length would lead to a proportional *increase* in its weight (or a *decrease* in its length would lead to a proportional *decrease* in its weight).

Suppose a steel beam is L_1 feet long and weighs W_1 pounds, and a similar steel beam is L_2 feet long and weighs W_2 pounds. When setting up the four terms that describe this *direct proportion*, we must be certain that if the length L_1 is the first term of the proportion, then its corresponding weight W_1 is the third term of the proportion. This places L_1 *directly opposite* W_1 and L_2 *directly opposite* W_2.

$$\text{first} \rightarrow \frac{L_1}{L_2} = \frac{W_1}{W_2} \begin{array}{l} \leftarrow \text{third} \\ \leftarrow \text{fourth} \end{array}$$
$$\text{second} \rightarrow \qquad\qquad$$

EXAMPLE 3

An 18-ft steel beam weighs 1125 lb. How many feet of a similar section weigh 625 lb?

Solution:

1. Formula

$$\frac{L_1}{L_2} = \frac{W_1}{W_2}, \text{ where } \frac{L_1}{L_2} \text{ is the ratio of the lengths}$$

$$\frac{W_1}{W_2} \text{ is the ratio of the weights}$$

2. Setup

$$L_1 = 18 \text{ ft} \qquad W_1 = 1125 \text{ lb}$$
$$L_2 = ? \quad \text{unknown} \qquad W_2 = 625 \text{ lb}$$

3. Equation

$$\frac{L_1}{L_2} = \frac{W_1}{W_2}$$

$$\frac{18 \text{ ft}}{L_2} = \frac{1125 \text{ lb}}{625 \text{ lb}} \qquad \text{substituting and cancelling lb}$$

$$(18 \text{ ft})(625) = 1125 \, L_2 \qquad \text{cross product property}$$

$$11{,}250 \text{ ft} = 1125 \, L_2 \qquad \text{simplifying}$$

$$\frac{1}{1125}(11{,}250 \text{ ft}) = \frac{1}{1125}(1125 \, L_2) \qquad \text{multiplying each side by } \frac{1}{1125}$$

$$10 \text{ ft} = L_2 \qquad \text{simplifying}$$

4. Conclusion

$$L_1 = 18 \text{ ft} \qquad W_1 = 1125 \text{ lb}$$
$$L_2 = 10 \text{ ft} \qquad W_2 = 625 \text{ lb} \quad \blacksquare$$

5. Check: The check is left for you.

⬝⬝ Inverse Proportions

We can say that the time required to make a trip from Boston to Montreal is *inversely proportional* to the average rate of speed at which we travel, since an *increase* in the rate of speed leads to a proportional *decrease* in the time of travel (or a *decrease* in the rate of speed leads to a proportional *increase* in the time of travel).

Suppose it takes t_1 hours to travel from Boston to Montreal at an average speed of r_1 miles per hour (mph), and it takes t_2 hours to make the same trip at an average speed of r_2 miles per hour. When setting up the four terms that describe this **inverse proportion**, we must set the ratio of the times equal to the *inverse* ratio of the rates of speed. This places t_1 *diagonally opposite* r_1 and t_2 *diagonally opposite* r_2.

$$\text{first} \rightarrow \frac{t_1}{t_2} = \frac{r_2}{r_1} \begin{array}{l} \leftarrow \text{third} \\ \leftarrow \text{fourth} \end{array}$$
$$\text{second} \rightarrow$$

EXAMPLE 4

It takes two hours longer to make the trip from Boston to Montreal when traveling at an average speed of 40 mph than it does when traveling at an average speed of 50 mph. How long does it take to make the trip at each speed?

Solution:

1. Formula

$$\frac{t_1}{t_2} = \frac{r_2}{r_1}, \text{ where } \frac{t_1}{t_2} \text{ is the ratio of the times}$$

$$\frac{r_2}{r_1} \text{ is the } \textit{inverse} \text{ ratio of the rates of speed}$$

2. Setup

$$t_1 = x \qquad\qquad r_1 = 50 \text{ mph}$$
$$t_2 = x + 2 \text{ hr} \qquad r_2 = 40 \text{ mph}$$

3. Equation

$$\frac{t_1}{t_2} = \frac{r_2}{r_1}$$

$\dfrac{x}{x + 2 \text{ hr}} = \dfrac{40 \text{ mph}}{50 \text{ mph}}$	substituting and cancelling mph
$(50)(x) = 40(x + 2 \text{ hr})$	cross product property
$50x = 40x + 80 \text{ hr}$	simplifying and distributing
$50x + (-40x) = 40x + (-40x) + 80 \text{ hr}$	adding $-40x$ to each side
$10x = 80 \text{ hr}$	simplifying
$\dfrac{1}{10}(10x) = \dfrac{1}{10}(80 \text{ hr})$	multiplying each side by $\dfrac{1}{10}$
$x = 8 \text{ hr}$	simplifying

4. Conclusion

$$t_1 = x = 8 \text{ hr} \qquad\qquad r_1 = 50 \text{ mph}$$
$$t_2 = x + 2 \text{ hr} = 10 \text{ hr} \qquad r_2 = 40 \text{ mph} \quad \blacksquare$$

5. Check: The check is left for you.

EXERCISES 4.7

Use the cross product property to solve each proportion. Be sure to check your results.

1. $\dfrac{x}{51} = \dfrac{4}{3}$

2. $\dfrac{15}{8} = \dfrac{y}{6}$

3. $\dfrac{4}{3p} = \dfrac{-2}{7}$

4. $\dfrac{-6}{7} = \dfrac{-9}{5x}$

5. $\dfrac{5}{4} = \dfrac{x+1}{6}$

6. $\dfrac{8}{3z+1} = \dfrac{8}{5}$

7. $\dfrac{10}{3} = \dfrac{-5}{4-2z}$

8. $\dfrac{6-5t}{6} = \dfrac{2}{-3}$

9. $\dfrac{5}{9} = \dfrac{6x+2}{9x}$

10. $\dfrac{5x-3}{2} = \dfrac{5x}{-8}$

11. $\dfrac{-3t+4}{8} = \dfrac{3-4t}{12}$

12. $\dfrac{8x-3}{2x-4} = \dfrac{-1}{3}$

13. The electrical resistance of a wire is directly proportional to the length of the wire.

$$\frac{R_1}{R_2} = \frac{L_1}{L_2}, \text{ where } \frac{R_1}{R_2} \text{ is the ratio of the resistances and}$$

$$\frac{L_1}{L_2} \text{ is the ratio of the lengths.}$$

If 75.0 m of a certain wire has a resistance of 1.85 Ω, then what length has a resistance of 1.02 Ω?

14. The electrical resistance of a wire is inversely proportional to the cross-sectional area of the wire.

$$\frac{R_1}{R_2} = \frac{A_2}{A_1}, \text{ where } \frac{R_1}{R_2} \text{ is the ratio of the resistances}$$

$$\frac{A_2}{A_1} \text{ is the inverse ratio of the cross-sectional area}$$

A certain length of wire, having a diameter of .286 mm and a resistance of 268.5 Ω, is to be replaced with a new wire of the same length whose diameter is 1.024 mm. What is the resistance of this new wire?

15. The speed of a pulley is inversely proportional to the diameter of the pulley.

$$\frac{S_1}{S_2} = \frac{d_2}{d_1}, \text{ where } \frac{S_1}{S_2} \text{ is the ratio of the speeds}$$

$$\frac{d_2}{d_1} \text{ is the inverse ratio of the diameters}$$

Two pulleys are connected by a belt. One has a diameter of $3\frac{1}{4}$ in. and rotates at 1840 revolutions per minute (rpm). How fast is the other pulley moving if its diameter is $11\frac{1}{2}$ in?

16. When the temperature is constant, the pressure of a gas is inversely proportional to the volume occupied by the gas.

$$\frac{P_1}{P_2} = \frac{V_2}{V_1}, \text{ where } \frac{P_1}{P_2} \text{ is the ratio of the pressures}$$

$$\frac{V_2}{V_1} \text{ is the inverse ratio of the volumes}$$

The volume occupied by a gas is 75.0 cu in. when the pressure is 35.0 pounds per square inch (psi). What is its volume when the pressure is increased to 105.0 psi?

17. The *length of an arc* of a circle is directly proportional to the measure of the central angle (see Fig. 4.14).

$$\frac{S_1}{S_2} = \frac{\theta_1}{\theta_2}, \text{ where } \frac{S_1}{S_2} \text{ is the ratio of the arc lengths}$$

$$\frac{\theta_1}{\theta_2} \text{ is the ratio of the central angles.}$$

If the length of an arc is 15.0 cm when the central angle is 30°, what is the length of the arc whose central angle is 70°?

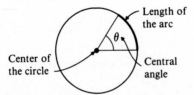

Length of the arc

θ

Center of the circle

Central angle

Figure 4.14

18. The *area of a sector* of a circle is directly proportional to the measure of the central angle (see Fig. 4.15).

$$\frac{A_1}{A_2} = \frac{\theta_1}{\theta_2}, \text{ where } \frac{A_1}{A_2} \text{ is the ratio of the sector areas}$$

$$\frac{\theta_1}{\theta_2} \text{ is the ratio of the central angles.}$$

If the area of the sector is 32.0 sq cm when the central angle is 80°, what is the area of the sector when the central angle is 135°?

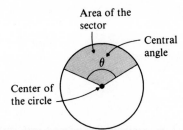

Area of the sector

Central angle

θ

Center of the circle

Figure 4.15

19. The speed of a gear is inversely proportional to the number of teeth in the gear.

$$\frac{S_1}{S_2} = \frac{T_2}{T_1}, \text{ where } \frac{S_1}{S_2} \text{ is the ratio of the speeds}$$

$$\frac{T_2}{T_1} \text{ is the inverse ratio of the number of teeth}$$

Two gears are meshed together. The smaller gear travels at 72 rpm and has 12 teeth fewer than the larger gear. How many teeth does the larger gear have if it travels at 27 rpm?

20. Under equivalent driving conditions, the amount of fuel consumed is directly proportional to the distanced traveled.

$$\frac{F_1}{F_2} = \frac{D_1}{D_2}, \text{ where } \frac{F_1}{F_2} \text{ is the ratio of the amounts of fuel consumed}$$

$$\frac{D_1}{D_2} \text{ is the ratio of the distances traveled}$$

If it takes 2 gal more fuel to travel 225 mi than it does to travel 150 mi, how many gallons of fuel are required for each trip?

21. For a series circuit, the voltage drop across a resistor is directly proportional to its resistance (see Fig. 4.16).

$$\frac{V_1}{V_2} = \frac{R_1}{R_2}, \text{ where } \frac{V_1}{V_2} \text{ is the ratio of the voltage drops}$$

$$\frac{R_1}{R_2} \text{ is the ratio of the resistances}$$

In a series circuit, if the voltage drop across a 100-Ω resistor is 5 V more than twice the voltage drop across a 40-Ω resistor, what is the voltage drop across each resistor?

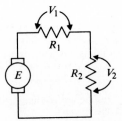

V_1

R_1

E

R_2 V_2

Figure 4.16

22. For a parallel circuit, the current through a resistor is inversely proportional to its resistance (see Fig. 4.17).

$$\frac{I_1}{I_2} = \frac{R_2}{R_1}, \text{ where } \frac{I_1}{I_2} \text{ is the ratio of the currents}$$

$$\frac{R_2}{R_1} \text{ is the inverse ratio of the resistances}$$

In a parallel circuit, if the current through a 28-Ω resistor is 1 A less than four times the current through a 98-Ω resistor, then what is the current through each resistor?

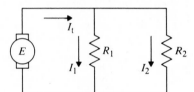

Figure 4.17

Decide whether each of the following is best described by a direct or an inverse proportion. Set up each proportion and solve. Be sure to check your results.

23. If the amount of heat lost through 6 in. of fiberglass insulation is 3750 BTU per hour, then how much insulation is needed to reduce the heat loss to 2500 BTU per hour?

24. If a man 6'0" tall casts a shadow of 4'0", how tall is a flagpole whose shadow length is 25'6".

25. A 75-N force will stretch a certain spring 7 cm farther than a 40-N force. How many centimeters will the 40-N force stretch the spring?

26. It takes six painters six more hours to paint a building than it does when nine painters are assigned the same job. How long does it take the six painters to complete the job?

Refer to exercise 17. When the central angle is 360°, the length of the arc becomes the circumference of the entire circle. Thus,

$$\frac{S_1}{S_2} = \frac{\theta_1}{\theta_2} \text{ can be written as } \frac{S_1}{2\pi r} = \frac{\theta_1}{360°}.$$

Using this proportion, find the lengths of the following arcs.

27.   50° $r = 4.00$ cm

28.   125° $r = 18.5$ in.

Refer to exercise 18. When the central angle is 360°, the area of the sector becomes the area of the entire circle. Thus,

$$\frac{A_1}{A_2} = \frac{\theta_1}{\theta_2} \text{ can be written as } \frac{A_1}{\pi r^2} = \frac{\theta_1}{360°}.$$

Using this proportion, find the area of the following sectors.

29. $r = 43.0$ mm

48°

30.

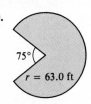

75°

$r = 63.0$ ft

31. Using the results of exercise 21, apply Kirchhoff's voltage law (Section 4.4) to find the voltage supply E.

32. Using the results of exercise 22, apply Kirchhoff's current law (Section 4.4) to find the total current I_t in the circuit.

33. Using the results of exercise 21, apply Ohm's law (Section 4.6) to find the current through each resistor.

34. Using the results of exercise 22, apply Ohm's law (Section 4.6) to find the voltage drop across each resistor.

4.8   **PERCENT**

Objectives

⊡ To use the percent proportion to solve a verbal problem involving percent.

⊡ To use the percent change proportion to solve a verbal problem involving percent change.

⊡ **Percent Proportion**

When the second term of a ratio is 100, the ratio can be expressed as a **percent**. For example, the ratio 37 to 100 can be written as

$$37:100 = \frac{37}{100} = 37 \underbrace{\text{ percent }} = 37\%$$

The word percent always means "out of every hundred."

DEFINITION

For any number P,

$$\frac{P}{100} = P\%.$$

All types of percent problems can be solved by using your knowledge of proportions.

THE PERCENT PROPORTION

$\dfrac{P}{100} = \dfrac{A}{T}$, where $\dfrac{P}{100}$ is the *percent* (written as a ratio whose second term is 100), and $\dfrac{A}{T}$ is the ratio of the *amount A* to the *total amount T.*

EXAMPLE 1

During a certain month, a wind-powered generator provided 380 kilowatt-hours (KWH) of the 550 KWH of electricity used in a home. What percentage of the electricity was provided by the wind-powered generator?

Solution:

1. Formula

$$\frac{P}{100} = \frac{A}{T}, \text{ where } \frac{P}{100} \text{ is the percent}$$

$$A \text{ is the amount}$$

$$T \text{ is the } total \text{ amount}$$

2. Setup

$$\frac{P}{100} = P\% = ? \quad \text{unknown}$$

$$A = 380 \text{ KWH}$$

$$T = 550 \text{ KWH}$$

3. Equation

$$\frac{P}{100} = \frac{A}{T}$$

$$\frac{P}{100} = \frac{380 \ \cancel{\text{KWH}}}{550 \ \cancel{\text{KWH}}} \qquad \text{substituting and cancelling KWH}$$

$$550P = 38,000 \qquad \text{cross product property}$$

$$\frac{1}{550}(550P) = \frac{1}{550}(38,000) \qquad \text{multiplying each side by } \frac{1}{550}$$

$$P = 69 \qquad \text{retain two significant digits}$$

4. Conclusion

$$\frac{P}{100} = \frac{69}{100} = 69\% \rightarrow \begin{cases} Note: \text{ Approximately 69 out} \\ \text{of every 100 KWH were} \\ \text{provided by the wind-} \\ \text{powered generator.} \end{cases}$$

$$A = 380 \text{ KWH}$$

$$T = 550 \text{ KWH} \quad \blacksquare$$

5. Check: The check is left for you.

EXAMPLE 2

A chemist has 24 L of a fuel mixture, of which 18 L are gasoline. How much pure gasoline must be added to make the fuel mixture 80% gasoline?

Solution:

1. Formula

$$\frac{P}{100} = \frac{A}{T}, \text{ where } \frac{P}{100} \text{ is the percent}$$

$$A \text{ is the amount}$$

$$T \text{ is the } total \text{ amount}$$

2. Setup

$$\frac{P}{100} = 80\% = \frac{80}{100}$$

Note: Let x be the amount of pure gasoline to be added.

$A = x + 18$ liters   amount of gasoline in the 80% mixture

$T = x + 24$ liters   *total* amount in the 80% mixture

3. Equation

$$\frac{P}{100} = \frac{A}{T}$$

$$\frac{80}{100} = \frac{x + 18}{x + 24} \qquad \text{substituting}$$

$$80(x + 24) = 100(x + 18) \qquad \text{cross product property}$$

$$80x + 1920 = 100x + 1800 \qquad \text{distributive law}$$

$$80x + (-80x) + 1920 = 100x + (-80x) + 1800 \qquad \text{adding } -80x \text{ to each side}$$

$$1920 = 20x + 1800 \qquad \text{simplifying}$$

$$1920 + (-1800) = 20x + 1800 + (-1800) \qquad \text{adding } -1800 \text{ to each side}$$

$$120 = 20x \qquad \text{simplifying}$$

$$\frac{1}{20}(120) = \frac{1}{20}(20x) \qquad \text{multiplying each side by } \frac{1}{20}$$

$$6 = x \qquad \text{simplifying}$$

4. Conclusion

$$\frac{P}{100} = 80\% = \frac{80}{100}$$

$$\left.\begin{array}{l} \text{amount of} \\ \text{gasoline to} \\ \text{be added} \end{array}\right\} \rightarrow x = 6 \text{ liters}$$

$$A = x + 18 \text{ liters} = 24 \text{ liters}$$

$$T = x + 24 \text{ liters} = 30 \text{ liters} \quad \blacksquare$$

5. Check: The check is left for you.

⬚ Percent Change Proportion

In many technical situations, it is necessary to work with a **percent change**. The percent change can be an *increase* or a *decrease* in that quantity. Again, your knowledge of proportions can help you solve problems that involve a percent change.

PERCENT CHANGE PROPORTION

$\dfrac{P}{100} = \dfrac{A_n - A_o}{A_o}$, where $\dfrac{P}{100}$ is the percent change (written as a ratio whose second term is 100), A_n is the *new* amount, A_o is the *original* amount, and $\dfrac{A_n - A_o}{A_o}$ is the ratio of the change $(A_n - A_o)$ to the original amount (A_o).

EXAMPLE 3

The retail price of a home computer recently decreased from \$899 to \$649. Find the percent change in price.

Solution:

1. Formula

$$\frac{P}{100} = \frac{A_n - A_o}{A_o}, \text{ where } \frac{P}{100} \text{ is the percent change}$$
$$A_n \text{ is the } \textit{new} \text{ amount}$$
$$A_o \text{ is the } \textit{original} \text{ amount}$$

2. Setup

$$\frac{P}{100} = P\% = ? \qquad \text{unknown}$$

$$A_n = \$649$$

$$A_o = \$899$$

3. Equation

$$\frac{P}{100} = \frac{A_n - A_o}{A_o}$$

$$\frac{P}{100} = \frac{\$649 - \$899}{\$899} \qquad \text{substituting}$$

$$\frac{P}{100} = \frac{-250}{899} \qquad \text{simplifying and cancelling \$}$$

$$899P = -25,000 \qquad \text{cross product property}$$

$$\frac{1}{899}(899P) = \frac{1}{899}(-25,000) \qquad \text{multiplying each side by } \frac{1}{899}$$

$$P = -27.8 \qquad \text{retain three significant digits}$$

Note: The negative sign tells us that the percent change is a percent *decrease*.

4. Conclusion

$$\frac{P}{100} = \frac{-27.8}{100} = -27.8\%$$

$$A_n = \$649$$

$$A_o = \$899 \quad \blacksquare$$

5. Check: The check is left for you.

EXAMPLE 4

The retail price of a home computer was recently decreased by 20%. If the home computer now sells for $210 more than half the original price, what was the original price?

Solution:

1. Formula

$$\frac{P}{100} = \frac{A_n - A_o}{A_o}, \text{ where } \frac{P}{100} \text{ is the percent change}$$

A_n is the *new* amount

A_o is the *original* amount

2. Setup

$$\frac{P}{100} = -20\% = \frac{-20}{100}$$

Note: The negative sign is attached since the percent change is a *decrease*.

$$A_n = \frac{x}{2} + \$210$$

$$A_o = x$$

3. Equation

$$\frac{P}{100} = \frac{A_n - A_o}{A_o}$$

$$\frac{-20}{100} = \frac{\left(\frac{x}{2} + \$210\right) - x}{x} \qquad \text{substituting}$$

$$\frac{-20}{100} = \frac{\$210 - \frac{x}{2}}{x} \qquad \text{simplifying}$$

$$-20x = 100\left(\$210 - \frac{x}{2}\right) \qquad \text{cross product property}$$

$$-20x = \$21,000 - 50x \qquad \text{distributive law}$$

$$-20x + 50x = \$21,000 - 50x + 50x \qquad \text{adding } 50x \text{ to each side}$$

$$30x = \$21,000 \qquad \text{simplifying}$$

$$\frac{1}{30}(30x) = \frac{1}{30}(\$21,000) \qquad \text{multiplying each side by } \frac{1}{30}$$

$$x = \$700 \qquad \text{simplifying}$$

4. Conclusion

$$\frac{P}{100} = -20\% = \frac{-20}{100}$$

$$A_n = \frac{x}{2} + \$210 = \$560$$

$$A_o = x = \$700 \text{ (the original price)} \quad \blacksquare$$

5. Check: The check is left for you.

EXERCISES 4.8

Solve the following verbal problems.

1. A home valued at \$115,000 is insured for 80% of its value. If the house were destroyed by fire, how much would the insurance company pay the owner?

2. Plumber's solder contains 65% lead. How much lead is contained in 48 lb of plumber's solder?

3. Presently, 6.7% is deducted from a person's weekly earnings for Social Security Taxes (FICA). If \$70.66 is deducted from a person's paycheck for this tax, what are his or her weekly earnings?

4. On a cold winter's day, 37% of the school buses owned by a certain company failed to start. If 29 buses failed to start, how many buses does the company own?

5. Of 1735 bolts manufactured during the day, 73 were defective. What percent were defective?

6. In a welding shop, 1872 welds were made and 1793 were acceptable. What percent were defective?

7. A house is insured for 80% of its value. If the house were destroyed by fire, the owner would receive \$25,000 less than the actual value of the home. What is the value of the home?

8. A dry mixture of concrete contains 35% sand. The amount of sand in the mixture is 19.8 cu yd less than twice the total amount of the mixture. How much sand is in the mixture?

9. Twenty-four pounds of solder contain 13 lb of lead. How much pure lead must be added to the solder to make it 60% lead?

10. How many cubic centimeters of water must be added to a 300 cm^3 solution containing 40% acid to obtain a solution that is 30% acid?

11. How much water must be evaporated from a 25-lb brine solution that contains 3 lb of salt to leave a solution that is 15% salt?

12. How much water must be evaporated from a 35-lb brine solution containing 8% salt to obtain a solution that is 10% salt?

13. The pressure in a water tank changed from 44 psi to 55 psi. What was the percent change in pressure?

14. The speed of a pulley changed from 1840 rpm to 1320 rpm. What was the percent change in speed?

15. The resistance of a variable resistor was decreased by $12\frac{1}{2}\%$ to $1400\,\Omega$. What was the original resistance of the resistor?

16. Due to expansion, the length of a 24.00-ft steel beam increased by .25%. What is the expanded length?

17. The temperature of a coolant increased by 30%. If the temperature of the coolant is now 7°C less than twice the original temperature, what was the original temperature?

18. The voltage drop across a resistor was decreased by 78%. If the voltage drop across the resistor is now 15.0 volts less than half the original voltage drop, what is the present voltage drop across the resistor?

19. A dealer buys an automobile wholesale for $3250 and tries to sell it retail at a 40% markup. During a sale he reduces the retail price by 20%. What is the selling price of the automobile when it is on sale?

20. A dealer buys an automobile wholesale and tries to sell it retail at a 30% markup. During a sale he reduces the retail price by 30%. Is the automobile now priced above, below, or at the wholesale price? Explain.

Chapter Review

REVIEW EXERCISES

The following review exercises are grouped according to the objectives that should have been mastered in Chapter 4. Work each problem carefully. If any weaknesses appear, immediately refer to and read the subsection that matches that objective.

4.1 THE MULTIPLICATION AND ADDITION PROPERTIES OF EQUALITY

Objectives

⊡ To use the Multiplication Property of Equality to solve a simple equation.

Solve.

1. $6x = -20$ **2.** $-\dfrac{5}{3}y = 25$ **3.** $\dfrac{n}{7} = -3.8$ **4.** $\dfrac{8}{t} = -32$

Solve.

⊡ To use the Addition Property of Equality to solve a simple equation.

5. $x + 19 = -32$ **6.** $\dfrac{5}{9} = m - \dfrac{2}{3}$ **7.** $1.82 - p = .06$

⊡ To apply either property to solve a formula for a given variable.

8. Moment of inertia of a triangle, $I = \dfrac{bd^3}{12}$. Solve for the base, b.

9. Perimeter of an isosceles triangle, $P = 2a + b$. Solve for the nonequal side b.

4.2 USING THE PROPERTIES TOGETHER TO SOLVE SIMPLE EQUATIONS

Objectives

⊡ To solve an equation of the form $ax + b = c$, for x.

Solve.

10. $\dfrac{9x}{4} - 5 = 13$

□ To solve a simple equation in which the variable appears on each side of the equation.

Solve.

11. $3w - 10 = 6w + 17$

12. $3(4x + 4) - 2(x + 5) = 6(x - 1)$

13. $\dfrac{4 - 6t}{3} = 3 - t$

□ To apply the properties together to solve a formula for a given variable.

14. Surface area of a cylinder, $A = 2\pi r(h + r)$. Solve for the height, h.

15. Secondary current in a transformer, $I_s = \dfrac{I_p \omega M}{Z_s + Z_L}$. Solve for the secondary impedance, Z_s.

4.3 AN INTRODUCTION TO SOLVING VERBAL PROBLEMS

Objectives

□ To solve a verbal problem that contains one unknown.

16. The perimeter of an isosceles triangle is 62.5 ft. One of the equal sides measures 27.8 ft. What is the length of the nonequal side?

17. A student's average score on three exams is 65 points. The first two exam scores were 53 points and 67 points respectively. What was the third exam score?

□ To solve a verbal problem that contains more than one unknown.

18. The perimeter of an isosceles triangle is 62.5 ft. One of the equal sides measures 27.8 ft more than the nonequal side. What is the length of each side?

19. A student's average score on three exams is 65 points. The score on the first exam was 5 points less than the second exam score, and the score on the third exam was 15 points more than half the second exam score. Find the score on each exam.

4.4 VERBAL PROBLEMS INVOLVING CURRENTS AND VOLTAGES (OPTIONAL)

Objectives

□ To determine the currents in a parallel circuit such that the circuit conforms to Kirchhoff's current law.

20. Given the circuit in Fig. 4.18, determine the currents I_2 and I_3, if I_2 is six times as large as I_3, and $I_1 = .161$ A.

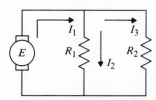

Figure 4.18

⊡ To determine the voltage drops in a series circuit such that the circuit conforms to Kirchhoff's voltage law.

21. Given the circuit in Fig. 4.19, determine the voltage drops V_1 and V_2 if V_1 is 3.00 volts less than three times V_2, and $E = 12.0$ V.

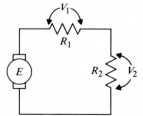

Figure 4.19

4.5 VERBAL PROBLEMS INVOLVING PARALLEL FORCES (OPTIONAL)

Objectives

⊡ To determine the forces acting on an object such that the object conforms to the force law of equilibrium.

22. If the beam in Fig. 4.20 conforms to the force law of equilibrium, determine the forces F_1 and F_3 if F_3 is 1200 lb more than twice F_1, and $F_2 = 1675$ lb.

⊡ To determine the distances to the forces acting on an object such that the object conforms to the moment law of equilibrium.

23. If the beam in Fig. 4.20 conforms to the moment law of equilibrium, determine the distances d_1, d_2, and d_3, if d_1 is 4.00 ft less than d_3 and d_2 is 4.00 ft less than twice d_3. Assume the forces F_1, F_2, and F_3 are the values found in exercise 22.

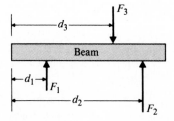

Figure 4.20

4.6 RATIO

Objectives

⊡ To write the ratio of one measurement to another using various forms.

24. Write a ratio for 9 in. to 3 ft using

  a) fractional form

  b) colon form

  c) decimal-to-one form

☐ To solve a verbal problem that involves ratios.

25. The ratio of the number of male students to the number of female students at a certain college is 9:2. If there are 1925 students altogether, how many of them are male?

26. The perimeter of a quadrilateral is 133 cm. The lengths of the sides are in the ratio 1:1:2:3. What are the lengths of the sides?

4.7 PROPORTIONS

Objectives

☐ To use the cross product property to solve a proportion.

Solve.

27. $\dfrac{4}{x} = \dfrac{6}{7}$

28. $\dfrac{4y - 1}{9} = \dfrac{27 - 3y}{2}$

☐ To solve a verbal problem that involves a direct proportion.

29. The voltage across a coil of a transformer is directly proportional to the number of turns of wire in the coil.

$$\dfrac{V_p}{V_s} = \dfrac{N_p}{N_s}, \text{ where } V_p \text{ is the primary voltage}$$

V_s is the secondary voltage
N_p is the number of turns in the primary coil
N_s is the number of turns in the secondary coil

A step-up transformer has 550 turns of wire in the secondary coil and 150 turns of wire in the primary coil. The voltage across the primary coil is 100 V less than the voltage across the secondary coil. Find the voltage across the primary coil.

☐ To solve a verbal problem that involves an inverse proportion.

30. For a *lever*, the forces and lever arm distances follow an inverse proportion (see Fig. 4.21). If the fulcrum is located 1.00 ft from an 1800-lb boulder, then how many pounds of force must be applied to a 6.00-ft lever in order to budge the boulder?

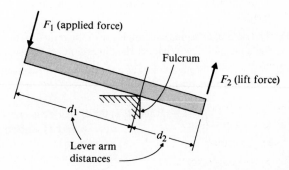

Figure 4.21

4.8 PERCENT

Objectives

⊡ To use the percent proportion to solve a verbal problem involving percent.

31. Forty cubic yards of cement contains 12 cu yd of sand. What percentage of the mixture is sand?

32. A chemist has 32 gal of a fuel mixture, of which 25% is alcohol. How much pure alcohol must she add to make the fuel mixture 40% alcohol?

⊡ To use the percent change proportion to solve a verbal problem involving percent change.

33. The charge on a capacitor increased from .124 coulombs to .140 coulombs. Find the percent change in the charge.

34. The speed of a pulley decreased by 40%. If the speed is now 700 rpm less than twice the original speed, what is its present speed?

CHAPTER TEST
Time
50 minutes

Score
15, 14 excellent
13, 12 good
11, 10, 9 fair
below 8 poor

If you have worked through the Review Exercises and corrected any weaknesses, then you are ready to take the following Chapter Test.

Solve.

1. $8 - y = -20$

2. $8 = \dfrac{-16}{t}$

3. $\dfrac{5h}{4} - 4 = 11$

4. $4m - 13 = 8 - 5m$

5. $3x - (5 - 2x) = 17 - 4(3 + 5x)$

6. $\dfrac{-5y}{8} = \dfrac{3y - 2}{-5}$

7. $\dfrac{7p + 6}{3 - 2p} = -5$

8. $X_c = \dfrac{1}{2\pi f C}$, for f

9. $E = 1 - \dfrac{H_2}{H_1}$, for H_2

10. The length of a rectangle is twice its width. If each side of this rectangle were increased by 7 cm, the new perimeter would be 76 cm. What are the dimensions of the original rectangle?

11. The sides of an isosceles triangle are in the ratio $3:5:5$. The perimeter is 80 in. longer than the nonequal side. What are the dimensions of the isosceles triangle?

12. The current through a coil of a transformer is inversely proportional to the number of turns of wire in the coil.

$$\frac{I_p}{I_s} = \frac{N_s}{N_p}, \text{ where } I_p \text{ is the primary current}$$

I_s is the secondary current
N_p is the number of turns in the primary coil
N_s is the number of turns in the secondary coil

The current through the primary coil of a step-down transformer is .050 A, while the current through the secondary coil is 4.00 A. There are 3160 more turns of wire in the primary coil than there are in the secondary coil. How many turns of wire are there in the primary coil?

13. The horsepower of a certain type of automobile engine is directly proportional to its displacement. If a 1.8-liter engine develops 110 H.P., what size engine will develop 130 H.P.?

14. Sixty cubic yards of concrete contain 21.6 cu yd of gravel. How much gravel must be added to the concrete mixture to make it 40% gravel?

15. The pressure inside a storage tank decreased 25%. If the pressure is now 240 psi more than half the original pressure, what is the present pressure?

5

EXPONENTS, ROOTS, AND RADICALS

5.1 PRODUCTS AND EXPONENTS

Objectives

- ☐ To simplify products of the form $a^m a^n$, where m and n are positive integers.
- ☐ To simplify algebraic expressions of the form $(a^m)^n$, where m and n are positive integers.
- ☐ To simplify products of the form $(a^m b^n)^k$, where m, n, and k are positive integers.

☐ Product Rule for Exponents

In Chapter 1 you learned that

$$a^n = \underbrace{(a)(a)(a)\dots(a)}_{n \text{ factors}}$$

In the expression a^n, a is called the **base**, and n the **exponent**.

Suppose we wished to find the product $(x^5)(x^3)$. We could proceed as follows:

$$(x^5)(x^3) = \underbrace{\overbrace{(x)(x)(x)(x)(x)}^{\text{five factors}}\overbrace{(x)(x)(x)}^{\text{three factors}}}_{\text{eight factors}} = x^8.$$

Is there something we can do to the exponents to quickly obtain x^8?

$$(x^5)(x^3) = x^8 \quad \text{— add —}$$

As you can see, to multiply expressions with the *same* base, simply *add their exponents and retain the same base*.

PRODUCT RULE FOR EXPONENTS

$$a^m a^n = a^{m+n}$$

EXAMPLE 1

Simplify $y^{12} y^3$.

Solution:

$$y^{12} y^3 = y^{12+3} = y^{15} \quad \blacksquare \qquad a^m a^n = a^{m+n}$$

If a base has no exponent written, then the exponent is understood to be one.

$$\text{For any number } a, a = a^1.$$

EXAMPLE 2

Simplify $(xy^3)(x^9y^4)$.

Solution:

$$(xy^3)(x^9y^4) = (x^1y^3)(x^9y^4) \qquad a = a^1$$

$$= (x^1x^9)(y^3y^4) \qquad \begin{array}{l}\text{commutative and} \\ \text{associative laws of} \\ \text{multiplication}\end{array}$$

$$= (x^{10})(y^7) \qquad a^ma^n = a^{m+n}$$

$$= x^{10}y^7 \quad \blacksquare$$

Note: Do not attempt to add the exponents for $x^{10}y^7$. The bases, x and y, are *different*. Therefore, the product rule for exponents does not apply.

EXAMPLE 3

Simplify $(-3x^2yz^5)(-5x^7)(-y^8z)$.

Solution: There are three negative factors in this product. Thus, the result must be negative.

$$(-3x^2yz^5)(-5x^7)(-y^8z) = -(3\cdot 5)(x^2x^7)(yy^8)(z^5z) \qquad \begin{array}{l}\text{commutative and} \\ \text{associative laws} \\ \text{of multiplication}\end{array}$$

$$= -15x^9y^9z^6 \quad \blacksquare \qquad a^ma^n = a^{m+n}$$

Power Rule for Exponents

Suppose we wished to raise x^3 to the *fourth power*. We could proceed as follows:

$$(x^3)^4 = \overbrace{(x^3)(x^3)(x^3)(x^3)}^{\text{four factors}}$$

$$= (x^6) \quad (x^6) \qquad a^ma^n = a^{m+n}$$

$$= x^{12}$$

Is there a shortcut?

$$(x^3)^4 \;\; = \;\; x^{12}$$

multiply

As you can see, to raise a base and exponent to another exponent, simply *multiply the exponents and retain the same base.*

POWER RULE FOR EXPONENTS

$$(a^m)^n = a^{mn}$$

EXAMPLE 4

Simplify $(y^{12})^3$.

Solution:

$$(y^{12})^3 = y^{(12)(3)} = y^{36} \quad \blacksquare \qquad (a^m)^n = a^{mn}$$

EXAMPLE 5

Simplify $((x - 3)^3)^2$.

Solution: The base is the difference $(x - 3)$. Therefore,

$$((x - 3)^3)^2 = (x - 3)^6 \quad \blacksquare \qquad (a^m)^n = a^{mn}$$

⚃ Power of a Product

Suppose we wished to raise the product xy to the *fourth power*. We could proceed as follows:

$$
\begin{aligned}
(xy)^4 &= \overbrace{(xy)(xy)(xy)(xy)}^{\text{four factors}} \\
&= (\underbrace{xxxx})(\underbrace{yyyy}) \qquad \text{commutative and} \\
& \qquad\qquad\qquad\qquad \text{associative laws} \\
& \qquad\qquad\qquad\qquad \text{of multiplication} \\
&= \;\; (x^4) \quad (y^4) \\
&= \quad\;\; x^4 y^4
\end{aligned}
$$

As you can see, *the power of a product becomes the product of the powers.*

POWER OF A PRODUCT RULE FOR EXPONENTS

$$(ab)^n = a^n b^n$$

The power of a sum is *not* the same as the sum of the powers. $(a + b)^n$ and $a^n + b^n$ are *not* equivalent. For example, $(2 + 3)^2 = (5)^2 = 25$, while

$$(2)^2 + (3)^2 = 4 + 9 = 13.$$

EXAMPLE 6

Simplify $(-3y)^4$.

Solution:

$$(-3y)^4 = (-3)^4(y)^4 \qquad (ab)^n = a^n b^n$$

$$= 81y^4 \quad \blacksquare \qquad \text{simplifying}$$

Note: Remember, a negative number to an *even* power is always positive.

EXAMPLE 7

Simplify $(-5xy^4z^5)^3$.

Solution:

$$(-5xy^4z^5)^3 = (-5)^3(x)^3(y^4)^3(z^5)^3 \qquad (ab)^n = a^n b^n$$

$$= (-5)^3(x)^3(y^{12})(z^{15}) \qquad (a^m)^n = a^{mn}$$

$$= -125x^3y^{12}z^{15} \quad \blacksquare \qquad \text{simplifying}$$

Note: Remember, a negative number to an *odd* power is always negative.

EXAMPLE 8

Simplify $(2x^2y^3)^3(3x^4y)^2$.

Solution:

$$(2x^2y^3)^3(3x^4y)^2 = (2^3x^6y^9)(3^2x^8y^2) \qquad (ab)^n = a^nb^n, \text{ and } (a^m)^n = a^{mn}$$

$$= (2^3)(3^2)(\underbrace{x^6x^8})(\underbrace{y^9y^2}) \qquad \text{commutative and}$$
associative laws of
multiplication

$$= \underbrace{(8)(9)} \quad (x^{14}) \; (y^{11}) \qquad a^ma^n = a^{m+n}$$

$$= 72x^{14}y^{11} \quad \blacksquare \qquad \text{simplifying}$$

EXERCISES 5.1

Simplify the following.

1. $(2)^2(2)^4$

2. $(-3)^2(-3)^3$

3. $(2^2)^4$

4. $((-3)^2)^3$

5. x^4x^7

6. p^9p^7

7. $(x^4)^7$

8. $(p^9)^7$

9. $(m^2n)(m^5n^3)$

10. $t^8(-t^6w^4)$

11. $(-p^2q^5)(p^8q^6)$

12. $(-h^2k^7)(-h^4k)$

13. $(5x^2y^9)(4x^9y^3)$

14. $(2r^5s^5t^5)(-3rst^3)$

15. $(-3x^5y^7z)(xy)(-8x^9z)$

16. $(-\frac{1}{2}xy^2z^5)(-6xy)(-x^4y^6z^8)$

17. $((p-4)^2)^4$

18. $((9+8t)^4)^3$

19. $(p-4)^2(p-4)^4$

20. $(9+8t)^4(9+8t)^3$

21. $(4xy^4)^3$

22. $(-3t^2u^5)^3$

23. $(-3m^2n^4)^3(4m^3n^7)$

24. $-(2xy^3)(-5x^2y^5z^9)^2$

25. $(-m^2n^3)^4(4mn^2)^2(m^3n^3)^5$

26. $(3rst^2)^5(-2r^6t)^2(-s^7)^3$

27. $((3xy)^2)^2$

28. $((-2x^2yz^4)^3)^2$

Simplify the following. Assume that all variable exponents represent some positive integer.

29. $(a^{3t})(a^{2t})$

30. $(a^t)(a^3)$

31. $(a^{2t})^2$

32. $(a^t)^{5t}$

33. $(2^{5t})(2)(a^{4t})$

34. $(a^{3t})(2^{3t})(a)(2)$

5.2 QUOTIENTS AND EXPONENTS

Objectives

☐ To simplify quotients of the form $\dfrac{a^m}{a^n}$, where m and n are positive integers and $m = n$.

☐ To simplify quotients of the form $\dfrac{a^m}{a^n}$, where m and n are positive integers and $m > n$.

☐ To simplify quotients of the form $\dfrac{a^m}{a^n}$, where m and n are positive integers and $m < n$.

⊡ To simplify quotients of the form $\left(\dfrac{a^m}{a^n}\right)^k$, where m, n, and k are positive integers.

⊡ Quotient Rule for Exponents, Case 1

You already know that any number except zero divided by itself equals one $\left(\dfrac{a}{a} = 1\right)$. Does a number to a power divided by itself also equal one?

For example, does $\dfrac{x^4}{x^4} = 1$?

$$\frac{x^4}{x^4} = \overset{\text{four factors}}{\overbrace{\frac{(x)(x)(x)(x)}{\underbrace{(x)(x)(x)(x)}_{\text{four factors}}}}}$$

$$= \left(\frac{x}{x}\right)\left(\frac{x}{x}\right)\left(\frac{x}{x}\right)\left(\frac{x}{x}\right) \qquad \begin{array}{l}\text{multiplication rule for} \\ \text{fractions in reverse}\end{array}$$

$$= \underbrace{(1)\ (1)\ (1)\ (1)}_{} \qquad \frac{a}{a} = 1$$

$$= \qquad 1$$

If the bases are the same, and the exponent of the numerator *equals* the exponent of the denominator, then all of the factors will completely cancel, leaving an answer of 1.

QUOTIENT RULE FOR EXPONENTS, CASE 1

$$\frac{a^m}{a^m} = 1 \qquad (\text{provided } a \neq 0)$$

EXAMPLE 1

Simplify $\dfrac{16x^8 y^3}{18x^8 y^3}$.

Solution:

$$\frac{16x^8 y^3}{18x^8 y^3} = \left(\frac{2}{2}\right)\left(\frac{8}{9}\right)\left(\frac{x^8}{x^8}\right)\left(\frac{y^3}{y^3}\right) \qquad \begin{array}{l}\text{multiplication rule for} \\ \text{fractions in reverse}\end{array}$$

$$= (1)\left(\frac{8}{9}\right)\ (1)\ (1) \qquad \frac{a^m}{a^m} = 1$$

$$= \frac{8}{9} \qquad \blacksquare$$

⸪ Quotient Rule for Exponents, Case 2

Let us now consider quotients in which the exponent of the numerator is *greater than* the exponent of the denominator. For example, suppose we wished to find the quotient $\dfrac{x^5}{x^3}$? We could proceed as follows:

$$\frac{x^5}{x^3} = \frac{(x^3)(x^2)}{(x^3)} \qquad a^m a^n = a^{m+n}$$

$$= \left(\frac{x^3}{x^3}\right)\left(\frac{x^2}{1}\right) \qquad \text{multiplication rule for fractions in reverse}$$

$$= (1)\left(\frac{x^2}{1}\right) \qquad \frac{a^m}{a^m} = 1$$

$$= x^2$$

If the bases are the same, and the exponent of the numerator is *greater than* the exponent of the denominator, then all of the factors in the denominator will completely cancel.

Is there something we can do to the exponents to quickly obtain x^2?

$$\frac{x^5}{x^3} = x^2$$

As you can see, to simplify the quotient, simply *subtract the exponent of the denominator from the exponent of the numerator and retain the same base.*

QUOTIENT RULE FOR EXPONENTS, CASE 2

$$\text{For } m > n, \frac{a^m}{a^n} = a^{m-n} \qquad \text{(provided } a \neq 0)$$

EXAMPLE 2

Simplify $\dfrac{x^8 y^4}{x^5 y}$.

Solution:

$$\frac{x^8y^4}{x^5y} = \left(\frac{x^8}{x^5}\right)\left(\frac{y^4}{y^1}\right) \qquad a = a^1$$

$$= (x^{8-5})(y^{4-1}) \qquad \frac{a^m}{a^n} = a^{m-n}, m > n$$

$$= x^3y^3 \quad \blacksquare \qquad \text{simplifying}$$

EXAMPLE 3

Simplify $\dfrac{21x^3y^5z^4}{28x^3yz^3}$.

Solution:

$$\frac{21x^3y^5z^4}{28x^3yz^3} = \left(\frac{7}{7}\right)\left(\frac{3}{4}\right)\left(\frac{x^3}{x^3}\right)\left(\frac{y^5}{y^1}\right)\left(\frac{z^4}{z^3}\right)$$

$$= (1)\left(\frac{3}{4}\right)\ (1)\ (y^4)\ (z^1) \qquad \frac{a^m}{a^m} = 1 \text{ and}$$

$$\frac{a^m}{a^n} = a^{m-n}, m > n$$

$$= \frac{3}{4}y^4z \quad \text{or} \quad \frac{3y^4z}{4} \quad \blacksquare \qquad \text{simplifying}$$

Note: Be sure to simplify z^1 to z.

⋱ Quotient Rule for Exponents, Case 3

Finally, let's consider quotients in which the exponent of the numerator is *less than* the exponent of the denominator. To find the quotient $\dfrac{x^3}{x^5}$, we could proceed as follows:

$$\frac{x^3}{x^5} = \frac{(x^3)}{(x^3)(x^2)} \qquad a^ma^n = a^{m+n}$$

$$= \left(\frac{x^3}{x^3}\right)\left(\frac{1}{x^2}\right) \qquad \begin{array}{l}\text{multiplication rule for}\\ \text{fractions in reverse}\end{array}$$

$$= (1)\ \frac{1}{x^2} \qquad \frac{a^m}{a^m} = 1$$

$$= \frac{1}{x^2}$$

If the bases are the same, and the exponent of the numerator is *less than* the exponent of the denominator, then all of the factors in the numerator will completely cancel.

Do you see the shortcut? *Subtract the exponent of the numerator from the exponent of the denominator, retain the same base, and write the reciprocal of this result.*

QUOTIENT RULE FOR EXPONENTS, CASE 3

$$\text{For } m < n, \; \frac{a^m}{a^n} = \frac{1}{a^{n-m}} \qquad \text{(provided } a \neq 0\text{)}$$

EXAMPLE 4

Simplify $\dfrac{x^3 y}{x^9 y^5}$.

Solution:

$$\frac{x^3 y}{x^9 y^5} = \left(\frac{x^3}{x^9}\right)\left(\frac{y^1}{y^5}\right) \qquad a = a^1$$

$$= \left(\frac{1}{x^{9-3}}\right)\left(\frac{1}{y^{5-1}}\right) \qquad \frac{a^m}{a^n} = \frac{1}{a^{n-m}}, m < n$$

$$= \left(\frac{1}{x^6}\right)\left(\frac{1}{y^4}\right)$$

$$= \frac{1}{x^6 y^4} \quad \blacksquare \qquad\qquad \text{multiplying fractions}$$

EXAMPLE 5

Simplify $\dfrac{15 x^9 y^3 z^3}{25 x^6 y^3 z^5}$.

Solution:

$$\frac{15 x^9 y^3 z^3}{25 x^6 y^3 z^5} = \left(\frac{5}{5}\right)\left(\frac{3}{5}\right)\left(\frac{x^9}{x^6}\right)\left(\frac{y^3}{y^3}\right)\left(\frac{z^3}{z^5}\right)$$

$$= (1)\left(\frac{3}{5}\right)(x^3)\,(1)\left(\frac{1}{z^2}\right) \qquad \begin{array}{l}\text{quotient rules for}\\\text{exponents}\end{array}$$

$$= \frac{3x^3}{5z^2} \quad \blacksquare \qquad\qquad \text{multiplying fractions}$$

Note: Record the base and its new exponent in the numerator if the larger of the two original exponents is in the numerator. Record the base and its new exponent in the denominator if the larger of the two original exponents is in the denominator.

⠒ Power of a Quotient

Is the power of a quotient the same as the quotient of the powers? For example, does $\left(\dfrac{x}{y}\right)^4 = \dfrac{x^4}{y^4}$?

$$\left(\frac{x}{y}\right)^4 = \overbrace{\left(\frac{x}{y}\right)\left(\frac{x}{y}\right)\left(\frac{x}{y}\right)\left(\frac{x}{y}\right)}^{\text{four factors}}$$

$$= \frac{\overbrace{(x)(x)(x)(x)}^{\text{four factors}}}{\underbrace{(y)(y)(y)(y)}_{\text{four factors}}} \qquad \text{multiplying fractions}$$

$$= \frac{x^4}{y^4}$$

As you can see, *the power of a quotient becomes the quotient of the powers.*

POWER OF A QUOTIENT RULE FOR EXPONENTS

$$\left(\frac{a}{b}\right)^n = \frac{a^n}{b^n} \qquad (\text{provided } b \neq 0)$$

EXAMPLE 6

Simplify $\left(\dfrac{-3}{y^3}\right)^2$.

Solution:

$$\left(\frac{-3}{y^3}\right)^2 = \frac{(-3)^2}{(y^3)^2} \qquad \left(\frac{a}{b}\right)^n = \frac{a^n}{b^n}$$

$$= \frac{9}{y^6} \quad \blacksquare \qquad (a^m)^n = a^{mn}$$

EXAMPLE 7

Simplify $\left(\dfrac{2x^2y^5}{3x^5y}\right)^3$.

Solution:

$$\left(\frac{2x^2y^5}{3x^5y}\right)^3 = \frac{(2x^2y^5)^3}{(3x^5y)^3} \qquad \left(\frac{a}{b}\right)^n = \frac{a^n}{b^n}$$

$$= \frac{(2)^3(x^2)^3(y^5)^3}{(3)^3(x^5)^3(y^1)^3} \qquad (ab)^n = a^nb^n$$

$$= \frac{8x^6y^{15}}{27x^{15}y^3} \qquad (a^m)^n = a^{mn}$$

$$= \frac{8y^{12}}{27x^9} \quad \blacksquare \qquad \text{the quotient rules}$$
$$\qquad\qquad\qquad\qquad\qquad \text{for exponents}$$

EXERCISES 5.2

Simplify the following.

1. $\dfrac{15^9}{15^9}$

2. $\dfrac{(-81)^{12}}{(-81)^{12}}$

3. $\dfrac{2^5}{2^2}$

4. $\dfrac{(-3)^6}{(-3)^3}$

5. $\dfrac{-2^8}{-2^{13}}$

6. $\dfrac{(-3)^5}{(-3)^9}$

7. $\dfrac{8x^4y^9}{24x^4y^9}$

8. $\dfrac{96xy^8}{16xy^8}$

9. $\dfrac{8g^8h^5}{16g^7h}$

10. $\dfrac{-4t^{15}u^4}{-12tu^3}$

11. $\dfrac{4xyz}{-28x^4y^2z^8}$

12. $\dfrac{-3r^2s^4t^5}{18r^{15}s^9t^6}$

13. $\dfrac{-14rs^9t^{12}}{-16r^3s^9t^3}$

14. $\dfrac{24x^8y^5z^6}{16x^8y^6z^5}$

15. $\dfrac{(y+2)^8}{(y+2)^5}$

16. $\dfrac{(2y-3)^5}{(2y-3)^6}$

17. $\dfrac{(4x^2y^3z)^3}{(3x^4yz)^2}$

18. $\dfrac{(-x^5y^8)^9}{(-xy^7)^6}$

19. $\left(\dfrac{-4}{z^4}\right)^3$

20. $\left(\dfrac{y}{-5}\right)^2$

21. $\left(\dfrac{1}{2t^5}\right)^5$

22. $\left(\dfrac{-1}{x^3yz^4}\right)^8$

23. $\left(\dfrac{2r^4s}{r^8s^2t}\right)^4$

24. $\left(\dfrac{m^3n}{-2m^4n^3}\right)^5$

25. $\left(\dfrac{p^2q^3}{rs^2}\right)^5\left(\dfrac{-3r^3s^5}{2p^5q}\right)^2$

26. $\dfrac{(x^3y^4)^3(x^2y^6)^3}{(x^3y^6)^5}$

Simplify the following. Assume that all variable exponents represent some positive integer.

27. $\dfrac{a^{3t}}{a^t}$

28. $\dfrac{a^{2t}}{a^{6t}}$

29. $\dfrac{a^3}{a^{t+3}}$

30. $\dfrac{a^{2t+1}}{a^{2t}}$

31. $\left(\dfrac{-1}{a^{3t}}\right)^3$

32. $\left(\dfrac{1}{a^t}\right)^t$

5.3 ZERO AND NEGATIVE EXPONENTS

Objectives

⊡ To simplify algebraic expressions of the form a^0.

⊡ To simplify algebraic expressions of the form a^{-n} and $\dfrac{1}{a^{-n}}$.

⊡ To simplify algebraic expressions containing integral exponents.

⊡ Zero as an Exponent

All the exponents that we have worked with thus far have been positive integers. Can other types of exponents be defined? Can any meaning be given to an expression such as x^0 (read "x to the zero power")?

You already know that $\dfrac{x^4}{x^4} = 1\left(\dfrac{a^m}{a^m} = 1\right)$. Let us substitute $\dfrac{x^4}{x^4}$ for $\dfrac{a^m}{a^n}$ in the

Quotient Rule for Exponents, Case 2 $\left(\dfrac{a^m}{a^n} = a^{m-n}\right)$. Then

$$\frac{x^4}{x^4} = x^{4-4} \qquad \left(\frac{a^m}{a^n} = a^{m-n}\right)$$
$$\downarrow \qquad \downarrow$$
$$1 = x^0$$

It certainly seems reasonable to *define a base to the zero power as the number* 1.

For any number a, (except 0) $a^0 = 1$.

EXAMPLE 1

Does $(2 + 3)^0 = 2^0 + 3^0$?

Solution:

$$(2 + 3)^0 = 2^0 + 3^0?$$
$$(5)^0 = 1 + 1 ?$$
$$1 = 2? \quad \text{No.}$$

Therefore, $(2 + 3)^0 \neq 2^0 + 3^0$ ∎

Note: Even for an exponent of zero, the power of a sum is *not* the same as the sum of the powers.

⬚ Negative Exponents

Our exponents now include positive integers and zero. Can negative exponents be defined? Can any meaning be given to an expression such as x^{-2} (read "x to the negative two power")?

You already know that $\dfrac{x^3}{x^5} = \dfrac{1}{x^2}$ $\left(\dfrac{a^m}{a^n} = \dfrac{1}{a^{n-m}}, m < n\right)$. Let us substitute $\dfrac{x^3}{x^5}$ for $\dfrac{a^m}{a^n}$ in the Quotient Rule for Exponents, Case 2 $\left(\dfrac{a^m}{a^n} = a^{m-n}\right)$. Then

$$\frac{x^3}{x^5} = x^{3-5} \qquad \left(\frac{a^m}{a^n} = a^{m-n}\right)$$
$$\frac{1}{x^2} = x^{-2}.$$

It certainly seems reasonable to *define a base with a negative exponent as the reciprocal of that base with the corresponding positive exponent.*

For any number a (except 0) $a^{-n} = \dfrac{1}{a^n}$

EXAMPLE 2

Does $(2 \cdot 3)^{-2} = 2^{-2} \cdot 3^{-2}$?

Solution:

$$\underbrace{(2 \cdot 3)}^{-2} = 2^{-2} \cdot 3^{-2} \ ?$$

$$(6)^{-2} = \left(\frac{1}{2^2}\right)\left(\frac{1}{3^2}\right)?$$

$$\frac{1}{6^2} = \left(\frac{1}{4}\right)\left(\frac{1}{9}\right) \ ?$$

$$\frac{1}{36} = \frac{1}{36}? \quad \text{Yes.}$$

Therefore, $(2 \cdot 3)^{-2} = 2^{-2} \cdot 3^{-2}$ ■

Note: The power of a product is still the product of the powers, even if the exponents are negative.

Although we often work with negative exponents in technical mathematics and calculus, we generally leave a final answer to a problem in terms of positive exponents only.

EXAMPLE 3

Rewrite $3x^0 y^3 z^{-1}$ using only positive exponents.

Solution:

$$3x^0 y^3 z^{-1} = 3(1)y^3 \left(\frac{1}{z}\right) \qquad a^0 = 1 \text{ and } a^{-n} = \frac{1}{a^n}$$

$$= \frac{3y^3}{z} \quad ■ \qquad \text{multiplying fractions}$$

EXAMPLE 4

Rewrite $\dfrac{x^{-3}}{y^{-2}}$ using only positive exponents.

Solution:

$$\frac{x^{-3}}{y^{-2}} = \frac{\left(\dfrac{1}{x^3}\right)}{\left(\dfrac{1}{y^2}\right)} \qquad a^{-n} = \frac{1}{a^n}$$

$$= \left(\frac{1}{x^3}\right)\left(\frac{y^2}{1}\right) \qquad \frac{a}{b} \div \frac{c}{d} = \frac{a}{b} \cdot \frac{d}{c}$$

$$= \frac{y^2}{x^3} \quad \blacksquare \qquad \text{multiplying fractions}$$

Note: Notice that the original numerator, x^{-3}, is shifted to the denominator and written as x^3. Similarly, the original denominator, y^{-2}, is shifted to the numerator

$$\left(\underset{\text{down}}{\overset{\text{up}}{\frac{x^{-3}}{y^{-2}}}} = \frac{y^2}{x^3} \right) \text{ and written as } y^2.$$

EXAMPLE 5

Rewrite $\dfrac{x^2 y^{-5}}{m^{-4} n^3}$ using only positive exponents.

Solution:

$$\underset{\text{down}}{\overset{\text{up}}{\frac{x^2 y^{-5}}{m^{-4} n^3}}} = \frac{x^2 m^4}{n^3 y^5} \quad \blacksquare \qquad \text{shifting of negative exponents}$$

 CAUTION This shifting of negative exponents does not work if the numerator and/or denominator of a fraction contains sums or differences.

Therefore, do not write $\underset{\text{down}}{\overset{\text{up}}{\frac{3^{-2}}{2^{-2} + 5}}} = \frac{2^2}{3^2 + 5} = \frac{4}{14} = \frac{2}{7}$ *wrong*

Instead, $\dfrac{3^{-2}}{2^{-2} + 5} = \dfrac{\dfrac{1}{3^2}}{\dfrac{1}{2^2} + 5} = \dfrac{\dfrac{1}{9}}{\dfrac{1}{4} + 5} = \dfrac{\dfrac{1}{9}}{\dfrac{21}{4}} = \dfrac{1}{9} \cdot \dfrac{4}{21} = \dfrac{4}{189}$ *correct*

⠒ Integral Exponents

Having now defined the zero and negative exponents, we need only one *general* quotient rule for exponents. For your convenience, this rule and all the other rules and definitions for exponents are listed together below. They hold true for *any* integral exponents *m* and *n*.

THE RULES AND DEFINITIONS FOR EXPONENTS

Definitions:

1. $a^n = \underbrace{(a)(a)(a)\ldots(a)}_{n \text{ factors}}$

2. $a^1 = a$

3. $a^0 = 1 \qquad (a \neq 0)$

4. $a^{-n} = \dfrac{1}{a^n} \qquad (a \neq 0)$

Rules:

1. product rule: $a^m a^n = a^{m+n}$
2. power rule: $(a^m)^n = a^{mn}$
3. power of a product rule: $(ab)^n = a^n b^n$
4. quotient rule: $\dfrac{a^m}{a^n} = a^{m-n} \qquad (a \neq 0)$
5. power of a quotient rule: $\left(\dfrac{a}{b}\right)^n = \dfrac{a^n}{b^n} \qquad (a \neq 0)$

EXAMPLE 6

Simplify $x^{-8}y^{-4}x^{-3}y^6$, leaving an answer with all positive exponents.

Solution:

$$x^{-8}y^{-4}x^{-3}y^6 = x^{\overbrace{-8+(-3)}}y^{\overbrace{-4+6}} \qquad a^m a^n = a^{m+n}$$

$$= x^{-11}y^2$$

$$= \frac{y^2}{x^{11}} \quad \blacksquare \qquad a^{-n} = \frac{1}{a^n}$$

EXAMPLE 7

Simplify $((x-3)^{-2})^{-4}$, leaving an answer with all positive exponents.

Solution:

$$((x-3)^{-2})^{-4} = (x-3)^{\overbrace{(-2)(-4)}} \qquad (a^m)^n = a^{mn}$$

$$= (x-3)^8 \quad \blacksquare \qquad \text{simplifying}$$

EXAMPLE 8

Simplify $(-2x^0y^{-3}z^4)^{-3}$, leaving an answer with all positive exponents.

Solution:

$$(-2x^0y^{-3}z^4)^{-3} = (-2)^{-3}(x^0)^{-3}(y^{-3})^{-3}(z^4)^{-3} \qquad (ab)^n = a^nb^n$$

$$= (-2)^{-3} \ (x^0) \ (y^9) \ (z^{-12}) \qquad (a^m)^n = a^{mn}$$

$$= \frac{(1)y^9}{(-2)^3z^{12}} \qquad a^{-n} = \frac{1}{a^n}, \text{ and}$$

$$a^0 = 1$$

$$= \frac{y^9}{-8z^{12}} \ \blacksquare \qquad \text{evaluating } (-2)^3$$

EXAMPLE 9

Simplify $\dfrac{(y+6)^{-7}}{(y+6)^{-5}}$, leaving an answer with all positive exponents.

Solution:

$$\frac{(y+6)^{-7}}{(y+6)^{-5}} = (y+6)^{-7-(-5)} \qquad \frac{a^m}{a^n} = a^{m-n}$$

$$= (y+6)^{-7+5} \qquad \text{subtraction rule}$$

$$= (y+6)^{-2}$$

$$= \frac{1}{(y+6)^2} \ \blacksquare \qquad a^{-n} = \frac{1}{a^n}$$

EXAMPLE 10

Simplify $\dfrac{3^{-2}x^{-4}y^6}{3x^2y^{-1}}$, leaving an answer with all positive exponents.

Solution:

$$\frac{3^{-2} \ x^{-4} \ y^6}{3x^2 \ y^{-1}} \overset{\text{up}}{\underset{\text{down}}{}} = \frac{y^6y^1}{3^23x^2x^4} \qquad \begin{array}{l}\text{shifting negative}\\ \text{exponents to obtain}\\ \text{positive exponents}\end{array}$$

$$= \frac{y^7}{3^3x^6} \qquad a^ma^n = a^{m+n}$$

$$= \frac{y^7}{27x^6} \ \blacksquare \qquad \text{evaluating } 3^3$$

Note: You could also immediately apply the quotient rule and *subtract* the exponents. However, when subtracting negative integers, you must be careful not to make a careless mistake in changing signs. Try this problem by applying the quotient rule first. See if you can obtain the same answer.

EXERCISES 5.3

Indicate whether the following are true or false. Assume that no variable has a value of 0.

1. $2^{-1} + 3^{-1} = 5^{-1}$

2. $x^{-1} + y^{-1} = (x + y)^{-1}$

3. $2^{-1}3^{-1} = 6^{-1}$

4. $x^{-1}y^{-1} = (xy)^{-1}$

5. $2 \cdot 3^0 = (2 \cdot 3)^0$

6. $xy^0 = (xy)^0$

7. $\left(\dfrac{1}{3}\right)^{-2} = 3^2$

8. $\left(\dfrac{1}{x}\right)^{-2} = x^2$

9. $\dfrac{1}{(2 + 3)^{-2}} = (2 + 3)^2$

10. $\dfrac{1}{(x + y)^{-2}} = (x + y)^2$

11. $\dfrac{3^{-1}}{(2^{-1})(4^{-1})} = \dfrac{(2)(4)}{3}$

12. $\dfrac{x^{-1}}{y^{-1}z^{-1}} = \dfrac{yz}{x}$

13. $\dfrac{3^{-1}}{2^{-1} + 4^{-1}} = \dfrac{2 + 4}{3}$

14. $\dfrac{x^{-1}}{y^{-1} + z^{-1}} = \dfrac{y + z}{x}$

15. $(-2)^{-4} = \dfrac{1}{2^4}$

16. $(-x)^{-n} = \dfrac{1}{x^n}$ (n is an *even* integer)

17. $(-2)^{-3} = -\dfrac{1}{2^3}$

18. $(-x)^{-n} = -\dfrac{1}{x^n}$ (n is an *odd* integer)

Simplify the following, leaving an answer with all positive exponents. Assume that no variable has a value of 0.

19. $4x^{-1}y^0z^{-1}$

20. $-3x^{-2}yz^0$

21. $\dfrac{3^{-1}x^{-2}}{y^{-2}z^0}$

22. $\dfrac{4^{-2}x^2y^{-3}}{3x^0z^{-1}}$

23. $(-x)^{-4}$

24. $(-x)^{-5}$

25. $x^{-3}x^{-4}x^5x^0$

26. $x^8x^{-4}x^0x^4$

27. $\dfrac{y^{-4}}{y^{-3}}$

28. $\dfrac{y^6}{y^{-1}}$

29. $(3xyz)^{-2}$

30. $(-3xyz)^{-2}$

31. $((1 - y)^{-3})^2$

32. $((1 - y)^{-3})^{-3}$

33. $\left(\dfrac{3}{y}\right)^{-3}$

34. $\left(\dfrac{3x}{2yz}\right)^{-3}$

35. $(y + 7)^{-2}(y + 7)^2$

36. $(x + y)^4(x + y)^{-1}(x + y)^{-4}$

37. $\dfrac{(x + y)^4}{(x + y)^{-2}}$

38. $\dfrac{(z - 4)^{-5}}{(z - 4)^0}$

39. $x^{-5}y^{-9}x^5y$

40. $x^{-2}y^{-3}x^{-6}y^9x^7y^{-1}x^2$

41. $\dfrac{4^2x^{-4}y^{-7}}{4^{-2}x^4y}$

42. $\dfrac{-6^2x^{-5}y}{-6^{-3}x^4y^2}$

43. $(2x^{-4}y^{-1}z^5)^3$

44. $(-5x^{-1}y^{-4}z^6)^{-2}$

45. $\left(\dfrac{3x^{-1}}{y^2}\right)^{-3}$

46. $\left(\dfrac{2x^{-4}}{3y^4z^{-1}}\right)^4$

47. $\left(\dfrac{4x^{-3}y^8z}{8x^{-5}y^{-2}z^{-1}}\right)^{-2}$

48. $\left(\dfrac{-x^{-2}y^4}{y^{-4}x}\right)^6$

49. $\dfrac{x^{-3}(y-z)^4}{(y-z)^{-3}x^{-4}}$

50. $\dfrac{(x^{-1}(y-z)^2)^{-2}}{(y-z)^{-1}x}$

5.4   SCIENTIFIC NOTATION

Objectives

⊡ Given a number in scientific notation, to write it in decimal notation.

⊡ Given a number in decimal notation, to write it in scientific notation.

⊡ Scientific Notation to Decimal Notation

Technicians and engineers must frequently work with very large and very small numbers. To write and work with numbers like 5,370,000,000,000 and .0000000004562 is certainly cumbersome and inconvenient. Usually what is done is to write such numbers in a more compact form called **scientific notation**.

A number is written in **scientific notation** if it is in the form

$$
\underbrace{k}_{\substack{\text{a number} \\ \text{between} \\ \text{one and ten}}} \qquad \underbrace{\times}_{\text{times}} \qquad \underbrace{10^n}_{\substack{\text{a power} \\ \text{of ten}}}
$$

Scientific notation can contain either

$$
\text{positive powers of ten,} \begin{cases} 10^1 = (10) & = 10 \\ 10^2 = (10)(10) & = 100 \\ 10^3 = (10)(10)(10) & = 1000 \\ 10^4 = (10)(10)(10)(10) & = 10,000 \\ \vdots \qquad \vdots \qquad \vdots \end{cases}
$$

$$\left.\begin{array}{l} 10^{-1} = \dfrac{1}{10^1} = \dfrac{1}{10} \\[2ex] 10^{-2} = \dfrac{1}{10^2} = \dfrac{1}{100} \\[2ex] 10^{-3} = \dfrac{1}{10^3} = \dfrac{1}{1000} \\[2ex] 10^{-4} = \dfrac{1}{10^4} = \dfrac{1}{10{,}000} \\[2ex] \quad\vdots \qquad \vdots \qquad \vdots \end{array}\right\}$$ negative powers of ten,

or ten to the zero power, $\{10^0 = 1.$

An example of a number written in scientific notation is

$$\underbrace{1.65}_{\substack{\text{a number}\\\text{between}\\\text{one and ten}}} \quad \underbrace{\times}_{\text{times}} \quad \underbrace{10^3}_{\substack{\text{a power}\\\text{of ten}}}$$

What decimal number might 1.65×10^3 represent? You know that 10^3 represents 1000, and multiplying by 1000 has the effect of moving the decimal point three places to the right. Thus,

$$\underset{\substack{\text{scientific}\\\text{notation}}}{1.65} \times 10^3 = 1.65 \times 1000 = \underset{\substack{\text{three places}\\\text{to the right}}}{1.650} = \underset{\substack{\text{decimal}\\\text{notation}}}{1650.}$$

Notice that the *positive* exponent tells us how many places to the *right* the decimal point should be moved.

EXAMPLE 1

The modulus of elasticity for a certain type of steel is written in scientific notation as 2.97×10^7 pounds per square inch (psi). Write this number in decimal notation.

Solution: The positive exponent, 7, tells us to move the decimal point seven places to the right. Thus,

$$2.97 \times 10^7 \text{ psi} = \underset{\substack{\text{seven places}\\\text{to the right}}}{2.9700000} = 29{,}700{,}000 \text{ psi.} \quad \blacksquare$$

Another example of a number written in scientific notation is

$$\underbrace{1.65}_{\substack{\text{a number} \\ \text{between} \\ \text{one and ten}}} \quad \underbrace{\times}_{\text{times}} \quad \underbrace{10^{-3}}_{\substack{\text{a power} \\ \text{of ten}}}.$$

What decimal number might 1.65×10^{-3} represent? Of course, 10^{-3} represents $\dfrac{1}{10^3}$ or $\dfrac{1}{1000}$. You know that multiplying by $\dfrac{1}{1000}$ is the same as dividing by 1000, and dividing by 1000 has the effect of moving the decimal point three places to the left. Thus,

$$\underbrace{1.65 \times 10^{-3}}_{\substack{\text{scientific} \\ \text{notation}}} = 1.65 \times \frac{1}{1000} = \frac{1.65}{1000} = \underbrace{001.65}_{\substack{\text{three places} \\ \text{to the left}}} = \underbrace{.00165}_{\substack{\text{decimal} \\ \text{notation}}}$$

Notice that the *negative* exponent tells us how many places to the *left* the decimal point should be moved.

EXAMPLE 2

The value of a certain electrical capacitor is written in scientific notation as 1.5×10^{-6} farads (F). Write this number in decimal notation.

Solution: The negative exponent, -6, tells us to move the decimal point to the left six places. Thus,

$$1.5 \times 10^{-6} \text{ F} = \underbrace{000001.5}_{\substack{\text{six places} \\ \text{to the left}}} = .0000015 \text{ F} \quad \blacksquare$$

Decimal Notation to Scientific Notation

If you are given a number in decimal notation and wish to write it in scientific notation, then you must determine *a power of ten* and *a number between one and ten* whose product is the original number.

To determine the *power of ten*:

1. Place a vertical bar in the number so that there is *one* significant digit to its left.

2. Starting at the vertical bar, count the number of places you must move to get to the decimal point. If you move to the *right* n places use 10^n, and if you move to the *left* n places use 10^{-n}.

To determine the *number between one and ten*:
Drop all nonsignificant (place-holding) zeros, and locate the new decimal point at the vertical bar.

(handwritten, left margin) Vertical Bar Between one & ten value

EXAMPLE 3

The value of a certain electrical resistor is $68,\overline{0}00,000\,\Omega$. Write this number in scientific notation.

Solution: The power of ten is $\rightarrow 10^7$.

$$6 \mid 8,\overline{0}00,000.$$

Note: The vertical bar with *one* significant digit to its left

seven places to the right

The number between one and ten is $\rightarrow 6.80$.

$$6 \mid 8,\overline{0}00,000.$$

Note: Drop the nonsignificant zeros, and locate the new decimal point at the vertical bar.

Thus, $68,\overline{0}00,000\,\Omega = 6.80 \times 10^7\,\Omega.$ ∎

Note: In scientific notation, the number between one and ten contains the significant digits of the number, while the power of ten represents the nonsignificant (place-holding) zeros.

EXAMPLE 4

The time needed by a computer to perform an addition is .0000320 sec. Write this number in scientific notation.

Solution: The power of ten is $\rightarrow 10^{-5}$

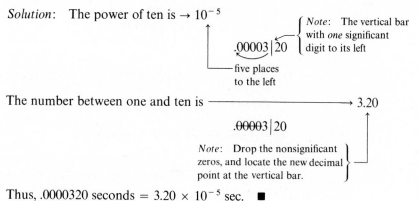

Note: The vertical bar with *one* significant digit to its left

$$.00003 \mid 20$$

five places to the left

The number between one and ten is $\longrightarrow 3.20$

$$.00003 \mid 20$$

Note: Drop the nonsignificant zeros, and locate the new decimal point at the vertical bar.

Thus, $.0000320$ seconds $= 3.20 \times 10^{-5}$ sec. ∎

EXERCISES 5.4

Write the number in each of the following statements in decimal notation.

1. The coefficient of linear expansion for steel is 6.5×10^{-6} per °F.

2. The value of an electrical resistor is $4.70 \times 10^4\,\Omega$.

3. The strength of a riveted joint is 1.74×10^6 lb.

4. The value of an electrical capacitor is 7.50×10^{-9} F.

5. The modulus of elasticity for eastern white pine is 7.6×10^9 Pa.

6. The diameter of AWG #36 wire is 5.00×10^{-3} in.

7. The charge of an electron is 1.6×10^{-19} C.

8. The voltage drop across a resistor is 1.70×10^{-6} V.

9. The memory unit of a computer contains 6.25×10^8 bits.

10. The distance light travels in one year (called a light year) is 5.87×10^{12} mi.

Write the number in each of the following statements in scientific notation.

11. The coefficient of linear expansion for steel is .0000117 per °C.

12. The value of an electrical resistor is 100,000,000 Ω.

13. The bending moment of a steel beam is 144,000 lb-in.

14. The value of an electrical inductor is .00035 H.

15. The modulus of elasticity for Douglas fir is 8,300,000,000 Pa.

16. The diameter of AWG #32 wire is .007950 in.

17. A coulomb is an electrical unit of quantity representing 6,240,000,000,000,000,000 electrons.

18. The amount of current flowing in an electrical circuit is .000350 A.

19. The thickness of a microprocessor chip is .00100 cm.

20. The distance from earth to Alpha Centauri (our nearest star) is 25,500,000,000,000 mi.

5.5 CALCULATING WITH SCIENTIFIC NOTATION

Objectives

☐ To multiply and divide numbers using scientific notation.

☐ To add and subtract numbers using scientific notation.

⊡ Multiplication and Division

Suppose we had to multiply 7,350,000,000 by .0000000264. Since most calculators only have an eight-digit display, we could not enter these decimal numbers into a calculator. We could multiply out the long way,

$$
\begin{array}{r}
7,350,000,000 \\
\times \quad .0000000264 \\
\hline
29400000000 \\
44100000000 \\
14700000000 \\
\hline
194.0400000000 = 194, \quad \text{three significant digits}
\end{array}
$$

but keeping track of so many zeros is quite an inconvenience. However, if we wrote the numbers in scientific notation, we could use a calculator to multiply the numbers between one and ten (the significant digits), and use the *product rule for exponents* to multiply the powers of ten (the place-holding zeros).

EXAMPLE 1

Find the product $(7,350,000,000)(.0000000264)$ using scientific notation. Write your answer in scientific notation.

Solution:

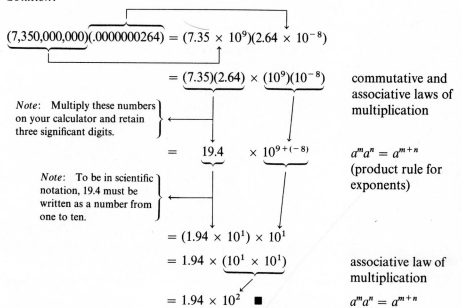

$$(7,350,000,000)(.0000000264) = (7.35 \times 10^9)(2.64 \times 10^{-8})$$

$$= (7.35)(2.64) \times (10^9)(10^{-8}) \qquad \text{commutative and}$$
associative laws of multiplication

Note: Multiply these numbers on your calculator and retain three significant digits.

$$= 19.4 \times 10^{9+(-8)} \qquad a^m a^n = a^{m+n}$$
(product rule for exponents)

Note: To be in scientific notation, 19.4 must be written as a number from one to ten.

$$= (1.94 \times 10^1) \times 10^1$$

$$= 1.94 \times (10^1 \times 10^1) \qquad \text{associative law of}$$
multiplication

$$= 1.94 \times 10^2 \qquad \blacksquare \qquad a^m a^n = a^{m+n}$$

Note: Of course, $1.94 \times 10^2 = 194$, which agrees with the longer method previously described.

Dividing large and small numbers can also be easily performed with scientific notation. Once the numbers are written in scientific notation, use your calculator to divide the numbers between one and ten (the significant digits), and use the *quotient rule for exponents* to divide the powers of ten (the place-holding zeros).

EXAMPLE 2

Find the quotient $\dfrac{22,000,000}{.000000882}$ using scientific notation. Write your answer in scientific notation.

Solution:

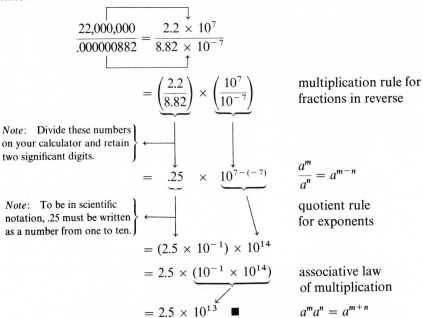

$$\frac{22,000,000}{.000000882} = \frac{2.2 \times 10^{7}}{8.82 \times 10^{-7}}$$

$$= \left(\frac{2.2}{8.82}\right) \times \left(\frac{10^{7}}{10^{-7}}\right) \qquad \text{multiplication rule for fractions in reverse}$$

Note: Divide these numbers on your calculator and retain two significant digits.

$$= \quad .25 \quad \times \quad 10^{7-(-7)} \qquad \frac{a^{m}}{a^{n}} = a^{m-n}$$

quotient rule
for exponents

Note: To be in scientific notation, .25 must be written as a number from one to ten.

$$= (2.5 \times 10^{-1}) \times 10^{14}$$

$$= 2.5 \times (10^{-1} \times 10^{14}) \qquad \text{associative law of multiplication}$$

$$= 2.5 \times 10^{13} \quad \blacksquare \qquad a^{m}a^{n} = a^{m+n}$$

⊡ Addition and Subtraction

To add or subtract decimal numbers, we must arrange the numbers in columns making sure that the decimal points line up. Only then can we add or subtract the numbers. For example, to add 9,274,000,000 and 898,000,000 we would write

$$
\begin{array}{r}
9,274,000,000.\\
+ \quad 898,000,000. \quad \leftarrow \text{decimal points in line}\\
\hline
10,172,000,000.
\end{array}
$$

Consequently, when using scientific notation to find a sum or a difference, we must be certain that all the powers of ten have exactly the same exponent before we add the significant digits.

EXAMPLE 3

Find the sum 9,274,000,000 + 898,000,000 using scientific notation. Write your answer in scientific notation.

Solution:

$$9,274,000,000 + 898,000,000 = (9.274 \times 10^{9}) + (8.98 \times 10^{8})$$

Note: Make the powers of ten the same before you add.

$$= (9.274 \times 10^{9}) = ((.898 \times 10^{1}) \times 10^{8})$$

$$= (9.274 \times 10^9) + (.898 \times 10^9)$$

Note: Since the powers of ten are now the same, you can add 9.274 and .898.

$$a^m a^n = a^{m+n}$$

$$= (9.274 + .898) \times 10^9$$

distributive law in reverse (like adding similar terms)

$$= \quad 10.172 \times 10^9$$

Note: To be in scientific notation, 10.172 must be written as a number from one to ten.

$$= (1.0172 \times 10^1) \times 10^9$$

$$= 1.0172 \times 10^{10} \quad \blacksquare \qquad a^m a^n = a^{m+n}$$

Note: Alternately, we could have written 9.274×10^9 as 92.74×10^8, and added $(92.74 \times 10^8) + (8.98 \times 10^8)$. Try it. See if you can obtain the same answer.

EXAMPLE 4

Find the difference $(2.73 \times 10^{-6}) - (3.9 \times 10^{-8})$ using scientific notation. Write your answer in scientific notation.

Solution:

$$(2.73 \times 10^{-6}) - (3.9 \times 10^{-8}) = ((273 \times 10^{-2}) \times 10^{-6}) - (3.9 \times 10^{-8})$$

Note: Make the powers of ten the same before you subtract.

$$= (273 \times 10^{-8}) - (3.9 \times 10^{-8}) \qquad a^m a^n = a^{m+n}$$

Note: Since the powers of ten are now the same, you can subtract 3.9 from 273.

$$= (273 - 3.9) \times 10^{-8}$$

distributive law in reverse (like subtracting similar terms.)

Note: Precise to the nearest whole number

$$= \quad 269 \times 10^{-8}$$

Note: To be in scientific notation, 269 must be written as a number from one to ten.

$$= (2.69 \times 10^2) \times 10^{-8}$$

$$= 2.69 \times 10^{-6} \quad \blacksquare \qquad a^m a^n = a^{m+n}$$

Note: Alternately, we could use $a^{-n} = \dfrac{1}{a^n}$ to write $\dfrac{2.73}{10^6} - \dfrac{3.9}{10^8}$. Try this problem by finding a least common denominator (10^8) and subtracting these fractions. See if you can obtain the same answer.

Application

ELECTRICAL POWER

When working with formulas, it is often required that you substitute the numerical quantities in their *basic unit size* directly into the formula. For example, the power dissipated by an electrical resistor can be found by using the formula $P = \dfrac{V^2}{R}$.

FORMULA: Electrical Power

$P = \dfrac{V^2}{R}$, where P is the power in watts, V is the voltage in volts, and R is the resistance in ohms.

However, the voltage drop V across the resistor must be in volts, and the resistance R must be in ohms, if the power P is to be in its usual units of watts. If *prefixes* are attached to the basic units (volts, ohms, watts), then *these prefixes must be converted to powers of ten before the formula can be used.*

You may recall that in Chapter 3, a table was given that listed the most commonly used prefixes. For your convenience, this table is given again. However, their meanings are now given as *powers of ten* (see Table 5.1).

EXAMPLE 5

A 470 MΩ (megohm) resistor has a voltage drop of 2.7 kV (kilovolts) across it. What power is being dissipated by the resistor?

TABLE 5.1

Prefixes	Symbol	Meaning
mega	M	10^6
kilo	k	10^3
centi	c	10^{-2}
milli	m	10^{-3}
micro	μ	10^{-6}

Solution: The prefixes must first be converted to powers of ten.

$$R = 470 \text{ M}\Omega = 470 \times 10^6 \ \Omega, \text{ and } V = 2.7 \text{ kV} = 2.7 \times 10^3 \text{V}.$$

Therefore,

$$P = \frac{V^2}{R} = \frac{(2.7 \times 10^3)^2}{470 \times 10^6} \qquad \text{substituting}$$

$$= \frac{(2.7)^2 \times (10^3)^2}{470 \times 10^6} \qquad (ab)^n = a^n b^n$$

$$= \frac{7.29 \times 10^6}{470 \times 10^6} \qquad (a^m)^n = a^{mn}$$

$$= .016 \times 10^0 \qquad \frac{a^m}{a^n} = a^{m-n}$$

$$= .016 \times 1 \qquad a^0 = 1$$

$$= .016 \text{ watts}$$

$$P = 16 \times 10^{-3} \text{ watts} = 16. \ milliwatts \text{ (mW)} \quad \blacksquare$$

Application

THERMAL EXPANSION AND THERMAL STRESS

When the temperature of an object changes, the dimensions of the object will also change. Nearly all objects *expand* when heated and *contract* when cooled. The number of units the length expands or contracts per degree of temperature change is called the **coefficient of thermal expansion** of the material, and is denoted by the Greek letter α (alpha). Listed in Table 5.2 are some average values of α for various structural materials.

TABLE 5.2   Average Coefficients of Thermal Expansion

Material	α (per degree Fahrenheit)
Brass	$1.04 \times 10^{-5}/°\text{F}$
Cast Iron	$5.9 \times 10^{-6}/°\text{F}$
Steel	$6.5 \times 10^{-6}/°\text{F}$
Concrete	$5.5 \times 10^{-6}/°\text{F}$
Douglas Fir	$3.0 \times 10^{-6}/°\text{F}$

The *total change in length* due to thermal expansion or contraction is denoted by the Greek letter δ (delta). Delta (δ) depends on the material's *coefficient of expansion* (α), the amount of *temperature change* (t), and the *original length* (L) of the object.

FORMULA: **Thermal Expansion**

$$\delta = \alpha t L$$

EXAMPLE 6

Fifty-foot steel railroad tracks are laid with a small space between the end of one section and the beginning of the next. If the tracks are laid at 40°F, what space should be left between them so that they will just touch when the temperature reaches 105°F (see Fig. 5.1)?

Solution:

$$\alpha = \frac{6.5 \times 10^{-6}}{°F} \quad \text{from Table 5.2}$$

$$t = (105°F - 40°F) = 65°F = 6.5 \times 10^1 \ °F$$

$$L = 5\bar{0}. \text{ ft} = \left(\frac{5\bar{0}. \text{ ft}}{1}\right)\left(\frac{12 \text{ in.}}{1 \text{ ft}}\right) = 6\bar{0}0. \text{ in.} = 6.0 \times 10^2 \text{ in.}$$

Thus

$$\delta = \alpha t L = \frac{6.5 \times 10^{-6}}{°F}(6.5 \times 10^1 \ °F)(6.0 \times 10^2 \text{ in.})$$

$$= \underbrace{(6.5)(6.5)(6.0)}_{\downarrow} \times \underbrace{(10^{-6})(10^1)(10^2)}_{\downarrow} \text{ in.}$$

$$= \qquad 250. \qquad \times \qquad 10^{-3} \text{ in.}$$

$$\delta = .25 \text{ in. or about } \frac{1}{4} \text{ in.} \quad \blacksquare$$

If the railroad tracks were laid with *no* space between them, they could not expand. When thermal expansion is restricted in this manner, the object will be subjected to *thermal stresses*. If these stresses become severe enough, the object will become deformed.

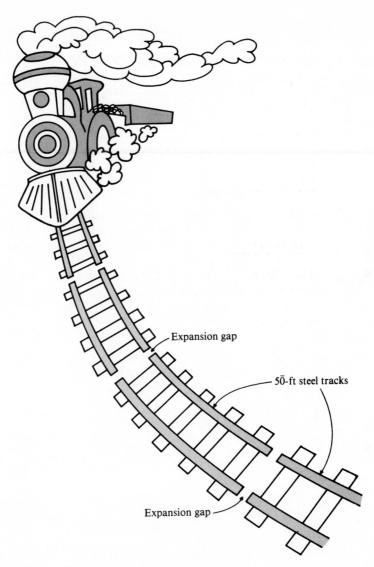

Expansion gap

5$\overline{0}$-ft steel tracks

Expansion gap

Figure 5.1

The *thermal stress* (*s*) depends on the material's *coefficient of expansion* (*α*), the amount of *temperature change* (*t*), and the material's *modulus of elasticity* (*E*). The modulus of elasticity is a measure of the stiffness of the material.

FORMULA: **Thermal Stress**

$$s = \alpha t E$$

EXAMPLE 7

Fifty-foot steel railroad tracks are laid with *no* space between the end of one section and the beginning of the next. If the tracks are laid at 40°F, what stress will be induced in a steel track when the temperature reaches 105°F? (For steel, $E = 2.9 \times 10^7$ psi.)

Solution:

$$\alpha = \frac{6.5 \times 10^{-6}}{°F} \quad \text{from Table 5.2}$$

$$t = (105°F - 40°F) = 65°F = 6.5 \times 10^1 \, °F$$

$$E = 2.9 \times 10^7 \text{ psi}$$

Thus,

$$s = \alpha t E = \left(\frac{6.5 \times 10^{-6}}{°F}\right)(6.5 \times 10^1 \, °F)(2.9 \times 10^7 \text{ psi})$$

$$= \underbrace{(6.5)(6.5)(2.9)}_{\downarrow} \times \underbrace{(10^{-6})(10^1)(10^7)}_{\downarrow} \text{ psi}$$

$$= \quad\quad 120 \quad\quad \times \quad\quad 10^2 \text{ psi}$$

$$s = 12{,}000 \text{ psi} \quad \blacksquare$$

Note: At approximately 30,000 psi, steel will start to perform. A serious situation will certainly arise when the thermal stress of 12,000 psi is added to the stress developed in the rails as the train goes over the tracks. It is easy to see why engineers must account for thermal expansion in the design and construction of railways, highways, buildings, bridges, and so on.

EXERCISES 5.5

Simplify each of the following. Write your answer in decimal notation.

1. $(10^{13})(10^{-14})$ **2.** $(10^{-12})(10^{15})$

3. $(10^{-3})^2$ **4.** $(10^{-3})^{-2}$

5. $\dfrac{10^9}{10^7}$

6. $\dfrac{10^{-8}}{10^{-5}}$

7. $10^4 + 10^5$

8. $10^5 - 10^{-7}$

9. $\left(\dfrac{10^3 10^{-7}}{10^{-5}}\right)^2$

10. $(10^0 - 10^{-1})^2$

Perform the indicated operations using scientific notation. Write your answer in scientific notation.

11. $(3 \times 10^6)(5 \times 10^{-4})$

12. $(36,000,000,000)(.0000092)$

13. $\dfrac{.00000795}{.0000000924}$

14. $\dfrac{8.7 \times 10^{19}}{3.2 \times 10^{16}}$

15. $(9.84 \times 10^7) + (8.1 \times 10^6)$

16. $(7.1 \times 10^{-7}) - (7.1 \times 10^{-6})$

17. $(4.31 \times 10^{-9}) - (1.8 \times 10^{-11})$

18. $(8.00 \times 10^8) + 1,900,000$

19. $(3.92 \times 10^{-8})^2$

20. $(4.6 \times 10^{-8})^{-3}$

21. $\dfrac{(42,000,000)(6.2 \times 10^{-3})}{(5.8 \times 10^5)}$

22. $\dfrac{(8.95 \times 10^{12})(167,000,000)}{.000000792}$

23. $(1.83 \times 10^9)(.0000000265)^3$

24. $(1.83 \times 10^9) + (1.9 \times 10^4)^2$

For the following exercises, convert all prefixes to powers of ten before applying the given formula. Attach an appropriate prefix, along with the basic unit, to your answer.

25. The current (I) through a resistor is 16.1 μA and the voltage drop (V) across it is 330 V. What is the value of the resistor (R)? $\left(\text{Use Ohm's law, } R = \dfrac{V}{I}.\right)$

26. Find the voltage drop across a 22.0 MΩ resistor if the current through the resistor is .325 mA. $\left(\text{Use Ohm's law, } R = \dfrac{V}{I}.\right)$

27. Referring to exercise 25, what power is being dissipated by the resistor? $\left(\text{Use } P = \dfrac{V^2}{R}.\right)$

28. Referring to exercise 26, what power is being dissipated by the resistor? $\left(\text{Use } P = \dfrac{V^2}{R}.\right)$

29. Resistances of 330 kΩ and 1.33 MΩ are connected in series. What is the total resistance? (Use $R_t = R_1 + R_2$.)

30. Resistances of 330 kΩ and 1.33 MΩ are connected in parallel. What is the total resistance? $\left(\text{Use } \dfrac{R_1 R_2}{R_1 + R_2}.\right)$

31. A concrete roadway on a bridge is being laid at a temperature of 35°F. How large an expansion gap should be allowed between the 30.0-ft long concrete slabs in order that the slabs will just touch each other at a temperature of 85°F (see Fig. 5.2)?

32. Referring to exercise 31, if the concrete is laid with *no* expansion gap, what would be the thermal stress in a concrete slab if the temperature rose to 120°F? (For concrete, $E = 2.0 \times 10^6$ psi. Concrete will start to break up at approximately 1000 psi).

33. A surveyor's steel tape is exactly 100 ft long when used at 70°F. What error is made in measuring a distance of 12,000 ft when the temperature is 0°F.

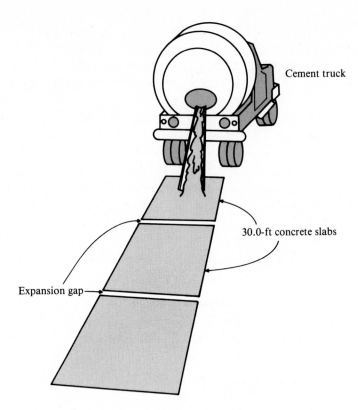

Cement truck

30.0-ft concrete slabs

Expansion gap

Figure 5.2

5.6 ROOTS AND RADICALS

Objectives

☐ To find the square root of a number or a perfect square variable expression.

☐ To find the cube root (or higher root) of a number or a perfect cube variable expression.

☐ Square Roots

The numbers 9 and 16 are examples of *perfect squares* since there exist integers that we can *square* to obtain 9 and 16.

$$(3)^2 = (3)(3) = 9$$
$$(-3)^2 = (-3)(-3) = 9$$
$$(4)^2 = (4)(4) = 16$$
$$(-4)^2 = (-4)(-4) = 16$$

The numbers 3 and -3 are called the **square roots** of 9, since either 3 or -3 when squared equals 9. Similarly, 4 and -4 are the square roots of 16, since either one when squared equals 16. As you can see, there are always *two* square roots of a *positive* number. One of the square roots will be *positive* while the other square root will be *negative*. They will always be *opposites*.

Mathematicians use the symbol $\sqrt[2]{}$ to indicate the positive square root of a number. The *2* is called the **root index** and the $\sqrt{}$ is called the **radical sign**. For the *positive square root*, the index *2* is usually omitted and you'll simply see $\sqrt{}$. We write $\sqrt{a}$ for the positive, or *principal*, square root of *a*.

Thus, the *positive* square root of 9 is written $\sqrt{9}$, and $\sqrt{9} = 3$. Similarly, the *positive* square root of 16 is written $\sqrt{16}$, and $\sqrt{16} = 4$.

For the *negative* square root of a number, mathematicians use the symbol $-\sqrt[2]{}$, or simply $-\sqrt{}$.

The *negative* square root of 9 is written $-\sqrt{9}$, and $-\sqrt{9} = -3$. The *negative* square root of 16 is written $-\sqrt{16}$, and $-\sqrt{16} = -4$.

CAUTION

$\sqrt{9}$ does *not* equal both 3 and -3. It is *only* equal to 3. Only if you see $\pm\sqrt{9}$ (read "plus or minus the square root of 9") does it mean you want *both* 3 and -3.

EXAMPLE 1

Find the indicated square roots.

a) $\sqrt{81}$ **b)** $-\sqrt{25}$ **c)** $\pm\sqrt{16}$ **d)** $\sqrt{.04}$ **e)** $\sqrt{\dfrac{1}{4}}$

Solution:

a) $\sqrt{81} = 9$, since 9 is the positive number we can square to obtain 81.

b) $-\sqrt{25} = -5$, since -5 is the negative number we can square to obtain 25.

c) $\pm\sqrt{16} = \pm4$, since we want both square roots.

d) $\sqrt{.04} = .2$ (*not* .02)

e) $\sqrt{\dfrac{1}{4}} = \dfrac{1}{2}$ ∎

Note: Notice that the positive square root of a number that is greater than 1 (Example 1a) is smaller than the original number, while the positive square root of a number that is between 0 and 1 (Examples 1d and 1e) is larger than the original number.

The positive square root of 2 is written $\sqrt{2}$, and the negative square root of 2 is written $-\sqrt{2}$. However, since 2 is *not* a perfect square, there does not exist any integer that we can square to obtain 2. Therefore, we have no other *exact* representation for the square roots of 2 other than $\pm\sqrt{2}$. Numbers like $\sqrt{2}$ and $-\sqrt{2}$ are called **irrational numbers** since they can never be written exactly as terminating repeating decimals. Irrational numbers can only be written as nonrepeating decimals that continue forever. When grouped together, the rational and irrational numbers form what is called the **real number system**. Your calculator and the $\sqrt{}$ key can be used to approximate the values of such irrational numbers.

EXAMPLE 2

Using a calculator, approximate the following square roots. Retain four significant digits in your answer.

a) $\sqrt{2}$

b) $-\sqrt{2}$

Solution:

a) Enter

$$\boxed{2}\ \boxed{\sqrt{}}\ \overbrace{1.4142136.}^{\text{Display}}$$

Thus,

$$\sqrt{2} \approx 1.414. \quad \blacksquare$$

b) Enter

$$\boxed{2}\ \boxed{\sqrt{}}\ \boxed{+/-}\ \underbrace{-1.4142136.}_{\text{Display}}$$

Thus,

$$-\sqrt{2} \approx -1.414. \quad \blacksquare$$

Thus far we have only talked about the square roots of a positive number. Now, suppose you were asked to find the square roots of negative nine. What number can you square to obtain -9? You know that a positive number squared is positive and a negative number squared is also positive. Therefore, it is impossible for you to square any type of number that we have discussed so far (any *real number*) and obtain a negative result. Thus, no *real number* can be assigned to the square root of a negative number. In more advanced technical mathematics courses, it is shown that the square roots of a negative number are **imaginary numbers**. Imaginary numbers will not be discussed in this book. We will work entirely with real numbers.

CAUTION Don't get $-\sqrt{9}$ and $\sqrt{-9}$ confused. $-\sqrt{9}$ designates the negative square root of nine, which equals -3. However, $\sqrt{-9}$ designates the square root of negative nine, which *does not exist in the real number system*.

In this book, we will assume that when a variable appears within the radical symbol that the variable represents a *nonnegative* number. When a variable base is raised to an *even* power, it is called a **perfect square variable expression**. As illustrated in the next example, the positive square root of such an expression can easily be found.

EXAMPLE 3

Find the indicated square root of the variable expressions.

a) $\sqrt{x^2}$ **b)** $\sqrt{y^8}$ **c)** $\sqrt{z^{16}}$

Solution:

a) $\sqrt{x^2} = x$, since x^1 is the algebraic expression we can square to obtain x^2. $(x^1)^2 = x^2$, $(a^m)^n = a^{mn}$.

b) $\sqrt{y^8} = y^4$, since y^4 is the algebraic expression we can square to obtain y^8. $(y^4)^2 = y^8$, $(a^m)^n = a^{mn}$.

c) $\sqrt{z^{16}} = z^8$, since z^8 is the algebraic expression we can square to obtain z^{16}. $(z^8)^2 = z^{16}$, $(a^m)^n = a^{mn}$. Be careful, $\sqrt{z^{16}} \neq z^4$. ∎

Note: Notice that the exponent in each result is exactly one-half that of the exponent within the radical sign.

From Example 3, it should be clear that *the square root of a base to an even power is that base raised to half the even power.*

> If m is even and a is nonnegative, $\sqrt{a^m} = a^{m/2}$.

In Section 5.7 we will investigate the square root of a base to an *odd* power.

⊡ Cube Roots and Higher Roots

The number 8 is an example of a **perfect cube** since there exists an integer that we can *cube* (raise to the third power) to obtain 8.

$$(2)^3 = (2)(2)(2) = 8$$

The number 2 is called the **cube root** of 8, since 2 when cubed equals 8. *There is only one cube root of a positive number.* Negative two is *not* a cube root of 8, since

$$(-2)^3 = (-2)(-2)(-2) = -8.$$

Negative two is the *cube root* of -8.

Mathematicians use the symbol $\sqrt[3]{}$ to designate the cube root of a number. Thus, the cube root of 8 is written $\sqrt[3]{8}$, and $\sqrt[3]{8} = 2$. Similarly, the cube root of -8 is written $\sqrt[3]{-8}$, and $\sqrt[3]{-8} = -2$. Although the square root of a negative number does not exist in the real number system, the cube root of a negative number certainly does.

The *fourth roots* of 16 are 2 and -2, since

$$2^4 = (2)(2)(2)(2) = 16$$

and

$$(-2)^4 = (-2)(-2)(-2)(-2) = 16.$$

The positive fourth root of 16 is written $\sqrt[4]{16}$, and $\sqrt[4]{16} = 2$. The negative fourth root of 16 is written $-\sqrt[4]{16}$, and $-\sqrt[4]{16} = -2$. The fourth root of -16, $\sqrt[4]{-16}$, *does not exist* in the real number system, since it is impossible to raise any real number to the fourth power (an even power) and obtain a negative result.

The *fifth root* of 32 is 2 only, since

$$2^5 = (2)(2)(2)(2)(2) = 32, \text{ thus } \sqrt[5]{32} = 2.$$

The *fifth root* of -32 exists and equals -2, since

$$(-2)^5 = (-2)(-2)(-2)(-2)(-2) = -32, \text{ thus } \sqrt[5]{-32} = -2.$$

Table 5.3 serves as a summary of the nature of the nth root(s) of a.

TABLE 5.3 Nature of the *n*th root(s) of *a*

n	*a*	*n*th root(s) of *a*
even	positive	one positive root one negative root
even	negative	does not exist in the real number system
odd	positive	one positive root only
odd	negative	one negative root only

EXAMPLE 4

Find the indicated *n*th roots.

a) $\sqrt[3]{-27}$ b) $\sqrt[4]{-81}$

Solution:

a) $\sqrt[3]{-27} = -3$, since -3 is the number we can cube to obtain -27. $(-3)^3 = (-3)(-3)(-3) = -27$.

b) $\sqrt[4]{-81}$ *does not exist* in the real number system. $3^4 = +81$, and $(-3)^4 = +81$. ∎

Your calculator and the $\boxed{\text{inv}}\,\boxed{y^x}$ keys (or $\boxed{\sqrt[x]{y}}$ key) can be used to find the *n*th root of a number that is not a perfect *n*th power. However, with most calculators, you cannot enter a negative base. If you do, your calculator will either print ERROR or it will blink some number. Your calculator must be used in conjunction with the information in Table 5.3 to properly evaluate the *n*th root of a number.

EXAMPLE 5

Using your calculator, evaluate $\sqrt[3]{-20}$. Retain four significant digits.

Solution: According to Table 5.3, if *n* is odd and *a* is negative, there should be *one negative root*.

Enter $\boxed{20}\,\boxed{+/-}\,\boxed{\text{inv}}\,\boxed{y^x}\,\boxed{3}\,\boxed{=}$ 2.7144176 *blinking*, or *érror*.
                                        Display   Display

The *blinking* 2.7144176 indicates that it is up to you to attach the correct sign. Thus, $\sqrt[3]{-20} \approx -2.714$.

    If your calculator displays *error*, you must clear the calculator and proceed as follows:

Enter $\boxed{20}\,\boxed{\text{inv}}\,\boxed{y^x}\,\boxed{3}\,\boxed{=}$ 2.7144176   (no $\boxed{+/-}$ entered)
                                        Display

Again, it is up to you to attach the correct sign. Thus, $\sqrt[3]{-20} \approx -2.714$. ∎

A variable base raised to a multiple of 3 power is a **perfect cube variable expression**. The cube root of such an expression can be easily found.

EXAMPLE 6

Find the indicated cube root of the variable expressions.

a) $\sqrt[3]{x^3}$ b) $\sqrt[3]{x^{27}}$

Solution:

a) $\sqrt[3]{x^3} = x$, since x^1 is the algebraic expression we can cube to obtain x^3. $(x^1)^3 = x^3$, $(a^m)^n = a^{mn}$.

b) $\sqrt[3]{x^{27}} = x^9$, since x^9 is the algebraic expression we can cube to obtain x^{27}. $(x^9)^3 = x^{27}$, $(a^m)^n = a^{mn}$. ∎

Note: Notice that the exponent in each result is exactly one-third that of the exponent within the radical sign.

From Example 6 it should be clear that *the cube root of a variable base to a multiple of 3 (3, 6, 9, 12, 15, . . .) power is that base raised to one-third the multiple of 3 power.*

If *m* is a multiple of 3, then

$$\sqrt[3]{a^m} = a^{m/3}.$$

In Section 5.7, we will investigate the cube root of a base to a nonmultiple of 3 power.

EXERCISES 5.6

Find the indicated root(s), if they exist.

1. $\sqrt{36}$ **2.** $-\sqrt{36}$ **3.** $-\sqrt{49}$ **4.** $\sqrt{49}$

5. $\pm\sqrt{100}$ **6.** $\pm\sqrt{9}$ **7.** $\sqrt{-64}$ **8.** $\sqrt{-25}$

9. $\sqrt{.09}$ **10.** $\sqrt{.25}$ **11.** $\sqrt{\dfrac{1}{9}}$ **12.** $\sqrt{\dfrac{1}{25}}$

13. $\sqrt[3]{64}$ **14.** $\sqrt[3]{-64}$ **15.** $\sqrt[3]{-125}$ **16.** $\sqrt[3]{125}$

17. $\sqrt[4]{-81}$ **18.** $\sqrt[4]{81}$ **19.** $\sqrt[5]{-1}$ **20.** $\sqrt[6]{-1}$

Find the indicated root. Assume that all variables represent nonnegative numbers.

21. $\sqrt{x^4}$ **22.** $\sqrt{y^{18}}$ **23.** $\sqrt{z^{36}}$ **24.** $\sqrt{p^{100}}$

25. $\sqrt{(x + y)^2}$ **26.** $\sqrt{(x + y)^{16}}$ **27.** $\sqrt[3]{x^9}$ **28.** $\sqrt[3]{y^6}$

29. $\sqrt[3]{z^{66}}$ **30.** $\sqrt[3]{z^{120}}$ **31.** $\sqrt[3]{(x + y)^3}$ **32.** $\sqrt[3]{(x + y)^{27}}$

Using your calculator, approximate the following roots, if they exist. Retain four significant digits in your answer.

33. $\sqrt{3}$ **34.** $\sqrt{10}$ **35.** $-\sqrt{41}$ **36.** $\sqrt{-109}$

37. $\sqrt[3]{55}$ **38.** $\sqrt[3]{731}$ **39.** $\sqrt[3]{-610}$ **40.** $\sqrt[3]{-104}$

41. $\sqrt[4]{-25}$ **42.** $\sqrt[4]{87}$ **43.** $\sqrt[5]{-8}$ **44.** $\sqrt[6]{-9}$

5.7   THE SQUARE ROOT AND CUBE ROOT OF A PRODUCT

Objectives

☐ To simplify the square root of a product.

☐ To simplify the cube root of a product.

☐ Square Root of a Product

In Section 5.1 you discovered that the power of a product is equivalent to the product of the powers. $(ab)^n = a^n b^n$. Might the positive square root of a product be equivalent to the product of the positive square roots? $\sqrt{ab} = \sqrt{a}\sqrt{b}$? Does

$$\sqrt{4 \cdot 9} = \sqrt{4} \cdot \sqrt{9}?$$
$$\sqrt{36} = 2 \cdot 3 ?$$
$$6 = 6 \quad \text{Yes.}$$

Does

$$\sqrt{25 \cdot 16} = \sqrt{25} \cdot \sqrt{16}?$$
$$\sqrt{400} = 5 \cdot 4 ?$$
$$20 = 20 \quad \text{Yes.}$$

As you can see, *the positive square root of a product is the same as the product of the positive square roots.*

SQUARE ROOT OF A PRODUCT RULE FOR RADICALS
For all nonnegative numbers a and b,
$$\sqrt{ab} = \sqrt{a}\sqrt{b}.$$

The positive square root of a sum is *not* the same as the sum of the positive square roots. $\sqrt{a + b}$ and $\sqrt{a} + \sqrt{b}$ are *not* equivalent. For example,

$$\sqrt{9 + 16} = \sqrt{25} = 5,$$

while

$$\sqrt{9} + \sqrt{16} = 3 + 4 = 7.$$

A square root is said to be simplified if all perfect square factors are removed from within the radical sign. To accomplish this task, we can use the square root of a product rule for radicals, $\sqrt{ab} = \sqrt{a}\sqrt{b}$.

EXAMPLE 1

Simplify $\sqrt{9x^4}$.

Solution:

$$\sqrt{9x^4} = \sqrt{9}\sqrt{x^4} \qquad \sqrt{ab} = \sqrt{a}\sqrt{b}$$
$$= 3x^2 \quad \blacksquare \qquad \sqrt{a^m} = a^{m/2}, \text{ if } m \text{ is even}$$

EXAMPLE 2

Simplify $\sqrt{9 + x^4}$.

Solution: *Be careful.* The square root of a sum is not the sum of the square roots. $\sqrt{9 + x^4} \neq 3 + x^2$. $\sqrt{9 + x^4}$ is already simplified. $\quad \blacksquare$

EXAMPLE 3

Simplify $\sqrt{81x^2y^{16}z^{36}}$.

Solution:

$$\sqrt{81x^2y^{16}z^{36}} = \sqrt{81}\sqrt{x^2}\sqrt{y^{16}}\sqrt{z^{36}} \qquad \sqrt{ab} = \sqrt{a}\sqrt{b}$$
$$= 9xy^8z^{18} \quad \blacksquare \qquad \sqrt{a^m} = a^{m/2}, \text{ if } m \text{ is even}$$

At times, the perfect square factors may be hidden. For example, $\sqrt{48}$ may seem to be already simplified, since 48 is *not* a perfect square. However, 4 is a factor of 48, and 4 is certainly a *perfect square factor* that can be removed from within the radical sign.

$$\sqrt{48} = \sqrt{4 \cdot 12} = \sqrt{4}\sqrt{12} \qquad \sqrt{ab} = \sqrt{a}\sqrt{b}$$
$$= 2\sqrt{12}$$

Note: But 4 is also a factor of 12.

$$= 2\sqrt{4 \cdot 3}$$
$$= 2\sqrt{4}\sqrt{3} \qquad \sqrt{ab} = \sqrt{a}\sqrt{b}$$
$$= 2(2)\sqrt{3}$$
$$= 4\sqrt{3}$$

Since all perfect square factors are now removed from within the radical sign, $\sqrt{48} = 4\sqrt{3}$ is *simplified*.

If you had thought of 48 as (16)(3), you could have removed the perfect square *16* from within the radical sign and immediately obtained the simplified answer as follows:

$$\sqrt{48} = \sqrt{16 \cdot 3} = \underbrace{\sqrt{16}}\sqrt{3} = 4\sqrt{3}.$$

It is always to your advantage to remove the largest possible perfect square factor hidden in the number. The perfect squares to look for are

4, 9, 16, 25, 36, 49, 64, 81, 100, 121, 144,

and so on.

EXAMPLE 4

Simplify $\sqrt{72}$.

Solution: The largest perfect square factor hidden in 72 is 36. Thus, $\sqrt{72} = \sqrt{36 \cdot 2} = \underbrace{\sqrt{36}}\sqrt{2} = 6\sqrt{2}.$ ■

In Section 5.6 you learned that the square root of a base to an even power is that base raised to half the even power $(\sqrt{a^m} = a^{m/2})$. Using this fact and $\sqrt{ab} = \sqrt{a}\sqrt{b}$, we can now simplify the square root of a base to an *odd* power.

EXAMPLE 5

Simplify $\sqrt{x^7}$.

Solution: The largest possible perfect square variable factor hidden in x^7 is x^6. Thus,

$$\sqrt{x^7} = \sqrt{x^6 x} = \underbrace{\sqrt{x^6}}\sqrt{x} \qquad \sqrt{ab} = \sqrt{a}\sqrt{b}$$

$$= x^3\sqrt{x} \quad \blacksquare \quad \sqrt{a^m} = a^{m/2}, \text{ if } m \text{ is even}$$

Note: The simplified form for the square root of a base x to an *odd* power will always contain $\sqrt{x}$.

EXAMPLE 6

Simplify $\sqrt{20xy^6z^5}$.

Solution:

$$\sqrt{20xy^6z^5} = \sqrt{4 \cdot 5x\,y^6\,z^4\,z}$$

$\left\{\begin{array}{l}Note\text{:} \quad \text{These are perfect} \\ \text{square factors.}\end{array}\right.$

$$= \sqrt{4y^6z^4}\sqrt{5xz} \qquad \sqrt{ab} = \sqrt{a}\sqrt{b}$$

$$= 2y^3z^2\sqrt{5xz} \quad \blacksquare \qquad \sqrt{a^m} = a^{m/2}, \text{ if } m \text{ is even}$$

To simplify a product of square roots, use the *square root of a product rule in reverse*. $\sqrt{a}\sqrt{b} = \sqrt{ab}$. Then simplify the radical expression that develops.

EXAMPLE 7

Simplify $\sqrt{2xy^3}\sqrt{8x^3y^2}$.

Solution:

$$\sqrt{2xy^3}\sqrt{8x^3y^2} = \sqrt{(2xy^3)(8x^3y^2)} \qquad \sqrt{a}\sqrt{b} = \sqrt{ab}$$

$$= \sqrt{16x^4y^5} \qquad a^m a^n = a^{m+n}$$

$$= \sqrt{16\;x^4\;y^4\;y}$$

$\left.\begin{array}{l}Note\text{:} \quad \text{These are perfect} \\ \text{square factors.}\end{array}\right\}$

$$= \sqrt{16x^4y^4}\sqrt{y} \qquad \sqrt{ab} = \sqrt{a}\sqrt{b}$$

$$= 4x^2y^2\sqrt{y} \quad \blacksquare \qquad \sqrt{a^m} = a^{m/2}, \text{ if } m \text{ is even}$$

⊡ Cube Root of a Product

The rules and ideas for simplifying square roots can easily be extended to cube roots. *A cube root is said to be simplified if all perfect cube factors are removed from within the radical sign.* To accomplish this task, we can use the cube root of a product rule for radicals, $\sqrt[3]{ab} = \sqrt[3]{a}\sqrt[3]{b}$.

EXAMPLE 8

Simplify $\sqrt[3]{27x^3y^{15}}$.

Solution:

$$\sqrt[3]{27x^3y^{15}} = \sqrt[3]{27}\sqrt[3]{x^3}\sqrt[3]{y^{15}} \qquad \sqrt[3]{ab} = \sqrt[3]{a}\sqrt[3]{b}$$

$$= 3xy^5 \quad \blacksquare \qquad \sqrt[3]{a^m} = a^{m/3}, \text{ if } m \text{ is a multiple of } 3$$

If the expression is not a perfect cube, *then remove the largest possible perfect cube factors that are hidden in the expression.* The perfect cubes to look for are 8, 27, 64, 125, 216, 343, 512, 729, 1000 , and so on.

EXAMPLE 9

Simplify $\sqrt[3]{24}$.

Solution: The largest perfect cube factor hidden in 24 is 8. Thus,

$$\sqrt[3]{24} = \sqrt[3]{8}\,\sqrt[3]{3} = 2\sqrt[3]{3} \quad \blacksquare$$

In Section 5.6 you learned that the cube root of a base to a multiple of 3 power is that base raised to one-third the multiple of 3 power ($\sqrt[3]{a^m} = a^{m/3}$). Using this fact and $\sqrt[3]{ab} = \sqrt[3]{a}\sqrt[3]{b}$, we can now simplify the cube root of a base to a non-multiple of 3 power.

EXAMPLE 10

Simplify.

a) $\sqrt[3]{x^4}$ **b)** $\sqrt[3]{x^8}$

Solution:

a) The largest perfect cube variable factor hidden in x^4 is x^3. Thus,

$$\sqrt[3]{x^4} = \sqrt[3]{x^3 x} = \sqrt[3]{x^3}\,\sqrt[3]{x} \qquad \sqrt[3]{ab} = \sqrt[3]{a}\,\sqrt[3]{b}$$
$$= x\sqrt[3]{x} \quad \blacksquare \qquad \sqrt[3]{a^m} = a^{m/3},\ \text{if } m \text{ is a multiple of 3}$$

b) The largest perfect cube variable factor hidden in x^8 is x^6. Thus,

$$\sqrt[3]{x^8} = \sqrt[3]{x^6 x^2} = \sqrt[3]{x^6}\,\sqrt[3]{x^2} \qquad \sqrt[3]{ab} = \sqrt[3]{a}\,\sqrt[3]{b}$$
$$= x^2\sqrt[3]{x^2} \quad \blacksquare \qquad \sqrt[3]{a^m} = a^{m/3},\ \text{if } m \text{ is a multiple of 3}$$

Note: The simplified form for the cube root of a base x to a nonmultiple of 3 power will always contain either $\sqrt[3]{x}$ or $\sqrt[3]{x^2}$.

EXAMPLE 11

Simplify $\sqrt[3]{64x^9 y^7 z^{14}}$.

Solution:

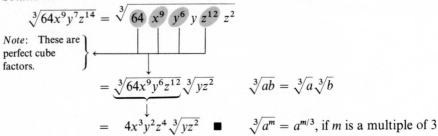

$$\sqrt[3]{64x^9y^7z^{14}} = \sqrt[3]{64 \; x^9 \; y^6 \; y \; z^{12} \; z^2}$$

Note: These are perfect cube factors. $\Bigg\}$ ←

$$= \underbrace{\sqrt[3]{64x^9y^6z^{12}}} \; \sqrt[3]{yz^2} \qquad \sqrt[3]{ab} = \sqrt[3]{a}\sqrt[3]{b}$$

$$= \quad 4x^3y^2z^4 \sqrt[3]{yz^2} \quad \blacksquare \qquad \sqrt[3]{a^m} = a^{m/3}, \text{ if } m \text{ is a multiple of } 3$$

To simplify a product of cube roots, use the cube root of a product rule in reverse. $\sqrt[3]{a}\sqrt[3]{b} = \sqrt[3]{ab}$. Then simplify the radical expression that develops.

EXAMPLE 12

Simplify $\sqrt[3]{2xy^2}\sqrt[3]{4x^2y^2}$.

Solution:

$$\sqrt[3]{2xy^2}\sqrt[3]{4x^2y^2} = \sqrt[3]{(2xy^2)(4x^2y^2)} \qquad \sqrt[3]{a}\sqrt[3]{b} = \sqrt[3]{ab}$$

$$= \sqrt[3]{8x^3y^4} \qquad\qquad a^ma^n = a^{m+n}$$

$$= \sqrt[3]{8 \; x^3 \; y^3 \; y}$$

Note: These are perfect cube factors. $\Bigg\}$ ←

$$= \underbrace{\sqrt[3]{8x^3y^3}} \; \sqrt[3]{y} \qquad \sqrt[3]{ab} = \sqrt[3]{a}\sqrt[3]{b}$$

$$= \quad 2xy\sqrt[3]{y} \quad \blacksquare \qquad \sqrt[3]{a^m} = a^{m/3}, \text{ if } m \text{ is a multiple of } 3$$

EXERCISES 5.7

Simplify the following radical expressions. Assume that all variables represent nonnegative numbers.

1. $\sqrt{9x^2y^2}$ **2.** $\sqrt{25x^4y^{10}}$

3. $\sqrt{100x^4y^8z^{100}}$ **4.** $\sqrt{64x^{16}y^{36}z^{64}}$

5. $\sqrt[3]{8x^{15}}$ **6.** $\sqrt[3]{64y^{12}}$

7. $\sqrt[3]{27x^6y^{27}}$ **8.** $\sqrt[3]{125x^3y^{27}}$

Simplify the following roots. Assume that all variables represent nonnegative numbers.

9. $\sqrt{24}$ **10.** $\sqrt{63}$ **11.** $\sqrt{45}$ **12.** $\sqrt{80}$

13. $\sqrt[3]{40}$ **14.** $\sqrt[3]{54}$ **15.** $\sqrt[3]{250}$ **16.** $\sqrt[3]{128}$

17. $\sqrt{x^9}$ **18.** $\sqrt{x^{11}}$ **19.** $\sqrt[3]{x^{13}}$ **20.** $\sqrt[3]{x^{20}}$

Simplify the following radical expressions. Assume that all variables represent nonnegative numbers.

21. $\sqrt{12x^4y^6}$

22. $\sqrt{64x^9y^4}$

23. $\sqrt{32x^3y^8z}$

24. $\sqrt{27x^5y^9z}$

25. $\sqrt[3]{16y^6}$

26. $\sqrt[3]{64x^8y^3}$

27. $\sqrt[3]{81x^9y^8z}$

28. $\sqrt[3]{54x^{11}y^{10}z^2}$

29. $\sqrt{25 + y^2}$

30. $\sqrt{(25 + y)^2}$

31. $\sqrt[3]{8 + y^3}$

32. $\sqrt[3]{(8 + y)^3}$

33. $\sqrt{9x^2(25 + y^2)}$

34. $\sqrt{9x^2(25 + y)^2}$

35. $\sqrt[3]{x^3z^3(8 + y^3)}$

36. $\sqrt[3]{x^3z^3(8 + y)^3}$

37. $\sqrt{5x}\sqrt{5x}$

38. $\sqrt{2x}\sqrt{12x}$

39. $\sqrt{7x^3y}\sqrt{7xy^2}$

40. $\sqrt{20x^4y^3}\sqrt{5x^2y}$

41. $\sqrt[3]{3xy}\sqrt[3]{9x^2y}$

42. $\sqrt[3]{16x^5y}\sqrt[3]{4xy^2}$

43. $(\sqrt{5xy})^2$

44. $(\sqrt[3]{5xy})^3$

5.8   THE SQUARE ROOT AND CUBE ROOT OF A QUOTIENT

Objectives

⊡ To simplify the square root of a quotient.

⊡ To simplify the cube root of a quotient.

⊡ Simplifying the Square Root of a Quotient

In Section 5.2 you discovered that the power of a quotient is equivalent to the quotient of the powers. $\left(\dfrac{a}{b}\right)^n = \dfrac{a^n}{b^n}$. Might the positive square root of a quotient be equivalent to the quotient of the positive square roots? $\sqrt{\dfrac{a}{b}} = \dfrac{\sqrt{a}}{\sqrt{b}}$? Does

$$\sqrt{\frac{36}{9}} \;\rbrack = \;\lbrack \frac{\sqrt{36}}{\sqrt{9}} \rbrack \; ?$$

$$\sqrt{4} \;\lrcorner = \;\lbrack \frac{6}{3} \;? $$

$$\downarrow \qquad\qquad \downarrow$$

$$2 \;\; = \;\; 2 \qquad \text{Yes.}$$

Does

$$\sqrt{\frac{100}{4}} \left.\right] = \left.\begin{array}{c}\sqrt{100} \\ \overline{\sqrt{4}}\end{array}\right] ?$$

$$\sqrt{25} \left.\right\downarrow = \left[\begin{array}{c}10 \\ \overline{2}\end{array}\right] ?$$

$$5 = 5 \quad \text{Yes.}$$

As you can see, *the positive square root of a quotient is the same as the quotient of the positive square roots.*

SQUARE ROOT OF A QUOTIENT RULE FOR RADICALS
For all nonnegative numbers a and b,

$$\sqrt{\frac{a}{b}} = \frac{\sqrt{a}}{\sqrt{b}}, b \neq 0.$$

Remember, a square root is said to be simplified if all perfect square factors are removed from within the radical sign.

EXAMPLE 1

Simplify $\sqrt{\dfrac{16x^{10}}{25y^2}}$.

Solution:

$$\sqrt{\frac{16x^{10}}{25y^2}} = \frac{\sqrt{16x^{10}}}{\sqrt{25y^2}} \qquad \sqrt{\frac{a}{b}} = \frac{\sqrt{a}}{\sqrt{b}}$$

$$= \frac{4x^5}{5y} \quad \blacksquare \qquad \sqrt{ab} = \sqrt{a}\sqrt{b} \text{ and } \sqrt{a^m} = a^{m/2}, \text{ if } m \text{ is even}$$

EXAMPLE 2

Simplify $\sqrt{\dfrac{63z^3}{64x^2y^8}}$.

Solution:

$$\sqrt{\frac{63z^3}{64x^2y^8}} = \frac{\sqrt{63z^3}}{\sqrt{64x^2y^8}} \qquad\qquad \sqrt{\frac{a}{b}} = \frac{\sqrt{a}}{\sqrt{b}}$$

Note: These are perfect square factors. $\left.\right\}$

$$= \frac{\sqrt{9 \cdot 7z^2\, z}}{\sqrt{64x^2y^8}}$$

$$= \frac{3z\sqrt{7z}}{8xy^4} \quad\blacksquare \qquad\qquad \sqrt{ab} = \sqrt{a}\sqrt{b} \text{ and } \sqrt{a^m} = a^{m/2}, \text{ if } m \text{ is even}$$

If you were asked to simplify $\sqrt{\dfrac{12}{2}}$, you could divide 12 by 2 and write

$$\sqrt{\frac{12}{2}} = \sqrt{6},$$

or you could apply the square root of a quotient rule $\sqrt{\dfrac{a}{b}} = \dfrac{\sqrt{a}}{\sqrt{b}}$ as follows:

$$\sqrt{\frac{12}{2}} = \frac{\sqrt{12}}{\sqrt{2}} = \frac{\sqrt{4 \cdot 3}}{\sqrt{2}} = \frac{2\sqrt{3}}{\sqrt{2}}$$

We know that $\sqrt{6}$ and $\dfrac{2\sqrt{3}}{\sqrt{2}}$ must be equivalent, but how do we algebraically show that they are equivalent? Suppose we multiplied both numerator and denominator of $\dfrac{2\sqrt{3}}{\sqrt{2}}$ by $\sqrt{2}$. This would eliminate the square root in the denominator as follows:

$$\frac{2\sqrt{3}}{\sqrt{2}} \cdot \left(\frac{\sqrt{2}}{\sqrt{2}}\right) = \frac{2\sqrt{3}\sqrt{2}}{\sqrt{2}\sqrt{2}} \qquad\qquad \text{multiplying fractions}$$

$$= \frac{2\sqrt{6}}{\sqrt{4}} \qquad\qquad \sqrt{a}\sqrt{b} = \sqrt{ab}$$

$$= \frac{2\sqrt{6}}{2} \qquad\qquad \text{simplifying } \sqrt{4}$$

$$= \sqrt{6} \qquad\qquad \text{cancelling}$$

This process of removing a square root from the denominator is called **rationalizing the denominator**. Mathematicians have agreed that a radical expression is not truly simplified unless its denominator is free of all radical signs.

A radical expression containing a square root is simplified only if

1. all perfect square factors are removed from within the radical sign, and
2. the denominator is rationalized.

To rationalize a denominator that contains a square root, *multiply both numerator and denominator by a* **rationalizing** *factor that will produce a perfect square within the radical sign of the denominator.* As a general rule, it is best to remove all perfect square factors first; then, if need be, rationalize the denominator.

EXAMPLE 3

Simplify $\sqrt{\dfrac{x^8}{24}}$.

Solution:

$$\sqrt{\dfrac{x^8}{24}} = \dfrac{\sqrt{x^8}}{\sqrt{24}} = \dfrac{\sqrt{x^8}}{\sqrt{4 \cdot 6}} \qquad\qquad \sqrt{\dfrac{a}{b}} = \dfrac{\sqrt{a}}{\sqrt{b}}$$

Note: This is a perfect square factor.

$$= \dfrac{x^4}{2\sqrt{6}} \qquad\qquad \sqrt{ab} = \sqrt{a}\sqrt{b} \text{ and } \sqrt{a^m} = a^{m/2}, \text{ if } m \text{ is even}$$

$$= \dfrac{x^4}{2\sqrt{6}}\left(\dfrac{\sqrt{6}}{\sqrt{6}}\right) \qquad\qquad \text{multiplying by a rationalizing factor } (\sqrt{6}\sqrt{6} = \sqrt{36} = 6)$$

$$= \dfrac{x^4\sqrt{6}}{2\sqrt{36}}$$

$$= \dfrac{x^4\sqrt{6}}{2(6)}$$

$$= \dfrac{x^4\sqrt{6}}{12} \qquad \blacksquare \qquad \text{simplifying}$$

EXAMPLE 4

Simplify $\sqrt{\dfrac{8x^3}{9y^5}}$.

Solution:

$$\sqrt{\frac{8x^3}{9y^5}} = \frac{\sqrt{4 \cdot 2\,x^2\,x}}{\sqrt{9\,y^4\,y}} \qquad \sqrt{\frac{a}{b}} = \frac{\sqrt{a}}{\sqrt{b}}$$

Note: These are perfect square factors.

Note: These are perfect square factors.

$$= \frac{2x\sqrt{2x}}{3y^2\sqrt{y}} \qquad \text{removing the perfect square factors}$$

$$= \frac{2x\sqrt{2x}}{3y^2\sqrt{y}}\left(\frac{\sqrt{y}}{\sqrt{y}}\right) \qquad \text{multiplying by a rationalizing factor}$$
$$\qquad\qquad\qquad\qquad (\sqrt{y}\sqrt{y} = \sqrt{y^2} = y)$$

$$= \frac{2x\sqrt{2xy}}{3y^2(y)}$$

$$= \frac{2x\sqrt{2xy}}{3y^3} \quad \blacksquare \qquad a^m a^n = a^{m+n}$$

⬚ Simplifying the Cube Root of a Quotient

The rules and ideas for simplifying square roots can easily be extended to cube roots. *A cube root is said to be simplified if all perfect cube factors are removed from within the radical sign and its denominator is rationalized.*

EXAMPLE 5

Simplify $\sqrt[3]{\dfrac{16x^2y^4}{z^{27}}}$.

Solution:

$\begin{cases} Note: & \text{These are} \\ \text{perfect cube factors.} \end{cases}$

$$\sqrt[3]{\frac{16x^2y^4}{z^{27}}} = \frac{\sqrt[3]{16x^2y^4}}{\sqrt[3]{z^{27}}} = \frac{\sqrt[3]{8 \cdot 2x^2 \, y^3 \, y}}{\sqrt[3]{z^{27}}} \qquad \sqrt[3]{\frac{a}{b}} = \frac{\sqrt[3]{a}}{\sqrt[3]{b}}$$

$$= \frac{2y\sqrt[3]{2x^2y}}{z^9} \quad \blacksquare \qquad \sqrt[3]{ab} = \sqrt[3]{a}\sqrt[3]{b} \text{ and } \sqrt[3]{a^m} = a^{m/3},$$
$$\text{if } m \text{ is a multiple of 3}$$

EXAMPLE 6

Simplify $\sqrt[3]{\dfrac{x^3}{24}}$.

Solution:

$$\sqrt[3]{\frac{x^3}{24}} = \frac{\sqrt[3]{x^3}}{\sqrt[3]{8 \cdot 3}} \qquad \sqrt[3]{\frac{a}{b}} = \frac{\sqrt[3]{a}}{\sqrt[3]{b}}$$

Note: This is a
perfect cube
factor.

$$= \frac{x}{2\sqrt[3]{3}} \qquad \qquad \sqrt[3]{ab} = \sqrt[3]{a}\sqrt[3]{b} \text{ and } \sqrt[3]{a^m} = a^{m/3},$$
$$\text{if } m \text{ is a multiple of 3}$$

$$= \frac{x}{2\sqrt[3]{3}} \left(\frac{\sqrt[3]{9}}{\sqrt[3]{9}} \right) \qquad \text{multiplying by a rationalizing factor}$$
$$(\sqrt[3]{3}\sqrt[3]{9} = \sqrt[3]{27} = 3)$$

Note: Be careful.
$\sqrt[3]{3}$ is *not* a
rationalizing factor.
$\sqrt[3]{3}\sqrt[3]{3} = \sqrt[3]{9}$, but
9 is *not* a perfect cube.

$$= \frac{x\sqrt[3]{9}}{2\sqrt[3]{27}}$$

$$= \frac{x\sqrt[3]{9}}{2(3)}$$

$$= \frac{x\sqrt[3]{9}}{6} \quad \blacksquare \qquad \text{simplifying}$$

EXAMPLE 7

Simplify $\sqrt[3]{\dfrac{x^8}{27y^5}}$.

Solution:

$$\sqrt[3]{\dfrac{x^8}{27y^5}} = \dfrac{\sqrt[3]{x^6\,x^2}}{\sqrt[3]{27\,y^3\,y^2}} \qquad \sqrt[3]{\dfrac{a}{b}} = \dfrac{\sqrt[3]{a}}{\sqrt[3]{b}}$$

⎰ *Note*:   This is a perfect
⎱ cube factor.

Note: These are
perfect cube
factors.

$$= \dfrac{x^2\sqrt[3]{x^2}}{3y\sqrt[3]{y^2}} \qquad \text{removing the perfect cube factors}$$

$$= \dfrac{x^2\sqrt[3]{x^2}}{3y\sqrt[3]{y^2}}\left(\dfrac{\sqrt[3]{y}}{\sqrt[3]{y}}\right) \qquad \text{multiplying by a rationalizing factor}$$
$$\qquad\qquad\qquad (\sqrt[3]{y^2}\,\sqrt[3]{y} = \sqrt[3]{y^3} = y)$$

$$= \dfrac{x^2\sqrt[3]{x^2 y}}{3y(y)}$$

$$= \dfrac{x^2\sqrt[3]{x^2 y}}{3y^2} \quad\blacksquare \qquad \text{simplifying}$$

Application

RESONANT FREQUENCY

Resonant circuits are used in electronic communication for selecting a particular
TV channel or radio station. For example, the **tuner** of a radio receiver is a simple
resonant circuit that consists of a capacitor C and an inductor L connected in
series (see Fig. 5.3).

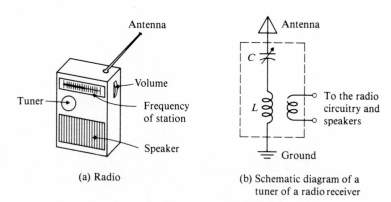

(a) Radio

(b) Schematic diagram of a
tuner of a radio receiver

Figure 5.3

The tuner is used to select a desired and specific radio signal frequency, which is called the **resonant frequency**. At the same time the tuner rejects the thousands of other radio waves that strike the antenna. The resonant frequency depends on the values of the variable capacitor C and the inductor L in the tuner.

FORMULA: **Resonant frequency**

$$f = \frac{1}{2\pi\sqrt{LC}}, \text{ where } f \text{ is the resonant frequency in Hertz (Hz)},$$
$$L \text{ is the inductance in henries (H)},$$
$$\text{and } C \text{ is the capacitance in farads (F).}$$

EXAMPLE 8

To what standard broadcast band (AM) frequency (550 kHz to 1610 kHz) are you tuned if the value of the variable capacitor is .0900μF (microfarads) and the value of the inductance is .250μH (microhenries)?

Solution:

$$.0900\mu F = .0900 \times 10^{-6}\,F = 9.00 \times 10^{-8}\,F$$

$$.250\mu H = .250 \times 10^{-6}\,H = 25.0 \times 10^{-8}\,H$$

$$f = \frac{1}{2\pi\sqrt{LC}} = \frac{1}{2\pi\sqrt{(25.0 \times 10^{-8})(9.00 \times 10^{-8})}}$$

$$= \frac{1}{2\pi\sqrt{225 \times 10^{-16}}} \qquad\qquad a^m a^n = a^{m+n}$$

$$= \frac{1}{2\pi\sqrt{225}\,\sqrt{10^{-16}}} \qquad\qquad \sqrt{ab} = \sqrt{a}\sqrt{b}$$

$$= \frac{1}{2\pi(15.0)(10^{-8})} \qquad\qquad \sqrt{a^m} = a^{m/2}$$

$$= \frac{10^8}{30.0\pi}\,Hz \qquad\qquad\qquad a^{-n} = \frac{1}{a^n}$$

$$= 1.06 \text{ MHz (megahertz)}$$

$$\text{or } 1060.\text{ kHz} \quad \blacksquare$$

EXERCISES 5.8

Simplify the following radical expressions. Assume that all variables represent positive numbers.

1. $\sqrt{\dfrac{4x^4}{9y^2}}$ **2.** $\sqrt{\dfrac{25x^{10}}{y^6}}$ **3.** $\sqrt{\dfrac{z^3}{16x^2y^{16}}}$ **4.** $\sqrt{\dfrac{8x^5y^6}{49z^2}}$

5. $\sqrt[3]{\dfrac{8x^3}{y^9}}$ **6.** $\sqrt[3]{\dfrac{27y^{27}}{64x^6}}$ **7.** $\sqrt[3]{\dfrac{8y^8}{125x^9}}$ **8.** $\sqrt[3]{\dfrac{3z^4}{8x^6y^3}}$

Simplify the following roots. Assume that all variables represent positive numbers.

9. $\sqrt{\dfrac{1}{20}}$ **10.** $\sqrt{\dfrac{1}{75}}$ **11.** $\sqrt{\dfrac{5}{48}}$ **12.** $\sqrt{\dfrac{4}{63}}$

13. $\sqrt[3]{\dfrac{1}{16}}$ **14.** $\sqrt[3]{\dfrac{1}{128}}$ **15.** $\sqrt[3]{\dfrac{8}{81}}$ **16.** $\sqrt[3]{\dfrac{3}{250}}$

17. $\sqrt{\dfrac{1}{x^3}}$ **18.** $\sqrt{\dfrac{1}{x^9}}$ **19.** $\sqrt[3]{\dfrac{1}{x^5}}$ **20.** $\sqrt[3]{\dfrac{1}{x^7}}$

Simplify the following radical expressions. Assume that all variables represent positive numbers.

21. $\sqrt{\dfrac{x^2}{8y^2}}$ **22.** $\sqrt{\dfrac{x^4}{16y^5}}$ **23.** $\sqrt{\dfrac{x^2y^2}{50z^3}}$ **24.** $\sqrt{\dfrac{5x^3}{27y^4z}}$

25. $\sqrt[3]{\dfrac{x^6}{16y^3}}$ **26.** $\sqrt[3]{\dfrac{8x^3}{27y^{10}}}$ **27.** $\sqrt[3]{\dfrac{x^3y^9}{32z^5}}$ **28.** $\sqrt[3]{\dfrac{z^4}{128xy^7}}$

29. $\sqrt{\dfrac{4+x^2}{4x^2}}$ **30.** $\sqrt[3]{\dfrac{8+x^3}{8x^3}}$ **31.** $\sqrt{9x^{-4}}$ **32.** $\sqrt{3x^{-6}}$

33. $\sqrt{44x^{-5}}$ **34.** $\sqrt{99x^{-3}y^{-2}}$ **35.** $\sqrt[3]{8x^{-5}}$ **36.** $\sqrt[3]{5x^{-27}}$

37. $\sqrt[3]{88x^{-5}}$ **38.** $\sqrt[3]{128x^{-4}y^{-9}}$ **39.** $\dfrac{\sqrt{18x^5}}{\sqrt{2x^3}}$ **40.** $\dfrac{\sqrt{24x^{11}}}{\sqrt{6x^5}}$

41. $\dfrac{\sqrt[3]{32x^{11}}}{\sqrt[3]{4x^8}}$ **42.** $\dfrac{\sqrt[3]{108x^{13}}}{\sqrt[3]{4x^4}}$

43. To what standard broadcast band frequency are you tuned if the inductance of the tuner measures .320 μH and the variable capacitor of the tuner measures .0800 μF?

44. To what standard broadcast band frequency are you tuned if the inductance of the tuner measures .320 μH and the variable capacitor of the tuner measures .180 μF?

5.9 USING SQUARE AND CUBE ROOTS TO SOLVE EQUATIONS

Objectives

☐ To solve an equation of the form $ax^2 = b$ for x.

☐ To solve an equation of the form $ax^3 = b$ for x.

☐ Equations of the Form $ax^2 = b$

Suppose you wished to solve the equation $x^2 = 25$, for x. What number(s) squared will equal 25? You know that $(5)^2 = 25$ and $(-5)^2 = 25$. Of course, the numbers 5 and -5 are the *square roots* of 25.

So if

$$x^2 = 25,$$

then

$$x = \pm \underbrace{\sqrt{25}}$$ read "plus or minus the square root of 25"

which simplifies to

$$x = \pm 5.$$ read "plus or minus 5"

Now consider the equation $x^2 = -25$. What number(s) squared will equal -25? You know that *no* real number multiplied by itself can ever be negative. Therefore, we say that the equation $x^2 = -25$ has *no real solution*.

SQUARE ROOT PROPERTY
If $x^2 = a$, and a is *positive*, then $x = \pm \sqrt{a}$.
If $x^2 = a$, and a is *negative*, then $x^2 = a$ has *no real solution*.

EXAMPLE 1

Solve $x^2 = 45$, for x.

Solution:

$$x^2 = 45$$
$$x = \pm\sqrt{45} \qquad x^2 = a, \text{ then } x = \pm\sqrt{a}$$
$$x = \pm 3\sqrt{5} \quad \blacksquare \qquad \text{simplifying radicals}$$

Note: Does $x^2 = 45$ when $x = \pm 3\sqrt{5}$?

$$(\pm 3\sqrt{5})^2 = 45?   \quad \text{replace the } x \text{ with } \pm 3\sqrt{5}$$

$$(\pm 3)^2(\sqrt{5})^2 = 45?   \quad (ab)^n = a^n b^n$$

$$\underbrace{(9)}\ \underbrace{(5)} = 45?   \quad \text{squaring}$$

$$45 = 45 \quad \text{true}$$

EXAMPLE 2

Solve $27x^2 = 3$ for x.

Solution: First, isolate the x^2 by dividing each side of the equation by 27.

$$27x^2 = 3$$

$$x^2 = \frac{3}{27}   \qquad \text{multiplying each side by } \frac{1}{27}$$

$$x^2 = \frac{1}{9}   \qquad \text{reducing}$$

$$x = \pm \sqrt{\frac{1}{9}}   \qquad x^2 = a,\ x = \pm\sqrt{a}$$

$$x = \pm \frac{1}{3}\ \blacksquare   \qquad \text{simplifying radicals}$$

Note: Check by replacing x with $\pm\dfrac{1}{3}$ in the original equation.

EXAMPLE 3

Solve $x^2 - 9 = 5x^2 + 7$, for x.

Solution: First, bring the x^2 terms together on one side of the equation.

$$x^2 - 9 = 5x^2 + 7$$

$$-9 = 4x^2 + 7   \qquad \text{adding } -x^2 \text{ to each side}$$

$$-16 = 4x^2   \qquad \text{adding } -7 \text{ to each side}$$

Note: A negative number ⟵——— $-4 = x^2$ $\qquad$ multiplying each side by $\dfrac{1}{4}$

NO REAL SOLUTION   ∎

Note: No real number can replace x and make $x^2 - 9 = 5x^2 + 7$ true.

EXAMPLE 4

Solve the area of a circle formula, $A = \pi r^2$, for the radius r. Do not rationalize the denominator.

Solution:

$$A = \pi r^2$$

$$\frac{A}{\pi} = r^2 \qquad \text{multiplying each side by } \frac{1}{\pi}$$

$$\pm\sqrt{\frac{A}{\pi}} = r \qquad x^2 = a, x = \pm\sqrt{a}$$

However, we know that the radius of a circle can never be a negative number. Therefore, only the *positive square root* is a possible answer.

Hence,

$$r = \sqrt{\frac{A}{\pi}} \quad \blacksquare$$

⬛ Equations of the Form $ax^3 = b$

Suppose you wished to solve the equation $x^3 = 8$, for x. What number(s) cubed will equal 8? You know that $(2)^3 = 8$ and that the number 2 is the *cube root* of 8.

So if

$$x^3 = 8,$$

then

$$x = \underbrace{\sqrt[3]{8}}_{},$$

which simplifies to

$$x = 2.$$

Next, consider the equation $x^3 = -8$. What number(s) cubed will equal -8? You know that $(-2)^3 = -8$, and that the number -2 is the *cube root* of -8.

So if

$$x^3 = -8,$$

then

$$x = \underbrace{\sqrt[3]{-8}}_{},$$

which simplifies to

$$x = -2.$$

CUBE ROOT PROPERTY

If $x^3 = a$ and a is *any* real number, then $x = \sqrt[3]{a}$.

It is important that you remember the following:

1. If a is *positive*, then the equation $x^2 = a$ has *two* solutions ($x = \pm\sqrt{a}$), while the equation $x^3 = a$ has only *one* solution ($x = \sqrt[3]{a}$).

2. If a is *negative*, then the equation $x^2 = a$ has *no* real solution, while the equation $x^3 = a$ still has *one* solution ($x = \sqrt[3]{a}$).

EXAMPLE 5

Solve $x^3 = 24$, for x.

Solution:

$$x^3 = 24$$
$$x = \sqrt[3]{24} \qquad x^3 = a, \text{ then } x = \sqrt[3]{a}$$
$$x = 2\sqrt[3]{3} \quad \blacksquare \qquad \text{simplifying radicals}$$

Note: Does $x^3 = 24$ when $x = 2\sqrt[3]{3}$?

$$(2\sqrt[3]{3})^3 = 24? \qquad \text{replace the } x \text{ with } 2\sqrt[3]{3}$$
$$(2)^3(\sqrt[3]{3})^3 = 24? \qquad (ab)^n = a^n b^n$$
$$\underbrace{(8)}\quad\underbrace{(3)} \quad = 24? \qquad \text{cubing}$$
$$24 = 24 \quad \text{true}$$

EXAMPLE 6

Solve the equation $6 + 5x^3 = 3x^3 - 48$ for x.

Solution:

$$6 + 5x^3 = 3x^3 - 48$$
$$6 + 2x^3 = -48 \qquad \text{adding } -3x^2 \text{ to each side}$$
$$2x^3 = -54 \qquad \text{adding } -6 \text{ to each side}$$
$$x^3 = -27 \qquad \text{multiplying each side by } \frac{1}{2}$$
$$x = \sqrt[3]{-27} \qquad x^3 = a, \text{ then } x = \sqrt[3]{a}$$
$$x = -3 \quad \blacksquare \qquad \text{simplifying radicals}$$

Note: Check by replacing x with -3 in the original equation.

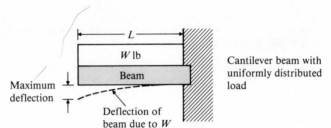

Cantilever beam with
uniformly distributed
load

Figure 5.4

EXAMPLE 7

The maximum deflection y of a cantilever beam carrying a uniformly distributed load of W pounds is given by the formula $y = \dfrac{WL^3}{8EI}$, where E is the modulus of elasticity and I is the moment of inertia for the beam. Solve for the length of the beam L (see Fig. 5.4). Do not rationalize the denominator.

Solution:

$$y = \frac{WL^3}{8EI}$$

$$\frac{8EIy}{W} = L^3 \qquad \text{multiplying each side by } \frac{8EI}{W}$$

$$\sqrt[3]{\frac{8EIy}{W}} = L \qquad x^3 = a, \text{ then } x = \sqrt[3]{a}$$

$$2\sqrt[3]{\frac{EIy}{W}} = L \quad \blacksquare \qquad \text{simplifying radicals}$$

EXERCISES 5.9

Solve the following equations for x. Write each answer in simplified radical form. Be sure to check your answer.

1. $x^2 = 49$
2. $x^2 = 121$
3. $x^3 = 64$
4. $x^3 = 216$
5. $x^2 = 75$
6. $x^2 = 72$
7. $x^3 = 56$
8. $x^3 = 108$
9. $3x^2 = 144$
10. $5x^2 = 250$
11. $4x^3 = 216$
12. $9x^3 = 288$
13. $-5x^2 = 125$
14. $2x^2 = -64$
15. $-5x^3 = 625$
16. $3x^3 = -81$
17. $2x^2 - 25 = x^2 - 9$
18. $3x^2 - 27 = 5 - x^2$
19. $36x^2 - 3 = 3$
20. $32x^2 - 7 = 4x^2$
21. $4x^3 - 5 = 3x^3 - 13$
22. $x^3 + 9 = 7x^3 - 153$
23. $40x^3 - 9 = 11$
24. $16x^3 + 9 = 8$
25. $6x^2 - 5 = 3x^2 - 9$
26. $x^3 - 8 = 2x^3 - 8$

Solve each formula for the indicated quantity. Do not rationalize any denominator.

27. Area of a square, $A = s^2$. Solve for the side, s.

28. Volume of a cube, $V = e^3$. Solve for the edge, e.

29. Volume of a sphere, $V = \dfrac{4}{3}\pi r^3$. Solve for the radius, r.

30. Volume of a cylinder, $V = \pi r^2 h$. Solve for the radius, r.

31. Electrical power, $P = \dfrac{V^2}{R}$. Solve for the voltage, V.

32. S.A.E. horsepower, H.P. $= \dfrac{D^2 N}{2.5}$. Solve for the diameter of the cylinder (the bore), D.

33. Maximum deflection of a simply supported beam carrying a uniformly distributed load, $y = \dfrac{5WL^3}{384EI}$. Solve for the length of the beam, L. (See Fig. 5.5).

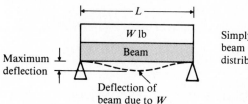

Figure 5.5

34. Maximum deflection of a simply supported beam with a concentrated load at midspan, $y = \dfrac{FL^3}{48EI}$. Solve for the length of the beam, L. (see Fig. 5.6).

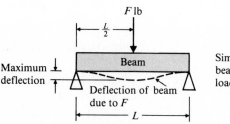

Figure 5.6

35. Referring to exercise 33, what is the length (in feet) of the longest $8'' \times 8''$ steel beam that can be used if the maximum deflection is not to exceed $\frac{1}{2}''$?

Assume that the modulus of elasticity is $E = 2.9 \times 10^7$ psi, the moment of inertia for an $8'' \times 8''$ beam is $I = 110.$ in^4, and the uniformly distributed load is $W = 19,500.$ lb.

36. Referring to exercise 34, what is the length (in feet) of the longest $8'' \times 8''$ steel beam that can be used if the maximum deflection is not to exceed $\frac{1}{2}''$?

Assume that the modulus of elasticity is $E = 2.9 \times 10^7$ psi, the moment of inertia for an $8'' \times 8''$ beam is $I = 110.$ in^4, and the concentrated load is $F = 19,500.$ lb.

Chapter Review

REVIEW
EXERCISES

The following review exercises are grouped according to the objectives that should have been mastered in Chapter 5. Work each problem carefully. If any weaknesses appear, immediately refer to and read the subsection that matches that objective.

5.1 PRODUCTS AND EXPONENTS

Objectives

☐ To simplify products of the form $a^m a^n$, where m and n are positive integers.

Simplify.

1. $y^8 y^5$

2. $(xy^7)(x^3 y^4)$

3. $(-2x^4 yz^6)(-3xz^4)(-x^6 y)$

☐ To simplify algebraic expressions of the form $(a^m)^n$, where m and n are positive integers.

Simplify.

4. $(y^8)^5$

5. $((2-x)^2)^4$

☐ To simplify products of the form $(a^m b^n)^k$, where m, n, and k are positive integers.

Simplify.

6. $(-2y)^6$

7. $(-3x^3 y^2 z^5)^3$

8. $(2x^3 y)^4 (-4xy^5)^2$

5.2 QUOTIENTS AND EXPONENTS

Objectives

☐ To simplify quotients of the form $\dfrac{a^m}{a^n}$, where m and n are positive integers and $m = n$.

Simplify.

9. $\dfrac{15x^6 y^2}{12x^6 y^2}$

☐ To simplify quotients of the form $\dfrac{a^m}{a^n}$, where m and n are positive integers and $m > n$.

Simplify.

10. $\dfrac{x^9 y^9}{xy^7}$

11. $\dfrac{18x^8 y^3 z^5}{32x^6 y^3 z^4}$

☐ To simplify quotients of the form $\dfrac{a^m}{a^n}$, where m and n are positive integers and $m < n$.

Simplify.

12. $\dfrac{xy^5}{x^4 y^6}$

13. $\dfrac{16x^4 y^5 z^6}{24x^5 yz^6}$

☐ To simplify quotients of the form $\left(\dfrac{a^m}{a^n}\right)^k$, where m, n, and k are positive integers.

Simplify.

14. $\left(\dfrac{-4}{y^4}\right)^2$

15. $\left(\dfrac{-x^6 y}{2x^2 y^4}\right)^3$

5.3 ZERO AND NEGATIVE EXPONENTS

Objectives

☐ To simplify algebraic expressions of the form a^0.

Simplify.   **16.** $(5^0 + 3^2)^2$

☐ To simplify algebraic expressions of the form a^{-n} and $\dfrac{1}{a^{-n}}$.

Simplify, leaving an
answer with all positive
exponents.

17. $\dfrac{1}{3^{-2}} + 3^{-2}$ **18.** $8^{-1}x^0y^{-3}$

19. $\dfrac{x^{-4}}{3y^{-2}}$ **20.** $\dfrac{6x^{-2}y^3}{8xy^{-3}}$

☐ To simplify algebraic expressions containing integral exponents.

Simplify, leaving an
answer with all positive
exponents.

21. $x^5y^{-4}x^{-6}y^{-1}$ **22.** $((x-1)^{-2})^3$

23. $(-2x^2y^{-3}z^0)^4$ **24.** $\dfrac{(x+y)^{-2}}{(x+y)^{-8}}$

25. $\dfrac{4^{-1}x^{-3}y}{4x^{-6}y^{-2}}$

5.4 SCIENTIFIC NOTATION

Objectives

☐ Given a number in scientific notation, to write it in decimal notation.

Write in decimal
notation.

26. 3.86×10^8 **27.** 7.30×10^{-5}

☐ Given a number in decimal notation, to write it in scientific notation.

Write in scientific
notation.

28. 220,000,000   **29.** .00000560

5.5 CALCULATING WITH SCIENTIFIC NOTATION

Objectives

☐ To multiply and divide numbers using scientific notation.

Perform the indicated
operation, writing your
answer in scientific
notation.

30. (624,000,000)(.000000573)

31. $\dfrac{.0000000775}{1,600,000,000}$

⬚ To add and subtract numbers using scientific notation.

Perform the indicated operation, writing your answer in scientific notation.

32. $(1.27 \times 10^{-8}) + (2.9 \times 10^{-9})$

33. $(7{,}395{,}000{,}000) - (5.0 \times 10^{7})$

5.6 ROOTS AND RADICALS

Objectives

⬚ To find the square root of a number or a perfect square variable expression.

Simplify.

34. a) $\sqrt{64}$ **b)** $-\sqrt{25}$ **c)** $\pm\sqrt{36}$

     d) $\sqrt{.09}$ **e)** $\sqrt{\dfrac{1}{9}}$

Using a calculator, approximate to four significant digits.

35. a) $\sqrt{3}$ **b)** $-\sqrt{3}$

Simplify.

36. a) $\sqrt{x^4}$ **b)** $\sqrt{y^{10}}$ **c)** $\sqrt{z^{36}}$

⬚ To find the cube root (or higher root) of a number or a perfect cube variable expression.

Simplify.

37. a) $\sqrt[3]{-64}$ **b)** $\sqrt[6]{-64}$

Using a calculator, approximate to four significant digits.

38. $\sqrt[3]{-16}$

Simplify.

39. a) $\sqrt[3]{x^9}$ **b)** $\sqrt[3]{x^{81}}$

5.7 THE SQUARE ROOT AND CUBE ROOT OF A PRODUCT

Objectives

⬚ To simplify the square root of a product.

Simplify.

40. $\sqrt{16x^6}$ **41.** $\sqrt{16 + x^6}$ **42.** $\sqrt{25x^8y^2z^{64}}$

43. $\sqrt{108}$ **44.** $\sqrt{x^{11}}$ **45.** $\sqrt{27x^{10}yz^7}$

46. $\sqrt{32xy^5}\sqrt{2x^3y}$

⬚ To simplify the cube root of a product.

Simplify.

47. $\sqrt[3]{64x^{12}y^3}$ **48.** $\sqrt[3]{54}$

49. a) $\sqrt[3]{x^{10}}$

     b) $\sqrt[3]{x^{11}}$

50. $\sqrt[3]{125x^{13}y^{12}z^{14}}$ **51.** $\sqrt[3]{32xy^5}\sqrt[3]{2x^3y}$

5.8 THE SQUARE ROOT AND CUBE ROOT OF A QUOTIENT

Objectives

⊡ To simplify the square root of a quotient.

Simplify.

52. $\sqrt{\dfrac{36x^2}{49y^6}}$ **53.** $\sqrt{\dfrac{99z^5}{81x^4y^{12}}}$ **54.** $\sqrt{\dfrac{x^{10}}{28}}$ **55.** $\sqrt{\dfrac{25x^5}{8y^7}}$

⊡ To simplify the cube root of a quotient.

Simplify.

56. $\sqrt[3]{\dfrac{81x^7y^2}{8z^9}}$ **57.** $\sqrt[3]{\dfrac{x^{27}}{128}}$ **58.** $\sqrt[3]{\dfrac{x^4}{64y^8}}$

5.9 USING SQUARE AND CUBE ROOTS TO SOLVE EQUATIONS

Objectives

⊡ To solve an equation of the form $ax^2 = b$ for x.

Solve.

59. $x^2 = 63$

60. $48x^2 = 3$

61. $2x^2 - 5 = 5x^2 + 22$

62. $P = I^2R$, for I (Do not rationalize the denominator.)

⊡ To solve an equation of the form $ax^3 = b$ for x.

Solve.

63. $x^3 = 40$

64. $6x^3 - 7 = 3x^3 - 31$

65. $y = \dfrac{WL^3}{120EI}$, for L. (Do not rationalize the denominator.)

CHAPTER TEST

Time
50 minutes

Score
30–27 excellent
26–24 good
23–18 fair
below 17 poor

If you have worked through the Review Exercises and corrected any weaknesses, then you are ready to take the following Chapter Test.

Simplify the following. Write all answers with positive exponents only.

1. x^4x^{-5}

2. $\dfrac{x^4}{x^{-5}}$

3. $(x - 1)^{-2}(x - 1)^{-1}$

4. $((x + y)^{-6})^0$

5. $(-2r^3st^4)^3$

6. $(-3x^{-1}y)^{-2}$

7. $(3xy^3z^5)(-4xy^4z^{-6})$

8. $\dfrac{42x^{-3}y}{7xy^{-1}}$

9. $\left(\dfrac{-3x^{-2}}{y^{-6}}\right)^3$

10. $(5^{-1} + 3^{-1})^2$

Perform the indicated operations using scientific notation. Write your answer in scientific notation.

11. $\dfrac{(6 \times 10^6)(3 \times 10^{-14})}{(3 \times 10^{-8})^2}$

12. $(5.74 \times 10^{12}) + (6.23 \times 10^{11})$

Simplify the following radical expressions. Assume that all variables represent positive numbers.

13. $\sqrt{60}$

14. $\sqrt[3]{72}$

15. $\sqrt{\dfrac{1}{20}}$

16. $\sqrt[3]{\dfrac{1}{48}}$

17. $\sqrt[3]{(x^3 + 8)^6}$

18. $\sqrt{40h^5}$

19. $\sqrt[3]{250v^5z^7}$

20. $\sqrt{12x^5}\sqrt{3x}$

21. $\sqrt{\dfrac{4x^4}{16y^{16}}}$

22. $\sqrt[3]{\dfrac{x^3}{32}}$

23. $\sqrt{\dfrac{45t^4}{24u^9}}$

24. $\sqrt[3]{8x^{-3}}$

25. $\sqrt[3]{\dfrac{27 + x^3}{27x^3}}$

26. $\sqrt{4^{-1} + 9^{-1}}$

Solve the following equations. Write your answer in simplified radical form.

27. $11x^2 - 35 = 6x^2 + 40$

28. $6x^3 + 1 = -2x^3$

29. $R = \dfrac{\alpha l}{d^2}$ for d, where d is the diameter of a wire. (Do not rationalize the denominator.)

30. $I = \dfrac{bh^3}{12}$ for h, where h is the height of a rectangle. (Do not rationalize the denominator.)

POLYNOMIALS
AND
FACTORING

6.1   ADDING AND SUBTRACTING POLYNOMIALS

Objectives

☐ To identify different types of polynomials.

☐ To add polynomials.

☐ To subtract polynomials.

☐ Polynomials

An algebraic expression such as $3x^5y + x^3y^2 - 2x - 5y^4 + y - 4$ is an example of a **polynomial**. Polynomials consist of a finite number of **terms**. Each term of a polynomial is separated from the next term by an addition sign or a subtraction sign. The polynomial

$$\underbrace{3x^5y}_{\substack{\text{first} \\ \text{term}}} + \underbrace{x^3y^2}_{\substack{\text{second} \\ \text{term}}} - \underbrace{2x}_{\substack{\text{third} \\ \text{term}}} - \underbrace{5y^4}_{\substack{\text{fourth} \\ \text{term}}} + \underbrace{y}_{\substack{\text{fifth} \\ \text{term}}} - \underbrace{4}_{\substack{\text{sixth} \\ \text{term}}}$$

consists of six terms. Notice that the terms of a polynomial can simply be constants (like the sixth term) or the *product* of a constant and one or more variables raised to *positive* integer powers (like the first through fifth terms).

Usually the numerical part of a term is referred to as its (numerical) coefficient. For example, 3 is the coefficient of the first term, and 1 is understood to be the coefficient of the second and fifth terms. The third, fourth, and sixth terms all have *negative* coefficients.

Some polynomials are given special names. A polynomial that contains only *one term* is called a **monomial**. A polynomial that contains *two* terms is called a **binomial**, and a polynomial that contains *three* terms is called a **trinomial**.

EXAMPLE 1

Identify each of the following as either a monomial, binomial, or trinomial.

a) $5xy^2z^3$

b) $5x - y^2 + z^3$

c) $5xy^2 + z^3$

d) $5 - x + y^2 - z^3$

e) $5xy^2z^{-3}$

Solution:

a) $5xy^2z^3 \rightarrow$ monomial (one term)   ■

b) $5x - y^2 + z^3 \rightarrow$ trinomial (three terms)   ■

c) $5xy^2 + z^3 \rightarrow$ binomial (two terms)   ■

d) $5 - x + y^2 - z^3 \rightarrow$ None of these. This is a *four*-term polynomial.   ■

e) $5xy^2z^{-3} \rightarrow$ None of these. This is *not* a polynomial, since polynomials do not contain variables with negative powers.   ■

⊡ Adding Polynomials

To add two polynomials, we simply add the similar terms of the polynomials. Remember, similar terms contain exactly the same variables and exactly the same exponents for these variables.

EXAMPLE 2

Simplify $(5x^5 - 3x^2) + (1 + 4x^2 - 6x^5)$.

Solution:

$$
\underbrace{(5x^5 - 3x^2) + (1 + 4x^2 - 6x^5)}_{\text{similar terms}} = 1 + x^2 - x^5 \quad\blacksquare \qquad \text{combining similar terms}
$$

Note: The order in which you write the terms is immaterial. Thus, $1 + x^2 - x^5 = x^2 - x^5 + 1 = -x^5 + x^2 + 1$, and so on.

⊡ Subtracting Polynomials

In Chapter 1 you learned that the opposite of a sum is the sum of the opposites,

$$-(a + b) = -a + (-b),$$

and you saw how a minus sign distributes over a sum. We can extend this rule from $-(a + b) = -a + (-b)$ to

$$-(a_1 + a_2 + a_3 + \cdots + a_n) = -a_1 + (-a_2) + (-a_3) + \cdots + (-a_n).$$

Simply remember to distribute the minus sign to each term of the polynomial.

Since subtraction is defined as adding the opposite, we can employ this rule when subtracting polynomials.

EXAMPLE 3

Simplify $(2xy^2 + 4y) - (5x^2y - 3y^2x + 4y)$.

Solution:

$$(2xy^2 + 4y) - (5x^2y - 3y^2x + 4y)$$

$$= 2xy^2 + 4y - 5x^2y + 3y^2x - 4y \qquad \begin{array}{l} -(a + b) = -a + (-b) \\ \text{distributing a negative} \end{array}$$

Note: $3y^2x = 3xy^2$, since multiplication is commutative

$$= 5xy^2 - 5x^2y \quad\blacksquare \qquad \text{combining similar terms}$$

If a problem contains more than one set of parentheses, start by simplifying the *innermost* parentheses first.

EXAMPLE 4

Simplify $3x - (5y - (3x - y))$.

Solution:

$$3x - (5y - (3x - y)) = 3x - (5y - 3x + y) \qquad -(a + b) = -a + (-b)$$

Note: Start here

$$= 3x - (6y - 3x) \qquad \text{combining similar terms}$$
$$= 3x - 6y + 3x \qquad -(a + b) = -a + (-b)$$
$$= 6x - 6y \quad \blacksquare \qquad \text{combining similar terms}$$

EXERCISES 6.1

Identify each of the following as either a monomial, binomial, or trinomial. If none of these, explain why.

1. $3xy^2 - zt^3$

2. $7a^2b^3 - c$

3. $3xy^2zt^3$

4. $7a^2b^3c$

5. $3x + y^2 - zt^3$

6. $7 - a^2b^3 + c$

7. $3 + x + y^2 - z + t^3$

8. $7 - a^2 + b^3 - c$

9. $3xy^{-2}zt^{-3}$

10. $\dfrac{7a^2b^3}{c}$

Simplify the following.

11. $x^3 + 4x^2y + xy^2 + 2yx^2 + 4xy^2 + 2y^3$

12. $27a^3 - 18a^2b + 3ab^2 - 9ba^2 + 6b^2a - b^3$

13. $(5x^2 - 3x + 4) + (4 - 3x - 5x^2)$

14. $(7m^2n^3 - 3mn + n^2m) + (3nm - 4mn^2)$

15. $(-p^4q - 4pq + 7pq^4) - (3pq + q^4p - qp^4)$

16. $-(rs + r^2s + 3) - (-6 + sr^2 + 3rs)$

17. $(x^2y^2 + 3xy) - (x^2y^2 + 2yx) + (-xy - x^2y)$

18. $-(8t + 2t^2) - (5t^2 + 3t) - (3 - 6t - 7t^2)$

19. $4 - x - (3 + (7 - 2x))$

20. $4mn^2 - (8mn - (1 - 5nm + 4m^2n))$

21. $-(xy - 3) - (yx + (4 - 3xy))$

22. $-(t^3 - (3 - t^2 + t^3)) - (3t^2 - (t^2 - 3))$

23. $4x - (3 - (2x + (2 - 6x)))$

24. $-(3x^2y - xy - (4yx^2 - (yx + 2x^2y)))$

6.2   MULTIPLYING POLYNOMIALS

Objectives

☐ To multiply a monomial by a polynomial.

☐ To multiply two binomials.

☐ To multiply two polynomials.

☐ The Product of a Monomial and a Polynomial

The distributive law, $a(b + c) = ab + ac$, and the product rule for exponents, $a^m a^n = a^{m+n}$, can be used together to multiply a monomial by a binomial.

EXAMPLE 1

Find the product $-3x^3(7 - 5x^4)$.

Solution:

$$-3x^3(7 - 5x^4) = (-3x^3)(7) + (-3x^3)(-5x^4) \qquad \text{distributive law}$$
$$= -21x^3 + 15x^7 \quad \blacksquare \qquad\qquad a^m a^n = a^{m+n}$$

We can extend the distributive law from $a(b + c) = ab + ac$ to

$$a(b_1 + b_2 + b_3 + \cdots + b_n) = ab_1 + ab_2 + ab_3 + \cdots + ab_n.$$

Simply remember to distribute the monomial to each term of the polynomial.

EXAMPLE 2

Find the product $xy^4(3xy^7 - 2y^5 + x - 1)$.

Solution:

$$xy^4(3xy^7 - 2y^5 + x - 1)$$
$$= 3x^2y^{11} - 2xy^9 + x^2y^4 - xy^4 \quad \blacksquare \qquad \begin{array}{l}\text{distributive law} \\ \text{and } a^m a^n = a^{m+n}\end{array}$$

To simplify expressions that combine the operations of addition, subtraction, and multiplication, be sure to always *multiply first* and then combine any similar terms that might appear.

EXAMPLE 3

Simplify $4xy^2 - 3xy^2(2x + 1)$.

Solution: You would violate the order of operations if you first subtracted $3xy^2$ from $4xy^2$ and then multiplied. *You must multiply first* by distributing $-3xy^2$ to the binomial $2x + 1$.

$$4xy^2 - 3xy^2(2x + 1) = 4xy^2 + (-3xy^2)(2x) + (-3xy^2)(1) \qquad \text{distributive law}$$
$$= 4xy^2 - 6x^2y^2 - 3xy^2 \qquad\qquad\qquad a^m a^n = a^{m+n}$$
$$\underbrace{\qquad\qquad}_{\substack{\text{similar} \\ \text{terms}}}$$

$$= xy^2 - 6x^2y^2 \quad \blacksquare \qquad\qquad \text{combining similar terms}$$

⊡ Multiplying Two Binomials

Suppose a rectangle has a length of $a + b$ and a width of $c + d$ (see Fig. 6.1).

Its area A can then be expressed as

$$A = lw = (a + b)(c + d).$$

The algebraic expression $(a + b)(c + d)$ represents *the product of two binomials.* How might we find the product of these two binomials?

Let's split the rectangular area in Fig. 6.1 into four smaller rectangular areas as shown in Fig. 6.2.

The area of each of these smaller rectangles is

$$A_1 = ac$$
$$A_2 = ad$$
$$A_3 = bc$$
$$A_4 = bd.$$

Certainly, the *sum* of these four areas, $ac + ad + bc + bd$, must be equal to the total area, $(a + b)(c + d)$.

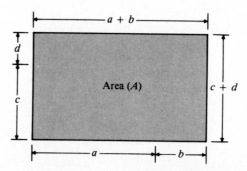

Figure 6.1

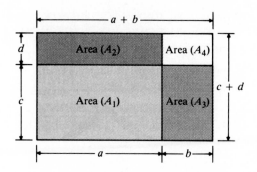

Figure 6.2

THE PRODUCT OF TWO BINOMIALS

$$(a + b)(c + d) = ac + ad + bc + bd$$

Algebraically, we can find the product of two binomials as follows:

$$(a + b)(c + d) = a(c + d) + b(c + d)$$

distributing each term
of the first binomial
to the second binomial

$$= ac + ad + bc + bd$$

distributing again twice

Or more quickly, we can multiply each term of the first binomial by each term of the second binomial as follows:

$$(a + b)(c + d) = ac + ad + bc + bd$$

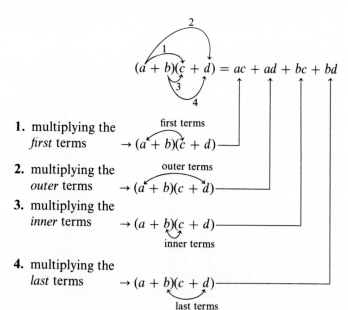

1. multiplying the
   first terms

   first terms

   $\rightarrow (a + b)(c + d)$

2. multiplying the
   outer terms

   outer terms

   $\rightarrow (a + b)(c + d)$

3. multiplying the
   inner terms

   $\rightarrow (a + b)(c + d)$

   inner terms

4. multiplying the
   last terms

   $\rightarrow (a + b)(c + d)$

   last terms

This method of multiplying the first (F), outer (O), inner (I), and last (L) terms of the binomials is called the FOIL method. The FOIL method provides a quick and easy way to find the product of any two binomials.

THE PRODUCT OF TWO BINOMIALS—FOIL METHOD

$$(a + b)(c + d) = ac + ad + bc + bd$$
$$\qquad\qquad\qquad\quad\text{F}\quad\ \text{O}\quad\ \text{I}\quad\ \text{L}$$

EXAMPLE 4

Find the product $(x + 7y)(4x + 3z)$.

Solution:

$$(x + 7y)(4x + 3z) = (x)(4x) + (x)(3z) + (7y)(4x) + (7y)(3z)$$
$$\qquad\qquad\qquad\qquad\ \text{F}\qquad\quad\ \text{O}\qquad\quad\ \text{I}\qquad\quad\ \text{L}$$
$$= 4x^2 + 3xz + 28xy + 21yz \quad \blacksquare$$

EXAMPLE 5

Find the product $(3m + 2n^2)(4 - 8n)$.

Solution:

$$(3m + 2n^2)(4 - 8n) = (3m)(4) + (3m)(-8n) + (2n^2)(4) + (2n^2)(-8n)$$
$$\qquad\qquad\qquad\qquad\quad\ \text{F}\qquad\quad\ \text{O}\qquad\qquad\ \text{I}\qquad\quad\ \text{L}$$
$$= 12m - 24mn + 8n^2 - 16n^3 \quad \blacksquare$$

Similar terms may appear after you multiply two binomials. If they do, be sure to combine them into a single term.

EXAMPLE 6

Find the product $(2t - 5)(t + 3)$.

Solution:

$$(2t - 5)(t + 3) = (2t)(t) + (2t)(3) + (-5)(t) + (-5)(3)$$

$$= 2t^2 + 6t - 5t - 15$$

similar terms

$$= 2t^2 + t - 15 \quad \blacksquare \qquad \text{combining similar terms}$$

Note: The product of these two binomials is a *trinomial*, since the outer and inner products combine into a single term.

EXAMPLE 7

Find the product $(x^2 - 9)(x^2 - 4)$.

Solution:

$$(x^2 - 9)(x^2 - 4) = (x^2)(x^2) + (x^2)(-4) + (-9)(x^2) + (-9)(-4)$$

$$= x^4 - 4x^2 - 9x^2 + 36$$

$$= x^4 - 13x^2 + 36 \quad \blacksquare \qquad \text{combining similar terms}$$

EXAMPLE 8

Simplify $5x - (3 - 2x)(2 + 3x)$.

Solution: First, find the product $(3 - 2x)(2 + 3x)$. Then subtract this product from $5x$.

$$(3 - 2x)(2 + 3x) = 6 + 9x - 4x - 6x^2 = 6 + 5x - 6x^2$$

Thus,

$$5x - \overline{(3 - 2x)(2 + 3x)} = 5x - (6 + 5x - 6x^2).$$ FOIL

Note: Retain this set of parentheses. It will help you remember to subtract *all* the terms of the trinomial.

$$= 5x \ominus 6 \ominus 5x \oplus 6x^2 \qquad -(a + b) = -a + (-b)$$
$$= 6x^2 - 6 \ \blacksquare \qquad \text{combining similar terms}$$

⫶ The Product of Two Polynomials

To multiply two binomials, we used the FOIL method and multiplied each term of the first binomial by each term of the second binomial. This procedure can be extended to the product of two polynomials.

THE PRODUCT OF TWO POLYNOMIALS
To multiply two polynomials, multiply each term of the first polynomial by each term of the second polynomial. Be sure to combine any similar terms.

EXAMPLE 9

Find the product $(x + y)(x^2 - xy + y^2)$.

Solution: There are actually *six* multiplications to perform.

$$(x + y)(x^2 - xy + y^2) = x^3 - x^2y + xy^2 + x^2y - xy^2 + y^3$$
$$= x^3 + y^3 \ \blacksquare \qquad \text{combining similar terms}$$

We could also multiply $(x + y)(x^2 - xy + y^2)$ by arranging the polynomials *vertically* and multiplying as follows:

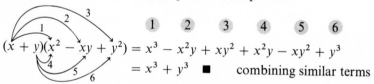

Multiplying each term of $(x^2 - xy + y^2)$ by x.

Multiplying each term of $(x^2 - xy + y^2)$ by y and aligning any similar terms that develop.

$$x^2 - xy + y^2$$
$$x + y$$
$$x^3 - x^2y + xy^2$$
aligning similar terms
$$x^2y - xy^2 + y^3$$
$$\underbrace{x^3 + 0 + 0 + y^3}_{\text{combining similar terms}} = x^3 + y^3$$

The advantage of the *vertical* method is that you group similar terms at the same time you perform the multiplications.

EXAMPLE 10

Use the vertical method to find the product $(x^2 + 2x + 3)(x^2 - 2x + 1)$.

Solution: There are *nine* multiplications to perform.

$$
\begin{array}{r}
x^2 + 2x + 3 \\
x^2 - 2x + 1 \\
\hline
\end{array}
$$

multiplying each term of $(x^2 + 2x + 3)$ by x^2 $\longleftarrow x^4 + 2x^3 + 3x^2$

aligning similar terms

multiplying each term of $(x^2 + 2x + 3)$ by $-2x$ $\longleftarrow \quad -2x^3 - 4x^2 - 6x$

aligning similar terms

multiplying each term of $(x^2 + 2x + 3)$ by 1 $\longleftarrow \qquad\qquad x^2 + 2x + 3$

$$x^4 + 0 + 0 - 4x + 3$$

combining similar terms

Thus, $(x^2 + 2x + 3)(x^2 - 2x + 1) = x^4 - 4x + 3$ ∎

EXERCISES 6.2

Perform the indicated operations. Be sure to combine any similar terms.

1. $4p^2(2p^3)$
2. $x^2y(-3xy^2)$
3. $4p^2(2 + p^3)$
4. $x^2y(-3 - xy^2)$
5. $4 + p^2(2 + p^3)$
6. $x^2 + y(-3 - xy^2)$
7. $(4 + p^2)(2 + p^3)$
8. $(x^2 + y)(-3 - xy^2)$
9. $-6t(t^3 - 4t^2 + 4t - 1)$
10. $-pq^3(q - 3p - 5p^3)$
11. $4xy^3 - 7xy(4 - x^2)$
12. $-6x^4 + 2x^2(7x^3 + 3x^2)$
13. $(4xy^3 - 7xy)(4 - x^2)$
14. $(-6x^4 + 2x^2)(7x^3 + 3x^2)$
15. $3p^2 + 4p^2(p^3 - 2)$
16. $7t^2u - 4t^2u(2 - t^2u)$
17. $(3p^2 + 4p^2)(p^3 - 2)$
18. $(7t^2u - 4t^2u)(2 - t^2u)$
19. $(x + 2)(x + 7)$
20. $(y + 3)(y + 5)$
21. $2(x^3 - 4) - 2x^2(x - 3)$
22. $5m^5(3m^4 + n) - 5n(m^5 - 6)$
23. $(4 - t)(2 + t)$
24. $(5 + 2x)(1 - 3x)$
25. $-9a^2b(-3ab^2)$
26. $-5x^3y^5(-x^2y^4)$
27. $x^4(3y - x^2) - (3x^4y - x^6)$
28. $8y^3(4y - y^4) - (-y^7 + y^4)$
29. $(3y - 5)(2y + 3)$
30. $(7t + u)(4t - u)$
31. $x - 4(x^2 - x + 2)$
32. $2x - 3y(x^2 - xy + y^2)$

33. $(x - 4)(x^2 - x + 2)$
34. $(2x - 3y)(x^2 - xy + y^2)$

35. $-(3x - 2)(x - 1)$
36. $-(2 + y)(5 - y)$

37. $4t - (6t + 1)(3t + 2)$
38. $t^2 - (3 + t)(7 + 2t)$

39. $(5 - m - 3m^2)(-6m^2)$
40. $(7 - m - m^2)(-2m^2)$

41. $(x^2 - 2x - 1)(x - 3)$
42. $(x^2 - 3x + 5)(x + 2)$

43. $11x + 24 - (x + 3)(x + 8)$
44. $6x^2 + x - (2x - 1)(3x + 2)$

45. $9(x - 2) + (x - 6)(x - 3)$
46. $-x(x + y) + (x + 3y)(x - 2y)$

47. $(x^2 + x + 3)(x^2 - x + 2)$
48. $(2x^2 - x + 3)(3x^2 + 4x - 1)$

49. $(x - 2)(3x^3 - x^2 - 2x - 1)$
50. $(x^2 + 2x - 3)(x^3 - x^2 + 3x + 4)$

51. $-x^3 - x(6x - 2x(x + 3))$
52. $-11a^2b - b(-18a + 6a(3 - 2a))$

53. $y^2 - (x(y + 2x) - (x + y)(2x - y))$
54. $t^2 - (s(3t + 4s) + (t - 4s)(t + s))$

6.3 SPECIAL PRODUCTS

Objectives

- ⊡ To multiply two binomials of the form $(a + b)(a - b)$.
- ⊡ To square a binomial.
- ⊡ To multiply three polynomials.

⊡ Products of the Form $(a + b)\,(a - b)$

A special product that occurs quite frequently is the product $(a + b)(a - b)$. The terms of the binomials are exactly the same, but one binomial contains the operation of addition, while the other contains the operation of subtraction.

The product can be found by using the FOIL method as follows:

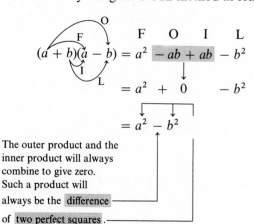

$$
\begin{array}{cccc}
 & \text{F} & \text{O} & \text{I} & \text{L} \\
(a + b)(a - b) = a^2 & - ab & + ab & - b^2 \\
= a^2 & + & 0 & - b^2 \\
= a^2 & - b^2 &
\end{array}
$$

The outer product and the inner product will always combine to give zero. Such a product will always be the difference of two perfect squares.

> **SPECIAL PRODUCT OF TWO BINOMIALS: The Difference of Two Perfect Squares**
>
> $$(a + b)(a - b) = (a - b)(a + b) = a^2 - b^2$$

EXAMPLE 1

Find the product $(3x + 4)(3x - 4)$.

Solution:

$$
\begin{aligned}
(3x + 4)(3x - 4) &= (3x)^2 - (4)^2 && (a + b)(a - b) = a^2 - b^2 \\
&= 3^2 x^2 - (4)^2 && (ab)^n = a^n b^n \\
&= 9x^2 - 16 \quad \blacksquare
\end{aligned}
$$

Note: Try this same problem by using the FOIL method. See if you can obtain the same result.

EXAMPLE 2

Find the product $(5y^3 - z^4)(5y^3 + z^4)$.

Solution:

$$
\begin{aligned}
(5y^3 - z^4)(5y^3 + z^4) &= (5y^3)^2 - (z^4)^2 && (a - b)(a + b) = a^2 - b^2 \\
&= 5^2 y^6 - z^8 && (ab)^n = a^n b^n \text{ and } (a^m)^n = a^{mn} \\
&= 25y^6 - z^8 \quad \blacksquare
\end{aligned}
$$

⊡ The Square of a Binomial

In Chapter 5 you were cautioned that the square of a sum is *not* the sum of the squares.

$$(a + b)^2 \neq a^2 + b^2$$

We know from the definition of an exponent that

$$(a + b)^2 = (a + b)(a + b)$$

→ two factors

Thus, we can use the FOIL method to find the square of the *sum* of two terms as follows:

$$(a + b)^2 = (a + b)(a + b) = a^2 + ab + ab + b^2$$

$$= a^2 + 2ab + b^2$$

The square of the *sum* of two terms is a trinomial whose middle term is positive and twice the product of the two terms in the sum.

THE SQUARE OF THE *SUM* OF TWO TERMS

$$(a + b)^2 = a^2 + 2ab + b^2$$

We can also use the FOIL method to find the square of the *difference* of two terms as follows:

$$(a - b)^2 = (a - b)(a - b) = a^2 - ab - ab + b^2$$

$$= a^2 - 2ab + b^2$$

The square of the difference of two terms is a trinomial whose middle term is negative and twice the product of the two terms in the sum.

THE SQUARE OF THE *DIFFERENCE* OF TWO TERMS

$$(a - b)^2 = a^2 - 2ab + b^2$$

Raising a binomial to a power is often called *expanding* the binomial.

EXAMPLE 3

Expand $(3x + 4)^2$.

Solution:

$$(3x + 4)^2 = (3x)^2 \boxed{+ \; 2(3x)(4)} + (4)^2 \qquad (a + b)^2 = a^2 + 2ab + b^2$$

$$\downarrow$$

$$= \quad 9x^2 + 24x + 16 \quad \blacksquare$$

Note: You could also write $(3x + 4)^2$ as $(3x + 4)(3x + 4)$ and use the FOIL method. Try it! See if you can obtain the same result.

EXAMPLE 4

Expand $(5y^3 - z^4)^2$.

Solution:

$$(5y^3 - z^4)^2 = (5y^3)^2 \boxed{- \; 2(5y^3)(z^4)} + (z^4)^2 \qquad (a - b)^2 = a^2 - 2ab + b^2$$

$$\downarrow$$

$$= \quad 25y^6 - 10y^3z^4 + z^8 \quad \blacksquare \qquad (a^m)^n = a^{mn}$$

⠒ **The Product of Three Polynomials**

To multiply three polynomials, find the product of any two of them and then multiply that product by the third polynomial. Because of the associative law of multiplication, $(ab)c = a(bc)$, we know that it doesn't matter which two are grouped together first.

EXAMPLE 5

Find the product $5x(x + 3)(x - 3)$.

Solution: First find the product $(x + 3)(x - 3)$. Then distribute the $5x$ to this product. Thus,

$$5x \boxed{(x + 3)(x - 3)} = 5x(x^2 - 9) \qquad (a + b)(a - b) = a^2 - b^2$$

$$= 5x^3 - 45x \quad \blacksquare \qquad \text{distributive law}$$

Note: Try finding the product $5x(x + 3)$ first. Then multiply this result by $(x - 3)$. See if you can obtain the same result.

EXAMPLE 6

Find the product $5x(x + 3)^2$.

Solution: First, expand $(x + 3)^2$. Then distribute the $5x$ to this product.

$$5x\,(x + 3)^2 = 5x(x^2 + 6x + 9) \qquad (a + b)^2 = a^2 + 2ab + b^2$$
$$= 5x^3 + 30x^2 + 45x \quad \blacksquare \qquad \text{distributive law}$$

 For products of the form $a(b + c)^2$, you must expand first, then distribute. You *cannot* distribute first, then expand.

$$a(b + c)^2 \neq (ab + ac)^2$$

For example,

$$4(2 + 3)^2 = 4(5)^2 = 4(25) = 100,$$

while

$$(4 \cdot 2 + 4 \cdot 3)^2 = (8 + 12)^2 = (20)^2 = 400.$$

EXAMPLE 7

Expand $(x - 2)^3$.

Solution: $(x - 2)^3 = (x - 2)\,(x - 2)^2 \qquad a^m a^n = a^{m+n}$
                                                       in reverse
$$= (x - 2)(x^2 - 4x + 4) \qquad (a - b)^2 = a^2 - 2ab + b^2$$

Note: To multiply a binomial by a trinomial, multiply each term of the binomial by each term of the trinomial. There are actually *six* multiplications to perform.

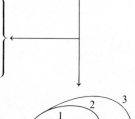

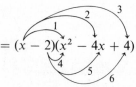

$$= x^3 - 4x^2 + 4x - 2x^2 + 8x - 8$$
$$= x^3 - 6x^2 + 12x - 8 \quad \blacksquare \quad \text{combining similar terms}$$

EXERCISES 6.3

Perform the indicated operations. Be sure to combine any similar terms.

1. $(x + 1)(x - 1)$

2. $(x - 3)(x + 3)$

3. $(x + 1)^2$

4. $(x - 3)^2$

5. $(2x - 3)(2x + 3)$

6. $(5x + 4)(5x - 4)$

7. $(2x - 3)^2$

8. $(5x + 4)^2$

9. $(4t^3 - y)(4t^3 + y)$

10. $(2x^2 + y^3)(2x^2 - y^3)$

11. $(4t^3 + y)^2$

12. $(2x^2 - y^3)^2$

13. $(4t^3y)^2$

14. $(2x^2y^3)^2$

15. $3x(x - 5)(x + 5)$

16. $5y(y + 8)(y - 8)$

17. $3x - (x - 5)(x + 5)$

18. $5y - (y + 8)(y - 8)$

19. $8m(2m - 1)^2$

20. $-3m(3m + 5)^2$

21. $8m - (2m - 1)^2$

22. $-3m - (3m + 5)^2$

23. $(5t - 2)^2(-2t)$

24. $(3p - 2q)^2(-2pq)$

25. $(5t - 2)^2 - 2t$

26. $(3p - 2q)^2 - 2pq$

27. $(x + 1)^3$

28. $(t + 4)^3$

29. $(3 - 2x)^3$

30. $(5 - 3x)^3$

31. $(2x^2 + y)^3$

32. $(4r^3 + t^4)^3$

33. $(2x^2y)^3$

34. $(4r^3t^4)^3$

35. $(x - 1)(x + 1)^2$

36. $(2x - 5)(2x + 5)^2$

37. $(x + 3)(x - 6)(x + 3)$

38. $(7x - 2)(2x + 7)(7x + 2)$

39. $(x + y)^4$

40. $(x - 2)^4$

41. $5(6p - 1)^2(3p)$

42. $-3y(y + 3)^2(-2)$

43. $((a + 4)(a - 4))^2$

44. $((4a - 1)(4a + 1))^2$

Find a polynomial that represents the area of the shaded region.

45.

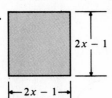

$2x - 1$

$2x - 1$

46.

$r + 1$

47.

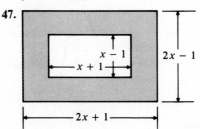

$x - 1$

$x + 1$

$2x - 1$

$2x + 1$

48.

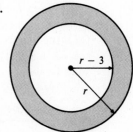

$r - 3$

r

Find a polynomial that represents the volume of the solid.

49.

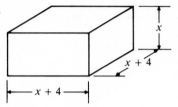

50.

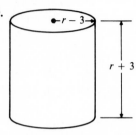

6.4   AN INTRODUCTION TO FACTORING

Objectives

⊡ To select the factors of a number that have a certain sum.

⊡ Given a monomial and one of its factors, to find the other factor.

⊡ To factor a polynomial that contains a common monomial factor.

⊡ To factor a polynomial by grouping terms.

⊡ Factoring a Number

In Chapter 2 you learned that to **factor** an integer meant to write that integer as a *product* of two (or more) other integers. For example, to factor the number 18, we could write:

number	factorizations
18	(1)(18)
	(2)(9)
	(3)(6)

and also

$$(-1)(-18)$$
$$(-2)(-9)$$
$$(-3)(-6)$$

When factoring certain trinomials (see Section 6.5), we will need to select the factors of a number that have a particular *sum*. For example, if we wished to select the factors of 18 whose *sum* is 11, we would select 2 and 9, since $(2)(9) = 18$ and $2 + 9 = 11$. Similarly, if we wished to select the factors of 18 whose *sum* is -9, we would select -3 and -6, since $(-3)(-6) = 18$ and $-3 + (-6) = -9$.

EXAMPLE 1

Find the factors of 24 whose sum is -10.

Solution: Since the number is *positive* (24) and the sum of the factors is to be *negative* (-10), we need only consider *two negative factors*.

number	factorizations to be considered	sum of the factors
24	$(-1)(-24)$	$-1 + (-24) = -25$
	$(-2)(-12)$	$-2 + (-12) = -14$
	$(-3)(-8)$	$-3 + (-8) = -11$
	$(-4)(-6)$	$-4 + (-6) = -10$

We should select -4 and -6, since
$(-4)(-6) = 24$, and $-4 + (-6) = -10$. ∎

EXAMPLE 2

Find the factors of -30 whose sum is 13.

Solution: Since the number is *negative* (-30) and the sum of the factors is to be positive (13), we need only consider factors of *opposite* signs in which the *larger of the factors is positive*.

number	factorizations to be considered	sum of the factors
-30	$(-1)(30)$	$-1 + 30 = 29$
	$(-2)(15)$	$-2 + 15 = 13$
	$(-3)(10)$	$-3 + 10 = 7$
	$(-5)(6)$	$-5 + 6 = 1$

We should select -2 and 15, since
$(-2)(15) = -30$, and $-2 + 15 = 13$. ∎

Factoring Monomials

To factor a monomial, we can use the product rule for exponents in reverse,

$$a^{m+n} = a^m a^n,$$

and write the monomial as a *product* of two other monomials. For example, to factor the monomial $18x^6$, we could write:

monomial	some possible factorizations
$18x^6$	$(3x)(6x^5)$
	$(9x^6)(2)$
	$(-x^2)(-18x^4)$
	$(-6x^3)(-3x^3)$
	$\vdots$

and so on

EXAMPLE 3

Find the missing factor of each monomial.

a) $48a^2 = (6a)(\ \ ?\ \ )$

b) $-56x^7y^3 = (\ \ ?\ \ )(7x^2y^3)$

c) $32y^{16}z = (-8y^4)(\ \ ?\ \ )$

Solution:

a) $48a^2 = (6a)(\boxed{8a})$, since $(6a)(8a) = 48a^2$. ∎

b) $-56x^7y^3 = (\boxed{-8x^5})(7x^2y^3)$, since $(-8x^5)(7x^2y^3) = -56x^7y^3$. ∎

c) $32y^{16}z = (-8y^4)(\boxed{-4y^{12}z})$, since $(-8y^4)(-4y^{12}z) = 32y^{16}z$. ∎

⸫ Common Term Factoring

To factor a polynomial means to write it as a *product* of other polynomials. Factoring polynomials is essential in order to extend your knowledge of solving equations (see Chapter 7). Factoring is also the foundation for simplifying algebraic fractions (see Chapter 8).

   If something is common to each term of a polynomial, then we can use the *distributive law in reverse* to factor out this common expression. This is known as *common term factoring*.

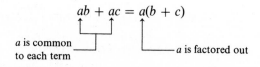

DISTRIBUTIVE LAW IN REVERSE—COMMON TERM FACTORING

$$ab + ac = a(b + c)$$

a is common to each term

a is factored out

When we factor a polynomial, we usually want to factor it *completely*. A polynomial is said to be factored completely if no further factoring is possible on any non-monomial factor (that is, if the nonmonomial factors are prime). When using common term factoring, this means that we must factor out the largest number and the largest variable expression common to each term of the polynomial.

EXAMPLE 4

Factor $24x^2 - 36y^3$ completely.

Solution: The largest number common to each term is 12. The terms do not have a common variable expression. Thus,

$$24x^2 - 36y^3 = (\boxed{12})(2x^2) + (\boxed{12})(-3y^3) \qquad \text{factoring monomials}$$
$$= 12(2x^2 - 3y^3) \quad \blacksquare \qquad \text{distributive law in reverse}$$

Note: We could also have factored out -12 as follows:

$$24x^2 - 36y^3 = (\boxed{-12})(-2x^2) + (\boxed{-12})(3y^3) \qquad \text{factoring monomials}$$
$$= -12(-2x^2 + 3y^3) \qquad \text{distributive law in reverse}$$
$$= -12(3y^3 - 2x^2) \qquad \text{commutative law of addition}$$

EXAMPLE 5

Factor $12x^3 + 8x^2 - 16x^5$ completely.

Solution: The largest number common to each term is 4, and the largest variable expression common to each term is x^2. Thus,

$$12x^3 + 8x^2 - 16x^5 = (\boxed{4x^2})(3x) + (\boxed{4x^2})(2) + (\boxed{4x^2})(-4x^3)$$
$$= 4x^2(3x + 2 - 4x^3) \quad \blacksquare \qquad \text{common term factoring}$$

Note: Notice that the largest variable expression that is common to each term is actually the one with the *smallest* exponent.

EXAMPLE 6

Factor $25x^2y^3 - 15x^2y^2 + 5xy^2$ completely.

Solution: The largest common monomial factor of each term is $5xy^2$. Thus,

$$25x^2y^3 - 15x^2y^2 + 5xy^2$$
$$= (\boxed{5xy^2})(5xy) + (\boxed{5xy^2})(-3x) + (\boxed{5xy^2})(1)$$

Note: Be sure to
write this 1.

$$= 5xy^2(5xy - 3x + 1) \quad \blacksquare \qquad \text{common term factoring}$$

Note: To check your factoring, simply distribute the common monomial factor to each term.

$$5xy^2(5xy - 3x + 1) = 25x^2y^3 - 15x^2y^2 + 5xy^2$$

Note: We need this 1, or
we would never get back
the $5xy^2$ term.

⠢ **Factoring by Grouping Terms**

Now, suppose we were given the polynomial

$$\underbrace{xy}_{\substack{\text{first} \\ \text{term}}} + \underbrace{3x}_{\substack{\text{second} \\ \text{term}}} + \underbrace{4y}_{\substack{\text{third} \\ \text{term}}} + \underbrace{12}_{\substack{\text{fourth} \\ \text{term}}}$$

and asked to find its factors. Notice that there is no common expression that can be factored out of *all* four terms. However, the first and second terms contain the common variable x, while the third and fourth terms contain the common number 4. Suppose we *grouped* these terms and tried to factor separately as follows:

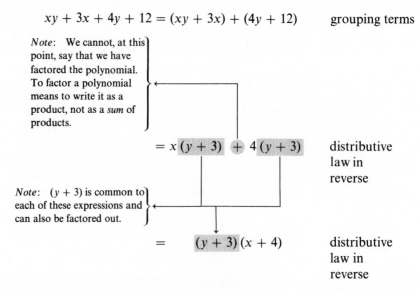

$$xy + 3x + 4y + 12 = (xy + 3x) + (4y + 12) \qquad \text{grouping terms}$$

Note: We cannot, at this point, say that we have factored the polynomial. To factor a polynomial means to write it as a product, not as a *sum* of products.

$$= x\,(y + 3) \; + \; 4\,(y + 3) \qquad \begin{array}{l}\text{distributive} \\ \text{law in} \\ \text{reverse}\end{array}$$

Note: $(y + 3)$ is common to each of these expressions and can also be factored out.

$$= \qquad (y + 3)\,(x + 4) \qquad \begin{array}{l}\text{distributive} \\ \text{law in} \\ \text{reverse}\end{array}$$

Thus, the polynomial $xy + 3x + 4y + 12$ is factored completely when we write it as $(y + 3)(x + 4)$. Of course, we can check our work by using the FOIL method to find the product $(y + 3)(x + 4)$.

$$\text{Check:} \quad (y + 3)(x + 4) = \overset{F}{xy} + \overset{O}{4y} + \overset{I}{3x} + \overset{L}{12}$$
$$= xy + 3x + 4y + 12 \qquad \begin{array}{l}\text{commutative law} \\ \text{of addition}\end{array}$$

EXAMPLE 7

Factor $6x^3 - 4x^2 + 9x - 6$ completely.

Solution: There is no common expression for *all* four terms. Therefore, group terms in pairs and factor each group separately as follows:

$$6x^3 - 4x^2 + 9x - 6 = (6x^3 - 4x^2) + (9x - 6) \qquad \text{grouping terms}$$

$$= 2x^2 (3x - 2) + 3 (3x - 2) \qquad \begin{array}{l} \text{distributive law} \\ \text{in reverse} \end{array}$$

$$= (3x - 2)(2x^2 + 3) \quad \blacksquare \qquad \begin{array}{l} \text{distributive law} \\ \text{in reverse} \end{array}$$

Note: There is usually more than one way to group the terms. Try grouping the first and third terms $(6x^3 + 9x)$ and the second and fourth terms $(-4x^2 - 6)$. See if you can obtain the same result.

Of course, the key to factoring by grouping is to develop common expressions so that we can factor them out.

EXAMPLE 8

Factor $8 + x^2y - 4x^2 - 2y$ completely.

Solution:

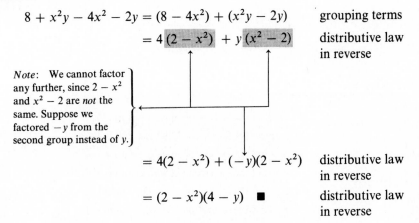

$$8 + x^2y - 4x^2 - 2y = (8 - 4x^2) + (x^2y - 2y) \qquad \text{grouping terms}$$

$$= 4 (2 - x^2) + y (x^2 - 2) \qquad \begin{array}{l} \text{distributive law} \\ \text{in reverse} \end{array}$$

Note: We cannot factor any further, since $2 - x^2$ and $x^2 - 2$ are *not* the same. Suppose we factored $-y$ from the second group instead of y.

$$= 4(2 - x^2) + (-y)(2 - x^2) \qquad \begin{array}{l} \text{distributive law} \\ \text{in reverse} \end{array}$$

$$= (2 - x^2)(4 - y) \quad \blacksquare \qquad \begin{array}{l} \text{distributive law} \\ \text{in reverse} \end{array}$$

Note: Check by finding the product $(2 - x^2)(4 - y)$.

Referring to Example 8, we could also have factored out -4 from the first group instead of 4 to obtain the following result:

$$(8 - 4x^2) + (x^2y - 2y) = -4(x^2 - 2) + y(x^2 - 2)$$
$$= (x^2 - 2)(y - 4)$$

Does this result agree with the previous answer? Does

$$(2 - x^2)(4 - y) = \underbrace{(x^2 - 2)}\ \underbrace{(y - 4)}?$$

$$= \underbrace{(-1)(2 - x^2)}\ \underbrace{(-1)(4 - y)}\quad \text{factoring} -1 \text{ from each factor}$$

$$= (2 - x^2)(4 - y)\quad \text{since } (-1)(-1) = 1$$

EXAMPLE 9

Factor $10x^3y^2 + 10x^2y^2 - 2xy^2 - 2y^2$ completely.

Solution: $2y^2$ is common to each term and should be factored out immediately.

$$10x^3y^2 + 10x^2y^2 - 2xy^2 - 2y^2 = 2y^2(5x^3 + 5x^2 - x - 1)\qquad \begin{array}{l}\text{common term}\\ \text{factoring}\end{array}$$

$$= 2y^2((5x^3 + 5x^2) + (-x - 1))\qquad \text{grouping terms}$$

$$= 2y^2(5x^2(x + 1) + \underbrace{(-x - 1)})\qquad \begin{array}{l}\text{distributive law}\\ \text{in reverse}\end{array}$$

Note: Factor -1
from this expression.

$$= 2y^2(5x^2\,(x + 1)\ + (-1)\,(x + 1)\,)$$

$$= 2y^2(x + 1)(5x^2 - 1)\quad \blacksquare\qquad \begin{array}{l}\text{distributive law}\\ \text{in reverse}\end{array}$$

EXERCISES 6.4

1. Find the factors of 10 whose sum is 7.

2. Find the factors of 32 whose sum is 18.

3. Find the factors of 36 whose sum is -15.

4. Find the factors of 72 whose sum is -22.

5. Find the factors of -48 whose sum is 2.

6. Find the factors of -12 whose sum is 1.

7. Find the factors of -56 whose sum is -1.

8. Find the factors of -27 whose sum is -6.

9. Find the factors of -25 whose sum is 0.

10. Find the factors of 81 whose sum is 0.

Find the missing factor.

11. $16x^2 = (8x)(\ \ ?\ \ )$

12. $32t^3 = (16t)(\ \ ?\ \ )$

13. $81xy^4 = (\ \ ?\ \ )(9y^2)$

14. $44x^2y^3 = (\ \ ?\ \ )(11xy)$

15. $56t^4u = (-7tu)(\ \ ?\ \ )$

16. $98p^{12}q^4 = (-7p^8q^4)(\ \ ?\ \ )$

17. $-36x^2z^{25} = ($? $)(6z^5)$

18. $-100t^3x^9 = ($? $)(-10x^3)$

19. $7y + 14x^2 = ($? $)(y + 2x^2)$

20. $8x^2y - 12x^2y^2 = ($? $)(2 - 3y)$

21. $25t^2 - 15t = -5t($? $)$

22. $24w^3z - 18w^3 = -6w^3($? $)$

23. $8 - 8t + 8t^2 = 8($? $)$

24. $-12x^3y^3 - 3xy = -3xy($? $)$

Using common term factoring, factor completely.

25. $8xy - 14x^2$

26. $12x^3y^3 + 16x^2y$

27. $16y^2 - 4xy^3 + 8xy^2$

28. $8x^2y - 12x^2 - 24x^3y$

29. $3x^3y^3 + 15x^3y^2 - 9x^2y^3 - 3x^2y^2$

30. $5x^5y - 5x^7y - 5x^4 + 10x^5y$

31. $6xy - 6x^2y - 12xy^2 + 18x^3y$

32. $2m^2n^3 + 6m^2n^4 - 10m^3n^3 + 2m^4n^6$

By grouping terms, factor each of the following completely.

33. $x^3 + 4x^2 + 3x + 12$

34. $y^3 + 4y^2 + 5y + 20$

35. $4xy^3 - 3x + 8y^3 - 6$

36. $6x^2y - 5y - 12x^2 + 10$

37. $15 + 2x^2y - 10x^2 - 3y$

38. $10 - 3x^2y - 5x^2 + 6y$

39. $t^3 + 2t^2 + t + 2$

40. $35mn^2 - 10n^2 + 7m - 2$

41. $9x^4 - 3x^3 - 54x^2 + 18x$

42. $4pq^3 - 8p^2q^2 - 12p^2q + 24p^3$

43. $45y^2 + 5wy^2 + 45y^4 + 5wy^4$

44. $6x^3y^4 - 6x^2y^3 + 6xy^2 - 6y$

Find a factored expression that represents each shaded area.

45.

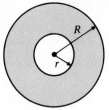

46.

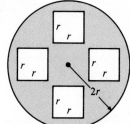

47.

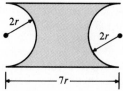

48.

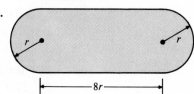

6.5   **MORE ON FACTORING**

Objectives

☐ To factor a trinomial of the form $ax^2 + bx + c$.

☐ To factor a trinomial of the form $x^2 + bx + c$.

☐ To factor the difference of two perfect squares.

⊡ Trinomials of the Form $ax^2 + bx + c$

In Section 6.2 we found that the product of two binomials can be a trinomial if the middle terms combine into a single term. For example,

$$(2x + 3)(x + 4) = 2x^2 \;\overset{\text{middle terms}}{\boxed{+\, 8x + 3x}}\; + 12$$

$$= 2x^2 + 11x + 12 \quad \text{a trinomial.}$$

Now, suppose we were given the trinomial, $2x^3 + 11x + 12$ and asked to find its factors. If somehow we knew to rewrite $11x$ as the sum $8x + 3x$, we could factor by grouping as follows:

$$2x^2 + 11x + 12 = 2x^2 \;\boxed{+\, 8x + 3x}\; + 12$$

$$= (2x^2 + 8x) + (3x + 12)$$

$$= 2x(x + 4) + 3(x + 4)$$

$$= (x + 4)(2x + 3)$$

But how will we know to write $11x$ as $8x + 3x$ and not as $6x + 5x$, or $2x + 9x$, or $4x + 7x$, and so on? Note the following:

$$2x^2 + 11x + 12 = \underset{\substack{\longrightarrow \\ \text{The product} \\ \text{of these} \\ \text{numbers is} \\ \text{also 24.}}}{\overset{\substack{\text{The product of} \\ \text{these numbers is 24.}}}{2\,x^2 + 8\,x + 3\,x + 12}}$$

Therefore, the correct numbers to select for the middle terms must have exactly the same product as the first and last coefficients.

FOUR-STEP PROCEDURE FOR FACTORING A TRINOMIAL OF THE FORM

$$ax^2 + bx + c$$

Step 1: Remove any common monomial factor.

Step 2: Record the product ac and the value of b. (Be sure to include any negative signs).

Step 3: Select two factors of the product ac whose sum is b.

Step 4: Write $ax^2 + bx + c$ as $ax^2 + \boxed{}x + \boxed{}x + c$ and fill in the boxes with the numbers selected in Step 3. Factor this expression by grouping terms.

EXAMPLE 1

Factor $6x^2 + 7x + 2$ completely.

Solution:

Step 1:   There is *no* common monomial factor for all three terms.

Step 2:

$$6\,x^2 + 7\,x + 2 \qquad \longrightarrow ac = (6)(2) = 12$$
$$b = 7$$

Step 3:   We must find two factors of 12 whose sum is 7. We should select 3 and 4, since $(3)(4) = 12$ and $3 + 4 = 7$.

Step 4: $6x^2 + 7x + 2 = 6x^2 + 3x + 4x + 2$

$$= (6x^2 + 3x) + (4x + 2) \qquad \text{grouping terms}$$
$$= 3x(2x + 1) + 2(2x + 1) \qquad \text{factoring by grouping terms}$$

$$= (2x + 1)(3x + 2) \quad \blacksquare$$

EXAMPLE 2

Factor $4x^2 - 20x + 25$ completely.

Solution:

Step 1:   There is no common monomial factor for all three terms.

Step 2:

$$4\,x^2 - 20\,x + 25 \qquad \longrightarrow ac = (4)(25) = 100$$
$$b = -20 \quad \text{(Be sure to include the negative sign.)}$$

Step 3:   We must find two factors of 100 whose sum is -20. We should select -10 and -10, since $(-10)(-10) = 100$ and $-10 + (-10) = -20$.

Step 4: $4x^2 - 20x + 25 = 4x^2 + -10\,x + -10\,x + 25$

$$= (4x^2 - 10x) + (-10x + 25) \qquad \text{grouping terms}$$
$$= 2x(2x - 5) + (-5)(2x - 5) \qquad \text{factoring by grouping terms}$$

$$= (2x - 5)(2x - 5)$$
$$= (2x - 5)^2 \quad \blacksquare \qquad\qquad a \cdot a = a^2$$

Note: Check by using $(a - b)^2 = a^2 - 2ab + b^2$.

EXAMPLE 3

Factor $8t^3 - 30t^2 - 8t$ completely.

Solution:

Step 1: The common monomial factor, $2t$, should be factored out immediately:

$$8t^3 - 30t^2 - 8t = 2t(4t^2 - 15t - 4)$$

Step 2: For the trinomial,

$$4t^2 - 15t - 4 \qquad ac = (4)(-4) = -16$$

$$b = -15.$$

Step 3: We must find two factors of -16 whose sum is -15. We should select 1 and -16, since $(1)(-16) = -16$ and $1 + (-16) = -15$.

Step 4:

$$8t^3 - 30t^2 - 8t = 2t(4t^2 - 15t - 4) \qquad \text{common term factoring}$$

$$= 2t(4t^2 + 1t + -16t - 4)$$

$$= 2t((4t^2 + t) + (-16t - 4))$$

$$= 2t(t(4t + 1) + (-4)(4t + 1)) \qquad \text{factoring by grouping terms}$$

$$= 2t(4t + 1)(t - 4) \qquad \blacksquare$$

EXAMPLE 4

Factor $8x^2 + 5xy - 3y^2$ completely.

Solution:

Step 1: There is no common monomial factor for all three terms.

Step 2:

$$8x^2 + 5xy - 3y^2 \qquad ac = (8)(-3y^2) = -24y^2$$

$$b = 5y$$

Step 3: We must find two factors of $-24y^2$ whose sum is $5y$. We should select $-3y$ and $8y$, since $(-3y)(8y) = -24y^2$ and $-3y + 8y = 5y$.

Step 4:

$$8x^2 + 5xy - 3y^2 = 8x^2 + \boxed{-3y}\,x + \boxed{8y}\,x - 3y^2$$
$$= (8x^2 - 3xy) + (8xy - 3y^2)$$
$$= x(8x - 3y) + y(8x - 3y) \qquad \text{factoring by}$$
$$ \text{grouping terms}$$
$$= (8x - 3y)(x + y) \quad \blacksquare$$

⊡ Trinomials of the Form $x^2 + bx + c$

To factor $x^2 + 9x + 14$, we could write it as $1x^2 + 9x + 14$ and use the four-step procedure for factoring trinomials as follows:

Step 1:   There is no common monomial factor.

Step 2:

$$\underset{\substack{\big\downarrow \\ b = 9}}{\boxed{1}\,x^2 + \boxed{9}\,x + \boxed{14}} \qquad\longrightarrow\quad ac = (1)(14) = 14$$

Step 3:   We must find two factors of 14 whose sum is 9. We should select 2 and 7, since $(2)(7) = 14$ and $2 + 7 = 9$.

Step 4:

$$x^2 + 9x + 14 = x^2 + \boxed{2}\,x + \boxed{7}\,x + 14$$
$$= (x^2 + 2x) + (7x + 14)$$
$$= x(x + 2) + 7(x + 2)$$
$$= (x + \boxed{2})(x + \boxed{7})$$

Notice that the factors we selected in Step 3 appear directly in the answer.

Therefore, to factor a trinomial of the form $x^2 + bx + c$, we can simply (1) select two factors of c whose sum is b, and (2) write $x^2 + bx + c$ as $(x + \boxed{})(x + \boxed{})$, and fill in the boxes with these factors.

This shorter version of factoring a trinomial only works if the coefficient of the x^2 term is 1.

EXAMPLE 5

Factor $x^2 - x - 20$ completely.

Solution:

$$x^2 - x - 20 = x^2 \underset{b\,=\,-1}{\underbrace{- 1}}\,x \overset{c\,=\,-20}{\underbrace{- 20}}$$

1. We must find two factors of -20 whose sum is -1. We should select -5 and 4, since $(-5)(4) = -20$ and $-5 + 4 = -1$.

2. $$x^2 - x - 20 = (x + \boxed{-5})(x + \boxed{4})$$
$$= (x - 5)(x + 4) \quad \blacksquare$$

EXAMPLE 6

Factor $3x^2y - 18xy + 27y$ completely.

Solution: The common monomial factor, $3y$, should be factored out immediately:

$$3x^2y - 18xy + 27y = 3y(x^2 - 6x + 9)$$

Now, for the trinomial,

$$x^2 \underset{b\,=\,-6}{\underbrace{- 6}}\,x + \overset{c\,=\,9}{\underbrace{9}}$$

1. We must find two factors of 9 whose sum is -6. We should select -3 and -3, since $(-3)(-3) = 9$ and $-3 + (-3) = -6$.

2. $3x^2y - 18xy + 27y = 3y(x^2 - 6x + 9)$ common term factoring
$$= 3y(x + \boxed{-3})(x + \boxed{-3})$$
$$= 3y(x - 3)(x - 3)$$
$$= 3y(x - 3)^2 \quad \blacksquare$$

Note: Remember, to check, you must expand first then distribute $3y$ to each term.

EXAMPLE 7

Factor $x^2 + 6x - 8$ completely.

Solution:

$$x^2 + 6\,x \overset{c\,=\,-8}{\underbrace{- 8}}$$
$$b = 6$$

We must find two factors of -8 whose sum is 6.

number	factorizations to be considered	sum of the factors
-8	$(-1)(8)$	$-1 + 8 = 7$
	$(-2)(4)$	$-2 + 4 = 2$

As you can see, it is impossible to find two factors of -8 whose sum is 6. Thus, we can say that $x^2 + 6x - 8$ is a *nonfactorable trinomial* (that is, it is prime). ■

Note: Not all trinomials can be factored into the product of two binomials. It is important that you keep this in mind.

⠒ The Difference of Two Perfect Squares

To factor $9x^2 - 4$, we could write it as $9x^2 + 0x - 4$ and use the four-step procedure for factoring a trinomial as follows:

Step 1:   There is no common monomial factor.

Step 2:

$$ac = (9)(-4) = -36$$
$$9\,x^2 + 0\,x - 4$$
$$b = 0$$

Step 3:   We must find two factors of -36 whose sum is 0. We should select 6 and -6, since $(6)(-6) = -36$ and $6 + (-6) = 0$.

Step 4:

$$
\begin{aligned}
9x^2 - 4 &= 9x^2 + 6x + -6x - 4 \\
&= (9x^2 + 6x) + (-6x - 4) \\
&= 3x(3x + 2) + (-2)(3x + 2) \\
&= (3x + 2)(3x - 2)
\end{aligned}
$$

In Section 6.3, you learned that the product $(a + b)(a - b) = a^2 - b^2$. Therefore, it isn't surprising that when factoring an expression of the form $a^2 - b^2$, we obtain $(a + b)(a - b)$.

For $9x^2 - 4$, observe the following:

$$\sqrt{9x^2} = 3x$$
$$9x^2 - 4 = (3x + 2)(3x - 2)$$
$$\sqrt{4} = 2$$

Therefore, to factor the *difference* of two perfect squares, we can simply write $a^2 - b^2$ as $(a + b)(a - b)$, where $a = \sqrt{a^2}$ and $b = \sqrt{b^2}$.

The *sum* of two perfect squares is nonfactorable (that is, $a^2 + b^2$ is prime).

$$a^2 + b^2 \neq \underbrace{(a + b)(a + b)}_{} \neq \underbrace{(a - b)(a - b)}_{}$$

$$\downarrow \qquad\qquad \downarrow$$

$$a^2 + 2ab + b^2 \qquad a^2 - 2ab + b^2$$

$$\text{or} \qquad\qquad \text{or}$$

$$(a + b)^2 \qquad\qquad (a - b)^2$$

EXAMPLE 8

Factor $25x^2 - 16y^2$ completely.

Solution:

$$a = \sqrt{25x^2} = 5x$$
$$25x^2 - 16y^2$$
$$b = \sqrt{16y^2} = 4y$$

Thus, $25x^2 - 16y^2 = (5x + 4y)(5x - 4y)$. ∎

EXAMPLE 9

Factor $3x^3 - 27x$ completely.

Solution: The common monomial factor, $3x$, should be factored out immediately:

$$3x^3 - 27x = 3x(x^2 - 9), \text{ but } x^2 - 9 \text{ is}$$
$$\text{the difference of two perfect squares.}$$

Thus, $3x^3 - 27x = 3x(x^2 - 9) = 3x(x + 3)(x - 3)$. ∎

EXAMPLE 10

Factor $t^6 - 81$ completely.

Solution: $t^6 - 81 = (t^3 + 9)(t^3 - 9)$ ∎

Note: The binomial $t^3 - 9$ is *not* the difference of two perfect squares because t^3 is not a perfect square variable expression ($\sqrt{t^3} = t\sqrt{t}$).

EXAMPLE 11

Factor $y^4 - 16$ completely.

Solution:

$$y^4 - 16 = \underbrace{(y^2 + 4)}\underbrace{(y^2 - 4)}$$

sum of two perfect squares (It's prime.)

difference of two perfect squares (It's factorable.)

Thus,

$$y^4 - 16 = (y^2 + 4)\underbrace{(y^2 - 4)} = (y^2 + 4)\underbrace{(y + 2)(y - 2)}. \quad \blacksquare$$

EXERCISES 6.5

Factor the following completely. If not factorable, write prime.

1. $2x^2 + 7x + 6$

2. $2x^2 + 11x + 15$

3. $3x^2 - 11x + 10$

4. $4x^2 - 16x + 15$

5. $x^2 + 7x + 6$

6. $x^2 + 8x + 15$

7. $6t^2 - 19t + 10$

8. $2w^2 + 5w - 12$

9. $2y^2 + y - 15$

10. $3p^2 + p - 10$

11. $x^2 - 8x + 12$

12. $x^2 - 13x + 40$

13. $4x^2 + 8x + 3$

14. $3x^2 - 10x + 3$

15. $y^2 - 81$

16. $t^2 - 25$

17. $f^2 - f + 42$

18. $v^2 - v - 42$

19. $x^2 - x - 56$

20. $x^2 - x + 56$

21. $4x^2 - 25y^2$

22. $81x^2 - 49y^2$

23. $4d^6 - 25$

24. $9g^{10} - 1$

25. $81x^2 + 4$

26. $25x^2 - y^2$

27. $10x^2 + 7xy + y^2$

28. $6x^2 + 31xy + 5y^2$

29. $p^2 + 2pq - 24q^2$

30. $t^2 - 3tu - 10u^2$

31. $25x^2 - 10xy + y^2$

32. $9h^2 - 12hk + 4k^2$

33. $8 - 2x^2$

34. $75 - 3x^2$

35. $32 - 4x - x^2$

36. $72 - y - y^2$

37. $24x^2 - 30x - 9$

38. $8x^2y + 2xy - y$

39. $5x^2y + 40xy + 80y$

40. $2x^3y + 12x^2y + 18xy$

41. $6t^5 + 3t^4 + 3t^3$

42. $6y^7 - 2y^6 - 10y^5$

43. $12x^3 + 48xy^2$

44. $3x^3y + 27xy^3$

45. $-2x^2y^3 + 8x^2y^2 + 10x^2y$

46. $-3x^5y^3 - 12x^4y^2 + 36x^3y$

47. $x^8 - y^8$

48. $x^4y^4 - z^4$

49. $4 - 4x^2 - y^2 + x^2y^2$

50. $x^2t^2 - x^2 - y^2t^2 + y^2$

Chapter Review

REVIEW EXERCISES

The following review exercises are grouped according to the objectives that should have been mastered in Chapter 6. Work each problem carefully. If any weaknesses appear, immediately refer to and read the subsection that matches that objective.

6.1 ADDING AND SUBTRACTING POLYNOMIALS

Objectives

☐ To identify different types of polynomials.

Identify.

1. a) $3x^2y + z$ ~binomial~ **b)** $3x^2yz$ ~monomial~ **c)** $3 + x^2y - z$ ~trinomial~

d) $3 + x^2 - y - z$ ~polynomial~   **e)** $3x^{-2}yz$ ~not a polynomial~

☐ To add polynomials.

Simplify.

2. $(3x^2 - x^5 + 2) + (3x^5 - 4x^2 - 2)$ ~= $1x^2 + 2x^5$ or $2x^5 - 1x^2$~

☐ To subtract polynomials.

Simplify.

3. $(2xy^2 - 3y) - (5y^2x + 2y + 3xy^2)$ ~dist.~ ~$-5y$~

4. $5xy - (3y - (2xy - y))$

6.2 MULTIPLYING POLYNOMIALS

Objectives

☐ To multiply a monomial by a polynomial.

Perform the indicated operations.

5. $-7y^2(3 - 2y^3)$ ~= $-21y^2 + 14y^5$ or $14y^5 - 21y^2$~

6. $xy^3(2xy^3 - 5x^3 + y - 1)$

7. $12x - 3x(3x^2y - 4)$ ~$12x \cdot 9x^3y + 12x = 9x^3y$~

☐ To multiply two binomials.

Perform the indicated operations.

8. $(x + 3y)(3x + 2z)$ ~FOIL $24x - 16xy^2 + 3y^3 -$~ ~$-2y^5$~

9. $(8x + y^3)(3 - 2y^2)$

10. $(4x - 3y)(x + 7y)$

11. $(5 - y^3)(4 - y^3)$ ~$20 5y^3 - 4y^3 + y^6$~ ~$20 - 9y^3 + y^6$~

12. $9x - (x - 5)(3x + 5)$

☐ To multiply two polynomials.

Find the product.

13. $(3x + y)(x^2 - 3xy + y^2)$

14. $(x^2 + 3x - 2)(x^2 - 3x - 2)$

6.3 SPECIAL PRODUCTS

Objectives

⊡ To multiply two binomials of the form $(a + b)(a - b)$.

Find the product.

15. $(2x + 5)(2x - 5)$

16. $(4x^2 - y^7)(4x^2 + y^7)$

⊡ To square a binomial.

17. $(2x + 5)^2$ $4x^2 + 20x + 25$

18. $(4x^2 - y^7)^2 = 16x^4 - 8x^2 y^7 + y^{14}$

$(4x^2 - y^7)(4x^2 + y^7)$

⊡ To multiply three polynomials.

Perform the indicated operations.

19. $2x(x + 5)(x - 5)$ **20.** $2x(x - 5)^2$ **21.** $(2x - 1)^3$

6.4 AN INTRODUCTION TO FACTORING

Objectives

⊡ To select the factors of a number that have a certain sum.

22. Find the factors of 28 whose sum is -11.

23. Find the factors of -64 whose sum is 12.

⊡ Given a monomial and one of its factors, to find the other factor.

Find the missing factor.

24. a) $54b^2 = (9b)(\ ?\ )$

b) $-28x^5y^2 = (\ ?\ )(4x^3y^2)$

c) $44y^9z = (-4y^3)(\ ?\ )$

⊡ To factor a polynomial that contains a common monomial factor.

Factor completely.

25. $16x^4 - 24y^3$

26. $8x^4 - 12x^3 - 20x^5$

27. $15xy^3 - 12x^2y + 9x^2y^2 + 3xy$

⊡ To factor a polynomial by grouping terms.

Factor completely.

28. $8x^3 + 6x^2 - 12x - 9$

29. $12 + 3x^2y - 9x^2 - 4y$

30. $12x^3y + 12x^2y - 3xy - 3y$

6.5 MORE ON FACTORING

Objectives

⊡ To factor a trinomial of the form $ax^2 + bx + c$.

Factor completely.

31. $6x^2 + 11x + 3$

32. $4x^2 + 20x + 25$

33. $9y^3 - 42y^2 - 15y$

34. $3x^2 + 5xy - 2y^2$

⊡ To factor a trinomial of the form $x^2 + bx + c$.

Factor completely.

35. $x^2 - x - 42$ **36.** $2x^2y - 16xy + 32y$ **37.** $x^2 + 8x - 12$

⊡ To factor the difference of two perfect squares.

Factor completely.

38. $36x^2 - 9y^2$ **39.** $5x^3 - 125x$

40. $x^6 - 16$ **41.** $y^4 - 81$

CHAPTER TEST

Time
50 minutes

Score
20, 19, 18 excellent
17, 16 good
15, 14, 13 fair
below 12 poor

If you have worked through the Review Exercises and corrected any weaknesses, then you are ready to take the following Chapter Test.

Perform the indicated operations.

1. $(3y^2 - 2xy) + (4yx - y^2)$ **2.** $3xy^2 - (4x^2y - (2y^2x + 3yx^2))$

3. $1 - 3xy^2(1 - 2x^3 + 3x^2y)$ **4.** $(x^2 + 2y^3)(3x + y^2)$

5. $(2x + 7)(x - 3)$ **6.** $(x - 2)(x^2 + 2x + 3)$

7. $4 - (5x - 2)(5x + 2)$ **8.** $2x(3x - 2)^2$

9. $3y(y - 2)(3 + y)$ **10.** $(2 - 3x)^3$

Factor the following completely.

11. $9xy - 18x^2y^3 - 6x$ **12.** $x^3 - 2x^2 + 3x - 6$

13. $4x^2 + 8x + 3$ **14.** $5x^2 - 16xy + 3y^2$

15. $x^2 + 5x - 36$ **16.** $2x^3y - 2x^2y - 112xy$

17. $6x - 24x^3$ **18.** $40 - 21x + 2x^2$

19. $x^2 + 10x - 21$ **20.** $12 - 3x^2 - 4x + x^3$

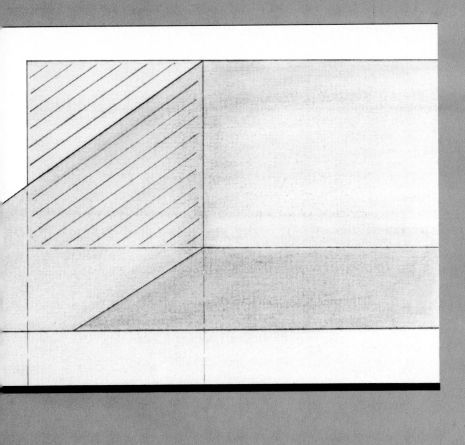

7

SOLVING OTHER TYPES OF EQUATIONS

7.1 SOLVING LITERAL EQUATIONS

Objectives

☐ To solve a literal equation in which the variable appears only once in the equation.

☐ To solve a literal equation or formula in which the variable appears more than once.

☐ Literal Equations in which the Variable Appears Once

In Chapter 4 you solved equations like $3x + 5 = 11$ for the variable x, by removing first the addition of 5 and then the multiplication by 3 as follows:

$$3x + 5 = 11$$

$$3x = 6 \qquad \text{adding } -5 \text{ to each side or}$$
$$\text{subtracting 5 from each side}$$

$$x = 2 \qquad \text{multiplying each side by } \frac{1}{3} \text{ or}$$
$$\text{dividing each side by 3}$$

We can say that $3x + 5 = 11$ is an equation of the form $ax + b = c$, where a, b, and c represent the **constants**. When we let letters (usually from the first part of the alphabet) represent constants in this way, we form what is called a **literal equation**. The equation $ax + b = c$ is an example of a literal equation.

If we solve this literal equation for the variable x, we will have a formula that can be used to find the solution of any equation of the form $ax + b = c$.

To solve $ax + b = c$ for x, we proceed in exactly the same way as if we were solving $3x + 5 = 11$. Thus,

$$ax + b = c$$

$$ax = c - b \qquad \text{adding } -b \text{ to each side or}$$
$$\text{subtracting } b \text{ from each side}$$

$$x = \frac{c - b}{a}. \qquad \text{multiplying each side by } \frac{1}{a} \text{ or}$$
$$\text{dividing each side by } a \ (a \neq 0)$$

To solve the equation $3x + 5 = 11$ using the formula $x = \dfrac{c - b}{a}$, we must first identify a, b, and c.

$$\overset{\displaystyle a}{\bigcirc} x + \overset{\displaystyle b}{\bigcirc} = \overset{\displaystyle c}{\bigcirc}$$
$$\downarrow \qquad \downarrow \qquad \downarrow$$
$$3x \ + \ 5 \ = \ 11$$

We can now substitute 3 for a, 5 for b, and 11 for c in this formula.

Thus, $x = \dfrac{c - b}{a} = \dfrac{11 - 5}{3} = \dfrac{6}{3} = 2$, which agrees with the previous result.

EXAMPLE 1

Solve the literal equation $a = \dfrac{bx + c}{d}$ for x.

Solution:

$$a = \frac{bx + c}{d}$$

$$ad = bx + c \qquad \text{multiplying each side by } d$$

$$ad - c = bx \qquad \begin{array}{l} \text{adding } -c \text{ to each side or} \\ \text{subtracting } c \text{ from each side} \end{array}$$

$$\frac{ad - c}{b} = x \quad \blacksquare \qquad \begin{array}{l} \text{multiplying each side by } \dfrac{1}{b} \text{ or} \\[2mm] \text{dividing each side by } b \ (b \neq 0) \end{array}$$

⊡ Literal Equations in which the Variable Appears More Than Once

The equation $ax + b = cx + d$ is an example of a literal equation in which the variable x appears twice. To solve a literal equation in which the variable appears more than once in the equation, you must

1. *Group* all the variable terms on one side of the equation and the nonvariable terms on the other side.

2. *Factor* out the variable you wish to solve for.

3. *Divide* each side of the equation by the expression being multiplied by the variable.

To solve $ax + b = cx + d$ for x, we must first bring the variable terms together on one side of the equation. To do this, we can either remove cx from the right side or ax from the left side.

Suppose we remove cx from the right side first.

Method 1:

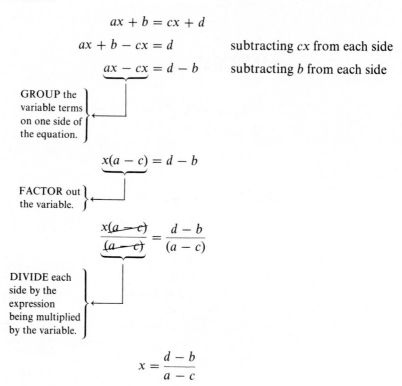

$$ax + b = cx + d$$

$$ax + b - cx = d \qquad \text{subtracting } cx \text{ from each side}$$

$$ax - cx = d - b \qquad \text{subtracting } b \text{ from each side}$$

GROUP the variable terms on one side of the equation.

$$x(a - c) = d - b$$

FACTOR out the variable.

$$\frac{x(a - c)}{(a - c)} = \frac{d - b}{(a - c)}$$

DIVIDE each side by the expression being multiplied by the variable.

$$x = \frac{d - b}{a - c}$$

Suppose we remove ax from the left side first.

Method 2:

$$ax + b = cx + d$$

$$b = cx + d - ax$$

$$b - d = cx - ax \leftarrow \text{GROUP}$$

$$b - d = x(c - a) \leftarrow \text{FACTOR}$$

$$\frac{b - d}{(c - a)} = \frac{x(c - a)}{(c - a)} \leftarrow \text{DIVIDE}$$

$$\frac{b - d}{c - a} = x$$

Are the two solutions the same? Does $\dfrac{d - b}{a - c} = \dfrac{b - d}{c - a}$? If we take $\dfrac{d - b}{a - c}$ and factor out -1 from the numerator and denominator, we would obtain

$$\frac{d-b}{a-c} = \frac{-1(b-d)}{-1(c-a)} = \frac{-1}{-1} \cdot \frac{(b-d)}{(c-a)} = 1 \cdot \frac{b-d}{c-a} = \frac{b-d}{c-a}.$$

Thus, $x = \dfrac{d-b}{a-c} = \dfrac{b-d}{c-a}$, and both answers are perfectly acceptable.

We can think of $x = \dfrac{d-b}{a-c}$ as a formula that can be used to solve equations of the form $ax + b = cx + d$. The formula will be valid except when $a = c$. If $a = c$, the denominator $(a - c)$ becomes zero, and division by zero is undefined.

EXAMPLE 2

Solve the literal equation $ax + b = \dfrac{cx}{d}$ for x.

Solution:

$$ax + b = \frac{cx}{d}$$

$$d(ax + b) = cx \qquad \text{multiplying each side by } d$$

$$adx + bd = cx \qquad \text{distributive law}$$

$$bd = cx - adx \qquad \text{group}$$

$$bd = x(c - ad) \qquad \text{factor}$$

$$\frac{bd}{(c - ad)} = \frac{x(c - ad)}{(c - ad)} \qquad \text{divide } (c - ad \neq 0)$$

$$\frac{bd}{c - ad} = x \quad \blacksquare$$

Note: If the variable terms had been grouped on the left side of the equation, we would have obtained

$$x = \frac{-bd}{ad - c}.$$

Although both answers are correct, $x = \dfrac{bd}{c - ad}$ is preferred, since it contains fewer negative signs.

At times it is necessary to solve a technical formula for a symbol that appears more than once.

EXAMPLE 3

Battery voltage $E = IR_i + IR_L$. Solve for the current, I.

Solution:

$$E = IR_i + IR_L$$
group (the symbol I is already grouped on the right side.)

$$E = I(R_i + R_L)$$ factor

$$\frac{E}{(R_i + R_L)} = \frac{I(\cancel{R_i + R_L})}{\cancel{(R_i + R_L)}}$$ divide

$$\frac{E}{R_i + R_L} = I \quad \blacksquare$$

EXAMPLE 4

Solve for the focal length of the lens, f, if the distance p of an object from a lens is $p = \dfrac{qf}{q - f}$.

Solution:

$$p = \frac{qf}{q - f}$$

$$p(q - f) = qf$$ multiplying each side by $(q - f)$

$$pq - pf = qf$$ distributive law

$$pq = qf + pf$$ group

$$pq = f(q + p)$$ factor

$$\frac{pq}{(q + p)} = \frac{f\cancel{(q + p)}}{\cancel{(q + p)}}$$ divide

$$\frac{pq}{q + p} = f \quad \blacksquare$$

EXERCISES 7.1

Solve each literal equation for x.

1. $\dfrac{ax}{b} = c$

2. $a = \dfrac{b}{cx + d}$

3. $ax = bx + c$

4. $a(b + cx) = dx$

5. $\dfrac{a + bx}{cx} = d$

6. $\dfrac{ax + b}{c} = dx + e$

7. $\dfrac{ax + b}{c} = \dfrac{dx}{e}$

8. $\dfrac{ax + b}{cx + d} = \dfrac{e}{f}$

9. $ax^2 + b = c$

10. $ax^3 + b = cx^3$

Solve each formula for the indicated variable.

11. Energy output of a drycell: $W = I^2 R_L t + I^2 R_i t$. Solve for the time the circuit operates, t.

12. Work done by an expanding gas: $W = PV_f - PV_i$. Solve for the pressure exerted by the gas, P.

13. Last term of an arithmetic progression: $L = a + nd - d$. Solve for the common difference, d.

14. Volume of concrete contained in a drainage pipe: $V = \pi R^2 L - \pi r^2 L$. Solve for the length of the pipe, L.

15. Resistance of a conductor: $R = \dfrac{\alpha L + \alpha L t}{d^2}$. Solve for the length of the conductor, L.

16. Deflection of a beam: $\delta = \dfrac{3LPx^2 - Px^3}{6EI}$. Solve for the concentrated load, P.

17. Thermal efficiency of a steam engine: $e = \dfrac{H_1 - H_2}{H_1}$. Solve for the heat delivered to the system, H_1.

18. Specific heat of a substance: $s = \dfrac{H}{Mt_2 - Mt_1}$. Solve for the mass of the body, M.

19. Total parallel resistance: $R_t = \dfrac{R_1 R_2}{R_1 + R_2}$. Solve for one of the resistances, R_1.

20. Volume expansion coefficient: $\beta = \dfrac{V_f - V_i}{V_i t}$. Solve for the initial volume, V_i.

21. Total power dissipated by two resistors connected in series: $P = I^2 R_1 + I^2 R_2$. Solve for the current, I (see Fig. 7.1).

22. Maximum deflection of a cantilever beam with a uniformly distributed load plus a concentrated load at the free end: $\delta_{max} = \dfrac{8PL^3 + 3WL^3}{24EI}$. Solve for the length of the beam, L (see Fig. 7.2).

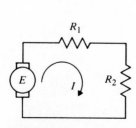

Figure 7.1

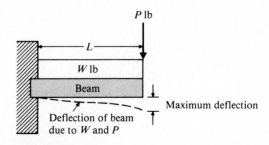

Figure 7.2

7.2 SOLVING EQUATIONS BY FACTORING

Objectives

⊡ To solve an equation that has a product of algebraic expressions equal to zero.

⊡ To solve a quadratic equation by factoring.

⊡ To solve other types of equations by factoring.

⊡ Products that Equal Zero

In Chapter 1 you learned that any number multiplied by zero equals zero.

$$0(a) = a(0) = 0$$

Now, suppose you know that the product of two numbers, a and b, equals zero.

$$ab = 0.$$

What does that tell you about either a or b? Surely *if the product of two numbers is zero, then at least one of those numbers must be zero.*

THE ZERO PRODUCT PROPERTY
If $ab = 0$, then either $a = 0$ or $b = 0$ (or both a and b are zero).

This property allows us to solve any equation that has a product of algebraic expressions equal to zero.

EXAMPLE 1

Solve $(x + 1)(x - 2) = 0$.

Solution:

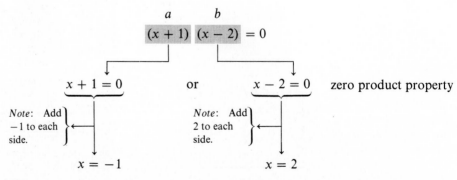

Thus, if $(x + 1)(x - 2) = 0$, then $x = -1$ or $x = 2$. ∎

Note: We can easily check these solutions as follows:

Does $(x + 1)(x - 2) = 0$, when $x = -1$?

$$(-1 + 1)(-1 - 2) = 0?$$
$$(0)(-3) \qquad = 0?$$
$$0 \qquad\qquad = 0 \quad \text{true}$$

Does $(x + 1)(x - 2) = 0$, when $x = 2$?

$$(2 + 1)(2 - 2) = 0?$$
$$(3)(0) \qquad = 0?$$
$$0 \qquad\quad = 0 \quad \text{true}$$

EXAMPLE 2

Solve $10x(x + 3)(2x - 5) = 0$.

Solution: In this equation, we have the product of three algebraic expressions that equal zero. Extending the zero product property, we can say that if $abc = 0$, then $a = 0$, $b = 0$, or $c = 0$.

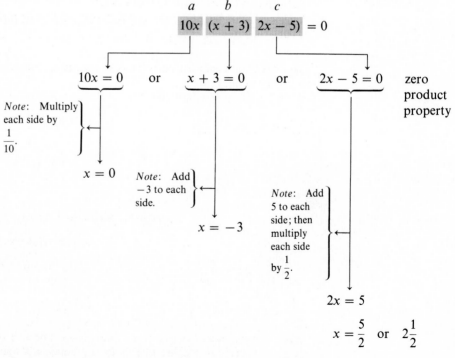

Thus, if $10x(x + 3)(2x - 5) = 0$, then $x = 0$, $x = -3$, or $x = \dfrac{5}{2}$.   ∎

⊡ Solving Quadratic Equations by Factoring

A literal equation of the form $ax^2 + bx + c = 0$, where x is the variable and a, b, and c are constants, is called a **quadratic equation in standard form**. All quadratic equations must contain a squared term (ax^2). Thus, for $ax^2 + bx + c = 0$, a cannot equal zero.

In Chapter 6 you learned to factor expressions like $ax^2 + bx + c$. Remember, factoring forms products, and if we can obtain a product that equals zero, we can apply the *zero product property*.

EXAMPLE 3

Solve the quadratic equation $x^2 + 2x - 8 = 0$.

Solution:

$$x^2 + 2x - 8 = 0$$

$$(x + 4)\ (x - 2) = 0 \qquad \text{factoring a trinomial}$$

$$x + 4 = 0 \quad \text{or} \quad x - 2 = 0 \qquad \text{zero product property}$$

$$x = -4 \quad \text{or} \quad x = 2 \quad \blacksquare$$

Note: We can easily check these solutions as follows:

Does $x^2 + 2x - 8 = 0$, when $x = -4$?

$$(-4)^2 + 2(-4) - 8 = 0?$$

$$16 + (-8) - 8 = 0?$$

$$0 = 0 \quad \text{true}$$

Does $x^2 + 2x - 8 = 0$, when $x = 2$?

$$(2)^2 + 2(2) - 8 = 0?$$

$$4 + 4 - 8 = 0?$$

$$0 = 0 \quad \text{true}$$

CAUTION Never start the factoring process unless one side of the equation equals zero. Remember, you are looking for a product that equals *zero*. Only then can you apply the zero product property. This means that quadratic equations must be in *standard form* before you begin to factor.

EXAMPLE 4

Solve $4x^2 - 8x = -3$.

Solution: We must first make one side of the equation equal zero.

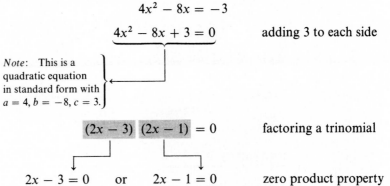

$$4x^2 - 8x = -3$$

$$4x^2 - 8x + 3 = 0 \qquad \text{adding 3 to each side}$$

Note: This is a quadratic equation in standard form with $a = 4, b = -8, c = 3.$

$$(2x - 3)\ (2x - 1) = 0 \qquad \text{factoring a trinomial}$$

$$2x - 3 = 0 \quad \text{or} \quad 2x - 1 = 0 \qquad \text{zero product property}$$

$$2x = 3 \qquad\qquad 2x = 1$$

$$x = \frac{3}{2} \quad \text{or} \quad\quad x = \frac{1}{2} \ \blacksquare$$

Note: Check by substituting $x = \dfrac{3}{2}$ and $x = \dfrac{1}{2}$ in the original equation, $4x^2 - 8x = -3$.

EXAMPLE 5

Solve $x^2 = 4x$.

Solution: We must first make one side of the equation equal zero.

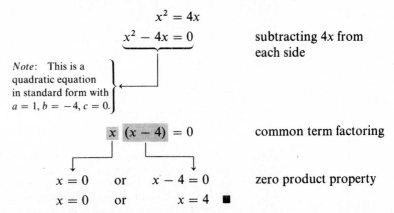

$$x^2 = 4x$$

$$x^2 - 4x = 0 \qquad \text{subtracting } 4x \text{ from each side}$$

Note: This is a quadratic equation in standard form with $a = 1, b = -4, c = 0.$

$$x\ (x - 4) = 0 \qquad \text{common term factoring}$$

$$x = 0 \quad \text{or} \quad x - 4 = 0 \qquad \text{zero product property}$$

$$x = 0 \quad \text{or} \quad\quad x = 4 \ \blacksquare$$

Often a student will immediately divide each side of an equation like $x^2 = 4x$ by x as follows:

$$x^2 = 4x$$

$$\frac{x^2}{x} = \frac{4x}{x}$$

$$x = 4.$$

Notice that if you take this route, you will lose the solution $x = 0$ (see Example 5).

CAUTION Avoid dividing each side of an equation by a variable expression. You may lose one of the solutions.

EXAMPLE 6

Solve $(t - 3)(t + 4) = t - 3$.

Solution: Although it is tempting to immediately divide each side of the equation by $(t - 3)$, we must *not* take this route since $(t - 3)$ is a variable expression. We should proceed as follows:

$$(t - 3)(t + 4) = t - 3$$
$$t^2 + t - 12 = t - 3 \qquad \text{FOIL}$$
$$t^2 + t - 9 = t \qquad \text{adding 3 to each side}$$
$$\underbrace{t^2 - 9 = 0} \qquad \text{adding } -t \text{ to each side}$$

Note: This is a quadratic equation in standard form with $a = 1, b = 0, c = -9$.

$$(t + 3)\ (t - 3) = 0 \qquad \text{factoring the difference of two perfect squares}$$

$$t + 3 = 0 \quad \text{or} \quad t - 3 = 0 \qquad \text{zero product property}$$

$$t = -3 \quad \text{or} \qquad t = 3 \quad \blacksquare$$

Note: To solve the quadratic equation $t^2 - 9 = 0$, we could also proceed as follows:

$$t^2 - 9 = 0$$
$$t^2 = 9 \qquad \text{adding 9 to each side}$$
$$t = \pm\sqrt{9} \qquad \text{if } x^2 = a, \text{ then } x = \pm\sqrt{a} \text{ provided } a \text{ is positive}$$
$$t = \pm 3 \qquad \text{simplifying radicals}$$

This is the *square root method* of solving an equation, which you learned in Chapter 5.

Solving Other Types of Equations by Factoring

Any equation that can be factored into a product of algebraic expressions that equals zero can be solved by using the zero product property.

EXAMPLE 7

Solve $x^3 + 9x = 6x^2$.

Solution: Although $x^3 + 9x = 6x^2$ is not a quadratic equation (since it contains an x^3 term), we might still be able to solve this equation if we can form a product that equals zero.

$$x^3 + 9x = 6x^2$$
$$x^3 - 6x^2 + 9x = 0 \qquad \text{subtracting } 6x^2 \text{ from each side}$$
$$x(x^2 - 6x + 9) = 0 \qquad \text{common term factoring}$$
$$x\,(x - 3)\,(x - 3) = 0 \qquad \text{factoring a trinomial}$$

$$x = 0 \quad \text{or} \quad x - 3 = 0 \qquad \text{zero product property}$$
$$x = 0 \quad \text{or} \quad x = 3 \quad \blacksquare$$

EXAMPLE 8

Solve $x^3 + 2x^2 + x + 2 = 0$

Solution:

$$x^3 + 2x^2 + x + 2 = 0$$
$$x^2(x + 2) + 1(x + 2) = 0 \qquad \text{factoring by grouping terms}$$
$$(x + 2)\,(x^2 + 1) = 0$$

$$x + 2 = 0 \quad \text{or} \quad x^2 + 1 = 0 \qquad \text{zero product property}$$
$$x = -2 \quad \text{or} \quad x^2 = -1$$

Note: If $x^2 = a$, and a is negative, then there is no real solution to the equation.

no real solution

Thus, the only real solution to $x^3 + 2x^2 + x + 2 = 0$ is $x = -2$. $\blacksquare$

Application

FREELY FALLING OBJECTS

Nearly four hundred years ago, Galileo discovered the laws of motion for freely falling objects. He found that the time (in seconds) it took an object to strike the ground was related to the distance (in feet) the object fell and the velocity (in feet per second) with which the object was projected downward.

FORMULA: **Freely Falling Objects**

$$d = vt + 16t^2$$

where d is the distance fallen (ft), v is the initial velocity $\left(\dfrac{\text{ft}}{\text{sec}}\right)$, and t is the time (sec).

EXAMPLE 9

A dive bomber projects a bomb downward with an initial velocity of 320 ft/s from a height of 4800 ft. How long will it take the bomb to reach the ground (see Fig. 7.3)?

Solution:

$$d = vt + 16t^2$$
$$4800 = 320t + 16t^2 \qquad \text{substituting}$$
$$-16t^2 - 320t + 4800 = 0 \qquad \text{adding } -16t^2 \text{ and } -320t$$
$$\text{to each side}$$

Note: This is a quadratic equation in standard form with $a = -16$, $b = -320$, $c = 4800$.

$$-16(t^2 + 20t - 300) = 0 \qquad \text{common term factoring}$$
$$-16(t + 30)(t - 10) = 0 \qquad \text{factoring a trinomial}$$

Note: The only possible factors that can be zero are $(t + 30)$ and $(t - 10)$.

$$t + 30 = 0 \qquad \text{or} \quad t - 10 = 0 \qquad \text{zero product property}$$
$$t = -30 \quad \text{or} \qquad t = 10$$

There appear to be two solutions to the problem. However, we cannot have a negative value for time. Thus, $t = 10$ s is the only meaningful solution. ∎

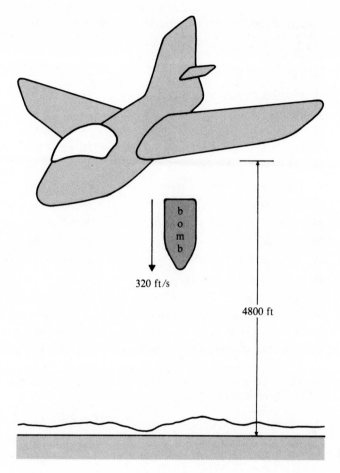

320 ft/s

4800 ft

Figure 7.3

EXERCISES 7.2

Using the zero product property, solve the following equations.

1. $(x - 4)(x + 3) = 0$

2. $(y + 5)(y - 11) = 0$

3. $(2x + 7)(3x + 1) = 0$

4. $(7x - 1)(2x - 9) = 0$

5. $3t(t - 3) = 0$

6. $x(3x - 5) = 0$

7. $y(9 - 2y)(4 - y) = 0$

8. $5n(2 + 3n)(2 + 3n) = 0$

9. $3m^2(5 - 7m)(2 + m) = 0$

10. $t^3(8 - 3t)(1 + t) = 0$

11. $(R^2 + 4)(R - 5) = 0$

12. $(3x - 1)(2x^2 + 3) = 0$

13. $(3x^2 - 4)(x + 9) = 0$

14. $3R^2(2R^2 - 5) = 0$

15. $(v^3 + 1)(v^2 + 1) = 0$

16. $(8y^3 + 1)(8y^2 - 1) = 0$

Using your knowledge of factoring and the zero product property, solve the following equations.

17. $x^2 + 3x - 10 = 0$

18. $x^2 - 7x + 12 = 0$

19. $2x^2 + 9x - 5 = 0$

20. $3x^2 + 5x + 2 = 0$

21. $y^2 - 10y = -16$

22. $t^2 + 3t = 28$

23. $4t^2 + 17t = 15$

24. $6y^2 = 17y - 5$

25. $3w^2 = 15w$

26. $8I^2 = 17I$

27. $(t + 2)(3t - 4) = t + 2$

28. $(2t - 5)(t - 3) = 2t - 5$

29. $2x^3 + 10x^2 + 8x = 0$

30. $6p^3 + 21p^2 = 12p$

31. $12h^4 + 3h^2 = 12h^3$

32. $m^5 = 9m^4 - 14m^3$

33. $2x^3 + 3x^2 - 8x - 12 = 0$

34. $3x^3 - x^2 - 27x + 9 = 0$

35. $x^3 - 2x^2 = 2 - x$

36. $y^3 - 3y = 5y^2 - 15$

37. $2x^4 = 16x^2$

38. $3v^6 = 54v^3$

39. $(x^3 + 1)(x^2 + 2) = x^2 + 2$

40. $(x^3 + 8)(x^2 - 2) = x^3 + 8$

41. $t^2 + (t + 8)(t - 2) = 2t$

42. $2y^2 - (3y + 2)(y - 1) = 2$

43. $x(3x + 2) = (x + 2)^2$

44. $y(3 - y) = (3 - y)^2$

45. A dive bomber projects a bomb downward with an initial velocity of 72 ft/s from a height of 1600 ft. How long will it take the bomb to reach the ground?

46. A baseball is simply dropped from the top of a 100-ft building. How long will it take the baseball to reach the ground?

47. The actual power delivered to the load by a generator with a voltage E and an internal resistance R_i is given by

$$P = EI - R_i I^2,$$

where I is the current flowing in the circuit (see Fig. 7.4). Find the current in the circuit if $P = 36$ watts, $E = 12$ volts, and $R_i = 1\,\Omega$.

48. Referring to exercise 47, find the current in the circuit if $P = 400$ W, $E = 80$ V, and $R_i = 4\,\Omega$.

49. The total surface area of a cylinder with radius r and height h is given by

$$S.A. = 2\pi r^2 + 2\pi rh$$

Find the radius of a cylinder if $S.A. = 48\pi$ cm^2 and $h = 5$ cm (see Fig. 7.5).

50. Referring to exercise 49, find the radius of a cylinder if $S.A. = 25\pi$ in^2 and $h = 2\frac{1}{2}$ in.

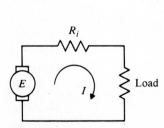

Figure 7.4

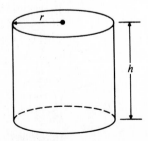

Figure 7.5

7.3 SOLVING QUADRATIC EQUATIONS BY COMPLETING THE PERFECT SQUARE

Objectives

⊡ To complete a perfect square trinomial.

⊡ To solve equations of the form $(x + a)^2 = b$ by using the square root property.

⊡ To solve a quadratic equation by completing the perfect square.

⊡ Completing a Perfect Square

The trinomials $x^2 + 6x + 9$ and $x^2 - 8x + 16$ are referred to as **perfect square trinomials**, since they can be written as $(x + 3)^2$ and $(x - 4)^2$ respectively.

Is there any special relationship between the coefficient of the second term and the last term in a perfect square trinomial? Observe the following:

$$x^2 + \boxed{6}x + 9$$
$\frac{1}{2}$ of 6 is
3, and $(3)^2$ is 9.

$$x^2 - \boxed{8}x + 16$$
$\frac{1}{2}$ of -8 is
-4, and $(-4)^2$ is 16.

Notice that *if we take one-half the coefficient of the second term and then square that result, we will obtain the last term of the perfect square trinomial.*

CAUTION The order in which the terms appear is critical. The trinomial $6x + 9 + x^2$ can be tested for being a perfect square trinomial only after it has been properly arranged as

second term
↑
first term | third term
$$x^2 + \boxed{6x} + \boxed{9}$$
x^2 term | constant term
↓
x term

We can use this idea to form our own perfect square trinomials. The process of finding the last term of an expression so that it becomes a perfect square trinomial is called **completing the perfect square**.

EXAMPLE 1

Complete each perfect square.

a) $x^2 - 12x + \underline{\quad?\quad}$

b) $x^2 + 3x + \underline{\quad?\quad}$

Solution:

a) $x^2 - \boxed{12}x + \boxed{36}$
$\frac{1}{2}$ of -12 is
-6, and $(-6)^2$
is 36.

b) $x^2 + \boxed{3}x + \boxed{\dfrac{9}{4}}$
$\frac{1}{2}$ of 3 is
$\frac{3}{2}$, and $(\frac{3}{2})^2$ is $\frac{9}{4}$.
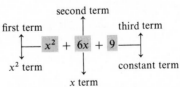

The trinomials $x^2 - 12x + 36$ and $x^2 + 3x + \dfrac{9}{4}$ are perfect square trinomials

since they can be written as $(x - 6)^2$ and $\left(x + \dfrac{3}{2}\right)^2$ respectively. ∎

 CAUTION   This procedure for completing the perfect square will only work if the coefficient of the x^2 term is 1.

⊡ Equations of the Form $(x + a)^2 = b$

In Chapter 5 you learned to solve equations of the form $x^2 = a$ for x by using the *square root property*.

SQUARE ROOT PROPERTY
If $x^2 = a$, and a is *positive*, then $x = \pm\sqrt{a}$.
If $x^2 = a$, and a is *negative*, then $x^2 = a$ has *no real solution*.

We can use this property to help us solve equations of the form $(x + a)^2 = b$ for x.

EXAMPLE 2

Solve $(x - 2)^2 = 9$.

Solution:

$$(x - 2)^2 = 9$$
$$x - 2 = \pm\sqrt{9} \qquad \text{square root property}$$
$$x - 2 = \pm 3 \qquad \text{simplifying radicals}$$
$$x = 2 \boxed{\pm} 3 \qquad \text{adding 2 to each side}$$

Note: There are two solutions.

$$x = \boxed{2 + 3} \quad \text{or} \quad x = \boxed{2 - 3}$$
$$x = 5 \qquad\quad \text{or} \quad x = -1 \quad ∎$$

Note: Alternately, we could have expanded first and solved as follows:

$$(x - 2)^2 = 9$$
$$x^2 - 4x + 4 = 9 \qquad (a - b)^2 = a^2 - 2ab + b^2$$
$$\underbrace{x^2 - 4x - 5 = 0} \qquad \text{adding } -9 \text{ to each side}$$

Note: This is a quadratic equation in standard form, with $a = 1$, $b = -4$ and $c = -5$.

$$(x - 5)(x + 1) = 0 \qquad \text{factoring a trinomial}$$

$$x - 5 = 0 \quad \text{or} \quad x + 1 = 0 \qquad \text{zero product property}$$

$$x = 5 \quad \text{or} \qquad x = -1$$

EXAMPLE 3

Solve $(x + 4)^2 = 12$.

Solution:

$$(x + 4)^2 = 12$$

$$x + 4 = \pm\sqrt{12} \qquad \text{square root property}$$

$$x + 4 = \pm 2\sqrt{3} \qquad \text{simplifying radicals}$$

$$x = -4 \ \boxed{\pm}\ 2\sqrt{3} \qquad \text{adding } -4 \text{ to each side}$$

Note: There are two solutions.

$$x = -4 + 2\sqrt{3} \quad \text{or} \quad x = -4 - 2\sqrt{3} \quad \blacksquare$$

Note: Be careful! Do not violate the order of operations. You *cannot* add or subtract before you multiply. Thus, to write

$$-4 + 2\sqrt{3} \text{ as } -2\sqrt{3} \text{ or}$$

$$-4 - 2\sqrt{3} \text{ as } -6\sqrt{3} \text{ is } \textit{wrong}.$$

Using a calculator, $x = -4 + 2\sqrt{3} \approx -.536$ three significant digits

$$\boxed{4}\,\boxed{+/-}\,\boxed{+}\,\boxed{2}\,\boxed{\times}\,\boxed{3}\,\boxed{\sqrt{\ }}\,\boxed{=}\ -.5358984$$

Display

Similarly,

$$x = -4 - 2\sqrt{3} \approx -7.46 \quad \text{three significant digits}$$

$$\boxed{4}\,\boxed{+/-}\,\boxed{-}\,\boxed{2}\,\boxed{\times}\,\boxed{3}\,\boxed{\sqrt{\ }}\,\boxed{=}\ -7.4641016$$

Display

⠒ Solving a Quadratic Equation by Completing the Perfect Square

Let's return to example 3, $(x + 4)^2 = 12$. If we expanded first, we would obtain

$$(x + 4)^2 = 12$$

$$x^2 + 8x + 16 = 12 \qquad (a + b)^2 = a^2 + 2ab + b^2$$

$$x^2 + 8x + 4 = 0 \qquad \text{adding } -12 \text{ to each side}$$

$$\begin{cases} \textit{Note}: & \text{This is a quadratic equation in} \\ & \text{standard form with } a = 1, b = 8, \text{and } c = 4. \end{cases}$$

Can we solve this quadratic equation by factoring? Notice that

$$x^2 + 8x + 4,$$

$$\downarrow c = 4$$

$$\downarrow b = 8$$

but there do not exist any factors of 4 whose sum is 8. Unlike the quadratic equations in Section 7.2, this quadratic equation *cannot* be solved by factoring. To solve a nonfactorable quadratic equation, we can write it in the form $(x + a)^2 = b$, and then use the square root property to solve for x. To go from $x^2 + 8x + 4$ to $(x + 4)^2 = 12$, you must *complete the perfect square* as follows:

FOUR-STEP PROCEDURE FOR SOLVING A QUADRATIC EQUATION BY COMPLETING THE PERFECT SQUARE
Using $x^2 + 8x + 4 = 0$ as an Example

Step 1: Isolate the constant on one side of the equation. $\quad \rightarrow x^2 + 8x = -4 \qquad$ adding -4 to each side

Step 2: Complete the perfect square and add this number to *each* side of the equation. $\quad \rightarrow x^2 + 8x + \boxed{16} = -4 + \boxed{16} \qquad$ adding 16 to each side

$\quad \downarrow \frac{1}{2}$ of 8 is 4 and $(4)^2$ is $\boxed{16}$.

Step 3: Factor the perfect square trinomial. $\quad \rightarrow (x + 4)^2 = 12$

Step 4: Solve for the variable by using the square root property. $\quad \rightarrow x = -4 \pm 2\sqrt{3} \qquad$ (See example 3.)

EXAMPLE 4

Solve $x^2 + 6x - 18 = 0$ by completing the perfect square.

Solution:

Step 1:

$$x^2 + 6x = 18 \qquad \text{isolating the constant}$$

Step 2:

$$x^2 + 6x + \boxed{9} = 18 + \boxed{9}$$

$\quad \downarrow \frac{1}{2}$ of 6 is 3 and $(3)^2$ is $\boxed{9}$.

completing the perfect square by adding 9 to each side

Step 3:

$$(x + 3)^2 = 27 \qquad \text{factoring the perfect square trinomial}$$

Step 4:

$$x + 3 = \pm\sqrt{27} \qquad \text{square root property}$$

$$x + 3 = \pm 3\sqrt{3} \qquad \text{simplifying radicals}$$

$$x = -3 \pm 3\sqrt{3} \qquad \text{adding } -3 \text{ to each side}$$

$$x = -3 + 3\sqrt{3} \approx 2.20 \quad \text{or} \quad x = -3 - 3\sqrt{3} \approx -8.20 \quad \blacksquare$$

EXAMPLE 5

Solve $y^2 - 3y - 9 = 0$ by completing the perfect square.

Solution:

Step 1:

$$y^2 - 3y = 9 \qquad \text{isolating the constant}$$

Step 2:

$$y^2 - 3y + \frac{9}{4} = 9 + \frac{9}{4} \qquad \begin{array}{l}\text{completing the perfect} \\ \text{square by adding } \dfrac{9}{4} \text{ to each side}\end{array}$$

$$\frac{1}{2} \text{ of } -3 \text{ is } -\frac{3}{2} \text{ and } \left(-\frac{3}{2}\right)^2 \text{ is } \frac{9}{4}.$$

$$\left(9 + \frac{9}{4} = \frac{36}{4} + \frac{9}{4} = \frac{45}{4}\right)$$

Step 3:

$$\left(y - \frac{3}{2}\right)^2 = \frac{45}{4} \qquad \text{factoring the perfect square trinomial}$$

Step 4:

$$y - \frac{3}{2} = \pm\sqrt{\frac{45}{4}} \qquad \text{square root property}$$

$$y - \frac{3}{2} = \pm\frac{3\sqrt{5}}{2} \qquad \text{simplifying radicals}$$

$$y = \frac{3}{2} \pm \frac{3\sqrt{5}}{2} \qquad \text{adding } \frac{3}{2} \text{ to each side}$$

$$y = \frac{3 \pm 3\sqrt{5}}{2} \quad \blacksquare \quad \text{adding fractions}$$

EXAMPLE 6

Solve $2x^2 - 3x - 2 = 0$ by completing the perfect square.

Solution:

Step 1:

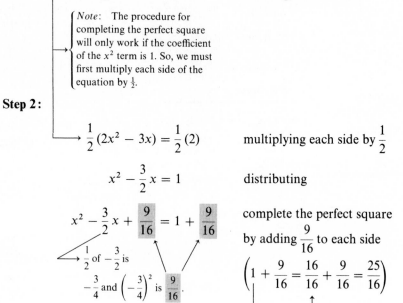

$$2x^2 - 3x = 2 \qquad \text{isolating the constant}$$

Note: The procedure for completing the perfect square will only work if the coefficient of the x^2 term is 1. So, we must first multiply each side of the equation by $\frac{1}{2}$.

Step 2:

$$\frac{1}{2}(2x^2 - 3x) = \frac{1}{2}(2) \qquad \text{multiplying each side by } \frac{1}{2}$$

$$x^2 - \frac{3}{2}x = 1 \qquad \text{distributing}$$

$$x^2 - \frac{3}{2}x + \frac{9}{16} = 1 + \frac{9}{16}$$

complete the perfect square by adding $\frac{9}{16}$ to each side

$$\left(1 + \frac{9}{16} = \frac{16}{16} + \frac{9}{16} = \frac{25}{16}\right)$$

$\frac{1}{2}$ of $-\frac{3}{2}$ is $-\frac{3}{4}$ and $\left(-\frac{3}{4}\right)^2$ is $\frac{9}{16}$.

Step 3:

$$\left(x - \frac{3}{4}\right)^2 = \frac{25}{16} \qquad \text{factoring the perfect square trinomial}$$

Step 4:

$$x - \frac{3}{4} = \pm\sqrt{\frac{25}{16}} \qquad \text{square root property}$$

$$x - \frac{3}{4} = \pm\frac{5}{4} \qquad \text{simplifying radicals}$$

$$x = \frac{3}{4} \pm \frac{5}{4} \qquad \text{adding } \frac{3}{4} \text{ to each side}$$

$$x = \frac{3}{4} + \frac{5}{4} \qquad \text{or} \qquad x = \frac{3}{4} - \frac{5}{4}$$

$$x = 2 \qquad \text{or} \qquad x = -\frac{1}{2} \quad \blacksquare$$

Note: Try solving the equation $2x^2 = 3x + 2$ by using factoring and the zero product property. See if you can obtain the same solutions.

EXAMPLE 7

Solve $t^2 + 12t + 40 = 0$ by completing the perfect square.

Solution:

Step 1:

$$t^2 + 12t = -40 \qquad \text{isolating the constant}$$

Step 2:

$$t^2 + 12t + \boxed{36} = -40 + \boxed{36}$$

$\frac{1}{2}$ of 12 is 6 and $(6)^2$ is $\boxed{36}$.

completing the perfect square by adding 36 to each side

Step 3:

$$(t + 6)^2 = \boxed{-4}$$

Note: A negative number

factoring the perfect square trinomial

Step 4:

$$\text{NO REAL SOLUTION} \quad \blacksquare \qquad \text{square root property}$$

As you have probably discovered, factorable quadratic equations will always have *rational solutions*, while nonfactorable quadratic equations will always have either *irrational solutions* or *no real solution* at all.

EXERCISES 7.3

Complete the perfect square, then write the perfect square trinomial in the form $(x + a)^2$.

1. $x^2 + 4x + \underline{\ ?\ }$

2. $x^2 + 10x + \underline{\ ?\ }$

3. $y^2 - 22y + \underline{\quad}$

4. $y^2 - 18y + \underline{\ ?\ }$

5. $t^2 + 9t + \underline{\ ?\ }$

6. $p^2 + 5p + \underline{\ ?\ }$

7. $x^2 - x + \underline{\ ?\ }$

8. $y^2 + y + \underline{\ ?\ }$

9. $y^2 + \frac{1}{2}y + \underline{\ ?\ }$

10. $w^2 + \frac{1}{4}w + \underline{\ ?\ }$

11. $t^2 - \dfrac{3}{5}t + \underline{}$

12. $t^2 + \dfrac{4}{3}t + \underline{}$

13. $x^2 + bx + \underline{}$

14. $x^2 + \dfrac{b}{a}x + \underline{}$

Solve the following equations by using the square root property. Write all irrational solutions in simplified radical form and as decimal approximations containing three significant digits.

15. $(x + 1)^2 = 16$

16. $(x - 4)^2 = 25$

17. $\left(t - \dfrac{1}{3}\right)^2 = \dfrac{1}{9}$

18. $\left(q + \dfrac{1}{7}\right)^2 = \dfrac{4}{49}$

19. $(x + 2)^2 = 17$

20. $(y - 5)^2 = 41$

21. $(n - 1)^2 = 20$

22. $(m + 6)^2 = 45$

23. $\left(x + \dfrac{1}{3}\right)^2 = \dfrac{16}{3}$

24. $\left(x - \dfrac{3}{5}\right)^2 = \dfrac{2}{5}$

Using the four-step procedure outlined in this section, solve the following quadratic equations by completing the perfect square. Write all irrational solutions in simplified radical form and as decimal approximations containing three significant digits.

25. $x^2 + 7x + 10 = 0$

26. $x^2 + 12x + 32 = 0$

27. $y^2 - 8y + 12 = 0$

28. $y^2 - 4y - 45 = 0$

29. $t^2 + 4t - 6 = 0$

30. $p^2 + 2p - 9 = 0$

31. $m^2 - 8m - 4 = 0$

32. $n^2 - 12n + 4 = 0$

33. $x^2 - x - 9 = 0$

34. $x^2 - 7x + 2 = 0$

35. $x^2 + 2x + 12 = 0$

36. $x^2 - 5x + 13 = 0$

37. $2t^2 + 8t - 17 = 0$

38. $3x^2 - 18x + 2 = 0$

39. $5y^2 - 8y - 21 = 0$

40. $4p^2 - 3p - 27 = 0$

41. $x^2 + bx = 0$

42. $ax^2 + bx = 0$

7.4   THE QUADRATIC FORMULA

Objective

⊡ To solve a quadratic equation by using the quadratic formula.

⊡ Deriving and Using the Quadratic Formula

In Section 7.2 we defined an equation of the form $ax^2 + bx + c = 0$ as a quadratic equation in standard form. If we could solve this literal equation for the variable x, we would have a *formula* that could be used to find the solution of *any* quadratic equation.

To solve $ax^2 + bx + c = 0$, we proceed as we did in Section 7.3 and apply the four-step procedure for solving a quadratic equation by completing the perfect square.

Step 1:

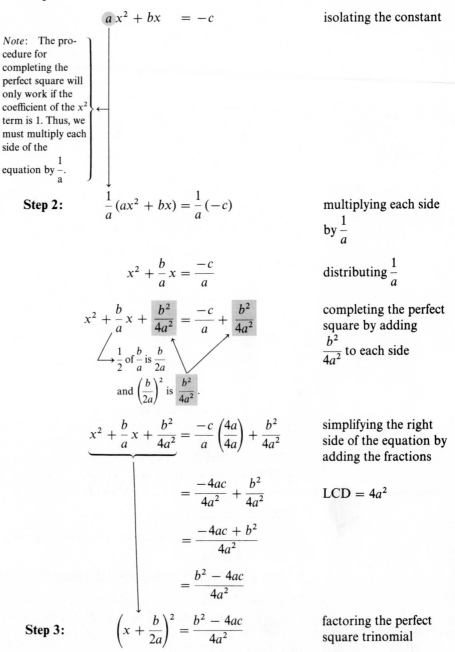

$$a\,x^2 + bx \quad = -c \qquad\qquad \text{isolating the constant}$$

Note: The procedure for completing the perfect square will only work if the coefficient of the x^2 term is 1. Thus, we must multiply each side of the equation by $\dfrac{1}{a}$.

Step 2:

$$\frac{1}{a}(ax^2 + bx) = \frac{1}{a}(-c) \qquad \text{multiplying each side by } \frac{1}{a}$$

$$x^2 + \frac{b}{a}x = \frac{-c}{a} \qquad \text{distributing } \frac{1}{a}$$

$$x^2 + \frac{b}{a}x + \boxed{\frac{b^2}{4a^2}} = \frac{-c}{a} + \boxed{\frac{b^2}{4a^2}} \qquad \begin{array}{l}\text{completing the perfect}\\\text{square by adding}\\ \frac{b^2}{4a^2} \text{ to each side}\end{array}$$

$$\frac{1}{2} \text{ of } \frac{b}{a} \text{ is } \frac{b}{2a}$$

$$\text{and } \left(\frac{b}{2a}\right)^2 \text{ is } \frac{b^2}{4a^2}.$$

$$x^2 + \frac{b}{a}x + \frac{b^2}{4a^2} = \frac{-c}{a}\left(\frac{4a}{4a}\right) + \frac{b^2}{4a^2} \qquad \begin{array}{l}\text{simplifying the right}\\\text{side of the equation by}\\\text{adding the fractions}\end{array}$$

$$= \frac{-4ac}{4a^2} + \frac{b^2}{4a^2} \qquad \text{LCD} = 4a^2$$

$$= \frac{-4ac + b^2}{4a^2}$$

$$= \frac{b^2 - 4ac}{4a^2}$$

Step 3:

$$\left(x + \frac{b}{2a}\right)^2 = \frac{b^2 - 4ac}{4a^2} \qquad \begin{array}{l}\text{factoring the perfect}\\\text{square trinomial}\end{array}$$

Step 4:

$$x + \frac{b}{2a} = \pm\sqrt{\frac{b^2 - 4ac}{4a^2}} \qquad \text{square root property}$$

$$x + \frac{b}{2a} = \pm\frac{\sqrt{b^2 - 4ac}}{\sqrt{4a^2}} \qquad \sqrt{\frac{a}{b}} = \frac{\sqrt{a}}{\sqrt{b}}$$

$$x + \frac{b}{2a} = \pm\frac{\sqrt{b^2 - 4ac}}{2a}$$

Note: We can simplify the square root of a product but we cannot simplify the square root of a sum or difference.

$$x = -\frac{b}{2a} \pm \frac{\sqrt{b^2 - 4ac}}{2a} \qquad \text{adding } -\frac{b}{2a} \text{ to each side}$$

$$x = \frac{-b \pm \sqrt{b^2 - 4ac}}{2a} \qquad \begin{array}{l}\text{adding fractions}\\ \text{LCD} = 2a\end{array}$$

Keep in mind that there are two solutions,

$$x = \frac{-b + \sqrt{b^2 - 4ac}}{2a} \quad \text{and} \quad x = \frac{-b - \sqrt{b^2 - 4ac}}{2a}.$$

QUADRATIC FORMULA

If $ax^2 + bx + c = 0$, then $x = \dfrac{-b \pm \sqrt{b^2 - 4ac}}{2a}$, $a \neq 0$.

When using the formula, the first part to evaluate is $b^2 - 4ac$. The quantity $b^2 - 4ac$ is called the **discriminant**. If the discriminant is a *perfect square*, the solutions to the quadratic equation will be rational numbers. If the discriminant is *not* a perfect square but is a *positive* number, the solutions will be *irrational numbers*. However, if the discriminant is a *negative* number, the quadratic equation will have *no real solution*, since we can't take the square root of a negative number.

EXAMPLE 1

Solve $x^2 + 4x - 12 = 0$ using the quadratic formula.

Solution: The equation $x^2 + 4x - 12 = 0$ is a quadratic equation in standard form. Before we can use the formula, we must identify the constants a, b, and c.

$$\underset{a = 1}{1}\,x^2 + \underset{b=4}{4}\,x \underset{c=-12}{-12} = 0$$

Be sure to include the negative sign.

First, evaluate the discriminant.

$$b^2 - 4ac = (4)^2 - \overbrace{4(1)(-12)} = 16 - (-48) = 16 + 48 = \boxed{64}.$$

Note: The discriminant is a *perfect square*. The solutions will be *rational numbers*.

$$x = \frac{-b \pm \sqrt{b^2 - 4ac}}{2a} = \frac{-(4) \pm \sqrt{64}}{2(1)} = \frac{-4 \pm 8}{2}$$

$$x = \frac{-4 + 8}{2} = \frac{4}{2} = 2 \quad \text{or} \quad x = \frac{-4 - 8}{2} = \frac{-12}{2} = -6 \quad \blacksquare$$

Note: Try solving $x^2 + 4x - 12 = 0$ by factoring. See if you can obtain the same results.

The quadratic equation must be in standard form before the formula is applied (that is, one side of the equation must equal zero). Make sure you don't mistake one coefficient for another when applying the quadratic formula.

EXAMPLE 2

Solve $y^2 = 7y - 3$ using the quadratic formula.

Solution: We must first make one side of the equation equal zero. Only then can we correctly identify a, b, and c.

$$b = -7$$

$$1\,y^2 - 7\,y + 3 = 0 \qquad \text{a quadratic equation}$$
$$a = 1 \qquad\qquad c = 3 \qquad \text{in standard form}$$

First, evaluate the discriminant.

$$b^2 - 4ac = (-7)^2 - \overbrace{4(1)(3)} = 49 - 12 = \boxed{37}$$

Note: The discriminant is *not* a perfect square, but it is *positive*. The solutions will be *irrational numbers*.

$$y = \frac{-b \pm \sqrt{b^2 - 4ac}}{2a} = \frac{-(-7) \pm \sqrt{37}}{2(1)} = \frac{7 \pm \sqrt{37}}{2}$$

$$y = \frac{7 + \sqrt{37}}{2} \quad \text{or} \quad y = \frac{7 - \sqrt{37}}{2} \quad \blacksquare$$

Note: Using a calculator,

$$y = \frac{7 + \sqrt{37}}{2} \approx 6.54. \quad \text{(three significant digits)}$$

$(\boxed{7} \, \boxed{+} \, \boxed{37} \, \boxed{\sqrt{}} \,) \, \boxed{\div} \, \boxed{2} \, \boxed{=} \, \underbrace{6.5413813}_{\text{Display}}$

Similarly,

$$y = \frac{7 - \sqrt{37}}{2} \approx .459. \quad \text{(three significant digits)}$$

$(\boxed{7} \, \boxed{-} \, \boxed{37} \, \boxed{\sqrt{}} \,) \, \boxed{\div} \, \boxed{2} \, \boxed{=} \, \underbrace{.4586187}_{\text{Display}}$

EXAMPLE 3

Solve $4t^2 - 1 = 8t$ using the quadratic formula.

Solution: We must first make one side of the equation equal zero. Only then can we correctly identify a, b, and c.

$$4t^2 - 8t - 1 = 0 \qquad \begin{array}{l}\text{a quadratic equation}\\\text{in standard form}\end{array}$$

with $b = -8$, $a = 4$, $c = -1$

First, evaluate the discriminant.

$$b^2 - 4ac = (-8)^2 - 4(4)(-1) = 64 - (-16) = 64 + 16 = \boxed{80}$$

Note: The discriminant is *not* a perfect square, but it is *positive*. The solutions will be *irrational numbers*.

$$t = \frac{-b \pm \sqrt{b^2 - 4ac}}{2a} = \frac{-(-8) \pm \sqrt{80}}{2(4)} = \frac{8 \pm \sqrt{80}}{8}$$

$$= \frac{8 \pm 4\sqrt{5}}{8} \qquad \begin{array}{l}\text{simplifying}\\\text{radicals}\end{array}$$

$$= \frac{\overset{1}{\cancel{4}}(2 \pm \sqrt{5})}{\underset{2}{\cancel{8}}} \qquad \begin{array}{l}\text{factoring and}\\\text{reducing}\end{array}$$

$$t = \frac{2 \pm \sqrt{5}}{2} \qquad \blacksquare$$

EXAMPLE 4

Solve $2x^2 - 4x + 3 = 0$ using the quadratic formula.

Solution:

$$b = -4$$

$$2x^2 - 4x + 3 = 0$$

$$a = 2 \qquad c = 3$$

$$b^2 - 4ac = (-4)^2 - \overline{4(2)(3)} = 16 - 24 = \boxed{-8}$$

Note: The discriminant is a *negative* number. The quadratic equation has *no real solution.*

For $2x^2 - 4x + 3 = 0$, there is *no real solution.* ∎

Application

BENDING MOMENT

When a uniformly distributed load is applied to a steel or wood beam, the force produced by the load creates a **bending moment**. Under this bending moment, the beam will deform slightly, resulting in a curvature of the beam. If the bending moment is great enough, the beam will break. To avoid disaster, structural engineers must investigate the bending moment at various points along the beam. This is necessary in order to design and select a beam of the proper size to safely carry a particular load.

For a simply supported beam of length L carrying a uniformly distributed load of w pounds per foot, the bending moment M at any distance x from one end of the beam is given by $M = \frac{1}{2}Lwx - \frac{1}{2}wx^2$ (see Fig. 7.6).

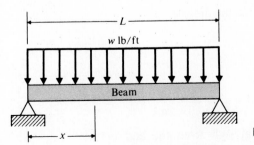

Figure 7.6

FORMULA: Bending Moment for a Simply Supported Beam

$$M = \tfrac{1}{2}Lwx - \tfrac{1}{2}wx^2$$

where M is the bending moment at any distance x from one end of the beam, L is the length of the beam, w is the uniformly distributed load.

EXAMPLE 5

A simply supported Douglas fir beam, 9 ft long, carries a uniformly distributed load of 800 lb/ft (including the weight of the beam). At what distance from the end of the beam is the bending moment 8100 lb-ft?

Solution:

$$M = \tfrac{1}{2}Lwx - \tfrac{1}{2}wx^2$$
$$8100 = \tfrac{1}{2}(9)(800)x - \tfrac{1}{2}(800)x^2 \qquad \text{substituting}$$
$$8100 = 3600x - 400x^2$$
$$400x^2 - 3600x + 8100 = 0 \qquad \text{writing the quadratic equation in standard form}$$

Note: We could apply the quadratic formula at this point, but these numbers are too large and cumbersome to use. It is best to remove any common factor before applying the formula.

$$100(4x^2 - 36x + 81) = 0 \qquad \text{common term factoring}$$

$$\frac{\cancel{100}(4x^2 - 36x + 81)}{\cancel{100}} = \frac{0}{100} \qquad \text{dividing each side by 100}$$

$$b = -36$$
$$4x^2 - 36x + 81 = 0$$
$$a = 4 \qquad c = 81$$

First, evaluate the discriminant.

$$b^2 - 4ac = (-36)^2 - 4(4)(81) = 1296 - 1296 = 0$$

$$x = \frac{-b \pm \sqrt{b^2 - 4ac}}{2a} = \frac{-(-36) \pm \sqrt{0}}{2(4)} = \frac{36}{8} = 4\tfrac{1}{2} \text{ ft}$$

Therefore, at $4\tfrac{1}{2}$ ft from the end of the beam, the bending moment will be 8100 lb-ft. ∎

Note: Notice that $4\frac{1}{2}$ ft is the midpoint of the 9-ft beam. For a simply supported beam, the *maximum bending moment* will occur at midspan. Thus, there is no greater bending moment along this beam than 8100 lb-ft.

EXERCISES 7.4

Solve the following equations by using the quadratic formula. Write all irrational solutions in simplified radical form and as decimal approximations containing three significant digits.

1. $x^2 - 3x - 10 = 0$ **2.** $x^2 + 8x - 9 = 0$

3. $y^2 + y = 6$ **4.** $y^2 - y = 20$

5. $6x^2 - 7x = 3$ **6.** $4x^2 + 4x = 3$

7. $4x^2 + 9 = 12x$ **8.** $9x^2 + 1 = -6x$

9. $t^2 + 3t = 7$ **10.** $t^2 + 5t = 9$

11. $n^2 + 8 = 5n$ **12.** $x^2 + 11 = 2x$

13. $m^2 - 3m = 9$ **14.** $n^2 + 5n = 5$

15. $2x^2 + 4x = 3$ **16.** $3x^2 + 6x = 5$

17. $w^2 = 4w + 11$ **18.** $p^2 = 2p + 12$

19. $x^2 + 8x + 4 = 0$ **20.** $x^2 + 6x - 18 = 0$

21. $3t^2 + 2t + 5 = 0$ **22.** $4x^2 + x + 6 = 0$

23. $3x^2 + 5x = 0$ **24.** $2y^2 - 7y = 0$

25. $6y^2 - 1 = 0$ **26.** $5y^2 - 2 = 0$

27. $x(2x - 3) = 7$ **28.** $2n(n - 5) = -9$

29. $(3t + 1)(t + 1) = 2$ **30.** $(2y - 1)(y + 2) = 3$

31. $(8 + 4x)(4 + 4x) = 128$ **32.** $(6 + 2x)(8 + 2x) = 96$

33. A simply supported Douglas fir beam, 10 ft long, carries a uniformly distributed load of 900 lb/ft (including the weight of the beam). At what distance from the end of the beam is the bending moment 10,000 lb-ft?

34. Referring to the beam in exercise 33, at what points along the beam is the bending moment zero?

7.5 DESIGN APPLICATIONS

Objective

⊡ To solve a design application using a quadratic equation.

⊡ Designing Structures

Engineers must often design structures to meet certain specifications, and many of the *design applications* they confront can be solved using *quadratic equations*. To help us solve the design applications in this section, we will set up each problem using the five-step procedure for solving verbal problems (see Chapter 4, Section 4.3). It may be helpful for you to review the five-step procedure.

EXAMPLE 1

An architect must design a fixed glass window to fit into a 6.00-ft by 8.00-ft opening. The plans call for 36.0 sq ft of glass to be encased in a solid wood frame of uniform width (see Fig. 7.7). What size glass and frame should be used to meet these specifications?

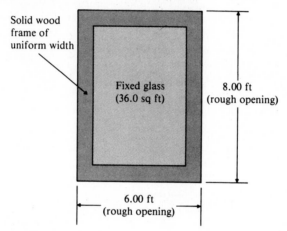

Solid wood frame of uniform width

Fixed glass (36.0 sq ft)

8.00 ft (rough opening)

6.00 ft (rough opening)

Figure 7.7

Solution:

1. FORMULA   Area of a rectangle

$$A = lw, \text{ where } A \text{ is the area}$$
$$l \text{ is the length}$$
$$w \text{ is the width}$$

2. SETUP   Let x be the uniform width of the solid wood frame (see Fig. 7.8).

$$\text{length of glass} = l = 6.00 - 2x$$
$$\text{width of glass} = w = 8.00 - 2x$$
$$\text{area of glass} = A = 36.0 \text{ sq ft}$$

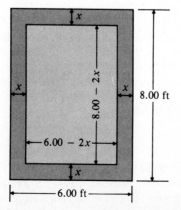

x

x x

$8.00 - 2x$

$6.00 - 2x$

x

8.00 ft

6.00 ft

Figure 7.8

3. EQUATION

$$A = lw$$

$36.0 = (6.00 - 2x)(8.00 - 2x)$ substituting

$36.0 = 48.0 - 28.0x + 4x^2$ FOIL

$-4x^2 + 28.0x - 12.0 = 0$ writing the quadratic equation in standard form

$-4(x^2 - 7.00x + 3.00) = 0$ common term factoring

$x^2 - 7.00x + 3.00 = 0$ dividing each side by -4

Note: A quadratic equation in standard form with $a = 1$
$b = -7.00$
$c = 3.00$

First, evaluate the discriminant.

$$b^2 - 4ac = (-7.00)^2 - 4(1)(3.00) = 49.0 - 12.0 = \boxed{37.0}$$

$$x = \frac{-b \pm \sqrt{b^2 - 4ac}}{2a} = \frac{-(-7.00) \pm \sqrt{37.0}}{2(1)} = \frac{7.00 \pm \sqrt{37.0}}{2}$$

Therefore,

$$x = \frac{7.00 + \sqrt{37.0}}{2} = 6.54 \text{ ft}$$

or

$$x = \frac{7.00 - \sqrt{37.0}}{2} = .459 \text{ ft.}$$

4. CONCLUSION   There appear to be two solutions. However, we must reject the solution $x = 6.54$ ft, since this value of x would force the length and width of the glass to be negative numbers. Thus, the only solution that makes sense in this application is $x = .459$ ft.

    The *uniform width of the solid wood frame* is .459 ft $\approx 5\frac{1}{2}$ in.

$$length \text{ of } glass = l = 6.00 - 2x$$
$$= 6.00 - 2(.459 \text{ ft})$$
$$= 5.08 \text{ ft} \approx 5 \text{ ft } 1 \text{ in.}$$

$$width \text{ of } glass = w = 8.00 - 2x$$
$$= 8.00 - 2(.459 \text{ ft})$$
$$= 7.08 \text{ ft} \approx 7 \text{ ft } 1 \text{ in.} \quad \blacksquare$$

5. CHECK

$$A = lw$$
$$36.0 \text{ ft}^2 = (5.08 \text{ ft})(7.08 \text{ ft})?$$
$$36.0 \text{ ft}^2 = 36.0 \text{ ft}^2 \quad \text{true}$$

EXERCISES 7.5

Solve the following design applications.

1. The length of a rectangular foundation is to be 6 ft more than its width. The area enclosed by the foundation is to be 720 sq ft. What should the dimensions of the foundation be?

2. A warehouse with rectangular sides and a flat roof (see Fig. 7.9) is to contain 1500 cu m of storage space. The length of the building is to be 5 m more than its width, and its height is to be 5 m. What should the dimensions of the warehouse be?

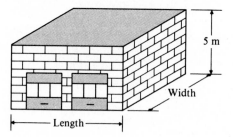

Figure 7.9

3. An architect is to design an A-frame for a chalet (see Fig. 7.10) to the following specifications.

a) The base of the frame is to be 4 m less than twice its height.

b) The total cross-sectional area of the frame is to be 48 sq m.

What should the dimensions of the frame be?

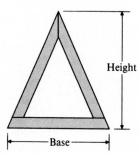

Figure 7.10

4. An architect is to design a pentagonal frame (see Fig. 7.11) for a house to the following specifications.

a) The base of the rectangular part of the frame is to be four times as long as the height.

b) The height of the triangular part of the frame is to be 4 ft less than the height of the rectangular part.

c) The total cross-sectional area of the frame is to be 320 sq ft.

What should the dimensions of the frame be?

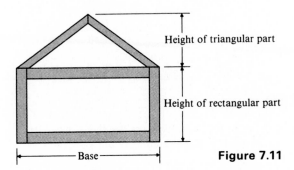

Height of triangular part

Height of rectangular part

Base

Figure 7.11

5. A parking lot is 20.0 m long and 18.0 m wide. The length and width are to be increased by the same amount until the parking area is twice that of the existing area (see Fig. 7.12). What will be the dimensions of the new lot?

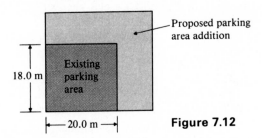

Proposed parking area addition

18.0 m

Existing parking area

20.0 m

Figure 7.12

6. An L-shaped angle iron (see Fig. 7.13) is to have a cross-sectional area of 10.0 cm². What must be the uniform width of the angle iron?

7. The area of a uniform walkway around a 20.0-ft by 40.0-ft rectangular swimming pool is to be equal to the area of the pool. What must be the width of the walkway? (See Fig. 7.14).

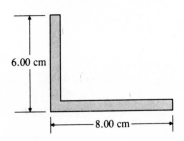

6.00 cm

8.00 cm

Figure 7.13

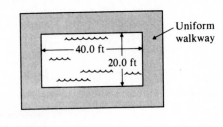

Uniform walkway

40.0 ft

20.0 ft

Figure 7.14

8. The area of a uniform walkway around the swimming pool shown in Fig. 7.15 is to be equal to twice the area of the pool. What must be the width of the walkway?

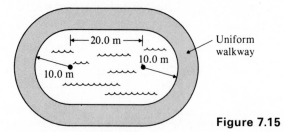

Figure 7.15

9. A square piece of sheet metal is to be made into a rectangular box with an open top by cutting out 2-in. by 2-in. squares from each corner and bending along the dotted lines as shown in Fig. 7.16. What are the dimensions of the original piece of sheet metal if the volume of the box is 50 cu in.?

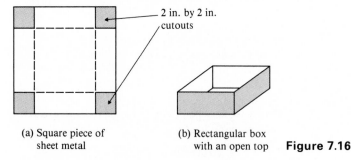

(a) Square piece of   (b) Rectangular box
sheet metal   with an open top   **Figure 7.16**

10. A rectangular piece of sheet metal 12 in. by 16 in. is to be made into a rectangular box with an open top by cutting out squares from each corner and folding up the sides. What is the height of the box if the area of the bottom of the box is to be 60 sq in.?

11. The perimeter of a rectangular foundation is to be 112 ft and the area enclosed by the foundation is to be 768 sq ft. What should the dimensions of the foundation be?

12. A rectangular heating duct is to be formed by bending a flat piece of sheet metal into a rectangle as shown in Fig. 7.17. The cross-sectional area of the duct is to be 32 sq. in. What should the dimensions of the duct be?

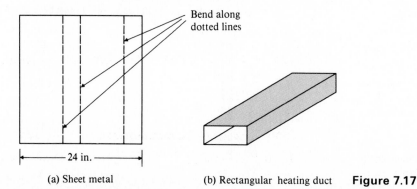

(a) Sheet metal   (b) Rectangular heating duct   **Figure 7.17**

Chapter Review

The following review exercises are grouped according to the objectives that should have been mastered in Chapter 7. Work each problem carefully. If any weaknesses appear, immediately refer to and read the subsection that matches that objective.

7.1 SOLVING LITERAL EQUATIONS

Objectives

⊡ To solve a literal equation in which the variable appears only once in the equation.

Solve.   **1.** $a(bx + c) = d$, for x

⊡ To solve a literal equation or formula in which the variable appears more than once.

Solve.   **2.** $\dfrac{ax + b}{c} = dx$, for x

3. $L = L_o + L_o \alpha t$, for L_o

4. $C_t = \dfrac{C_1 C_2}{C_1 + C_2}$, for C_2

7.2 SOLVING EQUATIONS BY FACTORING

Objectives

⊡ To solve an equation that has a product of algebraic expressions equal to zero.

Solve.   **5.** $(x - 5)(x + 3) = 0$ **6.** $7x(x - 4)(3x - 4) = 0$

⊡ To solve a quadratic equation by factoring.

Solve.   **7.** $y^2 + 5y - 24 = 0$ **8.** $3t^2 + 13t = 10$

9. $2z^2 = 16z$ **10.** $(2x - 3)(x + 5) = 2x - 3$

⊡ To solve other types of equations by factoring.

Solve.   **11.** $x^3 + 16x = 8x^2$ **12.** $v^3 - 3v^2 + 2v - 6 = 0$

7.3 SOLVING QUADRATIC EQUATIONS BY COMPLETING THE SQUARE

Objectives

⊡ To complete a perfect square trinomial.

Complete each perfect square.   **13. a)** $y^2 + 14y + \underline{\ ?\ }$ **b)** $x^2 - 5x + \underline{\ ?\ }$

⊡ To solve equations of the form $(x + a)^2 = b$ by using the square root property.

Solve.

14. $(x + 3)^2 = 16$ **15.** $(p - 5)^2 = 18$

⊡ To solve a quadratic equation by completing the perfect square.

Solve.

16. $x^2 - 8x - 34 = 0$ **17.** $y^2 + y - 11 = 0$

18. $3t^2 - 8t + 5 = 0$ **19.** $k^2 + 10k + 30 = 0$

7.4 THE QUADRATIC FORMULA

Objective

⊡ To solve a quadratic equation by using the quadratic formula.

Solve.

20. $x^2 + 3x - 28 = 0$ **21.** $y^2 = 9y - 4$

22. $3t^2 + 1 = 10t$ **23.** $4y^2 - 3y + 5 = 0$

7.5 DESIGN APPLICATIONS

Objective

⊡ To solve a design application using a quadratic equation.

Solve.

24. A rectangular swimming pool is surrounded by a wood deck of uniform width. The overall dimensions of the pool and deck are 56 ft by 40 ft. If the swimming area is to be 512 sq ft, what should the width of the deck be (see Fig. 7.18)?

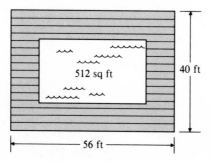

512 sq ft   40 ft

56 ft   **Figure 7.18**

CHAPTER TEST
Time
50 minutes

Score
15, 14 excellent
13, 12 good
11, 10, 9 fair
below 8 poor

If you have worked through the Review Exercises and corrected any weaknesses, then you are ready to take the following Chapter Test.

1. Solve for P. $A = P - Prt$

2. Solve for x. $4x(4x + 3)(x - 5) = 0$

3. Solve by factoring. $x^2 + x = 30$

4. Solve by factoring. $9x^2 + 1 = 6x$

5. Solve by factoring. $5y^2 = 25y$

6. Solve by factoring. $4t^3 + 7t^2 = 2t$

7. Complete the perfect square. $x^2 - 20x + \underline{\ ?\ }$

8. Complete the perfect square. $y^2 + 7y + \underline{\ ?\ }$

9. Solve by using the square root property. $(x - 2)^2 = 20$

10. Solve by completing the perfect square. $m^2 + 12m + 18 = 0$

11. Solve using the quadratic formula. $x^2 + 3x - 11 = 0$

12. Solve using the quadratic formula. $n^2 - 7 = 2n$

13. Solve using the quadratic formula. $3x^2 - x = 4$

14. Solve using the quadratic formula. $2k^2 - 6k + 7 = 0$

15. A parking lot 20.0 m by 30.0 m is to be tripled in size by adding the regions of uniform width shown in Fig. 7.19. What are the dimensions of the new lot?

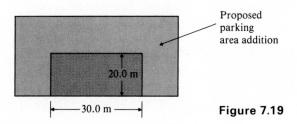

Proposed parking area addition

20.0 m

30.0 m

Figure 7.19

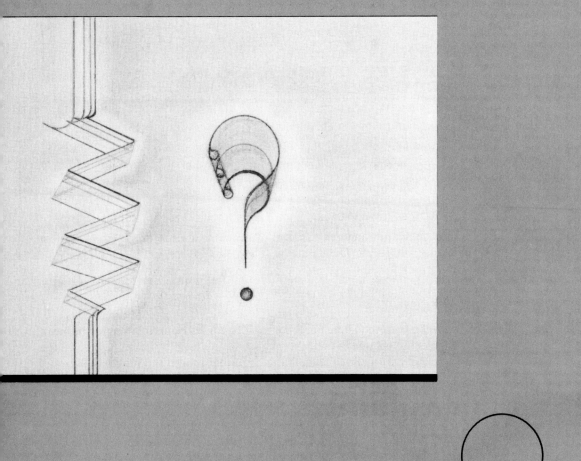

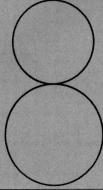

ALGEBRAIC FRACTIONS

$$8.1 \quad \textbf{BUILDING UP AND BREAKING DOWN ALGEBRAIC FRACTIONS}$$

Objectives

⊡ To find the restricted values of an algebraic fraction.

⊡ To find an equivalent algebraic fraction whose denominator is known.

⊡ To reduce an algebraic fraction to lowest terms.

⊡ Restricted Values

When the numerator and denominator of a fraction are polynomials, the fraction is referred to as an *algebraic fraction*. The following are examples of algebraic fractions:

$$\frac{x + 6}{3x^2} \quad \frac{3}{2y - 4} \quad \frac{x - 1}{x^2 + 3x - 10} \quad \frac{t^2 - 2t + 9}{t^2 + 1}$$

Since division by zero is undefined, we must be careful not to choose any value of the variable that will make the denominator of an algebraic fraction zero. Values of the variable that make the denominator zero are called the **restricted values** of the algebraic fraction.

EXAMPLE 1

Find the restricted values of the following algebraic fractions.

a) $\dfrac{x + 6}{3x^2}$ **b)** $\dfrac{3}{2y - 4}$ **c)** $\dfrac{x - 1}{x^2 + 3x - 10}$ **d)** $\dfrac{t^2 - 2t + 9}{t^2 + 1}$

Solution:

a) For $\dfrac{x + 6}{3x^2}$, the denominator will be zero if

$$3x^2 = 0$$
$$x^2 = 0 \qquad \text{dividing each side by 3}$$
$$x = 0. \qquad \text{square root property}$$

Thus, for $\dfrac{x + 6}{3x^2}$, $x \neq 0$. ■

b) For $\dfrac{3}{2y - 4}$, the denominator will be zero if

$$2y - 4 = 0$$
$$2y = 4 \qquad \text{adding 4 to each side}$$
$$y = 2. \qquad \text{dividing each side by 2}$$

Thus, for $\dfrac{3}{2y - 4}$, $y \neq 2$. ■

c) For $\dfrac{x-1}{x^2+3x-10}$, the denominator will be zero if

$$x^2+3x-10=0$$
$$(x+5)(x-2)=0 \qquad \text{factoring}$$
$$x+5=0 \quad \text{or } x-2=0 \qquad \text{zero product property}$$
$$x=-5 \text{ or} \qquad x=2.$$

Thus, for $\dfrac{x-1}{x^2+3x-10}$, $x\neq -5$ and $x\neq 2$. ∎

d) For $\dfrac{t^2-2t+9}{t^2+1}$, the denominator will be zero if

$$t^2+1=0$$
$$t^2=-1. \qquad \text{adding } -1 \text{ to each side}$$
$$\text{NO REAL SOLUTION} \qquad \text{square root property}$$

Thus, for $\dfrac{t^2-2t+9}{t^2+1}$, there are *no* restricted values for the variable. ∎

Note: Notice that there are no restrictions on the numerator of an algebraic fraction. It is only the denominator that can't be zero.

⠆ Fundamental Property of Fractions

The basic rules for working with rational numbers (see Chapter 2) can all be extended to algebraic fractions. In Chapter 2 you discovered that we could build up a fraction to higher terms by multiplying its numerator and denominator by the same quantity. In doing so, we obtained an *equivalent fraction*. For example, to change $\dfrac{3}{4}$ to an equivalent fraction whose denominator is 20, we proceeded as follows:

$$\frac{3}{4}=\frac{?}{20}$$

We must multiply 4 by 5 to obtain 20.

Thus,

$$\frac{3}{4}=\frac{5\,(3)}{5\,(4)}=\frac{15}{20}.$$

equivalent fractions

We referred to this property as the *fundamental property of fractions*. This same property applies to algebraic fractions as well and plays a critical part in the addition and subtraction of algebraic fractions (see Section 8.4).

FUNDAMENTAL PROPERTY OF FRACTIONS

$$\frac{a}{b} = \frac{ka}{kb}, \text{ provided } b \neq 0, k \neq 0.$$

EXAMPLE 2

Change $\dfrac{x + 2}{5x}$ to an equivalent algebraic fraction whose denominator is $10x^3y$.

Solution:

$$\frac{x + 2}{5x} = \frac{?}{10x^3y}$$

$\rightarrow$ We must multiply $5x$
by $2x^2y$ to obtain $10x^3y$. $\longrightarrow$

Thus,

$$\frac{x + 2}{5x} = \frac{2x^2y(x + 2)}{2x^2y(5x)} = \frac{2x^3y + 4x^2y}{10x^3y}. \quad \blacksquare$$

fundamental
property of
fractions

equivalent
fractions

Factoring is often necessary to determine an equivalent algebraic fraction whose denominator is known.

EXAMPLE 3

Change $\dfrac{x - 3}{x + 1}$ to an equivalent algebraic fraction whose denominator is $x^2 + 4x + 3$.

Solution:

$$\frac{x - 3}{x + 1} = \frac{?}{x^2 + 4x + 3}$$

Note: Factor this
denominator in order
to determine the
expression that must
be multiplied by $(x + 1)$
to obtain $x^2 + 4x + 3$.

$$\frac{x - 3}{x + 1} = \frac{?}{(x + 1)(x + 3)}$$

factoring

$\rightarrow$ We must multiply $(x + 1)$ by
$(x + 3)$ to obtain $(x + 1)(x + 3)$. $\longrightarrow$

Thus,

$$\frac{x-3}{x+1} = \frac{(x+3)(x-3)}{(x+3)(x+1)} = \frac{x^2-9}{x^2+4x+3}.$$ ■ fundamental property of fractions

equivalent fractions

EXAMPLE 4

Change $\dfrac{y+1}{2y^2-2y}$ to an equivalent algebraic fraction whose denominator is $6y^3 - 12y^2 + 6y$.

Solution:

$$\frac{y+1}{2y^2-2y} = \frac{?}{6y^3-12y^2+6y}$$

$$\frac{y+1}{2y(y-1)} = \frac{?}{6y(y-1)^2}$$ factoring

We must multiply $2y(y-1)$ by $3(y-1)$ to obtain $6y(y-1)^2$.

Thus,

$$\frac{y+1}{2y^2-2y} = \frac{y+1}{2y(y-1)} = \frac{3(y-1)(y+1)}{3(y-1)\,2y(y-1)}$$ fundamental property of fractions

$$= \frac{3(y^2-1)}{6y(y-1)^2}$$

$$= \frac{3y^2-3}{6y^3-12y^2+6y}$$ ■ simplifying

equivalent fractions

⁝ Cancellation Law

In Chapter 2 you discovered that we could break down a fraction to its lowest terms by cancelling common factors from the numerator and denominator. In doing so, we obtained a *reduced fraction*. For example, to reduce $\dfrac{15}{20}$ we proceeded as follows:

$$\frac{15}{20} = \frac{(5)3}{(5)4} = \frac{\overset{1}{(\cancel{5})3}}{\underset{1}{(\cancel{5})4}} = \frac{3}{4}$$

a reduced fraction

We referred to this property as the *cancellation law*. This same law applies to algebraic fractions as well and allows us to reduce an algebraic fraction to lowest terms.

CANCELLATION LAW

$$\frac{ka}{kb} = \frac{\overset{1}{\cancel{k}a}}{\underset{1}{\cancel{k}b}} = \frac{a}{b},$$

provided $b \neq 0$, $k \neq 0$.

 In Chapter 2, you were cautioned that the numerator and denominator of a fraction had to be written entirely as products if we were to apply the cancellation law and reduce a fraction correctly. The same caution applies for algebraic fractions.

EXAMPLE 5

Reduce $\dfrac{2t + 10t^2}{2t}$ to lowest terms.

Solution: Before we can cancel, we must write the numerator as a product. Factoring forms products, so begin by factoring the numerator.

$$\frac{2t + 10t^2}{2t} = \frac{2t(1 + 5t)}{2t} \qquad \text{common term factoring}$$

$$= \frac{\overset{1}{\cancel{2t}}(1 + 5t)}{\underset{1}{\cancel{2t}}} \qquad \text{cancellation law}$$

$$= \frac{1 + 5t}{1}$$

$$= 1 + 5t \quad \blacksquare \qquad \text{simplifying}$$

Note: Don't attempt to cancel as follows: $\dfrac{\overset{1}{\cancel{2t}} + 10t^2}{\underset{1}{\cancel{2t}}}$ *(Wrong)*.

Only common factors of the polynomials in the numerator and denominator will cancel, never their common terms.

EXAMPLE 6

Reduce $\dfrac{x-1}{x^2+3x-4}$ to lowest terms.

Solution: To cancel, we must have products, so begin by factoring the denominator.

$$\frac{x-1}{x^2+3x-4} = \frac{x-1}{(x+4)(x-1)} \qquad \text{factoring}$$

$$= \frac{\overset{1}{\cancel{x-1}}}{(x+4)(\underset{1}{\cancel{x-1}})} \qquad \text{cancellation law}$$

$$= \frac{1}{x+4} \quad \blacksquare \qquad \text{simplifying}$$

Note: The factor $x-1$ will cancel, provided $x \neq 1$. If $x = 1$, $\dfrac{x-1}{x-1} = \dfrac{0}{0}$; and we can never cancel zeros.

EXAMPLE 7

Reduce $\dfrac{y^2-2y-8}{8-2y^2}$ to lowest terms.

Solution:

$$\frac{y^2-2y-8}{8-2y^2} = \frac{(y-4)(y+2)}{2(2+y)(2-y)} \qquad \text{factoring completely}$$

$$= \frac{(y-4)\overset{1}{\cancel{(y+2)}}}{2\underset{1}{\cancel{(2+y)}}(2-y)} \qquad \text{cancellation law}$$

Note: Since addition is commutative, $y + 2 = 2 + y$. Thus we can cancel these factors.

$$= \frac{y-4}{2(2-y)}$$

$$= \frac{y-4}{4-2y} \quad \blacksquare \qquad \text{simplifying}$$

Remember, subtraction is *not* commutative, $a - b \neq b - a$. However, $a - b$ and $b - a$ differ only by sign. Therefore, $a - b$ is the *opposite* of $b - a$.

$$a - b = -(b - a)$$

Of course, any number (except 0) divided by its opposite equals -1.

$$\frac{a - b}{b - a} = \frac{-(b - a)}{b - a} = \frac{\overset{1}{-(b - a)}}{\underset{1}{b - a}} = -1$$

It is important when working with algebraic fractions that you are able to identify and cancel opposites.

To *identify opposites*, simply see if the expressions differ only by sign.

To *cancel opposites*, simply place a negative one (-1) in either the numerator or denominator of the algebraic fraction (but not in both).

EXAMPLE 8

Reduce $\dfrac{9 - x^2}{x^2 - 6x + 9}$ to lowest terms.

Solution:

$$\frac{9 - x^2}{x^2 - 6x + 9} = \frac{(3 + x)(3 - x)}{(x - 3)(x - 3)} \qquad \text{factoring}$$

Note: $3 - x$ and $x - 3$ are opposites since they differ only by sign. $3 - x = -(x - 3)$.

$$\frac{(3 + x)\overset{-1}{(3 - x)}}{(x - 3)\underset{1}{(x - 3)}} \quad \text{or} \quad \frac{(3 + x)\overset{1}{(3 - x)}}{(x - 3)\underset{-1}{(x - 3)}} \qquad \text{cancellation law}$$

$$\frac{-3 - x}{x - 3} \quad \text{or} \quad \frac{3 + x}{3 - x} \quad \blacksquare \qquad \text{simplifying}$$

Note: Although both answers are correct, $\dfrac{3 + x}{3 - x}$ is preferred since it contains fewer negative signs. Note also that $3 + x$ and $3 - x$ are *not* opposites. Although the *x*s differ in sign, the 3s do not. $3 + x \neq -(3 - x)$.

EXERCISES 8.1

Find the restricted values of the following algebraic fractions. *(use denominates)*

1. $\dfrac{3x - 7}{5x^3}$

2. $\dfrac{9 + 2x^2}{-x^2}$

3. $\dfrac{y - 1}{5y + 20}$

4. $\dfrac{y^2 + 2y + 3}{3y + 7}$

5. $\dfrac{2}{z^2 + 3z - 18}$ $\dfrac{(z^2 + 6z)(-3z - 18)}{z(z+10)-3(z-3)}$ **6.** $\dfrac{z - 3}{2z^2 + 7z - 4}$

7. $\dfrac{3x - 1}{2x^2 + 32}$ $z = -6 \quad z = 3$ **8.** $\dfrac{3x - 1}{2x^2 - 32}$

9. $\dfrac{8 - 3t}{t^3 + t^2 - 12t}$

10. $\dfrac{8 - 3t}{t^3 + t^2 + 12t}$

Find the missing numerator so that the fractions are equivalent.

11. $\dfrac{7y}{3x} = \dfrac{28x^4y^3}{12x^5y^2}$ $4x^4y^2$

12. $\dfrac{x - 1}{4x^2y^2} = \dfrac{?}{12x^5y^2}$

13. $\dfrac{y - 2}{y + 2} = \dfrac{?}{y^2 + 5y + 6}$

14. $\dfrac{y - 1}{y + 3} = \dfrac{?}{y^2 + 5y + 6}$

15. $\dfrac{z - 3}{z^2 + 3z} = \dfrac{?}{2z^3 + 12z^2 + 18z}$

16. $\dfrac{3z}{2z + 6} = \dfrac{?}{2z^3 + 12z^2 + 18z}$

17. $\dfrac{2t}{t + 5} = \dfrac{?}{3t^3 - 75t}$

18. $\dfrac{t - 2}{3t} = \dfrac{?}{3t^3 - 75t}$

19. $7w - 2 = \dfrac{?}{w - 3}$

20. $-6x = \dfrac{?}{x^2 + 3x + 2}$

Reduce the following algebraic fractions to lowest terms.

21. $\dfrac{3x - 9x^2}{3x}$

22. $\dfrac{5x}{10 - 5x}$

23. $\dfrac{6xy - 4xy^2 + 2x^2y}{2x^2y}$

24. $\dfrac{8xy^2}{24x^2y - 16xy + 4xy^2}$

25. $\dfrac{y + 3}{y^2 + 10y + 21}$

26. $\dfrac{2y^2 - 3y - 9}{2y + 3}$

27. $\dfrac{5n - 5m}{15m - 15n}$

28. $\dfrac{7x - 28y}{8y - 2x}$

29. $\dfrac{-36 - 36t}{-6}$

30. $\dfrac{-8t}{-16t - 16}$

31. $\dfrac{9y^2 - 9x^2}{6x + 6y}$

32. $\dfrac{12x + 12y}{6y^2 - 6x^2}$

33. $\dfrac{t^2 + 5t - 14}{2t^2 + 13t - 7}$

34. $\dfrac{3p^2 - 4p - 15}{p^2 + 6p + 9}$

35. $\dfrac{16 - w^2}{3w^2 - 10w - 8}$

36. $\dfrac{2v^2 - 7v - 15}{25 - v^2}$

37. $\dfrac{108 - 3x^2}{3x^3 + 36x^2 + 108x}$

38. $\dfrac{4x^3 + 8x^2 - 5x}{50x + 40x^2 + 8x^3}$

39. $\dfrac{2x^3 + x^2 - 8x - 4}{x + 2}$

40. $\dfrac{3y^2 - 4y - 4}{9y^3 - 18y^2 - 4y + 8}$

8.2 MULTIPLYING AND DIVIDING ALGEBRAIC FRACTIONS

Objectives

⊡ To multiply algebraic fractions.

⊡ To divide algebraic fractions.

⊡ Multiplying Algebraic Fractions

Recall from Chapter 2 that we multiplied fractions by simply multiplying their numerators and then multiplying their denominators.

MULTIPLICATION RULE FOR FRACTIONS

$$\frac{a}{b} \cdot \frac{c}{d} = \frac{ac}{bd},$$

provided $b \neq 0, d \neq 0$.

You also learned that it was advantageous to cancel first (if possible) before you actually found any products. This way the answer was always in reduced form. For example, to multiply $\dfrac{6}{7}$ times $\dfrac{3}{10}$, we proceeded as follows:

$$\frac{6}{7} \cdot \frac{3}{10} = \frac{6 \cdot 3}{7 \cdot 10} \qquad \text{multiplication rule for fractions}$$

$$= \frac{\overset{3}{\cancel{6}} \cdot 3}{7 \cdot \underset{5}{\cancel{10}}} \qquad \text{cancellation law} \left(\frac{6}{10} = \frac{2 \cdot 3}{2 \cdot 5} = \frac{3}{5} \right)$$

$$= \frac{9}{35} \qquad \text{simplifying}$$

We can multiply algebraic fractions exactly the same way.

EXAMPLE 1

Find the product $\dfrac{8x^2}{x-y} \cdot \dfrac{x-y}{10x^5}$.

Solution:

$$\dfrac{8x^2}{x-y} \cdot \dfrac{x-y}{10x^5} = \dfrac{(8x^2)\cdot(x-y)}{(x-y)\cdot(10x^5)} \qquad \text{multiplication rule for fractions}$$

$$= \dfrac{\overset{4}{\cancel{(8x^2)}}\cdot\overset{1}{\cancel{(x-y)}}}{\underset{1}{\cancel{(x-y)}}\cdot\underset{5x^3}{\cancel{(10x^5)}}} \qquad \text{cancellation law}$$

$$= \dfrac{4}{5x^3} \quad \blacksquare \qquad \text{simplifying}$$

To spot all the common factors that we can cancel, it is necessary to *completely factor* all numerators and denominators.

EXAMPLE 2

Find the product $\dfrac{y^2+y}{9-y^2} \cdot \dfrac{y+3}{2y^2}$.

Solution:

$$\dfrac{y^2+y}{9-y^2} \cdot \dfrac{y+3}{2y^2} = \dfrac{(y^2+y)\cdot(y+3)}{(9-y^2)\cdot(2y^2)} \qquad \text{multiplication rule for fractions}$$

$$= \dfrac{y(y+1)\cdot(y+3)}{(3+y)(3-y)\cdot(2y^2)} \qquad \text{factoring}$$

$$= \dfrac{\overset{1}{\cancel{y}}(y+1)\cdot\overset{1}{\cancel{(y+3)}}}{\cancel{(3+y)}(3-y)\cdot\underset{2y}{\cancel{(2y^2)}}} \qquad \text{cancellation law}$$

$$= \dfrac{y+1}{2y(3-y)}$$

$$= \dfrac{y+1}{6y-2y^2} \quad \blacksquare \qquad \text{simplifying}$$

EXAMPLE 3

Find the product $\dfrac{t^2 + 6t + 8}{2t^2 - 8} \cdot (2 - t)$.

Solution:

$$\dfrac{t^2 + 6t + 8}{2t^2 - 8} \cdot (2 - t) = \dfrac{t^2 + 6t + 8}{2t^2 - 8} \cdot \dfrac{2 - t}{1} \qquad a = \dfrac{a}{1}$$

$$= \dfrac{(t^2 + 6t + 8) \cdot (2 - t)}{(2t^2 - 8) \cdot (1)} \qquad \text{multiplication rule for fractions}$$

$$= \dfrac{(t + 2)(t + 4) \cdot (2 - t)}{2(t + 2)(t - 2) \cdot (1)} \qquad \text{factoring}$$

$$= \dfrac{\overset{1}{\cancel{(t + 2)}}(t + 4) \cdot \overset{-1}{\cancel{(2 - t)}}}{2\underset{1}{\cancel{(t + 2)}}\cancel{(t - 2)} \cdot (1)} \qquad \text{cancellation law}$$

$$= \dfrac{t + 4}{-2} \quad \blacksquare \qquad \text{simplifying}$$

Note: In cancelling the opposites, $2 - t$ and $t - 2$, we could have placed the -1 in the numerator and obtained the solution $\dfrac{-t - 4}{2}$. Both answers are correct.

Also, since $\dfrac{a}{-b} = -\dfrac{a}{b}$, you may wish to write $-\dfrac{t + 4}{2}$.

⠰ Dividing Algebraic Fractions

Recall from Chapter 2 that our procedure for dividing fractions was first to change the division symbol to a multiplication symbol and then to multiply the dividend by the *reciprocal* of the divisor.

DIVISION RULE FOR FRACTIONS

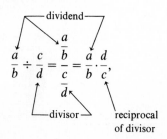

provided $b \neq 0$, $c \neq 0$, and $d \neq 0$.

For example, to divide $\dfrac{3}{8}$ by $\dfrac{5}{6}$, we proceeded as follows:

$$\frac{3}{8} \div \frac{5}{6} = \frac{3}{8} \cdot \frac{6}{5} \qquad \text{division rule for fractions}$$

$$= \frac{3 \cdot 6}{8 \cdot 5} \qquad \text{multiplication rule for fractions}$$

$$= \frac{3 \cdot \overset{3}{\cancel{6}}}{\underset{4}{\cancel{8}} \cdot 5} \qquad \text{cancellation law } \left(\frac{6}{8} = \frac{\cancel{2} \cdot 3}{\cancel{2} \cdot 4} = \frac{3}{4} \right)$$

$$= \frac{9}{20} \qquad \text{simplifying}$$

The same procedure applies for the division of two algebraic fractions.

EXAMPLE 4

Find the quotient $\dfrac{xy}{x+2} \div (x+2)$.

Solution:

$$\frac{xy}{x+2} \div (x+2) = \frac{xy}{x+2} \div \frac{x+2}{1} \qquad a = \frac{a}{1}$$

$$= \frac{xy}{x+2} \cdot \frac{1}{x+2} \qquad \text{division rule for fractions}$$

$$= \frac{(xy) \cdot (1)}{(x+2) \cdot (x+2)} \qquad \begin{array}{l}\text{multiplication rule}\\ \text{for fractions}\end{array}$$

Note: There are no cancellations. $\Big\}\leftarrow$

$$= \frac{xy}{x^2 + 4x + 4} \quad \blacksquare \qquad \text{simplifying}$$

EXAMPLE 5

Find the quotient $\dfrac{3t^2}{t^2 + 2t - 15} \div \dfrac{6t^3 - 6t^2}{t - 3}$.

Solution:

$$\frac{3t^2}{t^2 + 2t - 15} \div \frac{6t^3 - 6t^2}{t - 3} = \frac{3t^2}{t^2 + 2t - 15} \cdot \frac{t - 3}{6t^3 - 6t^2} \qquad \text{division rule for fractions}$$

$$= \frac{(3t^2) \cdot (t - 3)}{(t^2 + 2t - 15) \cdot (6t^3 - 6t^2)} \qquad \text{multiplication rule for fractions}$$

$$= \frac{(3t^2) \cdot (t - 3)}{(t + 5)(t - 3) \cdot (6t^2)(t - 1)} \qquad \text{factoring}$$

$$= \frac{\overset{1}{\cancel{(3t^2)}} \cdot \overset{1}{\cancel{(t - 3)}}}{(t + 5)\underset{1}{\cancel{(t - 3)}} \cdot \underset{2}{\cancel{(6t^2)}}(t - 1)} \qquad \text{cancellation law}$$

$$= \frac{1}{2(t + 5)(t - 1)}$$

$$= \frac{1}{2t^2 + 8t - 10} \quad \blacksquare \qquad \text{simplifying}$$

EXAMPLE 6

Find the quotient $\dfrac{\dfrac{x^2 - 16}{x^2 + x - 12}}{\dfrac{4 - x}{9 - x^2}}$.

Solution:

$$\frac{\dfrac{x^2 - 16}{x^2 + x - 12}}{\dfrac{4 - x}{9 - x^2}} = \frac{x^2 - 16}{x^2 + x - 12} \cdot \frac{9 - x^2}{4 - x} \qquad \text{division rule for fractions}$$

$$= \frac{(x^2 - 16) \cdot (9 - x^2)}{(x^2 + x - 12) \cdot (4 - x)} \qquad \text{multiplication rule for fractions}$$

$$= \frac{(x + 4)(x - 4) \cdot (3 - x)(3 + x)}{(x + 4)(x - 3) \cdot (4 - x)} \qquad \text{factoring}$$

$$= \frac{\overset{1}{\cancel{(x + 4)}}\overset{-1}{\cancel{(x - 4)}} \cdot \overset{-1}{\cancel{(3 - x)}}(3 + x)}{\underset{1}{\cancel{(x + 4)}}\underset{1}{\cancel{(x - 3)}} \cdot \underset{1}{\cancel{(4 - x)}}} \qquad \text{cancellation law}$$

$$= \frac{(-1)(-1)(3 + x)}{1}$$

$$= 3 + x \quad \blacksquare \qquad \text{simplifying}$$

EXERCISES 8.2

Simplify.

1. $\dfrac{8x^2y^3}{9z} \cdot \dfrac{15z^4}{10xy^4}$

2. $\dfrac{5z}{-6x^3y} \cdot 3xy^2$

3. $\dfrac{12m^5n^5}{5} \div \dfrac{18m^4n^6}{3n}$

4. $\dfrac{8}{m^3n^2} \div \dfrac{-2}{m^2n^5}$

5. $\dfrac{4x - 8}{x^3} \cdot \dfrac{x}{x - 2}$

6. $\dfrac{5x}{9 - 3x} \cdot \dfrac{12 - 4x}{25x^4}$

7. $\dfrac{5m - 10n}{4mn} \cdot \dfrac{-5}{6mn}$

8. $\dfrac{7p^2q^2}{-8} \div \left(\dfrac{21p^2}{16p - 8q}\right)$

9. $\dfrac{y^2 - 49}{y} \cdot \dfrac{y^3}{7 + y}$

10. $\dfrac{2z + 5}{15z} \cdot \dfrac{5z}{25 - 4z^2}$

11. $(9w^2 - 1) \div \left(\dfrac{3w - 1}{3w + 1}\right)$

12. $\dfrac{4t - 1}{4t + 1} \div (16t^2 - 1)$

13. $\dfrac{t^2 - 2t}{8t} \cdot \dfrac{4t^3}{1 - 4t^2}$

14. $\dfrac{9k^3}{12k - 2} \cdot \dfrac{1 - 36k^2}{36k^2}$

15. $\dfrac{(x + 6)^2}{14x^2} \cdot \dfrac{21x^3}{x^2 - 36}$

16. $\dfrac{42x^4y^3}{(x - 7)^2} \cdot \dfrac{x^2 - 49}{35xy^3}$

17. $\dfrac{(2 - x)^2}{3} \div \dfrac{4x^3 - 8x^2}{9}$

18. $\dfrac{10g - 2g^2}{8} \div \dfrac{(g - 5)^2}{16}$

19. $\dfrac{x^2 + 6x + 8}{x + 3} \cdot \dfrac{x^2 - 9}{x^2 + x - 12}$

20. $\dfrac{2y + 3}{y + 1} \cdot \dfrac{y^2 - 1}{2y^2 + y - 3}$

21. $\dfrac{x^2 + 5xy}{3y + 21} \div \dfrac{x^3 + 6x^2y + 5xy^2}{y^2 + 7y}$

22. $\dfrac{y^3 + 8y^2 + 16y}{32y^2} \div \dfrac{3y^3 - 48y}{16y^2 + 16}$

23. $\dfrac{2 + x}{2 - x} \cdot (x^2 - 4x + 4)$

24. $(x^2 - 10x + 25) \cdot \dfrac{5 + x}{5 - x}$

25. $\dfrac{t^2 - 3t - 10}{49 - 4t^2} \cdot \dfrac{2t^2 + t - 21}{25 - t^2}$

26. $\dfrac{16 - 6x - x^2}{9 - 4x^2} \cdot \dfrac{2x^2 + 3x - 9}{x^2 - 5x + 6}$

27. $\dfrac{\dfrac{2x + 6y}{x^2 - y^2}}{\dfrac{2}{x - y}}$

28. $\dfrac{\dfrac{16 - z^2}{z^2 + 5z - 24}}{\dfrac{z + 4}{z^2 + 8z}}$

29. $\dfrac{\dfrac{t^2 - 1}{t^2 + 8t + 7}}{\dfrac{3 - 4t + t^2}{t^2 - 49}}$

30. $\dfrac{\dfrac{4x^2 - 8x + 3}{4x^2 - 1}}{\dfrac{9x - 6x^2}{6x^2 + 5x + 1}}$

31. $\left(\dfrac{15x}{3x - 3y} \cdot \dfrac{7y - 7x}{5x^3}\right) \div \dfrac{-x^2}{14}$

32. $(50 - 8y^2) \div \left(\dfrac{9y^2 + 9y}{1 - y^2} \cdot \dfrac{2y^2 + 3y - 5}{18y^3}\right)$

8.3 ADDING AND SUBTRACTING ALGEBRAIC FRACTIONS WITH THE SAME DENOMINATOR

Objectives

⊡ To add algebraic fractions that have the same denominator.

⊡ To subtract algebraic fractions that have the same denominator.

⊡ To add or subtract algebraic fractions whose denominators are opposites.

⊡ Adding Algebraic Fractions with the Same Denominator

You learned in Chapter 2 that we could add fractions with the same denominator by simply adding their numerators and retaining the same denominator.

ADDITION RULE FOR FRACTIONS

$$\frac{a}{c} + \frac{b}{c} = \frac{a + b}{c},$$

provided $c \neq 0$.

For example, to add $\frac{3}{16}$ and $\frac{5}{16}$, we proceeded as follows:

$$\frac{3}{16} + \frac{5}{16} = \frac{3 + 5}{16} \qquad \text{addition rule for fractions}$$

$$= \frac{8}{16} \qquad \text{simplifying}$$

Note: Always write an answer in reduced form.

$$= \frac{\overset{1}{\cancel{8}} \cdot 1}{\underset{1}{\cancel{8}} \cdot 2} \qquad \text{cancellation law}$$

$$= \frac{1}{2} \qquad \text{reduced form}$$

In a similar manner, we can add algebraic fractions that have the same denominator.

EXAMPLE 1

Find the sum $\dfrac{3x + 2}{4x} + \dfrac{x + 6}{4x}$.

Solution:

$$\frac{3x + 2}{4x} + \frac{x + 6}{4x} = \frac{(3x + 2) + (x + 6)}{4x} \qquad \text{addition rule for fractions}$$

$$= \frac{4x + 8}{4x} \qquad \text{simplifying}$$

Note: After applying the addition rule for fractions, always look for common factors to see if you can *reduce*.

$$= \frac{4(x + 2)}{4x} \qquad \text{factoring}$$

$$= \frac{\overset{1}{\cancel{4}}(x + 2)}{\underset{x}{\cancel{4x}}} \qquad \text{cancellation law}$$

$$= \frac{x + 2}{x} \quad \blacksquare \qquad \text{reduced form}$$

Note: Just a reminder: you can never cancel common terms.

$$\frac{\overset{1}{\cancel{4x}} + 8}{\underset{1}{\cancel{4x}}} \qquad \textit{Wrong}$$

You can only cancel common factors.

EXAMPLE 2

Find the sum $\dfrac{5 - 4x}{x^2 - 4} + \dfrac{7 - 2x}{x^2 - 4}$.

Solution:

$$\frac{5 - 4x}{x^2 - 4} + \frac{7 - 2x}{x^2 - 4} = \frac{(5 - 4x) + (7 - 2x)}{x^2 - 4} \qquad \text{addition rule for fractions}$$

$$= \frac{12 - 6x}{x^2 - 4} \qquad \text{simplifying}$$

Note: Try factoring the numerator and denominator to see if this answer can be reduced.

$$= \frac{6(2 - x)}{(x + 2)(x - 2)} \qquad \text{factoring}$$

$$= \frac{\overset{-1}{6(2 - x)}}{\underset{1}{(x + 2)(x - 2)}} \qquad \text{cancellation law}$$

$$= \frac{-6}{x + 2} \quad \blacksquare \qquad \text{reduced form}$$

⊡ Subtracting Algebraic Fractions with the Same Denominator

You also learned in Chapter 2 that we could subtract fractions with the same denominator by simply subtracting their numerators and retaining the same denominator.

SUBTRACTION RULE FOR FRACTIONS

$$\frac{a}{c} - \frac{b}{c} = \frac{a - b}{c},$$

provided $c \neq 0$.

For example, to subtract $\dfrac{3}{16}$ from $\dfrac{5}{16}$, we proceeded as follows:

$$\frac{5}{16} - \frac{3}{16} = \frac{5-3}{16} \qquad \text{subtraction rule for fractions}$$

$$= \frac{2}{16} \qquad \text{simplifying}$$

Note: Always write an answer in reduced form.

$$= \frac{\overset{1}{\cancel{2}} \cdot 1}{\underset{8}{\cancel{2}} \cdot 8} \qquad \text{cancellation law}$$

$$= \frac{1}{8} \qquad \text{reduced form}$$

Similarly, we can subtract algebraic fractions that have the same denominator.

EXAMPLE 3

Find the difference $\dfrac{5x - y}{2y} - \dfrac{3x + 6y}{2y}$.

Solution:

$$\frac{5x - y}{2y} - \frac{3x + 6y}{2y} = \frac{(5x - y) - (3x + 6y)}{2y} \qquad \begin{array}{l}\text{subtraction rule}\\\text{for fractions}\end{array}$$

Note: We must subtract this entire second numerator. Note the switch in signs.

$$= \frac{5x - y - 3x - 6y}{2y} \qquad \begin{array}{l}\text{distributing a}\\\text{negative sign}\end{array}$$

$$= \frac{2x - 7y}{2y} \qquad \blacksquare \qquad \begin{array}{l}\text{simplifying}\\\text{(reduced form)}\end{array}$$

Note: A very common mistake is to forget to write parentheses in the numerator and to proceed as follows:

$$\frac{5x - y}{2y} - \frac{3x + 6y}{2y} = \frac{5x - y - 3x + 6y}{2y} = \frac{2x + 5y}{2y} \qquad \textit{Wrong}$$

Using parentheses will help remind you to subtract *all* of the terms in the second numerator.

 When subtracting algebraic fractions that have more than one term in their numerators, be sure to place parentheses around each numerator and to subtract the *entire* second numerator.

EXAMPLE 4

Find the difference $\dfrac{6+t}{t^2+t-6} - \dfrac{2-t}{t^2+t-6}$.

Solution:

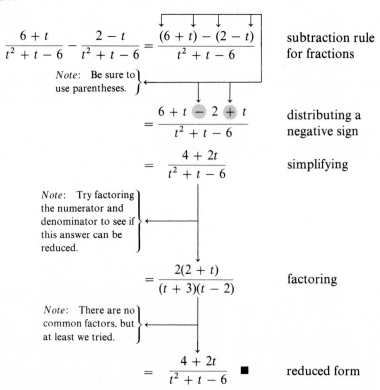

$$\frac{6+t}{t^2+t-6} - \frac{2-t}{t^2+t-6} = \frac{(6+t)-(2-t)}{t^2+t-6} \qquad \text{subtraction rule for fractions}$$

Note: Be sure to use parentheses.

$$= \frac{6+t-2+t}{t^2+t-6} \qquad \text{distributing a negative sign}$$

$$= \frac{4+2t}{t^2+t-6} \qquad \text{simplifying}$$

Note: Try factoring the numerator and denominator to see if this answer can be reduced.

$$= \frac{2(2+t)}{(t+3)(t-2)} \qquad \text{factoring}$$

Note: There are no common factors, but at least we tried.

$$= \frac{4+2t}{t^2+t-6} \quad \blacksquare \qquad \text{reduced form}$$

⬚ Adding and Subtracting Algebraic Fractions whose Denominators Are Opposites

In Chapter 2 we used the fact that $\dfrac{a}{-b} = -\dfrac{a}{b}$ to assist us in adding or subtracting fractions whose denominators are opposites.

We discovered the following:

$$\frac{a}{c} + \frac{b}{-c} = \frac{a}{c} + \left(-\frac{b}{c}\right) = \frac{a}{c} - \frac{b}{c} = \frac{a - b}{c}$$

To add fractions whose denominators are opposites, subtract the second numerator from the first and retain the denominator of the first fraction.

$$\frac{a}{c} - \frac{b}{-c} = \frac{a}{c} - \left(-\frac{b}{c}\right) = \frac{a}{c} + \frac{b}{c} = \frac{a + b}{c}$$

To subtract fractions whose denominators are opposites, add the numerators and retain the denominator of the first fraction.

Using this idea along with the fact that $a - b = -(b - a)$ allows us to add or subtract algebraic fractions whose denominators are opposites.

EXAMPLE 5

Simplify $\dfrac{y}{x - 6} - \dfrac{5y}{6 - x}$.

Solution: You should first note that the denominators are opposites, that is, $6 - x = -(x - 6)$.

$$\frac{y}{x - 6} - \frac{5y}{6 - x} = \frac{y}{x - 6} - \frac{5y}{-(x - 6)} \qquad a - b = -(b - a)$$

$$= \frac{y}{x - 6} + \frac{5y}{x - 6} \qquad \frac{a}{-b} = -\frac{a}{b}$$

$$= \frac{y + 5y}{x - 6} \qquad \text{addition rule for fractions}$$

$$= \frac{6y}{x - 6} \quad \blacksquare \qquad \text{simplifying (reduced form)}$$

EXAMPLE 6

Simplify $\dfrac{2w + 1}{w^2 - 9} + \dfrac{w - 2}{9 - w^2}$.

Solution: First, note that $9 - w^2$ is the opposite of $w^2 - 9$, that is, $9 - w^2 = -(w^2 - 9)$.

$$\frac{2w + 1}{w^2 - 9} + \frac{w - 2}{9 - w^2} = \frac{2w + 1}{w^2 - 9} + \frac{w - 2}{-(w^2 - 9)} \qquad a - b = -(b - a)$$

$$= \frac{2w + 1}{w^2 - 9} - \frac{w - 2}{w^2 - 9} \qquad \frac{a}{-b} = -\frac{a}{b}$$

Note: Remember to use parentheses and to subtract the entire second numerator.

$$= \frac{(2w + 1) - (w - 2)}{w^2 - 9} \qquad \text{subtraction rule for fractions}$$

$$= \frac{2w + 1 - w + 2}{w^2 - 9} \qquad \text{distributing a negative sign}$$

$$= \frac{w + 3}{w^2 - 9} \qquad \text{simplifying}$$

$$= \frac{w + 3}{(w + 3)(w - 3)} \qquad \text{factoring}$$

$$= \frac{\overset{1}{\cancel{w + 3}}}{\underset{1}{\cancel{(w + 3)}}(w - 3)} \qquad \text{cancellation law}$$

$$= \frac{1}{w - 3} \quad \blacksquare \qquad \text{reduced form}$$

EXERCISES 8.3

Perform the indicated operations. Be sure to write your answer in reduced form.

1. $\dfrac{7x}{9y^4} + \dfrac{2x}{9y^4}$

2. $\dfrac{14z}{3x^2y^2} - \dfrac{2z}{3x^2y^2}$

3. $\dfrac{3m + 2n}{6m^3n^4} + \dfrac{3m - 2n}{6m^3n^4}$

4. $\dfrac{4p - 3}{8pq^3} + \dfrac{4p - 5}{8pq^3}$

5. $\dfrac{6x}{2x - 5} - \dfrac{15}{2x - 5}$

6. $\dfrac{14}{3k + 7} + \dfrac{6k}{3k + 7}$

7. $\dfrac{5a - 2ab}{2ab} - \dfrac{7a + 2ab}{2ab}$

8. $\dfrac{3xy - 2y}{5xy} - \dfrac{8y - 2xy}{5xy}$

9. $\dfrac{4x - y}{3x - y} + \dfrac{7x - y}{3x - y}$

10. $\dfrac{13 - 2d}{7 - d} + \dfrac{4 + d}{7 - d}$

11. $\dfrac{7 + 8t}{2t + 3} - \dfrac{4 + 6t}{3 + 2t}$

12. $\dfrac{7m - 2n}{5m + n} - \dfrac{2m - 3n}{n + 5m}$

13. $\dfrac{2x^2}{x + 5} - \dfrac{50}{x + 5}$

14. $\dfrac{3x^2}{x - 3} + \dfrac{x^2 - 36}{x - 3}$

15. $\dfrac{x - y}{10xy^2} + \dfrac{3x - y}{-10xy^2}$

16. $\dfrac{3p - 4q}{4p^3q^2} - \dfrac{3p + 4q}{-4p^3q^2}$

17. $\dfrac{4x - 7y}{2x - y} - \dfrac{5y}{y - 2x}$

18. $\dfrac{3v - 6t^2}{3t^2 - v} + \dfrac{v}{v - 3t^2}$

19. $\dfrac{3t + 1}{t^2 - 4} + \dfrac{t + 7}{t^2 - 4}$

20. $\dfrac{3k - 16}{25 - k^2} + \dfrac{2k - 9}{25 - k^2}$

21. $\dfrac{9y}{4y^2 - 9} + \dfrac{5y - 6}{9 - 4y^2}$

22. $\dfrac{12}{16 - 9y^2} - \dfrac{6y - 4}{9y^2 - 16}$

23. $\dfrac{5x + 4}{3x^2 - 9x} - \dfrac{1 + 6x}{3x^2 - 9x}$

24. $\dfrac{10y - 3}{4y - 9y^2} - \dfrac{5 - 8y}{4y - 9y^2}$

25. $\dfrac{2n + 5}{n^2 + 6n + 9} + \dfrac{1 - 4n}{n^2 + 6n + 9}$

26. $\dfrac{5x + 6}{x^2 - 10x + 25} + \dfrac{9 - 2x}{x^2 - 10x + 25}$

27. $\dfrac{w^2 + 5w}{w^2 + w - 12} - \dfrac{2w + 4}{w^2 + w - 12}$

28. $\dfrac{t^2 - 8}{t^2 + 4t - 12} - \dfrac{2 - 3t}{t^2 + 4t - 12}$

29. $\dfrac{4x^2}{2x^2 - x - 15} - \dfrac{25}{2x^2 - x - 15}$

30. $\dfrac{1}{6x^2 + 7x - 3} - \dfrac{9x^2}{6x^2 + 7x - 3}$

31. $\dfrac{6x^2 - 5x - 4}{3x^2 - 5x - 2} - \dfrac{3x^2 + 5x - 1}{3x^2 - 5x - 2}$

32. $\dfrac{5x^2 + x - 1}{4x^2 + 8x - 5} - \dfrac{x^2 + 5x - 2}{4x^2 + 8x - 5}$

33. $\dfrac{2 + 3y}{2y} - \left(\dfrac{y + 1}{2y} + \dfrac{1 + 2y}{2y}\right)$

34. $\dfrac{x - 5}{7x^2} - \left(\dfrac{7x + 6}{7x^2} - \dfrac{6x + 11}{7x^2}\right)$

35. $\dfrac{x^2}{2 - x} - \left(\dfrac{8x}{2 - x} + \dfrac{12}{x - 2}\right)$

36. $\left(\dfrac{x^3}{4 - x^2} + \dfrac{6}{4 - x^2}\right) + \left(\dfrac{2x^2}{x^2 - 4} + \dfrac{3x}{x^2 - 4}\right)$

8.4 ADDING AND SUBTRACTING ALGEBRAIC FRACTIONS WITH DIFFERENT DENOMINATORS

Objectives

☐ Given algebraic fractions with different denominators, to find the least common denominator (LCD).

☐ To add or subtract algebraic fractions that have different denominators.

⊡ Finding the Least Common Denominator

Recall from Chapter 2 that to add or subtract fractions with different denominators, we first found a least common denominator (LCD) for the given fractions. If the LCD could not be found by inspection, we used the following two-step procedure to determine the LCD.

TWO-STEP PROCEDURE FOR DETERMINING THE LCD

Step 1: Completely factor each denominator, using exponents to indicate the number of times a factor appears.

Step 2: Multiply each different factor to the highest power that it appears in any one of the individual denominators.

For example, to find the LCD for $\dfrac{1}{24} + \dfrac{5}{42}$, we proceeded as follows:

Step 1:

$$\frac{1}{24} + \frac{5}{42} = \frac{1}{2^3 \cdot 3} + \frac{5}{2 \cdot 3 \cdot 7}$$

Note: The different factors are 2, 3, and 7. The highest powers to which they appear are 2^3, 3, and 7.

Step 2: Thus,

$$\text{LCD} = 2^3 \cdot 3 \cdot 7, \text{ or } 168.$$

We can use the same two-step procedure to determine the LCD of algebraic fractions with different denominators.

EXAMPLE 1

Find the LCD for $\dfrac{2x^2 + 1}{8x^3y} + \dfrac{1 - 3y}{12xy^2}$.

Solution:

Step 1:

$$\frac{2x^2 + 1}{8x^3y} + \frac{1 - 3y}{12xy^2} = \frac{2x^2 + 1}{2^3 \cdot x^3 \cdot y} + \frac{1 - 3y}{2^2 \cdot 3 \cdot x \cdot y^2}$$

Note: The different factors, that appear are 2, 3, x, and y. The highest powers to which they appear are 2^3, 3, x^3, and y^2.

Step 2:   Thus,

$$\text{LCD} = \quad 2^3 \cdot 3 \cdot x^3 \cdot y^2 \text{ or } 24x^3y^2. \quad \blacksquare$$

EXAMPLE 2

Find the LCD for $\dfrac{3}{x^2 + 2x} - \dfrac{2 - x}{x^2 + 4x + 4}$.

Solution:

Step 1:

$$\frac{3}{x^2 + 2x} - \frac{2 - x}{x^2 + 4x + 4} = \frac{3}{x\,(x + 2)} - \frac{2 - x}{(x + 2)^2}$$

Note: The different factors that appear are x and $x + 2$. The highest powers to which they appear are x and $(x + 2)^2$.

Step 2:   Thus,

$$\text{LCD} = \quad x(x + 2)^2 \text{ or } x^3 + 4x^2 + 4x. \quad \blacksquare$$

EXAMPLE 3

Find the LCD for $\dfrac{-12}{9 - y^2} - \dfrac{y - 1}{y - 3}$.

Solution:

Step 1:

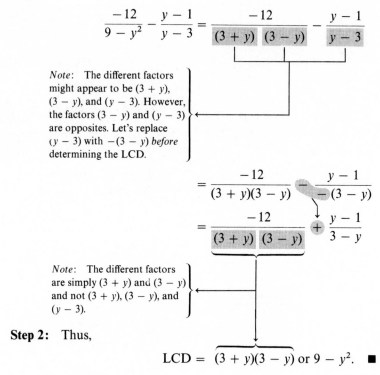

$$\frac{-12}{9 - y^2} - \frac{y - 1}{y - 3} = \frac{-12}{(3 + y)\ (3 - y)} - \frac{y - 1}{y - 3}$$

Note: The different factors might appear to be $(3 + y)$, $(3 - y)$, and $(y - 3)$. However, the factors $(3 - y)$ and $(y - 3)$ are opposites. Let's replace $(y - 3)$ with $-(3 - y)$ *before* determining the LCD.

$$= \frac{-12}{(3 + y)(3 - y)} - \frac{y - 1}{-(3 - y)}$$

$$= \frac{-12}{(3 + y)\ (3 - y)} + \frac{y - 1}{3 - y}$$

Note: The different factors are simply $(3 + y)$ and $(3 - y)$ and not $(3 + y)$, $(3 - y)$, and $(y - 3)$.

Step 2: Thus,

$$\text{LCD} = (3 + y)(3 - y) \text{ or } 9 - y^2. \quad \blacksquare$$

Note: When opposites appear as factors of the denominators, use the fact that $a - b = -(b - a)$ to rename the fractions *before* determining the LCD.

There are times when the LCD can be determined by inspection. *When the denominators are different and nonfactorable, the LCD is simply the product of the denominators.* For example,

the LCD for $\dfrac{1}{7} + \dfrac{3}{5}$ is simply $(7)(5)$, or 35.

— different, —
nonfactorable
denominators

EXAMPLE 4

Find the LCD for $\dfrac{t - 1}{t + 1} - \dfrac{t + 1}{t - 1}$.

Solution: The LCD for

$$\frac{t-1}{t+1} - \frac{t+1}{t-1}$$

different, nonfactorable denominators

is simply $(t+1)(t-1) = t^2 - 1$. ■

⬚ Adding and Subtracting Algebraic Fractions that Have Different Denominators

Once the LCD is chosen, we use the fundamental property of fractions to change each fraction to an equivalent fraction whose denominator is the LCD. Only then can we apply the addition or subtraction rule for fractions to find the sum or difference.

Let's return to the problem $\frac{1}{24} + \frac{5}{42}$ and find the sum.

$$\frac{1}{24} + \frac{5}{42} = \frac{1}{2^3 \cdot 3} + \frac{5}{2 \cdot 3 \cdot 7}$$ factoring the denominators

Note: LCD $= 2^3 \cdot 3 \cdot 7$. Now change to equivalent fractions that have $2^3 \cdot 3 \cdot 7$ as the denominator.

$$= \frac{7 \cdot 1}{7 \cdot 2^3 \cdot 3} + \frac{2^2 \cdot 5}{2^2 \cdot 2 \cdot 3 \cdot 7}$$ fundamental property of fractions

$$= \frac{7}{2^3 \cdot 3 \cdot 7} + \frac{20}{2^3 \cdot 3 \cdot 7}$$ simplifying

Note: Since the denominators are now the same, we can add the fractions.

$$= \frac{27}{2^3 \cdot 3 \cdot 7}$$ addition rule for fractions

$$= \frac{\overset{9}{\cancel{27}}}{2^3 \cdot \cancel{3} \cdot 7}$$ cancellation law

$$= \frac{9}{56}$$ reduced form

Using a similar procedure, we can add or subtract algebraic fractions that have different denominators.

EXAMPLE 5

Find the sum $\dfrac{2x^2 + 1}{8x^3y} + \dfrac{1 - 3y}{12xy^2}$.

Solution:

$$\frac{2x^2 + 1}{8x^3y} + \frac{1 - 3y}{12xy^2} = \frac{2x^2 + 1}{2^3 \; x^3 \; y} + \frac{1 - 3y}{2^2 \cdot 3 \; x \; y^2} \qquad \text{factoring the denominators}$$

Note: LCD $= 2^3 \cdot 3x^3y^2$. Now change to equivalent fractions that have $2^3 \cdot 3x^3y^2$ as the denominator.

$$= \frac{3y \cdot (2x^2 + 1)}{3y \cdot 2^3 x^3 y} + \frac{2x^2 \cdot (1 - 3y)}{2x^2 \cdot 2^2 \cdot 3xy^2} \qquad \text{fundamental property of fractions}$$

$$= \frac{6x^2y + 3y}{2^3 \cdot 3x^3y^2} + \frac{2x^2 - 6x^2y}{2^3 \cdot 3x^3y^2} \qquad \text{distributive law}$$

Note: Since the denominators are now the same, we can add the fractions.

$$= \frac{(6x^2y + 3y) + (2x^2 - 6x^2y)}{2^3 \cdot 3x^3y^2} \qquad \text{addition rule for fractions}$$

$$= \frac{2x^2 + 3y}{2^3 \cdot 3x^3y^2} \qquad \text{simplifying}$$

Note: Since $2x^2 + 3y$ is nonfactorable, we can't reduce further.

$$= \frac{2x^2 + 3y}{24x^3y^2} \qquad\blacksquare \qquad \text{reduced form}$$

EXAMPLE 6

Find the difference $\dfrac{3}{x^2 + 2x} - \dfrac{2 - x}{x^2 + 4x + 4}$.

Solution:

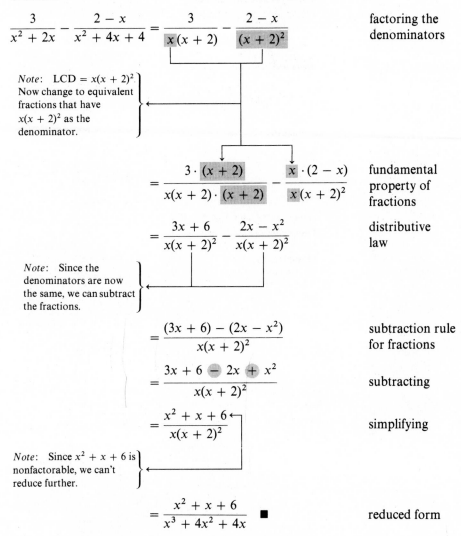

$$\frac{3}{x^2 + 2x} - \frac{2 - x}{x^2 + 4x + 4} = \frac{3}{\boxed{x(x+2)}} - \frac{2-x}{\boxed{(x+2)^2}}$$ factoring the denominators

Note: LCD = $x(x+2)^2$. Now change to equivalent fractions that have $x(x+2)^2$ as the denominator.

$$= \frac{3 \cdot (x+2)}{x(x+2) \cdot (x+2)} - \frac{x \cdot (2-x)}{x(x+2)^2}$$ fundamental property of fractions

$$= \frac{3x + 6}{x(x+2)^2} - \frac{2x - x^2}{x(x+2)^2}$$ distributive law

Note: Since the denominators are now the same, we can subtract the fractions.

$$= \frac{(3x + 6) - (2x - x^2)}{x(x+2)^2}$$ subtraction rule for fractions

$$= \frac{3x + 6 - 2x + x^2}{x(x+2)^2}$$ subtracting

$$= \frac{x^2 + x + 6}{x(x+2)^2}$$ simplifying

Note: Since $x^2 + x + 6$ is nonfactorable, we can't reduce further.

$$= \frac{x^2 + x + 6}{x^3 + 4x^2 + 4x} \quad \blacksquare$$ reduced form

EXAMPLE 7

Simplify $\dfrac{-12}{9 - y^2} - \dfrac{y - 1}{y - 3}$.

Solution:

$$\frac{-12}{9 - y^2} - \frac{y - 1}{y - 3} = \frac{-12}{(3 + y)(3 - y)} - \frac{y - 1}{y - 3}$$

opposites

$$a - b = -(b - a)$$

$$= \frac{-12}{(3 + y)\,(3 - y)} + \frac{y - 1}{3 - y}$$

Note: LCD = $(3 + y)(3 - y)$. Now change to equivalent fractions that have $(3 + y)(3 - y)$ as the denominator.

$$= \frac{-12}{(3 + y)(3 - y)} + \frac{(3 + y) \cdot (y - 1)}{(3 + y) \cdot (3 - y)}$$

fundamental property of fractions

$$= \frac{-12}{(3 + y)(3 - y)} + \frac{y^2 + 2y - 3}{(3 + y)(3 - y)}$$

FOIL

$$= \frac{y^2 + 2y - 15}{(3 + y)(3 - y)}$$

addition rule for fractions

Note: Factor the numerator to see if this answer can be reduced.

$$= \frac{\overset{-1}{(y + 5)\cancel{(y - 3)}}}{\underset{1}{(3 + y)\cancel{(3 - y)}}}$$

factoring and cancelling opposites

$$= \frac{-y - 5}{3 + y} \quad \blacksquare$$

reduced form

EXAMPLE 8

Simplify $\dfrac{t - 1}{t + 1} - \dfrac{t + 1}{t - 1}$.

Solution: LCD $= (t + 1)(t - 1)$. Now change to equivalent fractions that have $(t + 1)(t - 1)$ as a denominator.

$$\frac{t - 1}{t + 1} - \frac{t + 1}{t - 1} = \frac{(t - 1) \cdot (t - 1)}{(t - 1) \cdot (t + 1)} - \frac{(t + 1) \cdot (t + 1)}{(t + 1) \cdot (t - 1)}$$

fundamental
property
of fractions

$$= \frac{t^2 - 2t + 1}{(t - 1)(t + 1)} - \frac{t^2 + 2t + 1}{(t + 1)(t - 1)}$$

FOIL

Note: Be sure to remember the parentheses and subtract the *entire* second numerator.

$$= \frac{(t^2 - 2t + 1) - (t^2 + 2t + 1)}{(t + 1)(t - 1)}$$

subtraction
rule
for fractions

$$= \frac{t^2 - 2t + 1 \ominus t^2 \ominus 2t \ominus 1}{(t + 1)(t - 1)}$$

subtracting

$$= \frac{-4t}{(t + 1)(t - 1)}$$

simplifying

$$= \frac{-4t}{t^2 - 1} \quad \text{or} \quad \frac{4t}{1 - t^2} \quad \blacksquare$$

reduced form

EXERCISES 8.4

Find the LCD; then perform the indicated operation. Be sure your answer is in reduced form.

1. $\dfrac{5}{6x^2y} + \dfrac{3}{8xy^2}$

2. $\dfrac{7y}{12x^3} - \dfrac{5y}{16x}$

3. $\dfrac{8m - n}{6m^5} - \dfrac{7}{10m^4}$

4. $\dfrac{3 + s^2}{s^3t^4} + \dfrac{4 - t}{st^5}$

5. $\dfrac{b^2 - 1}{a^2b^3} + \dfrac{1 - a^2}{a^4b}$

6. $\dfrac{5 - n}{8n^7} + \dfrac{n - 3}{18n^8}$

7. $\dfrac{3 - y}{y} - \dfrac{7 - xy}{4xy}$

8. $\dfrac{3 - n}{n} - \dfrac{6mn - 1}{9mn}$

9. $\dfrac{4}{5y(y + 3)^2} - \dfrac{7}{10(y + 3)}$

10. $\dfrac{3}{2m(m - 1)} + \dfrac{m}{6(m - 1)^2}$

11. $\dfrac{5}{x + 3} + \dfrac{4}{x}$

12. $\dfrac{3}{y} + \dfrac{4}{2y + 5}$

13. $\dfrac{2k}{k + 2} + \dfrac{3k}{k - 2}$

14. $\dfrac{3x}{x - 5} - \dfrac{5x}{x + 5}$

15. $\dfrac{7p}{2p + 6} - \dfrac{5p}{3p + 9}$

16. $\dfrac{4x}{10 - 15x} - \dfrac{3x}{8 - 12x}$

17. $\dfrac{-4}{x^2 - 16} + \dfrac{1}{2x - 8}$

18. $\dfrac{1}{2x + 10} + \dfrac{x}{25 - x^2}$

19. $\dfrac{4}{t - 2} - \dfrac{t - 2}{t + 1}$

20. $\dfrac{3y}{2y - 3} + \dfrac{y - 1}{y + 2}$

21. $\dfrac{6 - 5x}{6x^2 - 9x} - \dfrac{x - 2}{3x}$

22. $\dfrac{1}{10x^2 - 5x} + \dfrac{2x + 1}{5x}$

23. $\dfrac{t}{t^2 + 6t + 9} - \dfrac{4}{t + 3}$

24. $\dfrac{4}{t^2 - 10t + 25} - \dfrac{t}{t - 5}$

25. $\dfrac{p + 2}{2p - 3} + \dfrac{12p + 3}{9 - 4p^2}$

26. $\dfrac{m + 1}{2m - 5} + \dfrac{8m + 15}{25 - 4m^2}$

27. $\dfrac{x - 3}{x^2 - 4} - \dfrac{x + 1}{x^2 + 4x + 4}$

28. $\dfrac{2 - k}{49 - k^2} - \dfrac{1 - k}{49 - 14k + k^2}$

29. $\dfrac{x - 1}{x + 3} + \dfrac{2x + 10}{x^2 + 7x + 12}$

30. $\dfrac{x + 3}{x + 2} - \dfrac{4x + 3}{2x^2 + 3x - 2}$

31. $\dfrac{-36}{2t^2 - 7t - 4} - \dfrac{t}{4 - t}$

32. $\dfrac{4p^2 - p}{3p^2 - 5p - 2} + \dfrac{p}{2 - p}$

33. $\dfrac{5}{6w + 12} + \dfrac{w - 1}{3w^3 + 12w^2 + 12w}$

34. $\dfrac{k + 3}{2k^3 - 12k^2 + 18k} - \dfrac{k - 1}{8k^3 - 24k^2}$

35. $\dfrac{2y}{y + 3} - \left(\dfrac{y + 1}{y - 3} + \dfrac{36}{y^2 - 9} \right)$

36. $\dfrac{x^2 + 29}{x^2 + x - 20} - \left(\dfrac{x - 1}{x + 5} - \dfrac{x + 1}{x - 4} \right)$

8.5   MIXED ALGEBRAIC EXPRESSIONS AND COMPLEX FRACTIONS

Objectives

☐ To change a mixed algebraic expression into a single algebraic fraction.

☐ To simplify a complex fraction.

☐ Mixed Algebraic Expressions

The sum or difference of a polynomial and an algebraic fraction is referred to as a **mixed algebraic expression**. The following are examples of mixed algebraic expressions:

$$2x^2 + \frac{2z}{3x^3 y^5} \qquad \frac{x^2 + 1}{2x} + 3x - 4 \qquad 2t - \frac{4}{t + 2} + 3$$

Recall from Chapter 2 that a *mixed number* represented the sum of an integer and a fraction. For example, $2\dfrac{1}{4} = 2 + \dfrac{1}{4}$. To change a mixed number to an improper fraction, we placed the integer over 1 and added the fractions. For example, to change $2\frac{1}{4}$ to an improper fraction, we proceeded as follows:

$$2\frac{1}{4} = 2 + \frac{1}{4} \qquad \text{definition of a mixed number}$$

$$= \frac{2}{1} + \frac{1}{4} \qquad a = \frac{a}{1}$$

$$= \frac{4 \cdot 2}{4 \cdot 1} + \frac{1}{4} \qquad \text{fundamental property of fractions}$$

$$= \frac{8}{4} + \frac{1}{4} \qquad \text{simplifying}$$

$$= \frac{9}{4} \qquad \text{addition rule for fractions}$$

In a similar way a mixed algebraic expression can be changed to a single algebraic fraction. Simply place the polynomial part of the mixed algebraic expression over 1 and add the fractions.

EXAMPLE 1

Change $2x^2 + \dfrac{2z}{3x^3y^5}$ to a single algebraic fraction.

Solution:

$$2x^2 + \frac{2z}{3x^3y^5} = \frac{2x^2}{1} + \frac{2z}{3x^3y^5} \qquad a = \frac{a}{1}$$

Note: When one of the denominators is 1, the LCD is the other denominator. Thus, LCD $= 3x^3y^5$.

$$= \frac{3x^3y^5 \cdot 2x^2}{3x^3y^5 \cdot 1} + \frac{2z}{3x^3y^5} \qquad \text{fundamental property of fractions}$$

$$= \frac{6x^5y^5}{3x^3y^5} + \frac{2z}{3x^3y^5} \qquad \text{simplifying}$$

$$= \frac{6x^5y^5 + 2z}{3x^3y^5} \quad \blacksquare \qquad \text{addition rule for fractions}$$

EXAMPLE 2

Change $\dfrac{x^2 + 1}{2x} + 3x - 4$ to a single algebraic fraction.

Solution:

$$\dfrac{x^2 + 1}{2x} + 3x - 4 = \dfrac{x^2 + 1}{2x} + \dfrac{3x - 4}{1} \qquad a = \dfrac{a}{1}$$

$$= \dfrac{x^2 + 1}{2x} + \dfrac{2x \cdot (3x - 4)}{2x \cdot 1} \qquad \text{fundamental property of fractions}$$

$$= \dfrac{x^2 + 1}{2x} + \dfrac{6x^2 - 8x}{2x} \qquad \text{distributive law}$$

$$= \dfrac{7x^2 - 8x + 1}{2x} \quad \blacksquare \qquad \text{addition rule for fractions}$$

EXAMPLE 3

Change $2t - \dfrac{4}{t + 2} + 3$ to a single algebraic fraction.

Solution:

$$2t - \dfrac{4}{t + 2} + 3 = 2t + 3 - \dfrac{4}{t + 2} \qquad \text{commutative law of addition}$$

$$= \dfrac{2t + 3}{1} - \dfrac{4}{t + 2} \qquad a = \dfrac{a}{1}$$

$$= \dfrac{(t + 2) \cdot (2t + 3)}{(t + 2) \cdot 1} - \dfrac{4}{t + 2} \qquad \text{fundamental property of fractions}$$

$$= \dfrac{2t^2 + 7t + 6}{t + 2} - \dfrac{4}{t + 2} \qquad \text{FOIL}$$

$$= \dfrac{2t^2 + 7t + 2}{t + 2} \quad \blacksquare \qquad \text{subtraction rule for fractions}$$

⊡ Complex Fractions

When the numerator and/or denominator of a fraction contain mixed algebraic expressions, the fraction is referred to as a **complex fraction**. The following are examples of complex fractions:

$$\dfrac{9 - \dfrac{1}{x^2}}{3 + \dfrac{1}{x}} \qquad \dfrac{y - 5 + \dfrac{6}{y + 2}}{y - 3 - \dfrac{5}{y + 1}}$$

To simplify such complex fractions, it is first necessary to change the mixed algebraic expressions into single algebraic fractions. After this is accomplished, the division rule for fractions can be applied. There are other methods for simplifying complex fractions. We choose this procedure now because it reinforces the use of *all* of the important properties and rules that have been previously discussed.

EXAMPLE 4

Simplify $\dfrac{9 - \dfrac{1}{x^2}}{3 + \dfrac{1}{x}}$.

Solution:

$$\frac{9 - \dfrac{1}{x^2}}{3 + \dfrac{1}{x}} = \frac{\dfrac{9}{1} - \dfrac{1}{x^2}}{\dfrac{3}{1} + \dfrac{1}{x}} \qquad a = \frac{a}{1}$$

$$= \frac{\dfrac{9x^2}{x^2} - \dfrac{1}{x^2}}{\dfrac{3x}{x} + \dfrac{1}{x}} \qquad \text{fundamental property of fractions}$$

$$= \frac{\dfrac{9x^2 - 1}{x^2}}{\dfrac{3x + 1}{x}} \qquad \text{addition and subtraction rules for fractions}$$

Note: Since we now have single algebraic fractions, we can apply the division rule for fractions.

$$= \frac{9x^2 - 1}{x^2} \cdot \frac{x}{3x + 1} \qquad \text{division rule for fractions}$$

$$= \frac{(9x^2 - 1) \cdot x}{x^2 \cdot (3x + 1)} \qquad \text{multiplication rule for fractions}$$

$$= \frac{(3x + 1)(3x - 1) \cdot x}{x^2 \cdot (3x + 1)} \qquad \text{factoring}$$

$$= \frac{\overset{1}{\cancel{(3x + 1)}}(3x - 1) \cdot \cancel{x}}{\underset{x}{\cancel{x^2}} \cdot \underset{1}{\cancel{(3x + 1)}}} \qquad \text{cancellation law}$$

$$= \frac{3x - 1}{x} \qquad \blacksquare \qquad \text{reduced form}$$

EXAMPLE 5

Simplify $\dfrac{y - 5 + \dfrac{6}{y + 2}}{y - 3 - \dfrac{5}{y + 1}}$.

Solution:

$$\frac{y - 5 + \dfrac{6}{y + 2}}{y - 3 - \dfrac{5}{y + 1}} = \frac{\dfrac{y - 5}{1} + \dfrac{6}{y + 2}}{\dfrac{y - 3}{1} - \dfrac{5}{y + 1}} \qquad a = \frac{a}{1}$$

$$= \frac{\dfrac{(y - 5)(y + 2)}{y + 2} + \dfrac{6}{y + 2}}{\dfrac{(y - 3)(y + 1)}{y + 1} - \dfrac{5}{y + 1}} \qquad \text{fundamental property of fractions}$$

$$= \frac{\dfrac{y^2 - 3y - 10}{y + 2} + \dfrac{6}{y + 2}}{\dfrac{y^2 - 2y - 3}{y + 1} - \dfrac{5}{y + 1}} \qquad \text{FOIL}$$

$$= \frac{\dfrac{y^2 - 3y - 4}{y + 2}}{\dfrac{y^2 - 2y - 8}{y + 1}} \qquad \text{addition and subtraction rules for fractions}$$

Note: Since we now have single algebraic fractions, we can apply the division rule for fractions.

$$= \frac{y^2 - 3y - 4}{y + 2} \cdot \frac{y + 1}{y^2 - 2y - 8} \qquad \text{division rule for fractions}$$

$$= \frac{(y^2 - 3y - 4) \cdot (y + 1)}{(y + 2) \cdot (y^2 - 2y - 8)} \qquad \text{multiplication rule for fractions}$$

$$= \frac{\overset{1}{\cancel{(y - 4)}}(y + 1) \cdot (y + 1)}{(y + 2) \cdot \underset{1}{\cancel{(y - 4)}}(y + 2)} \qquad \text{factoring and cancelling}$$

$$= \frac{(y + 1)^2}{(y + 2)^2} \text{ or } \frac{y^2 + 2y + 1}{y^2 + 4y + 4} \quad \blacksquare \qquad \text{reduced form}$$

Application

SERIES-PARALLEL CIRCUITS

Electric circuits are seldom purely series or purely parallel circuits. Often they contain a combination of series and parallel resistances like the circuit shown in Fig. 8.1.

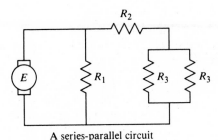

A series-parallel circuit　　　　　**Figure 8.1**

To analyze such series-parallel circuits, it is necessary to reduce them to simple series or simple parallel circuits. This can be accomplished by using the series and parallel resistance formulas and forming *equivalent circuits*.

For example, the two resistances labeled R_3 in Fig. 8.1 are connected in **parallel** and, according to the parallel resistance formula, their total resistance is

$$\frac{R_3 R_3}{R_3 + R_3} = \frac{R_3^{\,2}}{2R_3} = \frac{R_3}{2}.$$

Thus, the circuit in Fig. 8.1 can be considered equivalent to the circuit in Fig. 8.2.

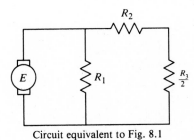

Circuit equivalent to Fig. 8.1　　　　**Figure 8.2**

The resistances R_2 and $\dfrac{R_3}{2}$ in Fig. 8.2 are connected in **series** and, according to the series resistance formula, their total resistance is

$$R_2 + \frac{R_3}{2}.$$

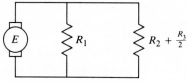

Circuit equivalent to Fig. 8.1   **Figure 8.3**

Thus, the circuit in Fig. 8.3 can be considered equivalent to the circuit in Fig. 8.1. The circuit in Fig. 8.3 is a **simple parallel circuit**. According to the parallel resistance formula, the total resistance in this circuit is

$$\frac{R_1 \cdot \left(R_2 + \dfrac{R_3}{2}\right)}{R_1 + \left(R_2 + \dfrac{R_3}{2}\right)}.$$

Thus, the total resistance in the circuit in Fig. 8.1 is

$$R_t = \frac{R_1 \cdot \left(R_2 + \dfrac{R_3}{2}\right)}{R_1 + \left(R_2 + \dfrac{R_3}{2}\right)}. \qquad Note: \quad \text{A complex fraction}$$

EXAMPLE 6

Write the total resistance of the circuit in Fig. 8.1 as a single algebraic fraction.

Solution:

$$R_t = \frac{R_1 \cdot \left(R_2 + \dfrac{R_3}{2}\right)}{R_1 + \left(R_2 + \dfrac{R_3}{2}\right)} = \frac{R_1 R_2 + \dfrac{R_1 R_3}{2}}{R_1 + R_2 + \dfrac{R_3}{2}} \qquad \text{distributing and removing parentheses}$$

$$= \frac{\dfrac{R_1 R_2}{1} + \dfrac{R_1 R_3}{2}}{\dfrac{R_1 + R_2}{1} + \dfrac{R_3}{2}} \qquad a = \dfrac{a}{1}$$

$$= \frac{\dfrac{2R_1 R_2}{2} + \dfrac{R_1 R_3}{2}}{\dfrac{2(R_1 + R_2)}{2} + \dfrac{R_3}{2}} \qquad \text{fundamental property of fractions}$$

$$= \frac{\dfrac{2R_1R_2 + R_1R_3}{2}}{\dfrac{2R_1 + 2R_2 + R_3}{2}} \qquad \text{addition rule for fractions}$$

$$= \frac{2R_1R_2 + R_1R_3}{2} \cdot \frac{2}{2R_1 + 2R_2 + R_3} \qquad \text{division rule for fractions}$$

$$= \frac{(2R_1R_2 + R_1R_3) \cdot \overset{1}{\cancel{2}}}{\underset{1}{\cancel{2}} \cdot (2R_1 + 2R_2 + R_3)} \qquad \begin{array}{l}\text{multiplication} \\ \text{rule for} \\ \text{fractions and} \\ \text{cancelling}\end{array}$$

$$= \frac{2R_1R_2 + R_1R_3}{2R_1 + 2R_2 + R_3} \quad \blacksquare \qquad \text{reduced form}$$

Note: Once the total resistance of a circuit is known, Ohm's law and Kirchhoff's current and voltage laws can be applied to determine the current through and the voltage drop across each resistor.

EXERCISES 8.5

Change each mixed algebraic expression to a single algebraic fraction.

1. $2t + \dfrac{4v}{3t^3}$

2. $\dfrac{3}{7x^2y^3} - 9x^3y^2$

3. $\dfrac{w^2 - 1}{5w} + 2w$

4. $8p - \dfrac{p^2 - 4}{2p}$

5. $y + 7 + \dfrac{5}{y}$

6. $\dfrac{3}{z} + 5 - 3z$

7. $5x - 1 - \dfrac{x^2 + 2x - 3}{5x}$

8. $\dfrac{x^2 + 5x + 4}{3x} - 2x - 3$

9. $\dfrac{16}{y - 4} - y + 4$

10. $2k - 3 + \dfrac{9}{2k + 3}$

11. $2 + \dfrac{4t}{2 - t} - t$

12. $t - \dfrac{14t}{t + 7} + 7$

13. $5x - 3 - \dfrac{2x - 1}{x + 1}$

14. $3x - 4 - \dfrac{x - 2}{x - 3}$

15. $\dfrac{3a - 18}{a^2 - 3a - 18} - 1$

16. $\dfrac{a^2 - 5a + 4}{2a^2 - a - 1} - 1$

Simplify each complex fraction.

17. $\dfrac{1 + 2m}{1 + \dfrac{1}{2m}}$

18. $\dfrac{n + 3m}{3 + \dfrac{n}{m}}$

19. $\dfrac{\dfrac{1}{x} - x}{\dfrac{1}{x} - 1}$

20. $\dfrac{\dfrac{1}{4x^2} - 1}{\dfrac{1}{x} - 2}$

21. $\dfrac{y - 3 + \dfrac{2}{y}}{\dfrac{4}{y} - y}$

22. $\dfrac{\dfrac{10}{y} - 7 + y}{y - \dfrac{25}{y}}$

23. $\dfrac{x + 3 - \dfrac{5}{x - 1}}{x + 1 + \dfrac{3}{x + 5}}$

24. $\dfrac{x - 2 - \dfrac{8}{x + 5}}{x + 4 + \dfrac{3x}{x - 3}}$

25. $\dfrac{t - \dfrac{t + 2}{t + 3} + 2}{t + \dfrac{3t + 3}{t + 3} + 5}$

26. $\dfrac{t - \dfrac{t + 8}{t - 2} + 2}{2t - \dfrac{2t + 10}{t - 2} + 1}$

27. $\dfrac{\dfrac{(2n + 1)^2}{5n + 1} + 1}{\dfrac{(2n - 1)^2}{3n + 3} - 1}$

28. $\dfrac{\dfrac{(4m - 3)^2}{12 - 26m} - 1}{\dfrac{(4m + 3)^2}{6 + 38m} - 1}$

29. a) Given the series-parallel circuit in Fig. 8.4(a), find the missing resistance in the equivalent circuit of Fig. 8.4(b).

   b) Using the simple series circuit in Fig. 8.4(b), find a single algebraic fraction that represents the total resistance in the circuit of Fig. 8.4(a).

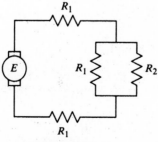

(a) Series-parallel
circuit

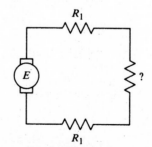

(b) Circuit equivalent to
Fig. 8.4(a) circuit

Figure 8.4

30. a) Given the series-parallel circuit in Fig. 8.5(a), find the missing resistances in the equivalent circuits of Figs. 8.5(b) and 8.5(c).

b) Using the simple parallel circuit in Fig. 8.5(c), find a single algebraic fraction that represents the total resistance in the circuit of Fig. 8.5(a).

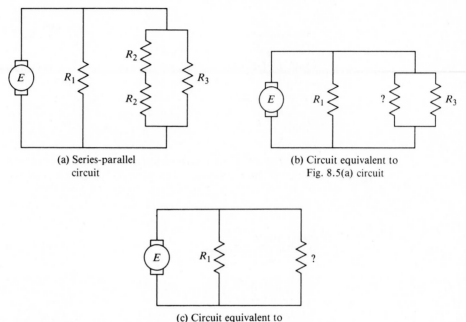

(a) Series-parallel
circuit

(b) Circuit equivalent to
Fig. 8.5(a) circuit

(c) Circuit equivalent to
Fig. 8.5(a) circuit

Figure 8.5

8.6   FRACTIONAL EQUATIONS

Objectives

⊡ To solve a fractional equation containing constants as denominators.

⊡ To solve a fractional equation containing variable expressions as denominators.

⊡ Constant Denominators

An equation that contains one or more algebraic fractions is referred to as a **fractional equation**. An example of a simple fractional equation is

$$\frac{t}{5} + \frac{3t}{10} = \frac{3}{2}.$$

How might we solve such an equation? It certainly would be nice to eliminate the denominators on each side of the equation. Recall the multiplication property of equality from Chapter 4.

MULTIPLICATION PROPERTY OF EQUALITY

If each side of an equation is *multiplied by* or *divided by* the same number (except 0), the new equation is equivalent to the original equation.

Is there a nonzero number that we can *multiply* each side of the equation by to eliminate *all* the denominators? The nonzero number we are looking for must be exactly divisible by each denominator if each denominator is to be completely cancelled and eliminated. Remember, the LCD was defined to be the *smallest* number exactly divisible by each denominator.

   Thus, if we wish to eliminate all the denominators of a fractional equation, we should multiply each side of the equation by the LCD of all the fractions in the equation.

   By eliminating the denominators we hope to obtain an equation that can be solved by methods that have been discussed previously.

EXAMPLE 1

Solve $\dfrac{t}{5} + \dfrac{3t}{10} = \dfrac{3}{2}$.

Solution:

$$\frac{t}{5} + \frac{3t}{10} = \frac{3}{2}$$

Note: First determine the LCD. LCD = 10.

$$\frac{10}{1}\left(\frac{t}{5} + \frac{3t}{10}\right) = \frac{10}{1}\left(\frac{3}{2}\right) \qquad \text{multiplying each side by the LCD}$$

$$\frac{\overset{2}{\cancel{10}} \cdot t}{\cancel{5}} + \frac{\overset{1}{\cancel{10}} \cdot 3t}{\cancel{10}} = \frac{\overset{5}{\cancel{10}} \cdot 3}{\cancel{2}} \qquad \text{distributing and cancelling}$$

Note: Each denominator is completely cancelled and eliminated.

$$2t + 3t = 15 \qquad \text{simplifying}$$

$$5t = 15 \qquad \text{combining similar terms}$$

$$t = 3 \ \blacksquare \qquad \text{dividing each side by 5.}$$

Note: As with any equation, you should check the solution.

$$\text{Does } \frac{t}{5} + \frac{3t}{10} = \frac{3}{2}, \text{ when } t = 3?$$

$$\frac{3}{5} + \frac{9}{10} = \frac{3}{2}?$$

$$\frac{6}{10} + \frac{9}{10} = \frac{3}{2}?$$

$$\frac{\overset{3}{\cancel{15}}}{\underset{2}{\cancel{10}}} = \frac{3}{2}?$$

$$\frac{3}{2} = \frac{3}{2} \quad \text{true}$$

EXAMPLE 2

Solve $\dfrac{2x + 1}{12} - \dfrac{x - 2}{8} = \dfrac{2}{5}$.

Solution:

$$\frac{2x + 1}{12} - \frac{x - 2}{8} = \frac{2}{5}$$

$$\frac{2x + 1}{2^2 \cdot 3} - \frac{x - 2}{2^3} = \frac{2}{5} \qquad \text{factoring the denominators}$$

Note: LCD $= 2^3 \cdot 3 \cdot 5$
or 120.

$$\frac{2^3 \cdot 3 \cdot 5}{1} \left(\frac{2x + 1}{2^2 \cdot 3} - \frac{x - 2}{2^3} \right) = \frac{2^3 \cdot 3 \cdot 5}{1} \left(\frac{2}{5} \right) \qquad \begin{array}{l}\text{multiplying} \\ \text{each side by} \\ \text{the LCD}\end{array}$$

$$\frac{\overset{2}{\cancel{2^3}} \cdot \overset{1}{\cancel{3}} \cdot 5(2x + 1)}{\underset{1}{\cancel{2^2}} \cdot \underset{1}{\cancel{3}}} - \frac{\overset{1}{\cancel{2^3}} \cdot 3 \cdot 5(x - 2)}{\underset{1}{\cancel{2^3}}} = \frac{2^3 \cdot 3 \cdot \overset{1}{\cancel{5}}(2)}{\underset{1}{\cancel{5}}} \qquad \begin{array}{l}\text{distributing} \\ \text{and cancelling}\end{array}$$

Note: Each
denominator
is completely
cancelled and
eliminated.

EXAMPLE 2 (*continued*)

$$10(2x + 1) - 15(x - 2) = 48 \qquad \text{simplifying}$$

$$20x + 10 - 15x + 30 = 48 \qquad \text{distributive law}$$

$$5x + 40 = 48 \qquad \begin{array}{l}\text{combining}\\ \text{similar terms}\end{array}$$

$$5x = 8 \qquad \begin{array}{l}\text{subtracting 40}\\ \text{from each side}\end{array}$$

$$x = \frac{8}{5} \qquad \begin{array}{l}\text{dividing each}\\ \text{side by 5}\end{array}$$

$$\text{or } x = 1\frac{3}{5} \quad \blacksquare$$

Note: Check by substituting $1\dfrac{3}{5} = \dfrac{8}{5}$ in the original equation.

⬚ Variable Denominators

We can never multiply each side of an equation by zero. This would violate the multiplication property of equality.

When working with fractional equations containing variable expressions as denominators, we must always *restrict the value of the variable* to be certain that we are not multiplying each side of the equation by an LCD that equals zero.

EXAMPLE 3

Solve $\dfrac{7}{6y} = \dfrac{2}{3y} + \dfrac{1}{y^2}$.

Solution: The LCD $= 6y^2$. We can multiply each side of the equation by $6y^2$, provided

$$6y^2 \neq 0.$$

$$y^2 \neq 0 \qquad \text{dividing each side by 6}$$

$$\boxed{y \neq 0} \qquad \text{square root property}$$

restriction on the variable ⟍

$$\frac{6y^2}{1}\left(\frac{7}{6y}\right) = \frac{6y^2}{1}\left(\frac{2}{3y} + \frac{1}{y^2}\right) \qquad \begin{array}{l}\text{multiplying each side}\\ \text{by } 6y^2, \ y \neq 0.\end{array}$$

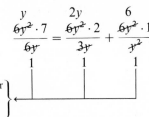

$$\frac{\cancel{6y^2} \cdot 7}{\cancel{6y}} = \frac{\cancel{6y^2} \cdot 2}{\cancel{3y}} + \frac{\cancel{6y^2} \cdot 1}{\cancel{y^2}}$$

distributing and cancelling

Note: Each denominator is completely cancelled and eliminated.

$$7y = 4y + 6$$

simplifying

$$3y = 6$$

subtracting 4y from each side

$$y = 2 \quad \blacksquare$$

dividing each side by 3

Note: This solution does not violate the restriction on the variable.

Note: Check by substituting $y = 2$ in the original equation.

EXAMPLE 4

Solve $\dfrac{3x}{x + 2} = 2 + \dfrac{5}{x}$.

Solution: The LCD $= x(x + 2)$. We can multiply each side of the equation by $x(x + 2)$, provided

$$x(x + 2) \neq 0.$$

$x \neq 0 \text{ or } x + 2 \neq 0$ zero product property

$x \neq 0 \text{ or } x \neq -2$ subtracting 2 from each side

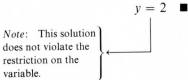

restrictions on the variable

$$\frac{x(x + 2)}{1} \left(\frac{3x}{x + 2}\right) = \frac{x(x + 2)}{1} \left(\frac{2}{1} + \frac{5}{x}\right)$$

multiplying each side by $x(x + 2)$, $x \neq 0, x \neq -2$

$$\frac{x\cancel{(x + 2)} \cdot 3x}{\cancel{x + 2}} = \frac{x(x + 2) \cdot 2}{1} + \frac{\cancel{x}(x + 2) \cdot 5}{\cancel{x}}$$

distributing and cancelling

EXAMPLE 4 (*continued*)

$$3x^2 = 2x(x + 2) + 5(x + 2)$$ simplifying

$$3x^2 = 2x^2 + 4x + 5x + 10$$ distributive law

Note: The x^2 terms signify that this equation is developing into a quadratic equation.

$$x^2 - 9x - 10 = 0$$ writing a quadratic equation in standard form

$$(x - 10)(x + 1) = 0$$ factoring

$$x - 10 = 0 \text{ or } x + 1 = 0$$ zero product property

$$x = 10 \text{ or } x = -1 \quad \blacksquare$$

Note: Neither of these solutions violates the restrictions on the variable.

Note: Check by substituting $x = 10$ and $x = -1$ in the original equation.

EXAMPLE 5

Solve $\dfrac{1}{x - 3} - \dfrac{6}{x^2 - 9} = 1$.

Solution:

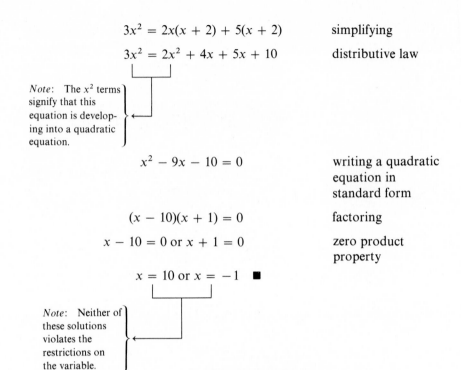

$$\frac{1}{x - 3} - \frac{6}{x^2 - 9} = 1$$

$$\frac{1}{x - 3} - \frac{6}{(x + 3)(x - 3)} = 1$$

The LCD $= (x - 3)(x + 3)$. We can multiply each side of the equation by $(x - 3)(x + 3)$, provided

$$(x - 3)(x + 3) \neq 0.$$

$$x - 3 \neq 0 \text{ or } x + 3 \neq 0 \quad \text{zero product property}$$

$$x \neq 3 \text{ or } x \neq -3$$

restrictions on the variable

$$\frac{(x-3)(x+3)}{1}\left(\frac{1}{x-3}-\frac{6}{(x+3)(x-3)}\right)=\frac{(x-3)(x+3)}{1}\left(\frac{1}{1}\right)$$

multiplying each side by $(x-3)(x+3)$
$x \neq 3,\ x \neq -3$

$$\frac{\overset{1}{\cancel{(x-3)}}(x+3)\cdot 1}{\cancel{x-3}}-\frac{\overset{1}{\cancel{(x-3)}}\cancel{(x+3)}\cdot 6}{\cancel{(x+3)}\cancel{(x-3)}}=\frac{(x-3)(x+3)\cdot 1}{1}$$

distributing and cancelling

$$(x+3)-6=(x-3)(x+3)$$

simplifying

$$x-3=x^2-9$$

combining similar terms and FOIL

Note: The x^2 term signifies that this equation is a quadratic equation.

$$x^2-x-6=0$$

writing a quadratic equation in standard form

$$(x-3)(x+2)=0$$

factoring

$$x-3=0 \text{ or } x+2=0$$

zero product property

$$x=3 \text{ or } x=-2$$

Note: This solution violates the restriction on the variable and must be discarded.

Therefore, the only solution to this equation is $x=-2$. ■

Note: Notice what happens if we attempt to check by substituting $x=3$ in the original equation:

$$\frac{1}{x-3}-\frac{6}{x^2-9}=1$$

$$\frac{1}{(3)-3}-\frac{6}{(3)^2-9}=1?$$

$$\frac{1}{0}-\frac{6}{0}=1?$$

Division by zero is undefined.

Do not confuse solving a fractional equation with adding fractions that have different denominators. Both start by finding the LCD. However, with fractional equations the denominators are cancelled and eliminated; with addition of fractions the LCD remains as part of the solution. Be sure you understand the difference between these two problems.

Solving a fractional equation	Adding fractions
$\dfrac{1}{2} + \dfrac{1}{3} = \dfrac{5}{x}$, LCD $= 6x$	$\dfrac{1}{2} + \dfrac{1}{3} + \dfrac{5}{x}$, LCD $= 6x$
$6x\left(\dfrac{1}{2} + \dfrac{1}{3}\right) = 6x\left(\dfrac{5}{x}\right)$	$\dfrac{3x}{6x} + \dfrac{2x}{6x} + \dfrac{30}{6x}$
$3x + 2x = 30$	$3x + 2x + 30$

Note: The denominators are cancelled and eliminated.

$5x = 30$

$x = 6$

Note: The denominator remains as part of the solution.

$\dfrac{5x + 30}{6x}$

Application

EYEGLASS LENSES

Nearsightedness (myopia) and farsightedness (hyperopia) are two common defects of the human eye. A nearsighted eye is capable of seeing nearby objects but cannot focus clearly on objects that are far away. A farsighted eye is capable of seeing distant objects but cannot focus clearly on objects that are close. Both nearsightedness and farsightedness can be corrected by the use of *eyeglass lenses* (Fig. 8.6). The purpose of the lens is to form an *image* of an object so that the image can be brought into focus. The *power P* of a lens needed to correct nearsightedness or farsightedness depends on the distance of the object d_1 and on the image distance d_2.

FORMULA: Power of a lens

$$P = \frac{1}{d_1} - \frac{1}{d_2},$$

where P is the power of the lens in diopters
d_1 is the distance to the object in meters
d_2 is the image distance in meters.

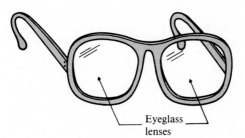

Eyeglass
lenses   **Figure 8.6**

Opticians use *negative numbers* to designate the power of a nearsighted lens and *positive numbers* to designate the power of a farsighted lens. Thus, a lens whose power is −3.0 diopters is a lens that corrects nearsightedness, while a lens whose power is +3.0 diopters is a lens that corrects farsightedness.

EXAMPLE 6

A person with nearsightedness cannot focus clearly on objects that are farther away than 50 cm. He is presently wearing eyeglasses with a power of −1.8 diopters. How far can he clearly see with these eyeglasses?

Solution: The purpose of the eyeglass lens is to form an image of an object at 50 cm. Thus, the image distance (d_2) is 50 cm, or .5 m.
   Using the power of a lens formula,

$$P = \frac{1}{d_1} - \frac{1}{d_2}.$$

$$-1.8 = \frac{1}{d_1} - \frac{1}{.5} \qquad \text{substituting}$$

Note: LCD $= .5d_1$

$$.5d_1(-1.8) = .5d_1\left(\frac{1}{d_1} - \frac{1}{.5}\right) \qquad \begin{array}{l}\text{multiplying each side}\\\text{by } .5d_1, d_1 \neq 0.\end{array}$$

$$-.9d_1 = .5 - d_1 \qquad \text{simplifying}$$

Note: If you don't wish to work with decimals, eliminate them by multiplying each side by 10.

$$-9d_1 = 5 - 10d_1 \qquad \text{multiplying each side by 10}$$

$$d_1 = 5 \text{ m} \qquad \text{adding } 10d_1 \text{ to each side}$$

Thus, this person can see objects clearly up to 5 m away.   ■

EXERCISES 8.6

Solve and check.

1. $\dfrac{x}{6} + \dfrac{x}{8} = \dfrac{7}{6}$

2. $\dfrac{2y}{3} - \dfrac{5y}{12} = 6$

3. $\dfrac{2y + 1}{2} - \dfrac{y}{3} = \dfrac{1}{2}$

4. $\dfrac{x + 1}{3} + \dfrac{x}{5} = 3$

5. $\dfrac{5t - 1}{3} - 2 = \dfrac{t + 7}{9}$

6. $\dfrac{3p - 1}{6} - 1 = \dfrac{p + 7}{9}$

7. $\dfrac{x}{6} - \dfrac{4x + 1}{4} = \dfrac{1}{6}$

8. $\dfrac{3}{4} - \dfrac{2x + 3}{2} = \dfrac{x}{8}$

9. $\dfrac{3y + 2}{5} - \dfrac{y - 3}{15} = \dfrac{y + 3}{3}$

10. $\dfrac{y + 2}{14} = \dfrac{2y - 5}{7} - \dfrac{3y - 1}{21}$

11. $\dfrac{5p + 3}{12} = \dfrac{p - 3}{15} + \dfrac{3p}{8}$

12. $\dfrac{11x + 1}{42} = \dfrac{x - 1}{9} + \dfrac{4x + 5}{24}$

13. $\dfrac{3w - 1}{28} - \dfrac{w + 1}{8} = 1 + \dfrac{5w + 3}{24}$

14. $1 + \dfrac{4w + 3}{48} = \dfrac{w + 12}{12} - \dfrac{w - 1}{40}$

Solve and check. Be sure to restrict the variable before multiplying each side of the equation by the LCD.

15. $\dfrac{4}{x} - \dfrac{3}{4} = \dfrac{1}{4x}$

16. $\dfrac{1}{3} - \dfrac{5}{8x} = \dfrac{1}{8}$

17. $\dfrac{2}{y^2} + \dfrac{1}{5y} = \dfrac{1}{3y}$

18. $\dfrac{3}{2y^2} - \dfrac{1}{4y} = \dfrac{5}{3y^2}$

19. $\dfrac{3t}{t - 1} - 3 = \dfrac{6}{t}$

20. $\dfrac{5t}{t + 3} - 5 = \dfrac{10}{t}$

21. $\dfrac{w - 2}{3w} - \dfrac{5}{w^2} = 0$

22. $\dfrac{2w - 1}{4w} - \dfrac{5}{2w^2} = 0$

23. $\dfrac{x + 1}{x - 5} - 1 = \dfrac{x + 2}{x}$

24. $\dfrac{2x + 1}{x + 2} - 2 = \dfrac{x + 7}{2x}$

25. $\dfrac{3}{x - 4} + \dfrac{x}{2} = \dfrac{3}{2x - 8}$

26. $\dfrac{5}{x + 3} + \dfrac{2x}{5} = \dfrac{21}{5x + 15}$

27. $\dfrac{k}{k - 4} = \dfrac{5}{k^2 - 16}$

28. $\dfrac{y}{2y + 1} = \dfrac{6}{4y^2 - 1}$

29. $\dfrac{m}{m - 3} = \dfrac{15}{m^2 - 5m + 6}$

30. $\dfrac{z}{z - 6} = \dfrac{8z}{z^2 - 4z - 12}$

31. $\dfrac{x}{x^2 + 3x - 4} - \dfrac{4}{x + 4} + \dfrac{7}{x - 1} = 0$

32. $\dfrac{1}{2x - 1} + \dfrac{3}{x + 3} - \dfrac{14}{2x^2 + 5x - 3} = 0$

33. $\dfrac{4}{x^2 - 4} = \dfrac{x}{2x + 4} - \dfrac{1}{x - 2}$

34. $\dfrac{3}{x^2 + 5x} + \dfrac{2}{x - 5} = \dfrac{6}{x^2 - 25}$

35. $\dfrac{y}{y - 5} - \dfrac{y - 10}{5 - y} = 1$

36. $\dfrac{2t}{t^2 - 1} = \dfrac{2}{t^2 + 2t + 1} - \dfrac{1}{1 - t}$

Solve the literal equations for x.

37. $\dfrac{x}{a} + \dfrac{x}{b} = c$

38. $\dfrac{1}{x} = \dfrac{1}{a} + \dfrac{1}{b}$

39. $\dfrac{1}{a} + \dfrac{1}{b + x} = \dfrac{1}{c}$

40. $\dfrac{x}{a} + b = \dfrac{x}{c} + d$

41. A person with farsightedness cannot focus clearly on objects that are nearer than 80 cm. She is presently wearing eyeglasses with a power of $+2.5$ diopters. How close can she see clearly with these eyeglasses?

42. A person with nearsightedness can focus clearly on an object 20 m away when wearing eyeglasses with a power of -1.5 diopters. How far can he see clearly without these eyeglasses?

Chapter Review

REVIEW EXERCISES

The following review exercises are grouped according to the objectives that should have been mastered in Chapter 8. Work each problem carefully. If any weaknesses appear, immediately refer to and read the subsection that matches that objective.

8.1 BUILDING UP AND BREAKING DOWN ALGEBRAIC FRACTIONS

Objectives

⊡ To find the restricted values of an algebraic fraction.

Restrict the variable.

1. a) $\dfrac{t - 3}{5t^2}$

c) $\dfrac{2x + 3}{x^2 + x - 20}$

b) $\dfrac{3 - 2y}{9 - 3y}$

d) $\dfrac{8 - 5z + z^2}{z^2 - 1}$

⊡ To find an equivalent algebraic fraction whose denominator is known.

Find the missing numerator.

2. $\dfrac{x + 3}{4x} = \dfrac{?}{24x^4y^2}$

3. $\dfrac{2y - 1}{y + 4} = \dfrac{?}{y^2 - 2y - 24}$

4. $\dfrac{t + 3}{3t^2 - 9t} = \dfrac{?}{6t^3 - 36t^2 + 54t}$

⊡ To reduce an algebraic fraction.

Reduce to lowest terms.

5. $\dfrac{5x - 10x^2}{5x}$

6. $\dfrac{t - 2}{t^2 + 4t - 12}$

7. $\dfrac{2y^2 + 5y - 3}{18 - 2y^2}$

8. $\dfrac{25 - x^2}{x^2 - 10x + 25}$

8.2 MULTIPLYING AND DIVIDING ALGEBRAIC FRACTIONS

Objectives

⊡ To multiply algebraic fractions.

Find the product.

9. $\dfrac{9y^2}{3x - y} \cdot \dfrac{3x - y}{6y^3}$

10. $\dfrac{2y^2 + y}{4 - y^2} \cdot \dfrac{y + 2}{2y^2}$

11. $\dfrac{t^2 - 3t - 28}{3t^2 - 48} \cdot (4 - t)$

⊡ To divide algebraic fractions.

Find the quotient.

12. $\dfrac{3x^2 y}{x + y} \div (x - y)$

13. $\dfrac{4t^2}{2t^2 + 3t - 5} \div \dfrac{8t^3 + 16t^2}{1 - t}$

14. $\dfrac{\dfrac{x^2 - 25}{x^2 - 6x + 5}}{\dfrac{5 + x}{1 - x^2}}$

8.3 ADDING AND SUBTRACTING ALGEBRAIC FRACTIONS WITH THE SAME DENOMINATOR

Objectives

⊡ To add algebraic fractions that have the same denominator.

Find the sum.

15. $\dfrac{5x + 1}{3x} + \dfrac{x + 5}{3x}$

16. $\dfrac{2x - 5}{9 - x^2} + \dfrac{x - 4}{9 - x^2}$

⊡ To subtract algebraic fractions that have the same denominator.

Find the difference.

17. $\dfrac{2y - 3x}{5y^2} - \dfrac{y + 3x}{5y^2}$

18. $\dfrac{3 + 3t}{t^2 + 3t - 40} - \dfrac{t - 7}{t^2 + 3t - 40}$

⊡ To add or subtract algebraic fractions whose denominators are opposites.

Simplify.

19. $\dfrac{5x}{x - 2} - \dfrac{4x}{2 - x}$

20. $\dfrac{7w + 9}{w^2 - 25} + \dfrac{2w - 16}{25 - w^2}$

8.4 ADDING AND SUBTRACTING ALGEBRAIC FRACTIONS WITH DIFFERENT DENOMINATORS

Objectives

⊡ Given algebraic fractions with different denominators, to find the least common denominator (LCD).

Find the LCD.

21. $\dfrac{5}{6xy^3} + \dfrac{y - 2}{10x^2 y}$

22. $\dfrac{5}{3y^2 + 9y} - \dfrac{4 - y}{y^2 + 6y + 9}$

23. $\dfrac{12}{4 - y^2} - \dfrac{y - 5}{y - 2}$

24. $\dfrac{t + 3}{t + 5} - \dfrac{t - 5}{t - 3}$

☐ To add or subtract algebraic fractions that have different denominators.

Simplify.

25. $\dfrac{5}{6xy^3} + \dfrac{y - 2}{10x^2y}$

26. $\dfrac{5}{3y^2 + 9y} - \dfrac{4 - y}{y^2 + 6y + 9}$

27. $\dfrac{12}{4 - y^2} - \dfrac{y - 5}{y - 2}$

28. $\dfrac{t + 3}{t + 5} - \dfrac{t - 5}{t - 3}$

8.5 MIXED ALGEBRAIC EXPRESSIONS AND COMPLEX FRACTIONS

Objectives

☐ To change a mixed algebraic expression into a single algebraic fraction.

Write as a single algebraic fraction.

29. $3y^2 + \dfrac{3}{5x^2y^5}$

30. $\dfrac{x^2 - 1}{x} + 2x - 5$

31. $2t - \dfrac{6}{3t - 1} + 1$

☐ To simplify a complex fraction.

Simplify.

32. $\dfrac{4 - \dfrac{1}{y^2}}{2 + \dfrac{1}{y}}$

33. $\dfrac{y - 2 - \dfrac{3y}{y + 6}}{y + 5 - \dfrac{2}{y + 6}}$

8.6 FRACTIONAL EQUATIONS

Objectives

☐ To solve a fractional equation containing constants as denominators.

Solve for the variable.

34. $\dfrac{t}{3} + \dfrac{2t}{9} = 5$

35. $\dfrac{5x - 5}{12} - \dfrac{3x - 1}{9} = \dfrac{x}{45}$

☐ To solve a fractional equation containing variable expressions as denominators.

Solve for the variable.

36. $\dfrac{3}{4y} = \dfrac{5}{6y} + \dfrac{1}{2y^2}$

37. $\dfrac{2y - 10}{y - 3} = 1 + \dfrac{3}{y}$

38. $\dfrac{1}{x - 5} - \dfrac{10}{x^2 - 25} = 1$

CHAPTER TEST
Time
50 minutes

Score
15, 14 excellent
13, 12 good
11, 10, 9 fair
below 8 poor

If you have worked through the Review Exercises and corrected any weaknesses, you are ready to take the following Chapter Test.

1. Restrict the variable. $\dfrac{x - 1}{x^2 - x - 2}$

2. Find the missing numerator. $\dfrac{3y + 2}{y - 3} = \dfrac{?}{y^2 - 9}$

3. Reduce to lowest terms. $\dfrac{16x^2 - 4x}{4x}$

4. Reduce to lowest terms. $\dfrac{36 - t^2}{t^2 - 3t - 18}$

5. Find the product. $\dfrac{4x^2 - 8x}{x^2 - 4} \cdot \dfrac{x^2 + 3x + 2}{8x^2}$

6. Find the quotient. $\dfrac{w^2 - 6w + 8}{2 + w} \div (16 - w^2)$

7. Find the sum. $\dfrac{3t - 1}{4t^2 - 25} + \dfrac{t - 9}{4t^2 - 25}$

8. Find the difference. $\dfrac{6 - x}{x^2 + 4x - 32} - \dfrac{x - 2}{x^2 + 4x - 32}$

9. Simplify. $\dfrac{3t}{2t - 1} + \dfrac{1 - 5t}{1 - 2t}$

10. State the LCD; then simplify. $\dfrac{y + 1}{3y} - \dfrac{y - 2}{3y - 1}$

11. State the LCD; then simplify. $\dfrac{2}{x^2 + 5x + 6} + \dfrac{x - 3}{3x^2 + 6x}$

12. Write as a single algebraic fraction. $k - 3 - \dfrac{3}{k + 2}$

13. Simplify. $\dfrac{y - \dfrac{8}{y - 2}}{2 - \dfrac{24}{y^2 - 4}}$

14. Solve for t. $\dfrac{3t - 1}{4} - \dfrac{t}{3} = \dfrac{1}{6}$

15. Solve for x. $\dfrac{8}{x^2 - 6x + 8} = \dfrac{x}{x - 4}$

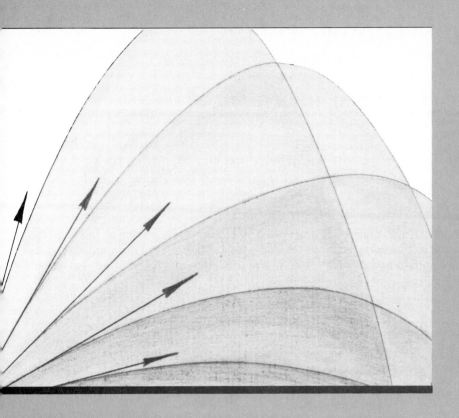

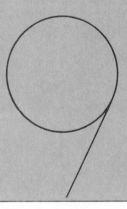

AN INTRODUCTION TO FUNCTIONS AND GRAPHS

9.1 THE FUNCTIONAL RELATION

Objectives

⊡ To express one quantity as a function of another.

⊡ To complete an ordered pair such that it satisfies a given function.

⊡ Functions

In technical mathematics, we are often interested in describing a relationship between two (or more) variables. For example, the area A of a square is related to the length of its side s (Fig. 9.1), and this *relation* can be described by the rule (formula)

$$A = s^2.$$

Area (A)

Side (s)   **Figure 9.1**

When we replace s with some value, we find a corresponding value of A. For example,

$$\text{if } s = 1 \text{ ft, then } A = s^2 = (1 \text{ ft})^2 = 1 \text{ ft}^2, \text{ or}$$

$$\text{if } s = 2 \text{ ft, then } A = s^2 = (2 \text{ ft})^2 = 4 \text{ ft}^2, \text{ and so on.}$$

In this way, the value of A *depends* upon the value chosen for s. Thus, we refer to A as the **dependent variable** and s as the **independent variable**. Listed in Table 9.1 are some arbitrarily chosen values of s and the corresponding values of A. Because s represents a length, we can select only positive values for s.

Observe in Table 9.1 that each value we select for the independent variable s yields *exactly one* value of the dependent variable A. Mathematicians refer to a rule with this characteristic as a **function**.

DEFINITION:

A **function** is a rule that describes how to find *exactly one* value of the dependent variable when a value of the independent variable is arbitrarily chosen.

TABLE 9.1

Independent variable	Dependent variable
s (ft)	*A* (ft²)
1	1
2	4
3	9
4	16
5	25

The equation $A = s^2$ defines A as a function of s. It tells us to *square* some value of s in order to determine a corresponding value of A. *In technical mathematics, we often try to define the functional relation between two quantities by writing an equation with the dependent variable isolated on the left side.* In Section 9.3 we will represent a function pictorially by graphing the function.

EXAMPLE 1

Write an equation that defines the side s of a square as a function of its area A.

Solution: We now want a rule that describes how to find *exactly one* value of s when a value of A is arbitrarily chosen. Therefore, using $A = s^2$, we isolate s as follows:

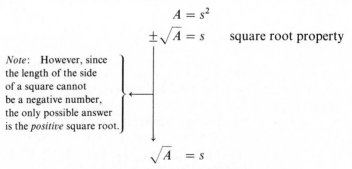

$$A = s^2$$
$$\pm\sqrt{A} = s \qquad \text{square root property}$$

Note: However, since the length of the side of a square cannot be a negative number, the only possible answer is the *positive* square root.

$$\sqrt{A} = s$$

Thus, the equation $s = \sqrt{A}$ defines s as a function of A. ∎

Note: It is important to note that the equation $y = \pm\sqrt{x}$ does *not* define y as a function of x, since *two* values of the dependent variable y would exist for each value of the independent variable x. However, when taken separately, $y = \sqrt{x}$ and $y = -\sqrt{x}$ each define a function.

EXAMPLE 2

A length of x ft is cut from the end of an 8-ft piece of copper pipe. Express the remaining length y as a function of x.

Solution: Referring to Fig. 9.2, it can be seen that the equation $y = 8 - x$ defines y as a function of x.   ∎

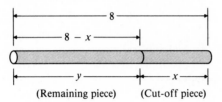

(Remaining piece)   (Cut-off piece)   **Figure 9.2**

Note: Because the length of the cut-off piece of pipe cannot be more than 8 ft, we can consider only those values of x between 0 ft and 8 ft. The collection of all numbers that can be chosen for the independent variable is called the **domain** of the function. Thus, the domain of this function is the collection of all real numbers between 0 and 8. The collection of all numbers that we obtain for the dependent variable is called the **range** of the function. The range of this function is also the collection of all real numbers between 0 and 8. For this example, the domain and range happen to be the same. This is *not* a characteristic of all functions.

⠒ Ordered Pairs

The function $y = 8 - x$ in Example 2 tells us to subtract the amount cut off the end (x) from 8, in order to determine the remaining length y. For example,

$$\text{if } x = 3 \text{ ft, then } y = 8 - x.$$

$$y = 8 - 3 \qquad \text{replace } x \text{ with 3}$$

$$y = 5 \text{ ft} \qquad \text{simplify}$$

The values $x = 3$ and $y = 5$ are said to *satisfy* the function. Another way of writing "if $x = 3$, then $y = 5$" is simply to write

$$(3, 5).$$

A pair of numbers written in this form is called an **ordered pair**.

In an ordered pair (x, y), the x-value (the independent variable) must be written first and the y-value (the dependent variable) written second. For example,

$$x = 3, y = 5 \text{ is written } (3, 5)$$

and

$$x = 5, y = 3 \text{ is written } (5, 3),$$

$$(3, 5) \neq (5, 3).$$

EXAMPLE 3

Complete the ordered pairs such that they satisfy the function $y = 1 - 4x^2$.

a) $(0, ?)$　　　　**b)** $(5, ?)$　　　　**c)** $(-5, ?)$　　　　**d)** $(-\frac{1}{2}, ?)$

Solution:

a) For $(0, ?)$, $x = 0$ so

$$y = 1 - 4x^2.$$
$$y = 1 - 4(0)^2 \qquad \text{replace } x \text{ with } 0$$
$$y = 1 - 4(0) \qquad \text{powers first}$$
$$y = 1 - 0 \qquad \text{multiplication next}$$
$$y = 1 \qquad \text{subtraction last}$$

Thus, the ordered pair is $(0, 1)$.　■

b) For $(5, ?)$, $x = 5$ so

$$y = 1 - 4x^2.$$
$$y = 1 - 4(5)^2 \qquad \text{replace } x \text{ with } 5$$
$$y = 1 - 4(25) \qquad \text{powers first}$$
$$y = 1 - 100 \qquad \text{multiplication next}$$
$$y = -99 \qquad \text{subtraction last}$$

Thus, the ordered pair is $(5, -99)$.　■

c) For $(-5, ?)$, $x = -5$ so

$$y = 1 - 4x^2.$$
$$y = 1 - 4(-5)^2 \qquad \text{replace } x \text{ with } -5$$
$$y = 1 - 4(25) \qquad \text{powers first}$$
$$y = 1 - 100 \qquad \text{multiplication next}$$
$$y = -99 \qquad \text{subtraction last}$$

Thus, the ordered pair is $(-5, -99)$.　■

d) For $(-\frac{1}{2}, ?)$, $x = -\frac{1}{2}$ so

$$y = 1 - 4x^2.$$
$$y = 1 - 4(-\tfrac{1}{2})^2 \qquad \text{replace } x \text{ with } -\tfrac{1}{2}$$
$$y = 1 - 4(\tfrac{1}{4}) \qquad \text{powers first}$$
$$y = 1 - 1 \qquad \text{multiplication next}$$
$$y = 0 \qquad \text{subtraction last}$$

Thus, the ordered pair is $(-\frac{1}{2}, 0)$.　■

Our previous definition of a function along with the concept of an ordered pair gives rise to an alternate definition of a function.

> **ALTERNATE DEFINITION:**
> A **function** is a collection of ordered pairs (x, y) in which no two distinct pairs have the same x-value.

For example, the collection of ordered pairs

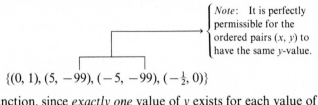

$\{(0, 1), (5, -99), (-5, -99), (-\frac{1}{2}, 0)\}$

defines a function, since *exactly one* value of y exists for each value of x.
However, the collection of ordered pairs

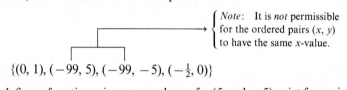

$\{(0, 1), (-99, 5), (-99, -5), (-\frac{1}{2}, 0)\}$

does *not* define a function, since *two* values of y (5 and -5) exist for a single value of x (-99).

The material in this section has introduced you to the idea of the functional relation. Probably the most important single concept in advanced mathematics is that of the function.

EXERCISES 9.1

1. Express the circumference C of a circle as a function of its diameter d.
2. Express the area A of a circle as a function of its radius r.
3. Express the diameter d of a circle as a function of its circumference C.
4. Express the radius r of a circle as a function of its area A.
5. Express the length l of a rectangle of width 4 as a function of its area A.
6. Express the height h of a cylinder of radius 3 as a function of its volume V.
7. Express the width w of a rectangle of length 5 as a function of its perimeter P.
8. Express the radius r of a cylinder of height 16 as a function of its volume V.
9. On a blueprint, the scale is 1 in. = 20 ft. Express the actual length y of a building as a function of the blueprint measurement x.
10. One out of every 500 transistors produced by a certain company is defective. Express the number n of defective transistors as a function of the total number t of transistors produced.
11. A 45,000-gal swimming pool is being filled at the rate of 15 gal/min. Express the amount A of water in the pool as a function of the number of minutes t the water has been filling the pool.

12. A chemist has 10 gal of water in an acid solution to which she adds x gallons of water. Express the total number of gallons y of water in the new solution as a function of x.

13. A 45,000-gal swimming pool is being drained at the rate of 15 gal/min. Express the amount A of water remaining in the swimming pool as a function of the number of minutes t the water has been draining from the pool.

14. A chemist has 10 gal of water in an acid solution from which she evaporates x gallons of water. Express the total number of gallons y of water in the new solution as a function of x.

15. Express in cents the daily cost y of renting a car as a function of the number x of miles driven, if the company charges \$30/day plus 10¢/mi.

16. Express in dollars the weekly earnings y of a commission salesperson as a function of the amount x of merchandise she sells, if she earns \$100/wk plus a 5% commission on sales.

17. State the domain and range of the function described in exercise 11.

18. State the domain and range of the function described in exercise 12.

19. State the domain and range of the function described in exercise 13.

20. State the domain and range of the function described in exercise 14.

Complete the ordered pairs such that they satisfy the given function.

Function	Ordered pairs
21. $y = x$	$(0, ?), (1, ?), (-1, ?), (2, ?), (-2, ?)$
22. $y = \dfrac{x}{5}$	$(0, ?), (5, ?), (-5, ?), (10, ?), (-10, ?)$
23. $y = 6 - 2x$	$(0, ?), (3, ?), (-3, ?), \left(\dfrac{1}{2}, ?\right), \left(-\dfrac{1}{2}, ?\right)$
24. $y = 3x + 5$	$(0, ?), \left(\dfrac{1}{3}, ?\right), \left(-\dfrac{5}{3}, ?\right), (2, ?), (-2, ?)$
25. $y = x^2$	$(0, ?), (1, ?), (-1, ?), (3, ?), (-3, ?)$
26. $y = 4 - x^2$	$(0, ?), (2, ?), (-2, ?), \left(\dfrac{1}{2}, ?\right), \left(-\dfrac{1}{2}, ?\right)$
27. $y = \dfrac{1}{x}$	$(3, ?), (-3, ?), \left(\dfrac{1}{3}, ?\right), \left(-\dfrac{1}{3}, ?\right)$
28. $y = \dfrac{1}{x^2}$	$(3, ?), (-3, ?), \left(\dfrac{1}{3}, ?\right), \left(-\dfrac{1}{3}, ?\right)$
29. $y = \dfrac{4}{\sqrt{x}}$	$(1, ?), (4, ?), (9, ?), (16, ?)$
30. $y = \sqrt{1 - x}$	$(1, ?), (-1, ?), \left(\dfrac{3}{4}, ?\right), \left(-\dfrac{3}{4}, ?\right), (0, ?)$
31. $y = 2x^2 - 5x + 2$	$(0, ?), \left(\dfrac{1}{2}, ?\right), (2, ?), (1, ?), (-1, ?)$
32. $y = 7 - 6x - x^2$	$(0, ?), (-7, ?), (-1, ?), (1, ?), (-3, ?)$

33. Which of the following expressions defines y as a function of x? Explain.

a) $y < x$ **b)** $y > x$ **c)** $y = x$ **d)** $y^2 = x$

34. Which of the following collection of ordered pairs defines a function? Explain.

a) $\{(0, 0), (1, 2), (-1, 2)\}$
b) $\{(0, 1), (1, 1), (2, 1), (3, 1)\}$
c) $\{(1, 0), (1, 1), (1, 2), (1, 3)\}$
d) $\{(0, 0), (1, 1), (1, -1), (4, 2), (4, -2)\}$

9.2 VARIATION

Objectives

⬚ To express one quantity as a function of another, if one of the quantities varies directly as the other.

⬚ To express one quantity as a function of another, if one of the quantities varies inversely as the other.

⬚ To express one quantity as a function of another, if one of the quantities varies directly or inversely as the square of the other.

⬚ Direct Variation

In engineering technology, the functional relation between two quantities is often stated in the language of **variation**. If we state that "y **varies directly** as x," we mean that the ratio of y to x is always the same. Thus,

$$\frac{y}{x} = k$$

or $y = kx$, where k is the **variation constant** $(k > 0)$.

dependent variable / independent variable

In direct variation, if x increases then y also increases; if x decreases then y also decreases.

EXAMPLE 1

Express y as a function of x, if y *varies directly* as x and $y = 6$ when $x = 3$.

Solution: To find k, the variation constant, replace x with 3 and y with 6 as follows:

$$y = kx \qquad \text{definition of a direct variation}$$
$$6 = k(3) \qquad \text{replacing } x \text{ with 3 and } y \text{ with 6}$$
$$2 = k \qquad \text{multiplying each side by } \frac{1}{3}$$

Thus, $y = kx$ becomes $y = 2x$. ∎

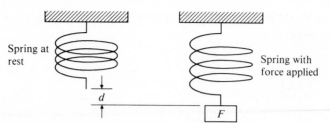

Spring at rest

Spring with force applied

d

F

Figure 9.3

EXAMPLE 2

Hooke's law states that the distance d a spring stretches varies directly as the force F applied to the spring (see Fig. 9.3). Express d as a function of F, if $d = 3$ in. when $F = 210$ lb.

Solution: To find the variation constant, replace d with 3 in. and F with 210 lb as follows:

$$d = kF \qquad \text{definition of a direct variation}$$

$$3 \text{ in.} = k(210 \text{ lb}) \qquad \text{replacing } d \text{ with 3 in. and } F \text{ with 210 lb}$$

$$\frac{3 \text{ in.}}{210 \text{ lb}} = k \qquad \text{dividing each side by 210 lb}$$

$$\frac{1 \text{ in.}}{70 \text{ lb}} = k \qquad \text{reducing}$$

Note: It is common practice to drop the units from the variation constant. $\left. \right\} \leftarrow \dfrac{1}{70} = k$

Note: It is also common practice to specify the units of the independent and dependent variables.

Thus, $d = \dfrac{1}{70} F$ or $d = \dfrac{F}{70}$, where F is in pounds and d is in inches. ∎

⠢ Inverse Variation

If we state that "y **varies inversely** as x," we mean that the product of x and y is always the same. Thus,

$$xy = k$$

or $y = \dfrac{k}{x}$, where k is the

variation constant $(k > 0)$.

dependent variable ⎵ ⎵ independent variable

In an inverse variation, if x increases then y decreases; if x decreases then y increases.

EXAMPLE 3

Express y as a function of x, if y *varies inversely* as x and $y = 6$ when $x = 3$.

Solution: To find k, the variation constant, replace x with 3 and y with 6 as follows:

$$y = \frac{k}{x} \qquad \text{definition of an inverse variation}$$

$$6 = \frac{k}{3} \qquad \text{replacing } x \text{ with 3 and } y \text{ with 6}$$

$$18 = k \qquad \text{multiplying each side by 3}$$

Thus,

$$y = \frac{k}{x} \text{ becomes } y = \frac{18}{x}. \quad \blacksquare$$

EXAMPLE 4

Boyle's law states that when the temperature of a confined gas remains constant, the pressure P it exerts varies inversely as the volume V it occupies. Express P as a function of V, if $P = 2$ psi when $V = 5 \text{ in}^3$.

Solution: To find k, the variation constant, replace P with 2 and V with 5 as follows:

$$P = \frac{k}{V} \qquad \text{definition of an inverse variation}$$

$$2 = \frac{k}{5} \qquad \text{replacing } P \text{ with 2 and } V \text{ with 5}$$

$$10 = k \qquad \text{multiplying each side by 5}$$

Thus, $P = \dfrac{10}{V}$, where V is in cubic inches (in^3) and P is in pounds per square inch (psi). $\quad \blacksquare$

Note: Be sure to specify the units of the independent and dependent variables.

⠇ **Quadratic Variation**

In **quadratic variation**, one quantity varies directly (or inversely) as the *square* of another quantity. If we state that "*y* varies directly as the *square* of *x*," we mean that the ratio of *y* to x^2 is always the same. Thus,

$$\frac{y}{x^2} = k$$

$$y = kx^2, \text{ where } k \text{ is the}$$

variation constant ($k > 0$).

dependent variable   independent variable

EXAMPLE 5

The distance *d* an object falls under the influence of gravity varies directly as the *square* of the duration of time *t* of the fall. Express *d* as a function of *t*, if $d = 144$ ft when $t = 3$ s.

Solution:

$$d = kt^2 \qquad \text{definition of direct quadratic variation}$$
$$144 = k(3)^2 \qquad \text{replacing } d \text{ with } 144 \text{ and } t \text{ with } 3$$
$$16 = k \qquad \text{dividing each side by } 3^2 \text{ or } 9.$$

Thus, $d = kt^2$ becomes $d = 16t^2$, where *t* is in seconds and *d* is in feet. ■

If we state that "*y* varies inversely as the *square* of *x*," we mean that the product of x^2 and *y* is always the same. Thus,

$$x^2y = k$$

or $$y = \frac{k}{x^2}, \text{ where } k \text{ is the variation constant } (k > 0).$$

dependent variable   independent variable

EXAMPLE 6

The intensity *I* of a light source varies inversely as the square of the distance *d* from the light source. Express *I* as a function of *d*, if $I = 3.75$ cd when $d = 1.35$ m.

Solution:

$$I = \frac{k}{d^2} \qquad \text{definition of inverse quadratic variation}$$

$$3.75 = \frac{k}{(1.35)^2} \qquad \text{replacing } I \text{ with 3.75 and } d \text{ with 1.35}$$

$$6.83 = k \qquad \text{multiplying each side by } (1.35)^2 \text{ and retaining three significant digits}$$

Thus, $I = \dfrac{6.83}{d^2}$, where d is in meters and I is in candelas. ∎

Note: We now have a rule (formula) that will tell us the intensity I of this light source at various distances d from the source. For example, the intensity at 2.00 m from this source can be found as follows:

$$I = \frac{6.83}{d^2} = \frac{6.83}{(2.00)^2} = 1.71 \text{ cd}$$

We can record this result as the ordered pair (2.00, 1.71).

independent ╲ ╱ dependent
variable   variable
first   second

EXERCISES 9.2

1. Express y as a function of x, if y varies directly as x and $y = 32$ when $x = 4$.
2. Express y as a function of x, if y varies directly as x and $y = 9$ when $x = 54$.
3. Express y as a function of x, if y varies inversely as x and $y = 18$ when $x = 5$.
4. Express y as a function of x, if y varies inversely as x and $y = 9$ when $x = 8$.
5. Express y as a function of x, if y varies directly as the square of x and $y = 9$ when $x = 6$.
6. Express y as a function of x, if y varies directly as the square of x and $y = 15$ when $x = 5$.
7. Express y as a function of x, if y varies inversely as the square of x and $y = 10$ when $x = 10$.
8. Express y as a function of x, if y varies inversely as the square of x and $y = 8$ when $x = 4$.
9. The weight W of a steel beam varies directly as its length L.
   a) Express W as a function of L, if $W = 1200$ lb when $L = 15$ ft.
   b) Find the weight of a similar beam whose length is 8 ft.
10. The electrical resistance R of a wire varies directly as the length L of the wire.
    a) Express R as a function of L, if $R = 1.5\ \Omega$ when $L = 75$ m.
    b) Find the resistance of a similar wire whose length is 80 m.
11. The circumference C of a circle varies directly as its diameter d.
    a) Express C as a function of d, if $C = 5.15$ ft when $d = 50$ cm.
    b) Find the circumference of a circle whose diameter is 70 cm.

12. The real estate tax T on a piece of property varies directly as its assessed valuation V.
    a) Express T as a function of V, if $T = \$3000$ when $V = \$120{,}000$.
    b) Find the tax on a piece of property valued at $84,000.

13. The safe load capacity S of a rectangular beam varies inversely as the distance d between its support columns.
    a) Express S as a function of d, if $S = 650$ lb when $d = 10$ ft.
    b) Find the safe load capacity if the distance between the supports is increased to 13 ft.

14. The time t it takes to travel a fixed distance varies inversely as the rate r of speed at which one travels.
    a) Express t as a function of r, if $t = 4$ hr when $r = 55$ mph.
    b) Find the time to travel the same distance if the rate of speed is 45 mph.

15. If the voltage in a circuit is constant, the current I through a resistor varies inversely as the resistance R.
    a) Express I as a function of R, if $I = 2$ A when $R = 50\ \Omega$.
    b) Find the current if the resistance is decreased to $20\ \Omega$.

16. The wavelength λ of a radio wave varies inversely as its frequency f.
    a) Express λ as a function of f, if $\lambda = 80$ m when $f = 3750$ kHz.
    b) Find the wavelength if the frequency is 4000 kHz.

17. The power P dissipated in an electrical resistor varies directly as the square of the current I through the resistor.
    a) Express P as a function of I, if $P = 80$ watts when $I = 4$ A.
    b) Find the power dissipated in the resistor if the current is 3 A.

18. The length L of a pendulum varies directly as the square of its period T.
    a) Express L as a function of T, if $L = 1.0$ m when $T = 2.0$ s.
    b) Find the length of a pendulum if its period is 1.0 s.

19. The gravitational force F between two objects varies inversely as the square of the distance d between them.
    a) Express F as a function of d, if $F = 200$ N when $d = 20$ m.
    b) Find the force between the objects if the distance is 40 m.

20. The electrical resistance R of a wire varies inversely as the square of its diameter d.
    a) Express R as a function of d, if $R = 2.5\ \Omega$ when $d = 0.20$ mm.
    b) Find the resistance of a wire having the same length if its diameter is 0.25 mm.

21. Fill in the table, given that y varies directly as x.

x	y
18	12
12	
9	
3	
1	

22. Fill in the table, given that y varies inversely as x.

x	y
18	12
12	
9	
3	
1	

23. Fill in the table, given that y varies inversely as the square of x.

x	y
18	12
12	
9	
3	
1	

24. Fill in the table, given that y varies directly as the square of x.

x	y
18	12
12	
9	
3	
1	

9.3 AN INTRODUCTION TO GRAPHING A FUNCTION

Objectives

   ⊡ To plot points associated with ordered pairs.

   ⊡ To graph a function of the form $y = kx$.

   ⊡ To graph a function of the form $y = kx^2$.

   ⊡ To graph a function of the form $y = \dfrac{k}{x}$.

   ⊡ To graph a function of the form $y = k\sqrt{x}$.

⊡ The Rectangular Coordinate System

Before we can graph a function, we need to form a *rectangular coordinate system*. To do this, we draw two number lines that intersect at their zero points as shown in Fig. 9.4. The horizontal number line is called the **horizontal axis**, and the vertical

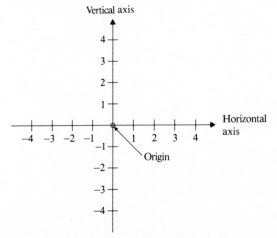

Figure 9.4

number line is called the **vertical axis**. The point where the axes intersect is called the *origin*.

A unique point in the rectangular coordinate system corresponds to every ordered pair (x, y). When graphing, we refer to the numbers in the ordered pair as **coordinates**. To plot

$$(3, 5)$$

first coordinate   second coordinate

we locate the *first coordinate*, 3, on the *horizontal axis* and draw a vertical line through 3 (see Fig. 9.5). Next, we locate the *second coordinate*, 5, on the *vertical*

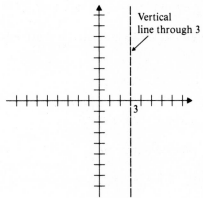

Figure 9.5

axis and draw a horizontal line through 5. The point where these lines intersect is (3, 5) (see Fig. 9.6). Notice that the lines and the axes form a rectangle; hence the name rectangular coordinate system.

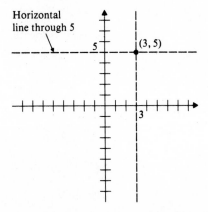

Figure 9.6

In the rectangular coordinate system, (3, 5) and (5, 3) do not represent the same point (see Fig. 9.7). The first coordinate always corresponds to a number on the horizontal axis, while the second coordinate always corresponds to a number on the vertical axis.

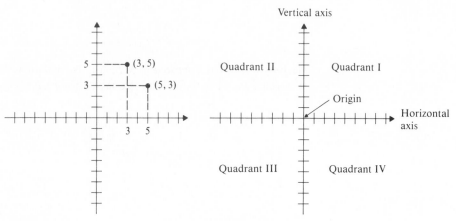

Figure 9.7   **Figure 9.8**

Notice that the rectangular coordinate system forms four regions called **quadrants,** which are numbered counterclockwise as shown in Fig. 9.8.

When plotting points, it will help to remember the following:

1. If both coordinates are positive, the point is located in the first quadrant.
2. If the first coordinate is negative and the second coordinate is positive, the point is located in the second quadrant.
3. If both coordinates are negative, the point is located in the third quadrant.
4. If the first coordinate is positive and the second coordinate is negative, the point is located in the fourth quadrant.
5. If the first coordinate is zero, the point is located on the vertical axis.
6. If the second coordinate is zero, the point is located on the horizontal axis.
7. If both coordinates are zero, the point is located at the origin.

EXAMPLE 1

Plot the points associated with each ordered pair.

a) $\left(3\frac{1}{2}, 2\frac{1}{3}\right)$ **b)** $(-4, 5)$ **c)** $(-1, -3)$ **d)** $(5, -6)$

e) $(0, 4)$ **f)** $(-7, 0)$ **g)** $(0, 0)$

Solution:

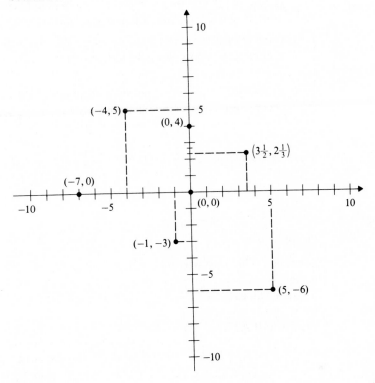

 Functions of the Form $y = kx$

It is impossible to actually list *all* the ordered pairs that satisfy a function if the domain of the function is the collection of all real numbers. However, we can pictorially show what the ordered pairs look like by *graphing* the function.

To graph a function, we will use the following four-step procedure:

FOUR-STEP PROCEDURE FOR GRAPHING A FUNCTION

1. TABULATE a few ordered pairs that satisfy the function. To do this, set up a table as follows:

Independent variable	Dependent variable	Ordered pair
x	y	(x, y)

2. PLOT and LABEL these ordered pairs in the rectangular coordinate system. Use the horizontal axis for the independent variable and the vertical axis for the dependent variable.

3. DISCOVER a pattern by imagining a thousand more ordered pairs being plotted. If you don't see a pattern, plot a few more ordered pairs that satisfy the function.

4. CONNECT the plotted points to form a smooth curve. Every point on the curve should satisfy the function.

First, let's consider functions of the form $y = kx$, $k \neq 0$.

EXAMPLE 2

Graph $y = x$.

Solution: The x-values are arbitrarily chosen and the y-values are found by using $y = x$. When possible, it is a good practice to choose both positive and negative values for x.

1. TABULATE

x	y	(x, y)
4	4	$(4, 4)$
2	2	$(2, 2)$
0	0	$(0, 0)$
-2	-2	$(-2, -2)$
-4	-4	$(-4, -4)$

$y = x$

2. PLOT and LABEL

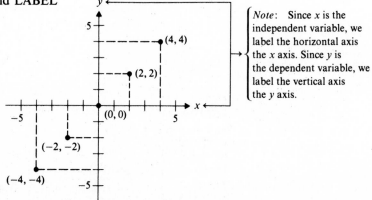

Note: Since x is the independent variable, we label the horizontal axis the x axis. Since y is the dependent variable, we label the vertical axis the y axis.

3. DISCOVER   It is obvious that if we were to plot a thousand more points, these points would begin to form a solid line.

4. CONNECT

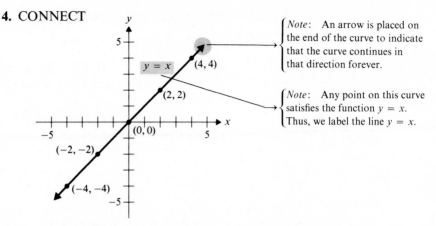

Note: An arrow is placed on the end of the curve to indicate that the curve continues in that direction forever.

Note: Any point on this curve satisfies the function $y = x$. Thus, we label the line $y = x$.

Note: The function $y = x$ is referred to as the **identity function**, since each value of x produces the same value of y.

EXAMPLE 3

The distance d a certain spring stretches when subjected to various forces F is given by $d = \dfrac{F}{70}$, where F is in pounds and d is in inches. Graph this function for $F \geq 0$.

Solution: The domain and range of the function are restricted to nonnegative numbers because we are talking about length and weight. For convenience, select F-values that are divisible by 70.

1. TABULATE

$$d = \frac{F}{70}$$

F	d	(F, d)
0	0	(0, 0)
70	1	(70, 1)
140	2	(140, 2)
280	4	(280, 4)

2. PLOT and LABEL

Note: Always identify the quantities and units of measure that are associated with each axis.

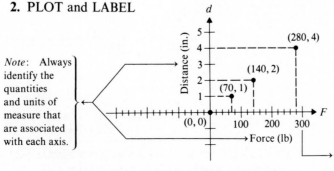

Note: Always choose a convenient scale for each axis. The scales need not be the same.

3. DISCOVER   The points are forming part of a straight line.

4. CONNECT

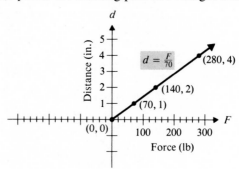

Note: The graph of $y = kx$ ($k \neq 0$) is always a straight line (or part of a straight line if the domain of the function is restricted).

⊡ Functions of the Form $y = kx^2$

Next, let's consider functions of the form $y = kx^2$, $k \neq 0$.

EXAMPLE 4

Graph $y = x^2$.

Solution: No matter what x-value is chosen the y-value can never be negative.

1. TABULATE

	x	y	(x, y)
	3	9	(3, 9)
	2	4	(2, 4)
	1	1	(1, 1)
$y = x^2$	0	0	(0, 0)
	-1	1	$(-1, 1)$
	-2	4	$(-2, 4)$
	-3	9	$(-3, 9)$

2. PLOT and LABEL

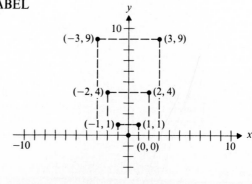

3. **DISCOVER**   If we were to plot a thousand more points, these points would begin to form a cup-shaped curve. No point on the curve would lie below the x-axis, and if (a, b) is on the curve then $(-a, b)$ is on the curve (that is, the curve is *symmetrical* with respect to the y-axis).

4. **CONNECT**

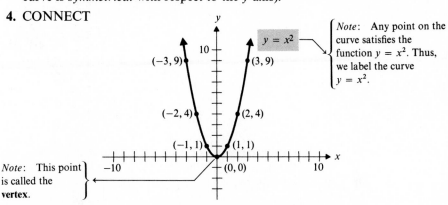

Note: Any point on the curve satisfies the function $y = x^2$. Thus, we label the curve $y = x^2$.

Note: This point is called the **vertex**.

Note: The function $y = x^2$ is referred to as the **squaring function**. The cup-shaped curve that is formed is called a **parabola**.

EXAMPLE 5

The distance d an object falls for a given time t is $d = 16t^2$, where t is in seconds and d is in feet. Graph this function for $t \geq 0$.

Solution: The domain and range of the function are restricted to nonnegative numbers because we are talking about distance and time.

1. **TABULATE**

$d = 16t^2$

t	d	(t, d)
0	0	$(0, 0)$
1	16	$(1, 16)$
2	64	$(2, 64)$
3	144	$(3, 144)$
4	256	$(4, 256)$

2. **PLOT and LABEL**

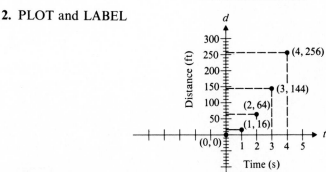

3. DISCOVER The points are forming part of a parabola.

4. CONNECT

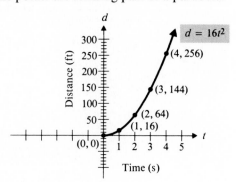

Note: The graph of $y = kx^2$ ($k \neq 0$) is always a parabola (or part of a parabola if the domain of the function is restricted).

⸬ Functions of the Form $y = \dfrac{k}{x}$

For functions of the form $y = \dfrac{k}{x}$ ($k \neq 0$), we must restrict $x = 0$ from the domain, since division by zero is undefined. Graphically, this tells us that $y = \dfrac{k}{x}$ cannot cross the y-axis (vertical axis), since every point on the y-axis has an x-value of zero.

EXAMPLE 6

Graph $y = \dfrac{1}{x}$.

Solution: Choose any x-value except $x = 0$.

1. TABULATE

$y = \dfrac{1}{x}$

x	y	(x, y)
4	$\frac{1}{4}$	$\left(4, \frac{1}{4}\right)$
2	$\frac{1}{2}$	$\left(2, \frac{1}{2}\right)$
1	1	$(1, 1)$
-1	-1	$(-1, -1)$
-2	$-\frac{1}{2}$	$\left(-2, -\frac{1}{2}\right)$
-4	$-\frac{1}{4}$	$\left(-4, -\frac{1}{4}\right)$

2. PLOT and LABEL

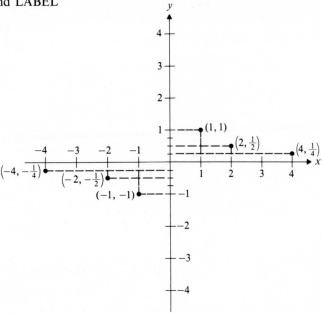

3. DISCOVER   If we were to plot a thousand more points, these points would always be located in the first or third quadrant. In the first quadrant, as we move along the x-axis in the positive direction, the y-value becomes smaller and smaller (that is, we start to approach the x-axis but never quite reach it). A similar pattern occurs in the third quadrant as we move along the x-axis in the negative direction. We must be careful not to join the points $(-1, -1)$ and $(1, 1)$. If we did, we would cross the y-axis, and this is impossible since x can't equal zero. What is happening between $x = 0$ and $x = 1$ (or between $x = 0$ and $x = -1$)? To determine this, let's plot some more points.

$$y = \frac{1}{x}$$

x	y	(x, y)
$\dfrac{1}{2}$	2	$\left(\dfrac{1}{2}, 2\right)$
$\dfrac{1}{4}$	4	$\left(\dfrac{1}{4}, 4\right)$
$-\dfrac{1}{2}$	-2	$\left(-\dfrac{1}{2}, -2\right)$
$-\dfrac{1}{4}$	-4	$\left(-\dfrac{1}{4}, -4\right)$

4. CONNECT

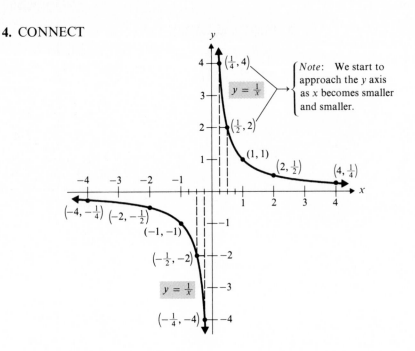

Note: We start to approach the y axis as x becomes smaller and smaller.

Note: The function $y = \dfrac{1}{x}$ is called the **reciprocal function**. The curve that is formed is called a **hyperbola**.

EXAMPLE 7

The pressure P exerted by a confined gas when occupying various volumes V is given by $P = \dfrac{10}{V}$, where P is in pounds per square inch (psi) and V is in cubic inches (in^3). Graph this function for $V > 0$.

Solution: The domain and range of the function are restricted to positive numbers because we are talking about volume and pressure.

1. TABULATE

$$P = \frac{10}{V}$$

V	P	(V, P)
10	1	(10, 1)
5	2	(5, 2)
2	5	(2, 5)
1	10	(1, 10)

2. PLOT and LABEL

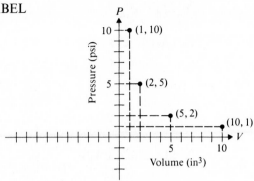

3. DISCOVER   The points are forming part of a hyperbola.

4. CONNECT

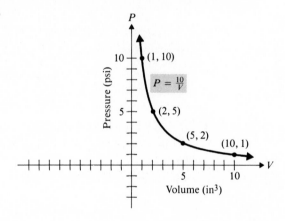

Note: The graph of $y = \dfrac{k}{x}$ $(k \neq 0)$ is always a hyperbola (or part of a hyperbola if the domain of the function is restricted).

⠠⠝ Functions of the Form $y = k\sqrt{x}$

For functions of the form $y = k\sqrt{x}$ $(k \neq 0)$, we must restrict all negative numbers from the domain, since the square root of a negative number is not a real number. Graphically, this tells us that no part of the function $y = k\sqrt{x}$ lies in the second or third quadrant.

EXAMPLE 8

Graph $y = \sqrt{x}$.

Solution: For convenience, choose *x*-values that are perfect squares.

1. TABULATE

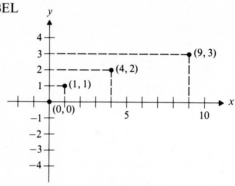

$y = \sqrt{x}$	x	y	(x, y)
	0	0	(0, 0)
	1	1	(1, 1)
	4	2	(4, 2)
	9	3	(9, 3)

2. PLOT and LABEL

3. DISCOVER If we were to plot a thousand more points, these points would always be located in the first quadrant. The value of *y* always increases as *x* increases.

4. CONNECT

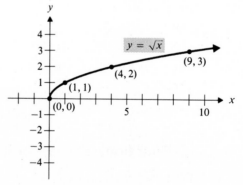

Note: The function $y = \sqrt{x}$ is referred to as the **square root** function.

EXERCISES 9.3

Plot each point on the same rectangular coordinate system.
1. $(4, 2), (-1, 5), (-3, -4), (2, -3), (0, -1), (4\frac{1}{4}, 0)$
2. $(5, 2), (-2\frac{2}{3}, 4), (-1, -2), (5, -4), (0, 6), (-5, 0)$

Name the coordinates that correspond to each point.

3.

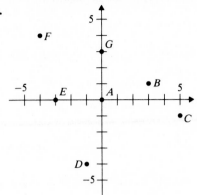

4.

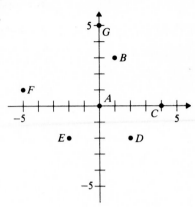

Graph each of the following functions.

5. $y = 4x$ **6.** $y = 5x$ **7.** $y = -4x$ **8.** $y = -5x$

9. $y = \dfrac{x}{4}$ **10.** $y = \dfrac{-x}{5}$ **11.** $y = 2x^2$ **12.** $y = 3x^2$

13. $y = -2x^2$ **14.** $y = -3x^2$ **15.** $y = \dfrac{x^2}{2}$ **16.** $y = \dfrac{-x^2}{3}$

17. $y = \dfrac{5}{x}$ **18.** $y = \dfrac{4}{x}$ **19.** $y = \dfrac{-5}{x}$ **20.** $y = \dfrac{-4}{x}$

21. $y = \dfrac{1}{5x}$ **22.** $y = \dfrac{-1}{4x}$ **23.** $y = 3\sqrt{x}$ **24.** $y = 2\sqrt{x}$

25. $y = -3\sqrt{x}$ **26.** $y = -2\sqrt{x}$ **27.** $y = \dfrac{-\sqrt{x}}{3}$ **28.** $y = \dfrac{-\sqrt{x}}{2}$

29. $y = \dfrac{1}{x^2}$ **30.** $y = \dfrac{2}{x^2}$ **31.** $y = \dfrac{-1}{x^2}$ **32.** $y = \dfrac{-2}{x^2}$

33. $y = x - 1$ **34.** $y = x + 2$ **35.** $y = x^2 - 1$ **36.** $y = x^2 + 2$

37. $y = \dfrac{1}{x - 1}$ **38.** $y = \dfrac{1}{x + 2}$ **39.** $y = \sqrt{x - 1}$ **40.** $y = \sqrt{x + 2}$

41. $y = \dfrac{1}{\sqrt{x - 1}}$ **42.** $y = \dfrac{1}{\sqrt{x + 2}}$

43. The weight W of steel beams of various lengths L is given by $W = 80L$, where W is in pounds and L is in feet. Graph this function for $L \geq 0$.

44. The electrical resistance R of wire of various lengths L is given by $R = L/50$, where R is in ohms and L is in meters. Graph this function for $L \geq 0$.

45. The safe load capacity S of rectangular beams with various distances d between their support columns is given by $S = \dfrac{6500}{d}$, where S is in pounds and d is in feet. Graph this function for $d > 0$.

46. The wavelength λ of a radio wave of frequency f is given by $\lambda = \dfrac{300,000}{f}$, where λ is in meters and f is in kilohertz. Graph this function for $f > 0$.

47. The power P dissipated in an electrical resistor for various currents I is given by $P = 5I^2$, where P is in watts and I is in amperes. Graph this function for $I \geq 0$.

48. The length L of a pendulum with period T is given by $L = \dfrac{T^2}{4}$, where L is in meters and T is in seconds. Graph this function for $T \geq 0$.

49. The gravitational force F between two objects placed various distances d apart is given by $F = \dfrac{80,000}{d^2}$, where F is in Newtons and d is in meters. Graph this function for $d > 0$.

50. The electrical resistance R of wire of various diameters d is given by $R = \dfrac{1}{10d^2}$, where R is in ohms and d is in millimeters. Graph this function for $d > 0$.

9.4 GRAPHING LINEAR FUNCTIONS AND LINEAR EQUATIONS

Objectives

⊡ To graph a function of the form $y = mx$ by using the slope.

⊡ To graph a function of the form $y = mx + b$ by using the slope and y-intercept.

⊡ To graph an equation of the form $Ax + By = C$ by using the x- and y-intercepts.

⊡ To graph equations of the form $x = a$ and $y = b$.

⊡ The Slope of a Line

In Section 9.3 you saw that the graph of $y = kx$ was a straight line that passed through the origin $(0, 0)$. How does the constant k affect the graph of the line? Consider the graphs of $y = \dfrac{2}{3}x$ and $y = -\dfrac{2}{3}x$ shown in Fig. 9.9.

Observe that

1. If k is *positive*, as in $y = \dfrac{2}{3}x$, the line slants *up* as we move to the *right*.

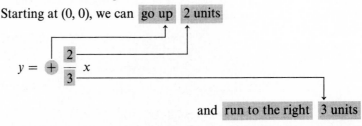

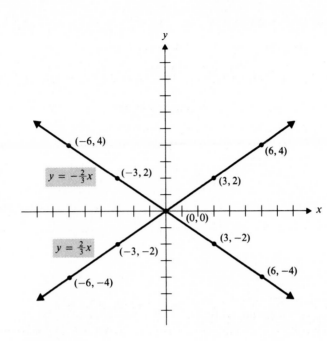

$$y = \frac{2}{3}x$$

x	y	(x, y)
6	4	(6, 4)
3	2	(3, 2)
0	0	(0, 0)
−3	−2	(−3, −2)
−6	−4	(−6, −4)

$$y = -\frac{2}{3}x$$

x	y	(x, y)
6	−4	(6, −4)
3	−2	(3, −2)
0	0	(0, 0)
−3	2	(−3, 2)
−6	4	(−6, 4)

Figure 9.9

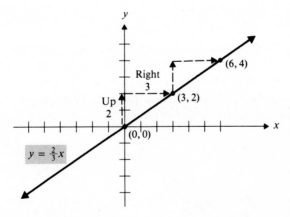

Figure 9.10

in order to locate the point (3, 2). Similarly, from (3, 2) we go up 2 units and run to the right 3 units in order to locate the point (6, 4) (see Fig. 9.10).

2. If k is *negative*, as in $y = -\frac{2}{3}x$, the line slants *up* as we move to the *left*, or *down* as we move to the *right*.

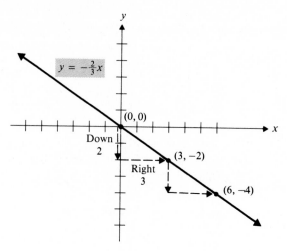

Figure 9.11

Starting at $(0, 0)$, we can go down 2 units

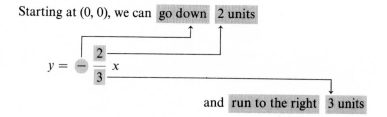

$$y = -\frac{2}{3}x$$

and run to the right 3 units

in order to locate the point $(3, -2)$. Similarly, from $(3, -2)$ we go down 2 units and run to the right 3 units in order to locate the point $(6, -4)$ (see Fig. 9.11). As you can see, the constant k determines the direction and steepness of the line. We refer to the direction and steepness of a line as its **slope**. A lowercase m is used to represent the slope.

$$\text{slope} = \frac{\text{vertical movement}}{\text{horizontal movement}}$$

$$y = \boxed{m}\, x$$

Since a graph is always read from left to right, we will consider the horizontal movement as moving to the *right* and the vertical movement as either *up* (if the slope is positive) or *down* (if the slope is negative).

EXAMPLE 1

Using the slope, graph $y = 3x$.

Solution: Since $y = 3x = +\dfrac{3}{1}x,$

$$\text{slope} = m = \boxed{+}\; \dfrac{\boxed{3}\quad \rightarrow \text{up } 3}{\boxed{1}\quad}$$

run right 1

Thus, starting at $(0, 0)$, we go up 3 units and run to the right 1 unit to locate another point on the line. We then connect these two points to form the graph of $y = 3x$ (Fig. 9.12).

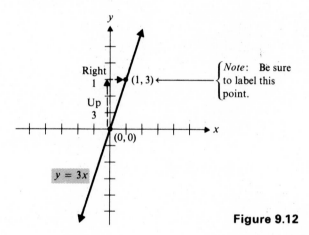

Figure 9.12

Note: You should check to see if the ordered pair $(1, 3)$ satisfies $y = 3x$.

$$y = 3x$$
$$3 = 3(1)? \quad \text{replacing } x \text{ with 1 and } y \text{ with 3}$$
$$3 = 3 \quad \text{true}$$

EXAMPLE 2

Using the slope, graph $y = \dfrac{-x}{4}$.

Solution: Since $y = \dfrac{-x}{4} = -\dfrac{1}{4}x,$

$$\text{slope} = m = \boxed{-}\; \dfrac{\boxed{1}\quad \rightarrow \text{down } 1}{\boxed{4}\quad}$$

run right 4

Thus, starting at $(0, 0)$, we do down 1 unit and run to the right 4 units to locate another point on the line. We then connect these two points to form the graph of $y = \dfrac{-x}{4}$ (Fig. 9.13).

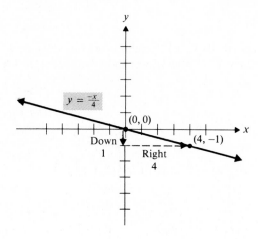

Figure 9.13

Note: Check to see if the ordered pair $(4, -1)$ satisfies $y = \dfrac{-x}{4}$.

⚃ The Linear Function

You have just seen that the graph of $y = mx$ is a straight line passing through the origin and having a slope of m. Suppose we add a constant b to mx to form $y = mx + b$. What effect does the constant b have on the line $y = mx$? Consider the graphs of $y = \dfrac{2}{3}x$, $y = \dfrac{2}{3}x + 4$, and $y = \dfrac{2}{3}x - 4$, shown in Fig. 9.14.

Observe that

1. The lines $y = \dfrac{2}{3}x$, $y = \dfrac{2}{3}x + 4$, and $y = \dfrac{2}{3}x - 4$ are parallel. They have the same direction and steepness. Thus, they have the same slope.

2. While the line $y = \dfrac{2}{3}x$ passes through the origin $(0, 0)$, the line $y = \dfrac{2}{3}x + \boxed{4}$ passes through the point $(0, \boxed{4})$, and the line $y = \dfrac{2}{3}x \boxed{- 4}$ passes through the point $(0, \boxed{-4})$.

$$y = \frac{2}{3}x + 4$$

x	y	(x, y)
6	8	(6, 8)
3	6	(3, 6)
0	4	(0, 4)
−3	2	(−3, 2)
−6	0	(−6, 0)

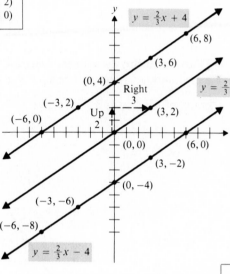

$$y = \frac{2}{3}x - 4$$

x	y	(x, y)
6	0	(6, 0)
3	−2	(3, −2)
0	−4	(0, −4)
−3	−6	(−3, −6)
−6	−8	(−6, −8)

Figure 9.14

As you can see, the graph of $y = mx + b$ is still a straight line of slope m. However, instead of passing through the origin as does $y = mx$, it passes through the point $(0, b)$. The point $(0, b)$ is called the **y-intercept**.

$$y = \underset{\substack{\uparrow \\ \text{y-intercept} = (0, b)}}{\boxed{m}} x + \boxed{b} \quad \overset{\text{slope}}{}$$

Since the graph of $y = mx + b$ is always a straight line, we refer to $y = mx + b$ as the **linear function**.

EXAMPLE 3

Using the slope and y-intercept, graph $y = \dfrac{3}{2}x - 3$.

Solution:

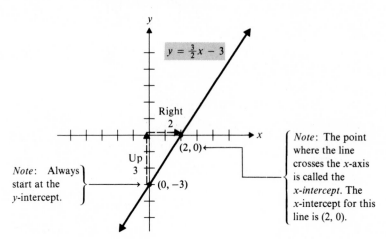

Thus, starting at the y-intercept $(0, -3)$, we go up 3 units and run right 2 units to locate another point on the line. We then connect these two points to form the graph of $y = \dfrac{3}{2}x - 3$ (Fig. 9.15).

Note: Always start at the y-intercept.

Note: The point where the line crosses the x-axis is called the *x-intercept*. The x-intercept for this line is $(2, 0)$.

Figure 9.15

EXAMPLE 4

Using the slope and y-intercept, graph $y = 2 - 3x$.

Solution: First, write $y = 2 - 3x$ in the form $y = mx + b$.

$$y = 2 - 3x = -\dfrac{3}{1}x + 2$$

slope

y-intercept $= (0, 2)$

Thus, starting at the y-intercept $(0, 2)$, we go down 3 units and run right 1 unit to locate another point on the line. We then connect these two points to form the graph of $y = 2 - 3x$ (Fig. 9.16).

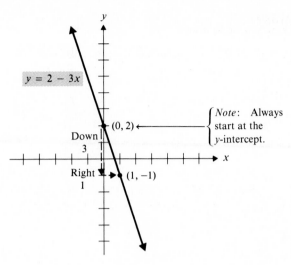

Figure 9.16

Note: Check to see if the ordered pair $(1, -1)$ satisfies the linear function $y = 2 - 3x$.

⦂ Linear Equations

Any equation of the form $Ax + By = C$, where A, B, and C are constants ($B \neq 0$), can be written in the form $y = mx + b$ as follows:

$$Ax + By = C$$

$By = C - Ax$	subtracting Ax from each side
$y = \dfrac{C - Ax}{B}$	dividing each side by B ($B \neq 0$)
$y = \dfrac{C}{B} - \dfrac{Ax}{B}$	subtraction rule for fractions in reverse
$y = -\dfrac{A}{B}x + \dfrac{C}{B}$	writing in the form $y = mx + b$

Thus, we know that the graph of $Ax + By = C$ is also a straight line with a slope of $-\dfrac{A}{B}$ and a y-intercept of $\left(0, \dfrac{C}{B}\right)$. We refer to an equation of the form $Ax + By = C$ as a **linear equation**, since its graph is a straight line.

One of the quickest ways to graph a linear equation is to find the x- and y-intercepts. To find the x-intercept of the equation $Ax + By = C$, let $y = 0$ and solve for x. Similarly, to find the y-intercept, let $x = 0$ and solve for y.

EXAMPLE 5

Using the x- and y-intercepts, graph $3x - 2y = 9$.

Solution:

To find the x-intercept, let $y = 0$ and solve for x.

$$3x - 2y = 9$$
$$3x - 2(0) = 9$$
$$3x = 9$$
$$x = 3$$

Thus, $(3, 0)$ is the x-intercept.

To find the y-intercept, let $x = 0$ and solve for y.

$$3x - 2y = 9$$
$$3(0) - 2y = 9$$
$$-2y = 9$$
$$y = -\frac{9}{2} \text{ or } -4\tfrac{1}{2}$$

Thus, $(0, -4\tfrac{1}{2})$ is the y-intercept.

We then connect these two points to form the graph of $3x - 2y = 9$ (Fig. 9.17).

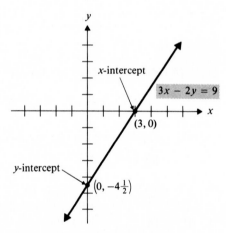

Figure 9.17

Note: When using this procedure, it is a good practice to find a third point as a *check.* For example if we arbitrarily choose $x = 1$, then

$$3x - 2y = 9$$
$$3(1) - 2y = 9 \qquad \text{replacing } x \text{ with } 1$$
$$3 - 2y = 9 \qquad \text{simplifying}$$
$$-2y = 6 \qquad \text{subtracting 3 from each side}$$
$$y = -3 \qquad \text{dividing each side by } -2$$

Thus, $(1, -3)$ should lie on the line drawn above. Check to see if it does.

⚃ **Horizontal and Vertical Lines**

In the equation $y = 2$, the x-variable appears to be missing. However, we could think of $y = 2$ as

$$0x + 1y = 2.$$
$$\downarrow \quad\quad \downarrow \quad\quad \downarrow$$
$$Ax + By = C$$

Notice that no matter what value we choose for x, $0x$ is always zero. This means that the ordered pairs that satisfy this equation can have any x-value as long as the y-value is 2. Thus, $(4, 2)$, $(2, 2)$, $(0, 2)$, $(-3, 2)$, $(-5, 2)$, and so on, all satisfy $0x + 1y = 2$ or $y = 2$. The graph of $y = 2$ is a *horizontal line* as shown in Fig. 9.18.

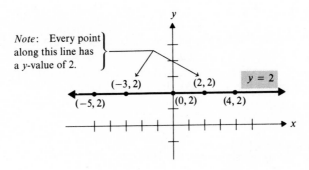

Note: Every point along this line has a y-value of 2.

Figure 9.18

The equation $y = 2$ can also be written as

$$y = 0x + 2.$$
$$\downarrow \quad\quad \downarrow$$
$$y = mx + b$$

The y-intercept is $(0, 2)$ and the slope is zero. All horizontal lines have a slope that equals zero. We refer to $y = b$ as the **constant function**, since every point on the line has the same y-value of b.

In the equation $x = 2$, the y-variable appears to be missing. However, we could think of $x = 2$ as

$$1x + 0y = 2$$
$$\downarrow \quad\quad \downarrow \quad\quad \downarrow$$
$$Ax + By = C$$

No matter what value we choose for y, $0y$ is always zero. This means that the ordered pairs that satisfy this equation can have any y-value as long as the x-value is 2. Thus, $(2, 4)$, $(2, 2)$, $(2, 0)$, $(2, -3)$, $(2, -5)$, and so on, all satisfy $1x + 0y = 2$

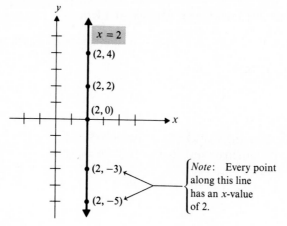

Figure 9.19

or $x = 2$. The graph of $x = 2$ is a *vertical line* as shown in Fig. 9.19. The equation $x = 2$ *cannot* be written in the form $y = mx + b$. The equation $x = 2$ does *not* have a y-intercept or a defined slope. Although every equation of the form $y = mx + b$ is a straight line, not all straight lines are of the form $y = mx + b$.

EXAMPLE 6

Graph

a) $y = -3$ **b)** $x = -\dfrac{5}{2}$

Solution:

a)

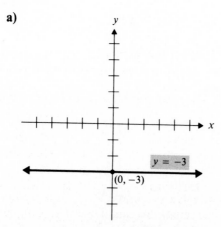

b)

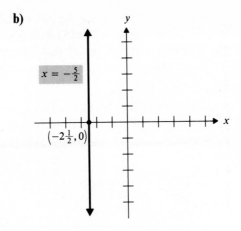

$$x = -\tfrac{5}{2}$$

$\left(-2\tfrac{1}{2}, 0\right)$

Application

THE COST OF ELECTRICITY

The *kilowatt-hour* (KWH) is the standard unit for measuring residential usage of electricity. The cost C of electricity is a function of the amount A of KWH used. Utility companies use a rate schedule to determine the cost of electricity.

EXAMPLE 7

Suppose the cost of electricity is given by the following rate schedule.

First 20 KWH or less	$2.00
Next 80 KWH	$.05 per KWH
Next 200 KWH	$.03 per KWH
Next 300 KWH	$.02 per KWH
All KWH over 600	$.01 per KWH

Graph the cost C against the amount A of KWH used.

Solution: The first 20 KWH cost a fixed amount of $2. Graphically, this is a horizontal line segment from (0, 2) to (20, 2). The next 80 KWH cost $.05 per KWH, for a total cost of (80)(.05), or $4. Thus, starting at (20, 2), we go up $4 and run right 80 KWH to locate the point (100, 6). The next 200 KWH cost $.03 per KWH, for a total cost of (200)(.03), or $6. Thus, starting at (100, 6), we go up $6 and run right 200 KWH to locate the point (300, 12). The next 300 KWH cost $.02 per KWH, for a total cost of (300)(.02), or $6. Thus, starting at (300, 12), we go up $6 and run right 300 KWH to locate the point (600, 18). <u>All KWH over 600</u> cost $.01 per KWH. In other words, the next 100 KWH would cost (100)(.01), or $1. Thus, starting at (600, 18), we go up $1 and run to the right 100 KWH to locate (700, 19). Connecting these points, we have the graph of the function (Fig. 9.20).

Note: Once the graph is drawn, it can be used to determine the cost for various KWH of usage. For example, the cost of 550 KWH of usage is $17 (see graph). Check this cost by using the rate schedule. See if you obtain $17.

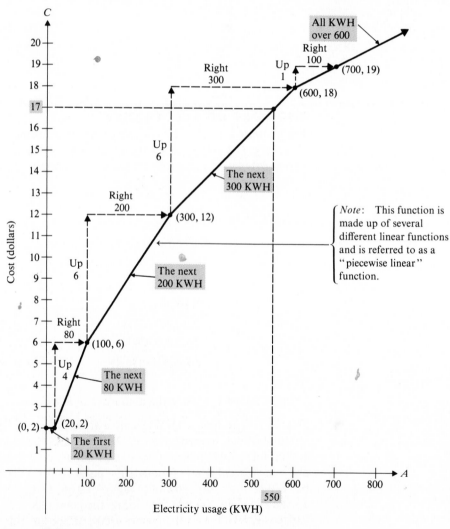

Figure 9.20

EXERCISES 9.4

Using the slope, graph the following:

1. $y = 4x$ **2.** $y = 6x$ **3.** $y = -2x$ **4.** $y = -3x$

5. $y = \dfrac{3}{4}x$ **6.** $y = \dfrac{x}{5}$ **7.** $y = \dfrac{-5x}{6}$ **8.** $y = -\dfrac{2}{3}x$

Using the slope and the y-intercept, graph the following:

9. $y = x + 4$ **10.** $y = x - 3$

11. $y = 4 - x$ **12.** $y = 3 - x$

13. $y = 2x + 3$ **14.** $y = 3x + 1$

15. $y = \dfrac{1}{2}x - 2$ **16.** $y = \dfrac{3}{5}x - 3$

17. $y = 3 - 2x$ **18.** $y = 6 - 5x$

19. $y = 4x - \dfrac{7}{2}$ **20.** $y = 3x + \dfrac{5}{3}$

21. $y = \dfrac{5}{2} - \dfrac{3}{2}x$ **22.** $y = -\dfrac{1}{4} - \dfrac{7}{4}x$

Using the x- and y-intercepts, graph the following:

23. $x + y = 8$ **24.** $x - y = 3$

25. $4x - 6y = 24$ **26.** $3x + 4y = 24$

27. $3x - 4y = 16$ **28.** $6x - 5y = 36$

29. $2x + 4y = -9$ **30.** $5x - 2y = -13$

31. $x + 2y = 0$ **32.** $3x - y = 0$

Graph the following:

33. $x = 4$ **34.** $y = -2$ **35.** $y = -\dfrac{5}{3}$ **36.** $x = \dfrac{7}{2}$

37. $x = 0$ **38.** $y = 0$ **39.** $y + 5 = 0$ **40.** $3x - 4 = 0$

41. The formula $T_f = \dfrac{9T_c}{5} + 32$ expresses the Fahrenheit temperature (T_f) as a function of the Celsius temperature (T_c).
a) Graph this function.
b) Using the graph, find T_f when $T_c = 20°C$.

42. The formula $v = 40 - \dfrac{5}{2}t$ expresses the speed (v, in feet per second) of an automobile as a function of the length of time (t, in seconds) it travels at a constant deceleration.
a) Graph this function.
b) Using the graph, find v after 16 seconds.

43. The cost C of electricity is a function of the amount A of kilowatt-hours used and is given by the following rate schedule.

First 15 KWH or less	$1.90
Next 85 KWH	$.06 per KWH
Next 200 KWH	$.04 per KWH
Next 400 KWH	$.02 per KWH
All KWH over 700	$.01 per KWH

a) Graph this function.
b) Using the graph, find C if $A = 450$ KWH.

44. The cost C of electricity is a function of the amount A of kilowatt-hours used and is given by the following rate schedule.

First 10 KWH or less	$1.40
Next 90 KWH	$.05 per KWH
Next 250 KWH	$.03 per KWH
Next 450 KWH	$.01 per KWH
All KWH over 800	$.005 per KWH

a) Graph this function.
b) Using the graph, find C if $A = 450$ KWH.

9.5 GRAPHING QUADRATIC FUNCTIONS

Objectives

 ⊡ To graph a function of the form $y = ax^2 + bx + c$.

⊡ The Parabolic Curve

Functions of the form $y = ax^2 + bx + c$, where a, b, and c are constants ($a \neq 0$) are called **quadratic functions**. The function $y = ax^2$ is an example of a quadratic function in which $b = 0$ and $c = 0$. In Section 9.3 we graphed functions like $y = ax^2$ and found that they formed a cup-shaped curve called a *parabola*. You may have noticed at that time that if $a > 0$, the parabola opens upward (see Fig. 9.21), and if $a < 0$, the parabola opens downward (see Fig. 9.22). The highest or lowest point on the graph of a parabola is called its **vertex**. For parabolas of the form $y = ax^2$, the vertex is always located at the origin $(0, 0)$. The vertical line that passes through the vertex and splits the parabola into two identical halves is

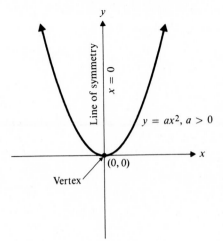

Figure 9.21

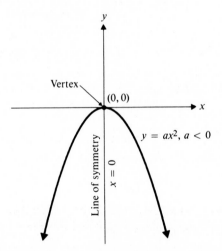

Figure 9.22

called its **line of symmetry**. For parabolas of the form $y = ax^2$, the line of symmetry is always $x = 0$ (the y-axis).

The graph of every quadratic function $y = ax^2 + bx + c$ is a parabola. If both b and c are *not* zero, the vertex is shifted away from the origin and the line of symmetry becomes $x = -\dfrac{b}{2a}$. Since the vertex of a parabola is always located along its line of symmetry, the x-coordinate of the vertex is $-\dfrac{b}{2a}$ (see Figs. 9.23 and 9.24). To find the y-coordinate of the vertex, simply replace x with $-\dfrac{b}{2a}$ in $y = ax^2 + bx + c$.

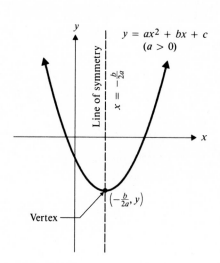

Figure 9.23   **Figure 9.24**

To graph a quadratic function of the form $y = ax^2 + bx + c$, we will use the following four-step procedure:

FOUR-STEP PROCEDURE FOR GRAPHING $y = ax^2 + bx + c$

Step 1: Determine the *line of symmetry* by using

$$x = -\frac{b}{2a}.$$

Step 2: Determine the location of the *vertex* $\left(-\dfrac{b}{2a}, y\right)$.

Step 3: Tabulate a few ordered pairs to the left and to the right of the line of symmetry that satisfy the function $y = ax^2 + bx + c$.

Step 4: Plot and connect the ordered pairs to form a parabola.

EXAMPLE 1

Graph $y = x^2 - 6x$.

Solution: $y = x^2 - 6x$ is a quadratic function with $a = 1$, $b = -6$, and $c = 0$. Since $a > 0$, we know that the parabola must open upward.

Step 1:

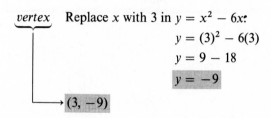

$$\underbrace{\textit{line of symmetry}}\; x = -\frac{b}{2a}$$

$$x = -\frac{(-6)}{2(1)}$$

$$\boxed{x = 3}$$

Step 2:

$$\underbrace{\textit{vertex}}$$ Replace x with 3 in $y = x^2 - 6x$:

$$y = (3)^2 - 6(3)$$

$$y = 9 - 18$$

$$\boxed{y = -9}$$

$$(3, -9)$$

Step 3:   Find a couple of ordered pairs to the *left* and to the *right* of the line of symmetry that satisfy the function.

	x	y	(x, y)
vertex	3	-9	$(3, -9)$
	0	0	$(0, 0)$
	1	-5	$(1, -5)$
	5	-5	$(5, -5)$
	6	0	$(6, 0)$

Step 4:

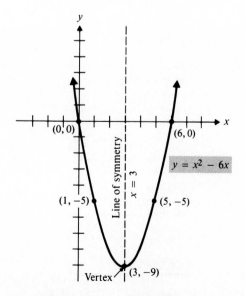

EXAMPLE 2

Graph $y = 3 - 2x - x^2$.

Solution: $y = 3 - 2x - x^2$ is a quadratic function with $a = -1$, $b = -2$, and $c = 3$. Since $a < 0$, we know that the parabola must open downward.

Step 1: $\underbrace{\textit{line of symmetry}}\ x = -\dfrac{b}{2a}$

$$x = -\frac{(-2)}{2(-1)}$$

$$x = -1$$

Step 2: $\underbrace{\textit{vertex}}$ Replace x with -1 in $y = 3 - 2x - x^2$.

$$y = 3 - 2(-1) - (-1)^2$$

$(-1, 4)$ $\qquad y = 4$

Step 3: Find a couple of ordered pairs to the *left* and to the *right* of the line of symmetry that satisfy the function.

	x	y	(x, y)
vertex	-1	4	$(-1, 4)$
	-2	3	$(-2, 3)$
	-3	0	$(-3, 0)$
	0	3	$(0, 3)$
	1	0	$(1, 0)$

Step 4:

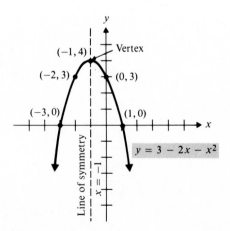

EXAMPLE 3

Graph $y = 4x^2 - 12x + 9$.

Solution: $y = 4x^2 - 12x + 9$ is a quadratic function with $a = 4$, $b = -12$, and $c = 9$. Since $a > 0$, we know that the parabola must open upward.

Step 1: <u>*line of symmetry*</u> $x = -\dfrac{b}{2a}$

$$x = -\dfrac{(-12)}{2(4)}$$

$$x = \dfrac{3}{2} \text{ or } 1\tfrac{1}{2}$$

Step 2: <u>*vertex*</u> Replace x with $\dfrac{3}{2}$ in $y = 4x^2 - 12x + 9$.

$\left(\dfrac{3}{2}, 0\right)$

$$y = 4\left(\dfrac{3}{2}\right)^2 - 12\left(\dfrac{3}{2}\right) + 9$$

$$y = 0$$

Step 3:

	x	y	(x, y)
vertex	$\dfrac{3}{2}$	0	$\left(\dfrac{3}{2}, 0\right)$
	0	9	$(0, 9)$
	1	1	$(1, 1)$
	2	1	$(2, 1)$
	3	9	$(3, 9)$

Step 4:

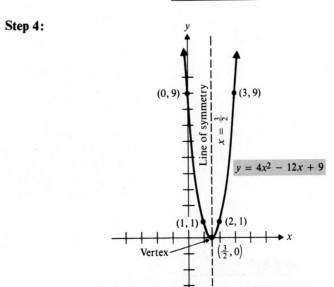

EXAMPLE 4

Graph $x^2 + 2y = -4$.

Solution: We must first write $x^2 + 2y = -4$ in the form $y = ax^2 + bx + c$. To do this, simply solve $x^2 + 2y = -4$ for y.

$$x^2 + 2y = -4$$

$$2y = -4 - x^2 \qquad \text{subtracting } x^2 \text{ from each side}$$

$$y = -2 - \tfrac{1}{2}x^2 \qquad \text{dividing each side by 2}$$

Note: This is a quadratic function with $a = -\tfrac{1}{2}$, $b = 0$, $c = -2$.

Step 1:   *line of symmetry* $x = -\dfrac{b}{2a} = -\dfrac{0}{2(-\frac{1}{2})} = 0$

$$x = 0$$

Step 2:   *vertex*   Replace x with 0 in $y = -2 - \tfrac{1}{2}x^2$.

$$y = -2 - \tfrac{1}{2}(0)^2$$

$(0, -2)$ $\qquad\qquad y = -2$

Step 3:

	x	y	(x, y)
vertex	0	-2	$(0, -2)$
	-2	-4	$(-2, -4)$
	-4	-10	$(-4, -10)$
	2	-4	$(2, -4)$
	4	-10	$(4, -10)$

Step 4:

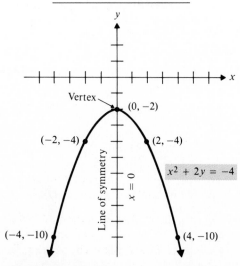

Application

THE FLIGHT OF A GOLF BALL

When a golf ball is driven with an initial velocity v at an angle θ with the horizontal, it will follow a *parabolic path* (Fig. 9.25). Shown in Fig. 9.25 are the flights of several golf balls, given the same initial velocity but projected at different angles. A golf

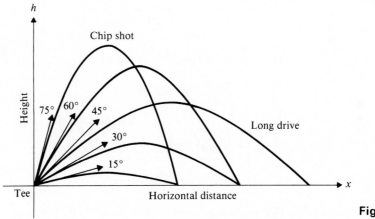

Figure 9.25

ball will travel the greatest distance when it is driven at an angle of 45° with the horizontal. Its path of flight is given by the following formula.

FORMULA: Flight of a Golf Ball When Driven at an Angle of 45°.

$$h = x - \frac{32x^2}{v^2},$$

where h is the height (ft), x is the horizontal distance (ft), v is the initial velocity (ft/s).

EXAMPLE 5

A golf ball is driven with an initial velocity of 160 ft/s at an angle of 45° with the horizontal.

a) What is the maximum height it will attain?
b) How far away from the tee will it strike the ground?
c) Graph the flight of the ball.

Solution:

$$h = x - \frac{32x^2}{v^2}$$

$$h = x - \frac{32x^2}{(160)^2} \qquad \text{replacing } v \text{ with } 160$$

$$h = x - .00125x^2 \qquad \text{simplifying}$$

Note: This is a quadratic function with $a = -.00125$, $b = 1$, and $c = 0$.

a) The maximum height will occur at the vertex of the parabolic flight.

$$\underbrace{line\ of\ symmetry}\ x = -\frac{b}{2a} = -\frac{1}{2(-.00125)} = 400$$

$$x = 400$$

$\underbrace{vertex}$ Replace x with 400 in $h = x - .00125x^2$.

$$h = (400) - .00125(400)^2$$

$$(400, 200) \qquad h = 200$$

The maximum height attained is 200 ft, which occurs at a distance of 400 ft from the tee. ■

b) When the ball is on the ground, the height is zero ($h = 0$).

$$h = \ \ x - .00125x^2$$

$$0 = \ \ x - .00125x^2 \qquad \text{replacing } h \text{ with } 0$$

Note: This is a quadratic equation, which we can solve by factoring and using the zero product property.

$$0 = x\ (1 - .00125x) \qquad \text{factoring}$$

$$x = 0 \text{ or } 1 - .00125x = 0 \qquad \text{zero product property}$$

Note: This is the x-value at the tee.

$$-.00125x = -1$$

$$x = \frac{-1}{-.00125}$$

$$x = 800$$

Note: This is the x-value when the ball strikes the ground.

The ball strikes the ground 800 ft from the tee. ■

c) Thus far, we have accumulated 3 points on the graph of the flight of the ball:

(0, 0) , when the ball is on the tee (*x*-intercept),

(400, 200) , when the ball attains its maximum height (vertex), and

(800, 0) , when the ball strikes the ground (*x*-intercept).

Actually, these three points are enough for us to *roughly* sketch the flight of the ball. However, for a more accurate graph, we will find two more ordered pairs that satisfy the function $h = x - .00125x^2$.

x	*h*	(*x, h*)
200	150	(200, 150)
600	150	(600, 150)

By plotting these five points and connecting them to form a parabolic curve, we have a graph of the flight of the golf ball (Fig. 9.26).

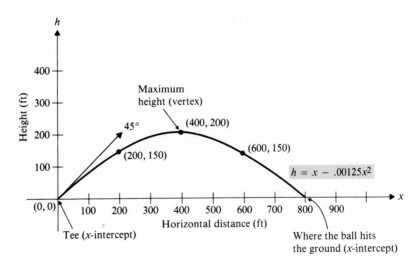

Figure 9.26

Note: Any projectile that has an initial velocity *v* at an angle θ with the horizontal will follow a parabolic path. A football thrown or kicked by a football player, a fly ball hit by a baseball player, a bullet fired from a gun, a rocket fired from a launch pad, and so on, all follow a parabolic path. The greatest distance that the object can travel always occurs when the object is projected at an angle θ of $45°$.

EXERCISES 9.5

Graph the following:

1. $y = x^2 + 4x$ 2. $y = x^2 - 8x$

3. $y = x^2 - 5$ 4. $y = x^2 + 4$

5. $y = 1 - x^2$ 6. $y = 3 - x^2$

7. $y = 5x - x^2$ 8. $y = 3x - x^2$

9. $y = x^2 - 2x + 1$ 10. $y = x^2 + 8x + 16$

11. $y = 5 - 4x - x^2$ 12. $y = 7 - 6x - x^2$

13. $y = 3x^2 - 6x + 1$ 14. $y = 2x^2 + 8x - 1$

15. $y = 5 + 3x - 2x^2$ 16. $y = 10 - x - 3x^2$

17. $x^2 - 3y = 6$ 18. $x^2 + 4y = -8$

19. $x^2 + 6x - 2y = 8$ 20. $x^2 + 12x + 4y = 1$

21. A golf ball is driven with an initial velocity of 100 ft/s at an angle of 45° with the horizontal.

    a) What is the maximum height it will attain?

    b) How far away from the tee will it strike the gound?

    c) Graph the flight of the ball.

22. A football is kicked with an initial velocity of 80 ft/s at an angle of 45° with the horizontal.

    a) What is the maximum height it will attain?

    b) How far away from the tee will it strike the ground?

    c) Graph the flight of the football.

Chapter Review

REVIEW EXERCISES

The following review exercises are grouped according to the objectives that should have been mastered in Chapter 9. Work each problem carefully. If any weaknesses appear, immediately refer to and read the subsection that matches that objective.

9.1 THE FUNCTIONAL RELATION

Objectives

⊡ To express one quantity as a function of another.

Express as a functional relation.

1. Write an equation that defines the side s of a square as a function of its perimeter P.

2. A length of x ft is cut from *each* end of an 8-ft piece of copper pipe. Express the remaining length y as a function of x.

☐ To complete an ordered pair such that it satisfies a given function.

Complete the ordered pairs.

3. $y = 1 - 9x^2$

    a) $(0, ?)$ **b)** $(2, ?)$

    c) $(-2, ?)$ **d)** $\left(-\dfrac{1}{3}, ?\right)$

9.2 VARIATION

Objectives

☐ To express one quantity as a function of another, if one of the quantities varies directly as the other.

Express as a functional relation.

4. Express y as a function of x, if y varies directly as x and $y = 12$ when $x = 2$.

5. Hooke's law states that the distance d a spring stretches varies directly as the force F applied to the spring. Express d as a function of F, if $d = 8$ cm when $F = 640$ N.

☐ To express one quantity as a function of another, if one of the quantities varies inversely as the other.

Express as a functional relation.

6. Express y as a function of x, if y varies inversely as x and $y = 12$ when $x = 2$.

7. Boyle's law states that when the temperature of a confined gas remains constant, the pressure P it exerts varies inversely as the volume V it occupies. Express P as a function of V, if $P = 6$ psi when $V = 20$ in^3.

☐ To express one quantity as a function of another, if one of the quantities varies directly or inversely as the square of the other.

Express as a functional relation.

8. The area A of a circle varies directly as the square of its diameter d. Express A as a function of d if $A = 28.27$ m^2 when $d = 6.00$ m.

9. The time t required to empty a swimming pool varies inversely as the square of the diameter d of the hose used. Express t as a function of d, if $t = 8$ hr when $d = 2$ in.

9.3 AN INTRODUCTION TO GRAPHING A FUNCTION

Objectives

☐ To plot points associated with ordered pairs.

Plot.

10. **a)** $(2\frac{1}{4}, 6\frac{2}{3})$ **b)** $(-3, 4)$
    c) $(-2, -8)$ **d)** $(4, -1)$
    e) $(0, 3)$ **f)** $(-6, 0)$
    g) $(0, 0)$

☐ To graph a function of the form $y = kx$.

Graph.

11. $y = -x$

12. $d = \dfrac{F}{80}$, where F is in Newtons
    d is in centimeters ($F \geq 0$)

☐ To graph a function of the form $y = kx^2$.

Graph.

13. $y = -x^2$

14. $A = .785d^2$, where A is in square feet
    d is in feet ($d \geq 0$)

⊡ To graph a function of the form $y = \dfrac{k}{x}$.

Graph.

15. $y = \dfrac{-1}{x}$

16. $P = \dfrac{120}{V}$, where P is in pounds per square inch
V is in cubic inches $(V > 0)$

⊡ To graph a function of the form $y = k\sqrt{x}$.

Graph.

17. $y = -\sqrt{x}$

9.4 GRAPHING LINEAR FUNCTIONS AND LINEAR EQUATIONS

Objectives

⊡ To graph a function of the form $y = mx$ by using the slope.

Graph using the slope.

18. $y = 5x$

19. $y = \dfrac{-x}{3}$

⊡ To graph a function of the form $y = mx + b$ by using the slope of y-intercept.

Graph using the slope
and y-intercept.

20. $y = \dfrac{2}{5}x - 2$

21. $y = 4 - 5x$

⊡ To graph an equation of the form $Ax + By = C$ by using the x- and y-intercepts.

Graph using the x- and
y-intercepts.

22. $4x - 3y = 15$

⊡ To graph equations of the form $x = a$ and $y = b$.

Graph.

23. a) $y = 4$

b) $x = -\dfrac{4}{3}$

9.5 GRAPHING QUADRATIC FUNCTIONS

Objectives

⊡ To graph a function of the form $y = ax^2 + bx + c$.

Graph.

24. $y = x^2 - 4x$

25. $y = 9 - 8x - x^2$

26. $y = 4x^2 - 4x + 1$

27. $x^2 + 4y = -4$

CHAPTER TEST
Time
50 minutes

Score
15, 14 excellent
13, 12 good
11, 10, 9 fair
below 8 poor

If you have worked through the Review Exercises and corrected any weaknesses, you are ready to take the following Chapter Test.

1. Express in cents the monthly cost C of a checking account as a function of the number n of checks written, if the bank charges $\$.10$ per check plus a service charge of $\$4$/mo.

2. Graph the function described in exercise 1 $(n \geq 0)$.

3. The lift L of an airplane wing varies directly as the square of the speed v of air flowing over it. Express L as a function of v if $L = 400$ lb/sq ft when $v = 200$ mph.

4. Graph the function described in exercise 2 ($v \geq 0$).

5. The force F necessary to remove a hubcap from a wheel varies inversely as the length L of the crowbar used. Express F as a function of L if $F = 60$ lb when $L = 2$ ft.

6. Graph the function described in exercise 5 ($L > 0$).

Graph the following lines by using the method described.

7. $y = \dfrac{-3x}{7}$ (slope)

8. $y = 6 - 4x$ (slope and y-intercept)

9. $y = \dfrac{2}{3} x - \dfrac{3}{4}$ (slope and y-intercept)

10. $5x - 3y = -14$ (x- and y-intercepts)

Graph the following

11. $y = 4\sqrt{x}$

12. $y = -5$

13. $2x = 3$

14. $y = 8x - x^2$

15. $y = x^2 + 8x + 12$

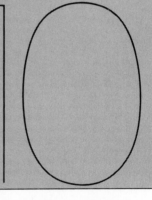

10

TWO BY TWO SYSTEMS OF LINEAR EQUATIONS

10.1 SOLVING SYSTEMS OF LINEAR EQUATIONS BY THE GRAPHING METHOD

Objectives

☐ To determine if a given ordered pair is a solution of a two by two linear system.

☐ To solve a two by two linear system by the graphing method.

☐ To determine if a two by two linear system has exactly one solution, no solution, or an Infinite number of solutions.

☐ Verifying a Solution

In Chapter 9 we referred to an equation of the form $Ax + By = C$, where A, B, and C are constants, as a *linear equation*. When *two* linear equations containing the same *two* unknowns (variables) are grouped together, a **two by two system of linear equations** is formed. Three examples of a two by two linear system are the following:

System I
$$3x + y = 5$$
$$2x - y = 5$$
Note: Both equations are of the form $Ax + By = C$.

System II
$$4x - 2y = 10$$
$$y = -3$$
Note: $y = -3$ is a linear equation since it can be written as $0x + y = -3$.

System III
$$x + 4y = 9$$
$$y = 6 - x$$
Note: $y = 6 - x$ is a linear equation since it can be written as $x + y = 6$.

A solution of a two by two linear system is an ordered pair (x, y) that satisfies BOTH linear equations.

EXAMPLE 1

Is $(1, 2)$ a solution for $\begin{array}{l} 3x + y = 5 \\ 2x - y = 5 \end{array}$?

Solution: The ordered pair $(1, 2)$ is a solution provided that it satisfies *both*

x-value ⤒ ⤒ y-value

linear equations.

Does $3x + y = 5$ when $x = 1$ and $y = 2$?

$$3(1) + (2) = 5?$$
$$3 + 2 = 5?$$
$$5 = 5 \quad \text{true}$$

Does $2x - y = 5$ when $x = 1$ and $y = 2$?

$$2(1) - (2) = 5?$$
$$2 - 2 = 5?$$
$$0 = 5 \quad \text{false}$$

Since $(1, 2)$ does not satisfy *both* equations, it is *not* a solution to the linear system. ■

EXAMPLE 2

Is $(2, -1)$ a solution for $\begin{matrix} 3x + y = 5 \\ 2x - y = 5 \end{matrix}$?

Solution: The ordered pair $(2, -1)$ is a solution provided that it satisfies *both*

x-value ⌐   ⌐ y-value

linear equations.

   Does $3x + y = 5$ when $x = 2$ and $y = -1$?

$$3(2) + (-1) = 5?$$
$$6 + (-1) = 5?$$
$$5 = 5 \quad \text{true}$$

Does $2x - y = 5$ when $x = 2$ and $y = -1$?

$$2(2) - (-1) = 5?$$
$$4 - (-1) = 5?$$
$$5 = 5 \quad \text{also true}$$

Since $(2, -1)$ satisfies *both* equations, it *is* a solution to the linear system. ■

⊡ The Graphing Method

In Chapter 9 you saw that the graph of a linear equation was a straight line. Suppose we were to graph $3x + y = 5$ and $2x - y = 5$ on the *same* axes (see Fig. 10.1). Observe that $(2, -1)$ is common to *both* lines and therefore satisfies *both* equations. The ordered pair $(2, -1)$ is the *only* such pair of numbers that is common to both lines, so $(2, -1)$ is the *only* solution to this system.

   As you can see, one method of finding the solution to a two by two linear system is to graph the lines on the same axes. *The ordered pair associated with the point where the two lines intersect is the solution to the linear system.* We refer to this method of solving a two by two linear system as the **graphing method**.

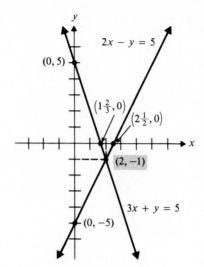

Figure 10.1

EXAMPLE 3

Use the graphing method to solve $\begin{aligned}4x - 2y &= 10 \\ y &= -3\end{aligned}$.

Solution: Graph the two lines on the same axes and note the point where they intersect (Fig. 10.2). The lines intersect at $(1, -3)$. Thus, the only solution is the ordered pair $(1, -3)$.   ∎

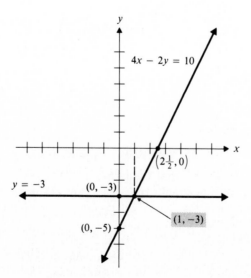

Figure 10.2

Note: Always check to see if the ordered pair satisfies *both* equations.

Does $4x - 2y = 10$ when $x = 1$ and $y = -3$?

$$4(1) - 2(-3) = 10?$$
$$4 - (-6) = 10?$$
$$10 = 10 \quad \text{true}$$

Does $0x + y = -3$ when $x = 1$ and $y = -3$?

$$0(1) + (-3) = -3?$$
$$0 + (-3) = -3?$$
$$-3 = -3 \quad \text{true}$$

EXAMPLE 4

Use the graphing method to solve $\begin{array}{l} x + 4y = 9 \\ y = 6 - x \end{array}$.

Solution: Graph the two lines on the same axes and note the point where they intersect (Fig. 10.3). The lines intersect at (5, 1). Thus, the only solution is the ordered pair (5, 1).   ■

Note: Check to see if (5, 1) satisfies *both* equations.

x-value ⤒ ⤒ y-value

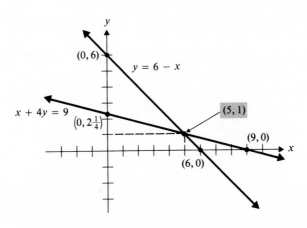

Figure 10.3

⠿ Types of Solutions

Does a two by two linear system always have exactly one solution? Can it have more than one solution or possibly no solution at all?

Consider the system $\begin{array}{l} 3x + y = 3 \\ 6x + 2y = -6 \end{array}$ (see Fig. 10.4). Observe that the lines appear to be parallel. There is no intersection point, so there is no ordered pair that satisfies *both* equations. Thus, there is *no solution* for this linear system.

To show that the lines are actually parallel, write them in the form $y = mx + b$. If their slopes are the *same* but the y-intercepts are *different*, the

slope ⤒ ⤒ y-intercept

lines are parallel.

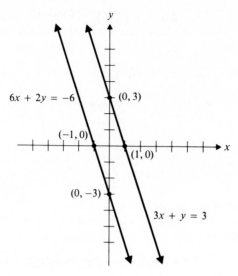

Figure 10.4

Solving $3x + y = 3$ for y we obtain $\qquad y = \boxed{-3}\,x + \boxed{3}$.

$$\underset{\substack{\text{same}\\\text{slope}}}{\uparrow} \qquad \underset{\substack{\text{different}\\\textit{y}\text{-intercept}}}{\uparrow}$$

$$\downarrow \qquad \downarrow$$

Solving $6x + 2y = -6$ for y we obtain $y = \boxed{-3}\,x - \boxed{3}$.

Thus, the lines are parallel and the system has no solution.

Next, consider the system $\begin{array}{r} 2x - 6y = 12 \\ -3x + 9y = -18 \end{array}$ (see Fig. 10.5). Observe that the lines appear to be the same lines (that is, they coincide). There are an infinite number of intersection points, and so there are an infinite number of ordered pairs that satisfy *both* equations. Thus, there are an *infinite number of solutions* to this system.

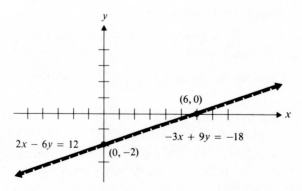

Figure 10.5

To show that the lines actually coincide, write them in the form $y = mx + b$. If their slopes and y-intercepts are the *same*, the lines coincide.

Solving $2x - 6y = 12$ for y we obtain $\qquad y = \dfrac{1}{3} x - 2$.

same   same
slope y-intercept

Solving $-3x + 9y = -18$ for y we obtain $y = \dfrac{1}{3} x - 2$.

Thus, the lines coincide and the system has an infinite number of solutions.

You have seen that there are three possibilities for the solution of a two by two linear system.

Case 1. If the lines have different slopes, the lines intersect once and the linear system has *exactly one solution.*

Case 2. If the lines have the same slope but different y-intercepts, the lines are parallel and the linear system has *no solution.*

Case 3. If the lines have the same slope and y-intercept, the lines coincide and the linear system has an *infinite number of solutions.*

EXAMPLE 5

Determine whether each linear system has exactly one solution, no solution, or an infinite number of solutions.

a) $x + 4y = 2$
$x - 4y = -2$

b) $3x - 12y = 6$
$-4x + 16y = 8$

c) $2x - 8y = -4$
$-5x + 20y = 10$

Solution:

a) $\qquad x + 4y = 2 \xrightarrow{\text{Solve for } y.} y = -\dfrac{1}{4} x + \dfrac{1}{2}$

$\qquad x - 4y = -2 \xrightarrow[\text{Solve for } y.]{} y = \dfrac{1}{4} x + \dfrac{1}{2}$

$\begin{cases} Note: \text{ The slopes} \\ \text{are } different. \end{cases}$

Since the slopes are different, this linear system must have *exactly one solution.* Since the y-intercepts are the same, we know the solution is $(0, \frac{1}{2})$. ■

$\begin{cases} Note: & \text{The } y\text{-intercepts} \\ & \text{are } \textit{different}. \end{cases}$

b) $3x - 12y = 6$ $\xrightarrow{\text{Solve for } y.}$ $y = \dfrac{1}{4}x - \dfrac{1}{2}$

$-4x + 16y = 8$ $\xrightarrow[\text{Solve for } y.]{}$ $y = \dfrac{1}{4}x + \dfrac{1}{2}$

$\begin{cases} Note: & \text{The slopes} \\ & \text{are the } \textit{same}. \end{cases}$

Since the slopes are the same and the y-intercepts are different, this system must have *no solution*. The lines are parallel. ■

c) $2x - 8y = -4$ $\xrightarrow{\text{Solve for } y.}$ $y = \dfrac{1}{4}x + \dfrac{1}{2}$

$\begin{cases} Note: & \text{The} \\ & y\text{-intercepts} \\ & \text{are the } \textit{same}. \end{cases}$

$-5x + 20y = 10$ $\xrightarrow[\text{Solve for } y.]{}$ $y = \dfrac{1}{4}x + \dfrac{1}{2}$

$\begin{cases} Note: & \text{The slopes} \\ & \text{are the } \textit{same}. \end{cases}$

Since the slopes and y-intercepts are the same, this system has an *infinite number of solutions*. Any ordered pair that satisfies one of the equations satisfies the other equation. The lines coincide. ■

EXERCISES 10.1

1. Is $(1, 3)$ a solution for $\begin{array}{l} 2x + 3y = 11 \\ x - 2y = 5 \end{array}$?

2. Is $(5, 2)$ a solution for $\begin{array}{l} 3x - y = 11 \\ x - 3y = 1 \end{array}$?

3. Is $(-1, 4)$ a solution for $\begin{array}{l} 9x + 2y = -1 \\ -x + 3y = 13 \end{array}$?

4. Is $(2, -3)$ a solution for $\begin{array}{l} 2x - 2y = 10 \\ 8x + 3y = 6 \end{array}$?

5. Is $(-2, -1)$ a solution for $\begin{array}{l} 7x - 3y = 11 \\ 4x - y = -7 \end{array}$?

6. Is $(-8, -3)$ a solution for $\begin{array}{l} x - 3y = 1 \\ x = -3 \end{array}$?

7. Is $\left(\dfrac{4}{3}, -4\right)$ a solution for $\begin{array}{l} 9x + y = 8 \\ y = -3x \end{array}$?

8. Is $\left(1, -\dfrac{3}{2}\right)$ a solution for $\begin{array}{l} x = 6y + 10 \\ 3x = -2y \end{array}$?

9. Is $\left(\dfrac{2}{3}, \dfrac{1}{4}\right)$ a solution for $\begin{array}{l} 3x + 4y = 3 \\ 6x - 20y = 1 \end{array}$?

10. Is $\left(\dfrac{7}{6}, -\dfrac{11}{12}\right)$ a solution for $\begin{array}{l} 5x + 2y = 4 \\ x - 2y = 3 \end{array}$?

Use the graphing method to solve the following two by two linear systems. Be sure to check your answers.

11. $\begin{array}{l} x + y = 3 \\ x - y = 7 \end{array}$

12. $\begin{array}{l} 2x + y = 6 \\ x - y = -3 \end{array}$

13. $\begin{array}{l} 3x - 2y = -8 \\ -x + 4y = 6 \end{array}$

14. $\begin{array}{l} 5x + 2y = 5 \\ 2x - y = 11 \end{array}$

15. $\begin{array}{l} y = 6 - 3x \\ 3x - 4y = 6 \end{array}$

16. $\begin{array}{l} 4x - 3y = 18 \\ y = 2x - 8 \end{array}$

17. $\begin{array}{l} 6x - 2y = 12 \\ x = 3 \end{array}$

18. $\begin{array}{l} 2x - y = 9 \\ y = -1 \end{array}$

19. $\begin{array}{l} 3x - 4y = 8 \\ y = \dfrac{x}{2} \end{array}$

20. $\begin{array}{l} 2x - 3y = 17 \\ y = -5x \end{array}$

21. $\begin{array}{l} 4x + 2y = 9 \\ 6x + 3y = 10 \end{array}$

22. $\begin{array}{l} 3y = 5 - x \\ 2x + 6y = 15 \end{array}$

23. $\begin{array}{l} x = 2y + 5 \\ 2x - 4y = 10 \end{array}$

24. $\begin{array}{l} 2x - 10y = 12 \\ 5x - 25y = 30 \end{array}$

Without graphing, determine whether each of the following linear systems has exactly one solution, no solution, or an infinite number of solutions.

25. $\begin{array}{l} 3x + y = 2 \\ 3x - y = -2 \end{array}$

26. $\begin{array}{l} 6x + 3y = 1 \\ -6x + 3y = -1 \end{array}$

27. $\begin{array}{l} 6x - 2y = -4 \\ -9x + 3y = 6 \end{array}$

28. $\begin{array}{l} -12x + 6y = -2 \\ 18x - 9y = 3 \end{array}$

29. $\begin{array}{l} -12x + 4y = 8 \\ 15x - 5y = 10 \end{array}$

30. $\begin{array}{l} 24x + 12y = 4 \\ -30x - 15y = 5 \end{array}$

10.2 SOLVING SYSTEMS OF LINEAR EQUATIONS BY THE SUBSTITUTION METHOD

Objectives

☐ To solve a two by two linear system by using the substitution method when one of the equations contains an isolated variable.

☐ To solve a two by two linear system by using the substitution method when neither of the equations contains an isolated variable.

⊡ **The Substitution Method**

Solving a two by two linear system by the graphing method has a serious limitation. Unless the coordinates of the solution are integers, it is nearly impossible to find the exact solution from the graph. Consider the linear system

$$7x - 3y = -14$$
$$y = -2.$$

We could try to find the solution by graphing (see Fig. 10.6). We might guess the solution to be $(-3, -2)$. If $(-3, -2)$ is the solution, it must satisfy *both* equations.

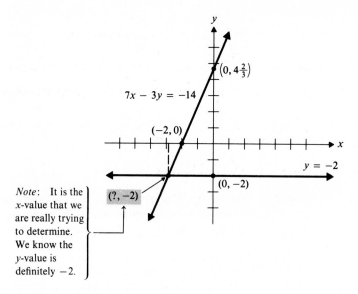

Note: It is the x-value that we are really trying to determine. We know the y-value is definitely -2.

Figure 10.6

Does $7x - 3y = -14$, when $x = -3$ and $y = -2$?

$$7(-3) - 3(-2) = -14?$$
$$-21 - (-6) = -14?$$
$$-15 = -14 \quad \text{false}$$

We can conclude that $(-3, -2)$ is not the exact solution of this system. The x-coordinate at the intersection point must be a rational number that is approximately equal to -3. The chances of guessing this rational number are slim at best. We need a more sophisticated method than graphing, one that will give us the exact solution every time.

Note that in the system

$$7x - 3y = -14 \qquad \text{first equation}$$
$$y = -2 \qquad \text{second equation}$$

the second equation states that y *is the same as* -2. Thus, we can *substitute* -2 for y in the first equation and solve for x algebraically as follows:

$$7x - 3\,\boxed{y} = -14 \qquad\qquad \text{adding } -6 \text{ to each side}$$

Note: We now have *one* equation and *one* unknown, which can be solved in the usual manner.

$$\rightarrow 7x - 3\,(\boxed{-2}) = -14 \qquad\qquad \text{substitution}$$

$$7x + 6 = -14 \qquad\qquad \text{simplifying}$$

$$7x = -20$$

$$x = -\frac{27}{7} \quad\text{or}\quad -2\frac{6}{7} \qquad\qquad \text{dividing each side by 7}$$

We now know that the x-coordinate at the intersection point is $-\dfrac{20}{7}$. Thus, the ordered pair $\left(-\dfrac{20}{7}, -2\right)$ is the solution to this system. This method of solving a two by two linear system is called the **substitution method**.

EXAMPLE 1

Use the substitution method to solve $\begin{array}{l} 6x + y = 12 \\ y = 2x \end{array}$.

Solution: The second equation states that y *is the same as* $2x$. Thus, we can *substitute* $2x$ for y in the first equation and solve for x algebraically as follows:

$$6x + \boxed{y} = 12$$

Note: We now have *one* equation and *one* unknown.

$$\rightarrow 6x + \boxed{2x} = 12 \qquad\qquad \text{substitution}$$

$$8x = 12 \qquad\qquad \text{simplifying}$$

$$x = \frac{12}{8} \qquad\qquad \text{dividing each side by 8}$$

$$x = \frac{3}{2} \quad\text{or}\quad 1\frac{1}{2} \qquad\qquad \text{cancellation law}$$

Now that we know the value of x, we can replace x with $\dfrac{3}{2}$ in the equation $y = 2x$ in order to find y.

$$y = \;\; 2\,\boxed{x}$$

$$y = 2\left(\boxed{\frac{3}{2}}\right) \qquad\qquad \text{replacing } x \text{ with } \frac{3}{2}$$

$$y = 3 \qquad\qquad \text{simplifying}$$

Thus, $\left(\dfrac{3}{2}, 3\right)$ is the solution to this system.

Note: Check to see if $\left(\dfrac{3}{2}, 3\right)$ satisfies *both* equations.   ■

x-value ——⤴ ⤴—— *y*-value

Remember, the solution to a two by two linear system is an ordered pair of numbers. Once you have used the substitution method to solve for one of the variables, be sure to complete the problem and find the value of the other variable.

EXAMPLE 2

Use the substitution method to solve $\begin{aligned} x &= 6 - 2y \\ 4x - 3y &= 2 \end{aligned}$.

Solution: The first equation states that *x is the same as* $(6 - 2y)$. Thus, we can *substitute* $(6 - 2y)$ for *x* in the second equation and solve for *y* algebraically as follows:

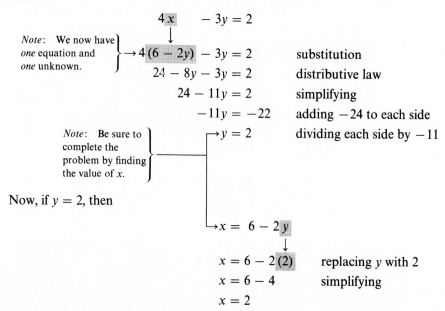

$$4\,x \quad - 3y = 2$$

Note: We now have *one* equation and *one* unknown.

$\rightarrow 4\,(6 - 2y) - 3y = 2$ substitution

$24 - 8y - 3y = 2$ distributive law

$24 - 11y = 2$ simplifying

$-11y = -22$ adding -24 to each side

Note: Be sure to complete the problem by finding the value of *x*.

$\rightarrow y = 2$ dividing each side by -11

Now, if $y = 2$, then

$\hookrightarrow x = 6 - 2\,y$

$x = 6 - 2\,(2)$ replacing *y* with 2

$x = 6 - 4$ simplifying

$x = 2$

Thus, $(2, 2)$ is the solution to this system.   ■

Note: Check to see if $(2, 2)$ satisfies *both* equations.

x-value ——⤴ ⤴—— *y*-value

⬚ More on the Substitution Method

If neither of the equations in a linear system contains an isolated variable, we must solve one of the equations for either *x* or *y* before the substitution method can be applied. If one of the variables in either of the equations has a coefficient of 1, it is easiest to solve for that variable and avoid unnecessary computation with fractions.

EXAMPLE 3

Use the substitution method to solve $\begin{array}{l} 6x + 9y = 5 \\ x + y = 1 \end{array}$.

Solution: It is best to solve the equation $x + y = 1$ for either x or y since each of these variables has a coefficient of 1. Let's solve for y.

$$x + y = 1$$
$$\boxed{y = 1 - x} \qquad \text{subtracting } x \text{ from each side}$$

Since y *and* $(1 - x)$ *are the same*, we can *substitute* $(1 - x)$ for y in the *other* equation, $6x + 9y = 5$, as follows:

$$6x + 9\,\boxed{y} = 5$$
$$\downarrow$$

$$6x + 9\,\boxed{(1 - x)} = 5 \qquad\qquad \text{substitution}$$

$$6x + 9 - 9x = 5 \qquad\qquad \text{distributive law}$$

$$9 - 3x = 5 \qquad\qquad \text{simplifying}$$

$$-3x = -4 \qquad\qquad \text{adding } -9 \text{ to each side}$$

$$x = \frac{4}{3} \quad \text{or} \quad 1\frac{1}{3} \qquad\qquad \text{dividing each side by } -3$$

Note: Be sure to complete the problem by finding the value of y.

Now, if $x = \dfrac{4}{3}$, then

$$y = 1 - x$$

$$y = 1 - \frac{4}{3} \qquad\qquad \text{replacing } x \text{ with } \frac{4}{3}$$

$$y = \frac{3}{3} - \frac{4}{3} \qquad\qquad \text{fundamental property of fractions}$$

$$y = -\frac{1}{3} \qquad\qquad \text{subtraction rule for fractions}$$

Thus, $\left(\dfrac{4}{3}, -\dfrac{1}{3}\right)$ is the solution to this system.   ∎

Note: Try solving the equation $x + y = 1$ for x and substituting this quantity in the other equation, $6x + 9y = 5$. See if you can obtain the same ordered pair.

EXAMPLE 4

Use the substitution method to solve $\begin{array}{l} 4x + 5y = -11 \\ 3x - 4y = 15 \end{array}$.

Solution: No variable has a coefficient of 1. Thus, no matter what variable we solve for, a *fractional substitution* will result. Let's solve $3x - 4y = 15$ for x.

$$3x - 4y = 15$$

$$3x = 15 + 4y \qquad \text{adding } 4y \text{ to each side}$$

$$x = \frac{15 + 4y}{3} \qquad \text{dividing each side by 3}$$

$$x = 5 + \frac{4y}{3} \qquad \text{writing as a mixed expression}$$

Since x *and* $\left(5 + \dfrac{4y}{3} \right)$ *are the same*, we can *substitute* $\left(5 + \dfrac{4y}{3} \right)$ for x in the *other* equation, $4x + 5y = -11$, as follows:

$$4\,x \qquad + 5y = -11$$

$$4\left(5 + \frac{4y}{3} \right) + 5y = -11 \qquad \text{substitution}$$

$$20 + \frac{16y}{3} + 5y = -11 \qquad \text{distributive law}$$

$$20 + \frac{16y}{3} + \frac{15y}{3} = -11 \qquad \text{fundamental property of fractions}$$

$$20 + \frac{31y}{3} = -11 \qquad \text{simplifying}$$

$$\frac{31y}{3} = -31 \qquad \text{adding } -20 \text{ to each side}$$

$$y = -3 \qquad \text{multiplying each side by } \frac{3}{31}$$

Now, if $y = -3$, then

$$x = 5 + \frac{4y}{3}$$

$$x = 5 + \frac{4(-3)}{3} \qquad \text{replacing } y \text{ with } -3$$

$$x = 5 + (-4) \qquad \text{simplifying}$$

$$x = 1$$

Thus, $(1, -3)$ is the solution to this system. ■

EXAMPLE 5

Use the substitution method to solve $\begin{array}{l} 6x - 2y = -1 \\ -9x + 3y = \dfrac{3}{2} \end{array}$.

Solution: No variable has a coefficient of 1. Thus, no matter what variable we solve for, a *fractional substitution* will result. Let's solve $6x - 2y = -1$ for y.

$$6x - 2y = -1$$

$$-2y = -1 - 6x \qquad \text{adding } -6x \text{ to each side}$$

$$y = \frac{-1 - 6x}{-2} \qquad \text{dividing each side by } -2$$

$$\boxed{y = \frac{1}{2} + 3x} \qquad \text{writing as a mixed expression}$$

Since y and $\left(\dfrac{1}{2} + 3x\right)$ *are the same*, we can *substitute* $\left(\dfrac{1}{2} + 3x\right)$ for y in the *other* equation, $-9x + 3y = \dfrac{3}{2}$, as follows:

$$-9x + \quad 3\,\underset{\downarrow}{y} \quad = \frac{3}{2}$$

$$-9x + 3\left(\frac{1}{2} + 3x\right) = \frac{3}{2} \qquad \text{substitution}$$

$$-9x + \frac{3}{2} + 9x = \frac{3}{2} \qquad \text{distributive law}$$

Note: The variable terms drop out, and we obtain a *true* statement.

$$\frac{3}{2} = \frac{3}{2} \qquad \text{simplifying}$$

To understand what's happening, write each equation in the form $y = mx + b$.

$$6x - 2y = -1 \xrightarrow{\text{Solve for } y.} y = \boxed{3}x + \boxed{\frac{1}{2}} \leftarrow \left\{ \begin{array}{l} Note: \quad \text{The} \\ y\text{-intercepts} \\ \text{are the } same. \end{array} \right.$$

$$-9x + 3y = \frac{3}{2} \xrightarrow[\text{Solve for } y.]{} y = \boxed{3}x + \boxed{\frac{1}{2}}$$

$$\left\{ \begin{array}{l} Note: \quad \text{The slopes} \\ \text{are the } same. \end{array} \right.$$

Since the slopes and the y-intercepts are the same, the lines coincide. Thus, this system has an *infinite number of solutions*. Any ordered pair that satisfies one of the equations satisfies the other. ∎

Note: Remember the following when solving a two by two linear system:

If *both variables drop out* and the resulting statement is *true*, $\xrightarrow{\text{then}}$ the lines coincide and the system has an *infinite number of solutions*.

If *both variables drop out* and the resulting statement is *false*, $\xrightarrow{\text{then}}$ the lines are parallel and the system has *no solution*.

Application

BEAM ANALYSIS

In Section 4.5 the force law and moment law of equilibrium were discussed. The force law of equilibrium states that if an object is at rest (in equilibrium), then the algebraic *sum* of all the forces acting on that object must be zero. The moment law of equilibrium states that the *sum* of all moments acting on an object at rest (in equilibrium) must also be equal to zero. The diagram showing the beam in Fig. 10.7 is accompanied by a force diagram. The load on the beam represents a downward force, and upward forces are exerted by each column in order to hold up this load. (We will assume that the beam is of negligible weight).

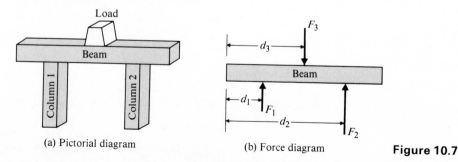

(a) Pictorial diagram   (b) Force diagram   **Figure 10.7**

If the beam in Fig. 10.7 is in equilibrium, two conditions must hold:

1. Force law of equilibrium $\Sigma F = 0$

    Note: Upward forces are considered positive and downward forces negative. $\Big\} \longrightarrow \quad F_1 + F_2 + (-F_3) = 0$

    or $F_1 + F_2 = F_3$

2. Moment law of equilibrium $\Sigma M = 0$

    Note: The moment of force is the product of the force and the distance to the force. $\Big\} \longrightarrow \quad F_1 d_1 + F_2 d_2 + (-F_3 d_3) = 0$

    or $F_1 d_1 + F_2 d_2 = F_3 d_3$

When the equations $F_1 + F_2 = F_3$ and $F_1 d_1 + F_2 d_2 = F_3 d_3$ are taken together, they form a *system* of equations. Such systems can be used to analyze the forces acting on a beam.

EXAMPLE 6

If the beam in Fig. 10.8 is in equilibrium, determine the forces F_1 and F_2.

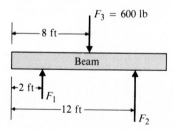

Figure 10.8

Solution:

$$F_1 + F_2 = F_3 \quad \rightarrow F_1 + F_2 \quad = 600 \qquad \text{replacing } F_3 \text{ with } 600$$
$$F_1 d_1 + F_2 d_2 = F_3 d_3 \rightarrow \underline{2F_1 + 12F_2 = 4800} \qquad \text{replacing } d_1 \text{ with } 2, d_2 \text{ with } 12,$$
$$\text{and } F_3 d_3 \text{ with } (600)(8) = 4800.$$

Note: This is a two by two linear system. $\Bigg\}$

Now, let's solve $F_1 + F_2 = 600$ for F_1 and use the substitution method.

$$F_1 + F_2 = 600$$
$$F_1 = 600 - F_2 \qquad \text{adding } -F_2 \text{ to each side}$$

Since F_1 and $(600 - F_2)$ *are the same,* we can *substitute* $(600 - F_2)$ for F_1 in the other equation, $2F_1 + 12F_2 = 4800$, as follows:

$$2\,\underset{\downarrow}{F_1} \quad + 12F_2 = 4800$$

$$2\,(600 - F_2) + 12F_2 = 4800 \qquad \text{substitution}$$

$$1200 - 2F_2 + 12F_2 = 4800 \qquad \text{distributive law}$$

$$1200 + 10F_2 = 4800 \qquad \text{simplifying}$$

$$10F_2 = 3600 \qquad \text{adding } -1200 \text{ to each side}$$

$$F_2 = 360 \text{ lb} \qquad \text{dividing each side by 10}$$

Now, if $F_2 = 360$ lb, then $F_1 = 600 - F_2$.

$$F_1 = 600 - (360) \qquad \text{replacing } F_2 \text{ with } 360$$

$$F_1 = 240 \text{ lb} \qquad \text{simplifying}$$

Thus, the upward forces that are exerted by each column are $F_1 = 240$ lb and $F_2 = 360$ lb. ∎

EXERCISES 10.2

Use the substitution method to solve the following linear systems. Be sure to check your answers.

1. $4x - 3y = 8$
  $y = 4$

2. $5x - 2y = 4$
  $x = -2$

3. $3x + y = 8$
  $x = 5y$

4. $3x - 2y = 14$
  $y = -2x$

5. $y = 3x - 5$
  $2x + y = 5$

6. $x = 4 - 3y$
  $x + 5y = 3$

7. $y = 9 - x$
  $4x - y = 6$

8. $y = 7x - 2$
  $5x - y = 8$

9. $x = 3y + 5$
  $2x - 6y = 10$

10. $x = 4 - 2y$
  $3x + 6y = 5$

11. $x - 5 = 0$
  $4x - 3y = -1$

12. $3x + 5y = 9$
  $x + y = -1$

13. $3x + 5y = 4$
  $x - y = -6$

14. $4x - 7y = 12$
  $x - y = 4$

15. $2x - 3y = 2$
  $8x - y = -3$

16. $-2x - y = -15$
  $x + 3y = 8$

17. $2x + 3y = 5$
  $3x - 2y = -12$

18. $5x + 2y = 11$
  $2x - 5y = 16$

19. $-3x + 4y = 12$
     $3x + 5y = 15$

20. $7x - 2y = 14$
     $9x + 2y = 18$

21. $3x - 6y = 1$
     $5x - 10y = 2$

22. $8x - 2y = -5$

     $-12x + 3y = \dfrac{15}{2}$

23. $3x + 7 = 2y + 4x$
     $3x + 5y = 7 + x$

24. $x - 6 = 4x - 3y$
     $5x + 8 = 4y + 2$

25. $3x - 4(x + y) = 0$
     $2x + 2(x - y) = 9$

26. $2(x - y) + 8y = 1$
     $4(2x + y) = 7x - y$

If each of the following beams is in equilibrium, determine the forces F_1 and F_2.

27.

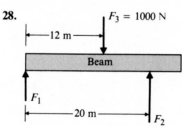

28.

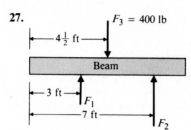

29.

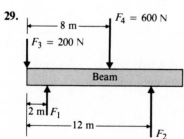

30.

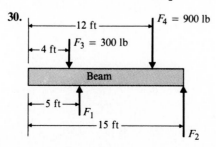

10.3   SOLVING SYSTEMS OF LINEAR EQUATIONS BY THE ADDITION METHOD

Objectives

▣ To solve a two by two linear system by using the addition method.

▣ To solve a two by two linear system by using the addition method when the multiplication property of equality must be applied first.

▣ The Addition Method

The substitution method (see Section 10.2) works well provided we can solve for x or y without introducing a *fractional substitution*. To avoid such fractional substitutions, we will develop another method for solving a two by two linear system. This new method, called the **addition method**, is a direct consequence of the *addition property of equality*.

The addition property of equality was introduced in Section 4.2. It states that if we add the *same* quantity to each side of an equation, the new equation formed is equivalent to the original equation. This property can also be stated as follows:

ADDITION PROPERTY OF EQUALITY
If
$$A = B,$$
and
$$C = D,$$
then
$$A + C = B + D.$$

The equation $C = D$ states that C is the *same* as D. Thus, by adding C to the left side and D to the right side of the equation $A = B$, we have actually added the *same* quantity to each side.

EXAMPLE 1

Use the addition method to solve $\begin{array}{l} 3x + 2y = 7 \\ 5x - 2y = 9 \end{array}$.

Solution: Think of $3x + 2y = 7$ as $A = B$,

and $5x - 2y = 9$ as $C = D$.

The addition property of equality states that

$$\text{if } A = B,$$
$$\text{and } C = D,$$

then A $+$ C $= B + D.$

$$(3x + 2y) + (5x - 2y) = 7 + 9 \qquad \text{addition property of equality}$$

Note: By using the addition method, we have eliminated the y terms. We now have *one* equation and *one* unknown, which can be solved in the usual manner.

$$8x + 0y = 16 \qquad \text{combining similar terms on the same side of the equation}$$

$$8x = 16 \qquad \text{simplifying}$$
$$x = 2 \qquad \text{dividing each side by 8}$$

Now, if we know $x = 2$, then we can replace x with 2 in *either of the original equations* in order to find the value of y. Suppose we choose to replace x with 2 in the equation $3x + 2y = 7$.

$$3x + 2y = 7$$
$$3(2) + 2y = 7 \qquad \text{replacing } x \text{ with 2}$$
$$6 + 2y = 7 \qquad \text{simplifying}$$
$$2y = 1 \qquad \text{adding } -6 \text{ to each side}$$
$$y = \frac{1}{2} \qquad \text{dividing each side by 2}$$

Thus, $\left(2, \dfrac{1}{2}\right)$ is the solution to this system. ∎

Note: The addition method can be speeded up by placing one equation under the other and adding *vertically* as follows:

Note: The addition property of equality is being applied *vertically* instead of horizontally.

$$3x + 2y = 7$$
$$5x - 2y = 9$$
$$8x + 0y = 16 \qquad \text{addition method}$$
$$8x = 16 \qquad \text{simplifying}$$
$$x = 2 \qquad \text{dividing each side by 8}$$

EXAMPLE 2

Use the addition method to solve $\begin{aligned} -5x + 3y &= 12 \\ 5x - 6y &= -10 \end{aligned}.$

Solution:

$$-5x + 3y = 12$$
$$5x - 6y = -10$$
$$\overline{0x - 3y = 2} \qquad \text{addition method}$$

$$-3y = 2 \qquad \text{simplifying}$$

$$y = -\frac{2}{3} \qquad \text{dividing each side by } -3$$

Note: Be sure to complete the problem by replacing y with $-\frac{2}{3}$ in either of the original equations and then solving for x.

$$-5x + 3y = 12$$

$$-5x + 3\left(-\frac{2}{3}\right) = 12 \qquad \text{replacing } y \text{ with } -\frac{2}{3} \text{ in one of the original equations}$$

$$-5x - 2 = 12 \qquad \text{simplifying}$$

$$-5x = 14 \qquad \text{adding 2 to each side}$$

$$x = -\frac{14}{5} \qquad \text{dividing each side by } -5$$

Thus, $\left(-\frac{14}{5}, -\frac{2}{3}\right)$ is the solution to this system. ∎

Note: Check to see if $\left(-\frac{14}{5}, -\frac{2}{3}\right)$ satisfies *both* equations.

⊡ More on the Addition Method

Consider the linear system $\begin{array}{l} 5x + 2y = 8 \\ 3x + \ y = 3 \end{array}$.

If the addition method is applied, neither of the variables is eliminated.

Note: Although we have *one* equation, it contains *two* unknowns. We can't determine the value of either x or y.

$$5x + 2y = 8$$
$$3x + \ y = 3$$
$$\rightarrow \overline{8x + 3y = 11} \qquad \text{addition method}$$

The addition method will eliminate a variable only if the coefficients of that variable are *opposites*. To force opposite coefficients, we can apply the *multiplication property of equality* to one or both of the equations.

EXAMPLE 3

Use the addition method to solve $\begin{array}{l} 5x + 2y = 8 \\ 3x + \ y = 3 \end{array}$.

Solution: If we multiply both sides of the equation $3x + y = 3$ by -2, the coefficients of the y terms will be *opposites*.

$$5x + 2y = 8 \xrightarrow{\text{Leave alone.}} 5x + 2y = 8$$

$$3x + \ y = 3 \xrightarrow[\substack{\text{Multiply each} \\ \text{side by } -2.}]{} \begin{array}{l} -6x - 2y = -6 \\ \hline -x + 0y = 2 \\ -x = 2 \\ x = -2 \end{array}$$

multiplication property of equality
addition method
simplifying
multiplying each side by -1

Now, if $x = -2$, then

$$\begin{array}{l} 3x + y = 3 \\ 3(-2) + y = 3 \\ -6 + y = 3 \\ y = 9 \end{array}$$

replacing x with -2 in one of the original equations
simplifying
adding 6 to each side

Thus, $(-2, 9)$ is the solution to this system. ∎

Note: Alternately, we could have eliminated x first and solved for y as follows:

$$5x + 2y = 8 \xrightarrow[\substack{\text{side by 3.}}]{\text{Multiply each}} 15x + 6y = 24$$

multiplication property of equality

$$3x + \ y = 3 \xrightarrow[\substack{\text{Multiply each} \\ \text{side by } -5.}]{} \begin{array}{l} -15x - 5y = -15 \\ \hline 0x + \ y = 9 \\ y = 9 \end{array}$$

multiplication property of equality
addition method
simplifying

EXAMPLE 4

Use the addition method to solve $\begin{array}{l} 9x - \ 8y = 4 \\ 12x - 12y = 5 \end{array}$.

Solution: Suppose we choose to eliminate x first and solve for y. The smallest number divisible by both 9 and 12 is *36*. Thus, multiplying the first equation by *4* and the second equation by -3 forces the x terms to have opposite coefficients.

$$9x - 8y = 4 \xrightarrow{\substack{\text{Multiply each} \\ \text{side by 4.}}} 36x - 32y = 16 \quad \text{multiplication property of equality}$$

$$12x - 12y = 5 \xrightarrow[\substack{\text{Multiply each} \\ \text{side by } -3.}]{} \underline{-36x + 36y = -15} \quad \substack{\text{multiplication property} \\ \text{of equality}}$$

$$0x + 4y = 1 \quad \text{addition method}$$

$$4y = 1 \quad \text{simplifying}$$

$$y = \frac{1}{4} \quad \text{dividing each side by 4}$$

Now, if $y = \dfrac{1}{4}$, then

$$9x - 8y = 4$$

$$9x - 8\left(\frac{1}{4}\right) = 4 \quad \text{replacing } y \text{ with } \frac{1}{4} \text{ in one}$$

of the original equations

$$9x - 2 = 4 \quad \text{simplifying}$$

$$9x = 6 \quad \text{adding 2 to each side}$$

$$x = \frac{6}{9} \quad \text{dividing each side by 9}$$

$$x = \frac{2}{3} \quad \text{cancellation law}$$

Thus, $\left(\dfrac{2}{3}, \dfrac{1}{4}\right)$ is the solution to this system. ∎

Note: Try eliminating y first and solving for x. See if you can obtain the same ordered pair.

EXAMPLE 5

Use the addition method to solve $\begin{array}{l} 6y = 3x - 5 \\ 2x - 4y = 1 \end{array}$.

Solution: We must first write $6y = 3x - 5$ in the form $Ax + By = C$.

$$6y = 3x - 5 \xrightarrow{\substack{\text{Add } -3x \text{ to} \\ \text{each side.}}} -3x + 6y = -5$$

$$2x - 4y = 1 \xrightarrow[\text{Leave alone.}]{} 2x - 4y = 1$$

Now, let's choose to eliminate x first and solve for y.

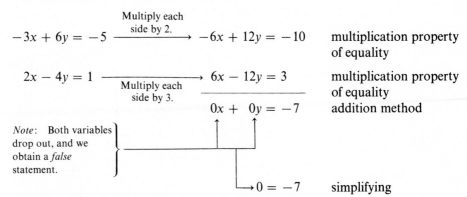

$$-3x + 6y = -5 \xrightarrow{\text{Multiply each side by 2.}} -6x + 12y = -10 \qquad \text{multiplication property of equality}$$

$$2x - 4y = 1 \xrightarrow[\text{Multiply each side by 3.}]{} 6x - 12y = 3 \qquad \text{multiplication property of equality}$$

$$\underline{\phantom{2x - 4y = 1 \xrightarrow{} 6x - 12y = 3}}$$

$$0x + 0y = -7 \qquad \text{addition method}$$

Note: Both variables drop out, and we obtain a *false* statement.

$$0 = -7 \qquad \text{simplifying}$$

If both variables drop out and the resulting statement is *false*, the lines are parallel and the system has *no solution*. ■

Application

CIRCUIT ANALYSIS

In Section 4.4 Kirchhoff's voltage law was discussed. It states that in a closed loop, the *sum* of the voltage drops across the electrical resistances must equal the voltage supply in that loop. The diagram showing the circuit in Fig. 10.9 is accompanied by a schematic diagram. Notice that there are *two* closed loops in this circuit. We will assume that the currents are flowing in the directions shown. Since both currents I_1 and I_2 are flowing in the *same* direction through R_3, the total current through R_3 must be the *sum* of the currents I_1 and I_2.

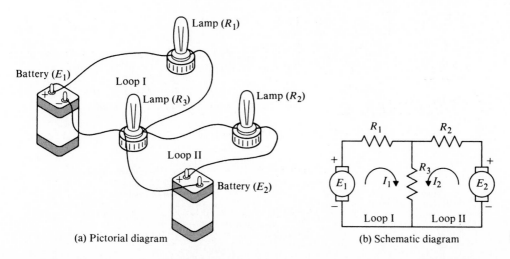

(a) Pictorial diagram

(b) Schematic diagram

Figure 10.9

According to Kirchhoff's voltage law, two conditions must hold for this circuit.

1. Loop I

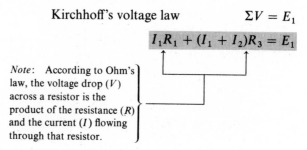

Kirchhoff's voltage law $\Sigma V = E_1$

$$I_1 R_1 + (I_1 + I_2)R_3 = E_1$$

Note: According to Ohm's law, the voltage drop (V) across a resistor is the product of the resistance (R) and the current (I) flowing through that resistor.

2. Loop II

Kirchhoff's voltage law $\Sigma V = E_2$

$$I_2 R_2 + (I_1 + I_2)R_3 = E_2$$

When the equations $I_1 R_1 + (I_1 + I_2)R_3 = E_1$ and $I_2 R_2 + (I_1 + I_2)R_3 = E_2$ are taken together, they form a *system* of equations. Such systems can be used to analyze the currents in multiloop circuits.

EXAMPLE 6

Given the circuit in Fig. 10.10, determine the currents I_1 and I_2.

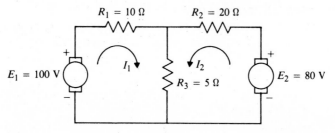

Figure 10.10

Solution: First, replace $R_1, R_2, R_3, E_1,$ and E_2 with their given values.

$$I_1 R_1 + (I_1 + I_2)R_3 = E_1 \rightarrow 10I_1 + 5(I_1 + I_2) = 100$$
$$I_2 R_2 + (I_1 + I_2)R_3 = E_2 \rightarrow 20I_2 + 5(I_1 + I_2) = 80$$

Next, write these equations in the form $Ax +\quad By = C.$

$$10I_1 + 5(I_1 + I_2) = 100 \rightarrow 15I_1 +\quad 5I_2 = 100$$
$$20I_2 + 5(I_1 + I_2) = 80\quad \rightarrow\quad 5I_1 + 25I_2 = 80$$

Note: This is a two by two linear system.

Let's choose to eliminate I_1 first and solve for I_2 by using the addition method.

$$15I_1 + 5I_2 = 100 \xrightarrow{\text{Leave alone.}} 15I_1 + 5I_2 = 100$$

$$5I_1 + 25I_2 = 80 \xrightarrow[\substack{\text{Multiply each} \\ \text{side by } -3.}]{} -15I_1 - 75I_2 = -240$$

$$-70I_2 = -140 \qquad \text{addition method}$$

$$I_2 = 2 \qquad \begin{array}{l}\text{dividing each side} \\ \text{by } -70\end{array}$$

Now, if $I_2 = 2$ A, then

$$5I_1 + 25I_2 = 80.$$

$$5I_1 + 25(2) = 80 \qquad \text{replacing } I_2 \text{ with 2}$$

$$5I_1 + 50 = 80 \qquad \text{simplifying}$$

$$5I_1 = 30 \qquad \text{adding } -50 \text{ to each side}$$

$$I_1 = 6 \qquad \text{dividing each side by 5}$$

Thus, the currents in each loop are $I_1 = 6$ A and $I_2 = 2$ A. ∎

EXERCISES 10.3

Use the addition method to solve the following linear systems. Be sure to check your answers.

1. $x + y = 9$
$x - y = 5$

2. $2x - y = 8$
$x + y = 7$

3. $2x - 3y = 5$
$4x + 3y = 7$

4. $7x + 8y = 3$
$x - 8y = 1$

5. $4x + 3y = -9$
$-4x + y = 3$

6. $-2x - 3y = 3$
$2x - 9y = -5$

7. $x - 5y = 12$
$7x + 5y = -2$

8. $3x - 2y = -7$
$6x + 2y = -5$

9. $3x + y = 7$
$5x + y = 3$

10. $4x - 3y = 8$
$4x - 2y = 7$

11. $2x - 5y = 8$
$6x + y = 0$

12. $2x + 7y = -9$
$5x + y = -6$

13. $3x - 3y = 10$
$5x + 6y = 2$

14. $2x + 4y = 17$
$4x - 2y = 9$

15. $8x + 3y = 1$
$6x + 2y = 0$

16. $7x - 9y = -3$
$3x - 6y = 3$

17. $15x - 8y = 20$
$9x - 10y = -14$

18. $6x + 8y = 7$
$16x + 12y = 0$

19. $y = 6x - 4$
$2x + 3y = 18$

20. $5x - 2y = 10$
$3x = 12 - 6y$

21. $12x - 3y = 8$
$y = 4x - 1$

22. $4y = x + 6$
$3x + 18 = 12y$

23. $3x - 2y = x + 5$
$2x + 4y = 4 - x$

24. $5x - 4y = y - 4$
$3x - 5y = 2(1 - y)$

25. $3x - (x - y) = 9$
$6x - 2(x + y) = 18 - 5y$

26. $3x + 5(x + y) = 12$
$2x - 4(x - y) = -3$

In each of the following circuits, determine the currents I_1 and I_2.

27.

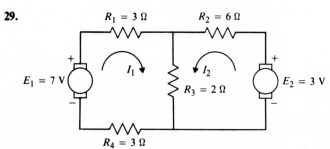

28.

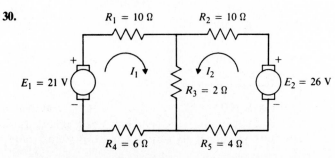

29.

30.

10.4 VERBAL PROBLEMS CONTAINING TWO UNKNOWNS

Objectives

- ☐ To solve various types of verbal problems that contain two unknowns by using a two by two linear system.

☐ Verbal Problems

In Chapter 4 you solved verbal problems containing two unknowns by assigning one of the unknowns the variable x and recording the other unknown in terms of x. Using this method, it was possible to develop *one* equation with *one* unknown and to solve for x. However, many verbal problems containing two unknowns can more naturally be set up and solved by using *two* equations and *two* unknowns.

FOUR-STEP PROCEDURE FOR SOLVING A VERBAL PROBLEM CONTAINING TWO UNKNOWNS

1. SETUP   Carefully read the problem at least twice: the first time for a general overview, and the second (and possibly third) time to determine which two quantities the problem is asking you to find. Let x represent one of these unknown quantities and y the other. Draw a diagram when appropriate.

2. SYSTEM   From the information given in the problem, develop *two* equations containing the *two* unknowns x and y.

3. SOLUTION   Use the *substitution method* or the *addition method* to solve the two by two linear system developed in Step 2.

4. CONCLUSION   Carefully state your conclusion. Be sure it seems reasonable and logically satisfies the conditions of the problem.

EXAMPLE 1

Two hundred twenty-five feet of fencing are needed to fence all four sides of a rectangular vegetable garden. The width of the garden is 42.7 ft less than the length. What are the dimensions of the garden?

Solution:

1. SETUP   The *two unknowns* are the length and width of the rectangular garden. Thus,

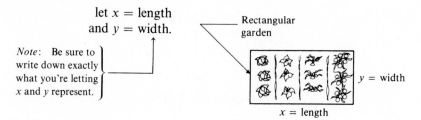

let x = length
and y = width.

Note: Be sure to write down exactly what you're letting x and y represent.

Rectangular garden

y = width

x = length

2. SYSTEM   If 225 ft of fencing are needed to fence all four sides of the rectangular garden, then

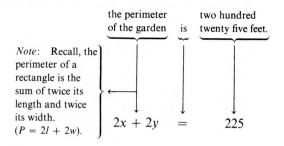

the perimeter of the garden   is   two hundred twenty five feet.

Note: Recall, the perimeter of a rectangle is the sum of twice its length and twice its width.
$(P = 2l + 2w)$.

$$2x + 2y \quad = \quad 225$$

Also,

the width of the garden   is   42.7 ft less than the length.

$$y \quad = \quad x - 42.7$$

Thus, the *system* is $\begin{aligned} 2x + 2y &= 225 \\ y &= x - 42.7 \end{aligned}$

3. SOLUTION   This system is easily solved by the substitution method.

$$2x + \quad 2\,y \quad\quad = 225$$

$$2x + 2\,(x - 42.7) = 225 \qquad \text{substitution}$$

$$2x + 2x - 85.4 = 225 \qquad \text{distributive law}$$

$$4x - 85.4 = 225 \qquad \text{simplifying}$$

$$4x = 310.4 \qquad \text{adding 85.4 to each side}$$

$$x = 77.6 \qquad \text{dividing each side by 4}$$

Now, if $x = 77.6$, then

$$y = x - 42.7$$
$$y = (77.6) - 42.7$$
$$y = 34.9$$

4. CONCLUSION

$$x = \text{length} = 77.6 \text{ ft}$$
$$y = \text{width} = 34.9 \text{ ft} \quad \blacksquare$$

Note: Be sure to label your answers with the correct units.

Note: In Section 4.3 example 3, this same problem was solved using *one* equation and *one* unknown. It is a good idea to look back at Section 4.3 and compare the two methods for solving a verbal problem containing two unknowns.

EXAMPLE 2

The ratio of the number of pounds of tin to the number of pounds of lead in a certain type of solder is $8:5$. How many pounds of each metal are contained in 52 lb of this solder?

Solution:

1. SETUP

Let x = the number of pounds of *tin* in the 52 lb of solder

and y = the number of pounds of *lead* in the 52 lb of solder.

2. SYSTEM   If the total weight of the solder is 52 lb, then

the number of pounds of tin	plus	the number of pounds of lead	is	fifty two pounds.
x	$+$	y	$=$	52

Also,

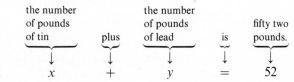

the ratio of the number of pounds of tin to the number of pounds of lead   is $8:5$

$$\frac{x}{y} = \frac{8}{5}$$

$$5x = 8y \qquad \text{cross product property}$$
$$5x - 8y = 0 \qquad \text{adding } -8y \text{ to each side}$$

Thus, the system is $\begin{array}{l} x + \;y = 52 \\ 5x - 8y = 0 \end{array}$.

3. SOLUTION   The *addition method* can be used to easily solve this system.

$$x + y = 52 \xrightarrow[\text{side by 8.}]{\text{Multiply each}} 8x + 8y = 416 \qquad \text{multiplication property of equality}$$

$$5x - 8y = 0 \xrightarrow{\text{Leave alone.}} 5x - 8y = 0$$

$$13x + 0y = 416 \qquad \text{addition method}$$
$$13x = 416 \qquad \text{simplifying}$$
$$x = 32 \qquad \text{dividing each side by 13}$$

Now, if $x = 32$, then

$$x + y = 52$$
$$(32) + y = 52$$
$$y = 20$$

4. CONCLUSION

$x =$ the number of pounds of tin in the 52 lb of solder $= 32$ lb

$y =$ the number of pounds of lead in the 52 lb of solder $= 20$ lb   ∎

Note: In Section 4.6 example 2, this same problem was solved using *one* equation and *one* unknown. Refer to Section 4.6 and compare the two methods for solving a verbal problem concerning ratios.

EXAMPLE 3

A chemist has a 5% sulfuric acid solution and a 9% sulfuric acid solution. How many liters of each must she use to make 400 L of a solution that is 6% sulfuric acid?

Solution:

1. SETUP   Let

$x =$ the amount of 5% sulfuric acid to be used

$y =$ the amount of 9% sulfuric acid to be used

2. SYSTEM   If the total amount is to be 400 L, then

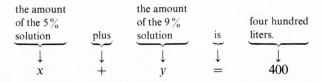

the amount of the 5% solution	plus	the amount of the 9% solution	is	four hundred liters.
↓	↓	↓	↓	↓
x	$+$	y	$=$	400

Also, the amount of *pure* sulfuric acid in the 400-L solution is

$$6\% \text{ of } 400 = \frac{6}{100} \cdot \frac{400}{1} = 24 \text{ L.}$$

Thus,

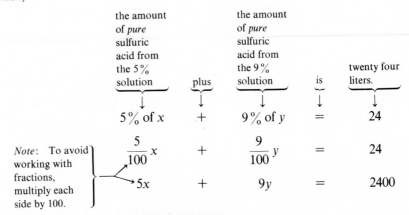

the amount of *pure* sulfuric acid from the 5% solution	plus	the amount of *pure* sulfuric acid from the 9% solution	is	twenty four liters.
↓	↓	↓	↓	↓
5% of x	$+$	9% of y	$=$	24
$\dfrac{5}{100}x$	$+$	$\dfrac{9}{100}y$	$=$	24
$5x$	$+$	$9y$	$=$	2400

Note: To avoid working with fractions, multiply each side by 100.

Therefore, the system is $\begin{array}{l} x + y = 400 \\ 5x + 9y = 2400 \end{array}$.

3. SOLUTION   The addition method can be used to easily solve this system.

$$x + y = 400 \xrightarrow[\text{side by } -5.]{\text{Multiply each}} -5x - 5y = -2000 \qquad \begin{array}{l}\text{multiplication}\\ \text{property of equality}\end{array}$$

$$5x + 9y = 2400 \xrightarrow[\text{Leave alone.}]{} \underline{5x + 9y = 2400}$$

$$\begin{array}{ll} 0x + 4y = 400 & \text{addition method} \\ 4y = 400 & \text{simplifying} \\ y = 100 & \text{dividing each side by 4} \end{array}$$

Now, if $y = 100$, then

$$x + y = 400$$
$$x + (100) = 400$$
$$x = 300$$

4. CONCLUSION

$x =$ the amount of 5% sulfuric acid to be used $= 300$ L

$y =$ the amount of 9% sulfuric acid to be used $= 100$ L ■

Note: You should always verify that the CONCLUSION satisfies the conditions of the problem.

1. Is the total amount 400 L?

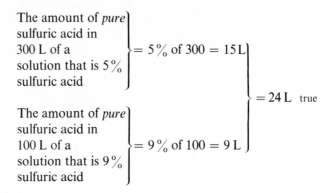

$$300\,\text{L} + 100\,\text{L} = 400\,\text{L} \quad \text{true}$$

amount of 5% solution to be used — amount of 9% solution to be used

2. Does the 400-L solution contain 24 L of *pure* sulfuric acid (that is, is it 6% sulfuric acid)?

The amount of *pure* sulfuric acid in 300 L of a solution that is 5% sulfuric acid $\Big\} = 5\%$ of $300 = 15\,\text{L}$

$\Big\} = 24\,\text{L}$ true

The amount of *pure* sulfuric acid in 100 L of a solution that is 9% sulfuric acid $\Big\} = 9\%$ of $100 = 9\,\text{L}$

EXAMPLE 4

It takes a canoeist 3 h to travel 15 mi upstream to campsite A. Paddling at the same rate, it takes her 2 h to travel 14 mi downstream to campsite B. Find the rate of speed at which she paddles and the rate of speed of the river's current.

Solution:

1. SETUP   Let

x = the rate of speed at which she paddles the canoe

y = the rate of speed of the river's current

2. SYSTEM   To organize the information in this problem it helps to set up a table as follows:

	Distance traveled (*d*)	Rate of speed (*r*)	Time (*t*)
upstream	15	$x - y$	3
downstream	14	$x + y$	2

Note: When traveling *upstream*, the current *hinders* the speed of the canoe. Thus, the actual speed is the speed at which she paddles (x) *minus* the speed of the current (y).

Note: When traveling *downstream*, the current *aids* the speed of the canoe. Thus, the actual speed is the speed at which she paddles (x) *plus* the speed of the current (y).

In order to develop the system of equations, it is necessary to know the relationship between distance d, rate of speed r, and time t. The relationship is given by the formula $d = rt$. Using the information gathered in the table and applying the formula

$$d \ = \ r \ \ t,$$

we have

$$15 = (x - y) \ 3 \rightarrow 3x - 3y = 15$$
$$14 = (x + y) \ 2 \rightarrow 2x + 2y = 14.$$

the system

3. SOLUTION   To solve, let's use the addition method.

$$3x - 3y = 15 \xrightarrow{\text{Multiply each side by 2.}} 6x - 6y = 30 \qquad \text{multiplication property of equality}$$

$$2x + 2y = 14 \xrightarrow[\text{Multiply each side by 3.}]{} 6x + 6y = 42$$

$$12x + 0y = 72 \qquad \text{addition method}$$
$$12x = 72 \qquad \text{simplifying}$$
$$x = 6 \qquad \text{dividing each side by 12}$$

Now, if $x = 6$, then

$$2x + 2y = 14$$
$$2(6) + 2y = 14$$
$$12 + 2y = 14$$
$$2y = 2$$
$$y = 1$$

4. CONCLUSION

x = the rate of speed at which she paddles the canoe = 6 mph

y = the rate of speed of the river's current = 1 mph   ∎

Note: Using the formula $d = rt$, see if you can verify that the conclusion satisfies the conditions of the problem.

EXERCISES 10.4

Solve each of the following verbal problems by using a two by two linear system.

1. The perimeter of an isosceles triangle is 59.6 cm. The nonequal side measures 15.4 cm less than each of the equal sides. What are the dimensions of the triangle?

2. The perimeter of a rectangle is 9820 km. The length is 2950 km more than the width. What are the dimensions of the rectangle?

3. An 8.0-ft board is cut into two pieces. The larger piece is 3.2 ft longer than twice the shorter piece. What are the lengths of each piece?

4. When two resistors are connected in series, their total resistance is 68.0 Ω. The smaller resistance is 6.4 Ω less than half the larger resistance. What is the value of each resistor?

5. The fuel mixture for a certain two-cycle chain saw engine requires that the ratio of gasoline to oil be 15:1. How many liters of gas and oil are there in a 4-L mixture?

6. In a step-up transformer, the ratio of the number of turns of wire in the primary coil to the number of turns of wire in the secondary coil is 1:160. If there are 15,295 turns of wire altogether, then how many turns of wire are there in each coil?

7. A dry mixture of concrete is made up of cement and gravel in the ratio 2:3 by volume. How many cubic yards of each must be used to make 42 yd^3 of this mixture?

8. The ratio of the number of male students to the number of female students at a certain college is 9:5. If there are 3990 students altogether, how many are male and how many are female?

9. A bartender has a vermouth containing 10% alcohol and a gin containing 40% alcohol. How many liters of each must be used to make a 4-L mixture of martinis that is 35% alcohol?

10. A chemist has a 20% insecticide solution and a 10% insecticide solution. How many gallons of each must be used to make 50 gal of a 16% insecticide solution?

11. An airplane travels 1200 mi in 3 h with a tail wind, and returns in 4 h traveling against a head wind. What is the speed of the airplane and the speed of the wind?

12. A motorboat can travel 20 mi upstream in 4 h and 22 mi downstream in 2 h. What is the speed of the motorboat and the speed of the river's current?

13. A builder purchased a 40-Ac piece of land for $120,000. Part of the land was forest and part of the land was swamp. He agreed to pay $10,000/Ac for buildable forest land and $2000/Ac for nonbuildable swamp land. How many acres of forest and how many acres of swamp did he purchase?

14. A total of $10,000 was invested in term certificates. Part of the money was invested at 10% and part of the money was invested at 12%. The total annual interest was $1090. How much was invested at each rate?

15. A car radiator contains 15 qt of a 20% antifreeze solution. To guard against freezing, it should be increased to a 60% antifreeze solution. How much 20% antifreeze should be drained and replaced with *pure* antifreeze to achieve a 60% solution?

16. A small aircraft leaves an airport at noon and flies due north at 200 mph. At 2 P.M. a jet leaves from the same airport and flies due north at 800 mph. How many hours is each plane in flight when the jet overtakes the small aircraft?

17. The graph of the line $y = mx + b$ passes through the point (x, y), provided that the coordinates of that point satisfy the equation $y = mx + b$. Determine the slope m and the y-intercept b for a line that passes through the points $(3, 2)$ and $(-6, 5)$.

18. Referring to exercise 17, determine the slope m and the y-intercept b for a line that passes through the points $(-7, -2)$ and $(3, 6)$.

10.5   AN INTRODUCTION TO DETERMINANTS

Objectives

- ☐ To evaluate a two by two determinant.
- ☐ To use Cramer's rule to solve a two by two linear system.

☐ Two by Two Determinants

In engineering technology, determinants are often used to solve systems of linear equations. A **determinant** is a square array of numbers flanked by two vertical lines. A **two by two determinant** has two rows and two columns (see Fig. 10.11).

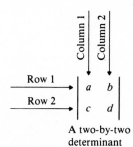

A two-by-two
determinant   **Figure 10.11**

To evaluate a two by two determinant, we multiply along the diagonals and subtract the results as follows:

$$\begin{vmatrix} a & b \\ c & d \end{vmatrix} = ad - cb$$

EXAMPLE 1

Evaluate the following determinants.

a) $\begin{vmatrix} 7 & -1 \\ 3 & 4 \end{vmatrix}$ b) $\begin{vmatrix} 2 & 0 \\ 4 & -5 \end{vmatrix}$ c) $\begin{vmatrix} -4 & 8 \\ 3 & -6 \end{vmatrix}$

Solution:

a) $\begin{vmatrix} 7 & -1 \\ 3 & 4 \end{vmatrix} = (7)(4) - (3)(-1) = 28 - (-3) = 31$ ∎

b) $\begin{vmatrix} 2 & 0 \\ 4 & -5 \end{vmatrix} = (2)(-5) - (4)(0) = -10 - 0 = -10$ ∎

c) $\begin{vmatrix} -4 & 8 \\ 3 & -6 \end{vmatrix} = (-4)(-6) - (3)(8) = 24 - 24 = 0$ ∎

Note: A determinant is simply a symbol that stands for a particular number.

⬛ **Cramer's Rule**

To show how determinants can be used to solve a two by two linear system, it is necessary to solve the general linear system

$$a_1 x + b_1 y = c_1$$
$$a_2 x + b_2 y = c_2$$

and observe the results. First, let's solve for x by eliminating y as follows:

$$a_1 x + b_1 y = c_1 \xrightarrow[\text{side by } b_2.]{\text{Multiply each}} a_1 b_2 x + b_1 b_2 y = c_1 b_2 \qquad \text{multiplication property of equality}$$

$$a_2 x + b_2 y = c_2 \xrightarrow[\text{side by } -b_1.]{\text{Multiply each}} -a_2 b_1 x - b_1 b_2 y = -c_2 b_1$$

$$\overline{\quad a_1 b_2 x - a_2 b_1 x = c_1 b_2 - c_2 b_1 \quad} \qquad \text{addition method}$$

$$x(a_1 b_2 - a_2 b_1) = c_1 b_2 - c_2 b_1 \qquad \text{factoring}$$

$$x = \frac{c_1 b_2 - c_2 b_1}{a_1 b_2 - a_2 b_1} \qquad \begin{array}{l}\text{dividing each}\\\text{side by}\\(a_1 b_2 - a_2 b_1)\end{array}$$

Next, solve for y by eliminating x as follows:

$$a_1 x + b_1 y = c_1 \xrightarrow[\text{side by } -a_2.]{\text{Multiply each}} -a_1 a_2 x - a_2 b_1 y = -a_2 c_1 \qquad \text{multiplication property of equality}$$

$$a_2 x + b_2 y = c_2 \xrightarrow[\text{side by } a_1.]{\text{Multiply each}} a_1 a_2 x + a_1 b_2 y = a_1 c_2$$

$$\overline{\quad a_1 b_2 y - a_2 b_1 y = a_1 c_2 - a_2 c_1 \quad} \qquad \text{addition method}$$

$$y(a_1 b_2 - a_2 b_1) = a_1 c_2 - a_2 c_1 \qquad \text{factoring}$$

$$y = \frac{a_1 c_2 - a_2 c_1}{a_1 b_2 - a_2 b_1} \qquad \begin{array}{l}\text{dividing each}\\\text{side by}\\(a_1 b_2 - a_2 b_1)\end{array}$$

Now, observe that

1. Both x and y have the same denominator, and this denominator can be represented by the determinant

$$\begin{vmatrix} a_1 & b_1 \\ a_2 & b_2 \end{vmatrix} = a_1 b_2 - a_2 b_1.$$

2. The numerator for x can be represented by the determinant

$$\begin{vmatrix} c_1 & b_1 \\ c_2 & b_2 \end{vmatrix} = c_1 b_2 - c_2 b_1.$$

3. The numerator for y can be represented by the determinant

$$\begin{vmatrix} a_1 & c_1 \\ a_2 & c_2 \end{vmatrix} = a_1 c_2 - a_2 c_1.$$

Putting these facts together, we have Cramer's rule.

CRAMER'S RULE

The solution of the system

$$a_1 x + b_1 y = c_1$$
$$a_2 x + b_2 y = c_2$$

is given by

$$x = \dfrac{\begin{vmatrix} c_1 & b_1 \\ c_2 & b_2 \end{vmatrix}}{\begin{vmatrix} a_1 & b_1 \\ a_2 & b_2 \end{vmatrix}} \quad \text{and} \quad y = \dfrac{\begin{vmatrix} a_1 & c_1 \\ a_2 & c_2 \end{vmatrix}}{\begin{vmatrix} a_1 & b_1 \\ a_2 & b_2 \end{vmatrix}}.$$

Cramer's rule is easy to remember provided that you note the following:

1.

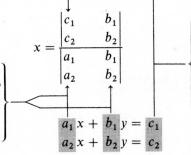

Note: The determinant in the denominator is made up of the coefficients of the x- and y-terms in the exact order they appear in the general system.

Note: The determinant in the numerator of x is formed by taking the determinant in the denominator and replacing the x-coefficients with the constants c_1 and c_2.

2.

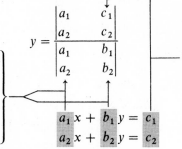

Note: The determinant in the denominator is made up of the coefficients of the x- and y-terms in the exact order they appear in the general system.

Note: The determinant in the numerator of y is formed by taking the determinant in the denominator and replacing the y-coefficients with the constants c_1 and c_2.

EXAMPLE 2

Using Cramer's rule solve $\begin{array}{r} 3x + 5y = -5 \\ -2x - y = 8 \end{array}$.

Solution:

$$x = \frac{\begin{vmatrix} -5 & 5 \\ 8 & -1 \end{vmatrix}}{\begin{vmatrix} 3 & 5 \\ -2 & -1 \end{vmatrix}} = \frac{(-5)(-1) - (8)(5)}{(3)(-1) - (-2)(5)} = \frac{-35}{7} = -5$$

$$3x + 5y = -5$$
$$-2x - 1y = 8$$

$$y = \frac{\begin{vmatrix} 3 & -5 \\ -2 & 8 \end{vmatrix}}{\begin{vmatrix} 3 & 5 \\ -2 & -1 \end{vmatrix}} = \frac{(3)(8) - (-2)(-5)}{(3)(-1) - (-2)(5)} = \frac{14}{7} = 2$$

$$3x + 5y = -5$$
$$-2x - 1y \quad 8$$

Thus, $(-5, 2)$ is the solution to this system. ■

Before Cramer's rule can be applied, the equations in the linear system must be in the general form $Ax + By = C$.

EXAMPLE 3

Using Cramer's rule solve $\begin{array}{l} 16y = 14x - 5 \\ 13x - 4 = 12y \end{array}$.

Solution:

First, write the equations in the form $Ax + By = C$.

$16y = 14x - 5 \longrightarrow 14x - 16y = 5$

$13x - 4 = 12y \longrightarrow 13x - 12y = 4$

$$x = \frac{\begin{vmatrix} 5 & -16 \\ 4 & -12 \end{vmatrix}}{\begin{vmatrix} 14 & -16 \\ 13 & -12 \end{vmatrix}} = \frac{(5)(-12) - (4)(-16)}{(14)(-12) - (13)(-16)} = \frac{4}{40} = \frac{1}{10}$$

$$14x - 16y = 5$$
$$13x - 12y = 4$$

$$y = \frac{\begin{vmatrix} 14 & 5 \\ 13 & 4 \\ 14 & -16 \\ 13 & -12 \end{vmatrix}} = \frac{(14)(4) - (13)(5)}{(14)(-12) - (13)(-16)} = \frac{-9}{40} = -\frac{9}{40}$$

$$14\,x - 16\,y = 5$$
$$13\,x - 12\,y = 4$$

Thus, $\left(\dfrac{1}{10}, -\dfrac{9}{40}\right)$ is the solution to this system. ∎

The material in this section has introduced you to the idea of a determinant and how it can be used to solve a two by two linear system. In more advanced technical mathematics courses you will discuss linear systems that contain three equations and three unknowns, four equations and four unknowns, and so on. You will discover at that time other methods to solve such multisystems.

EXERCISES 10.5

Evaluate the following two by two determinants.

1. $\begin{vmatrix} 1 & 2 \\ 6 & 4 \end{vmatrix}$　　　　**2.** $\begin{vmatrix} 9 & 4 \\ 1 & 3 \end{vmatrix}$

3. $\begin{vmatrix} 8 & -3 \\ 4 & -1 \end{vmatrix}$　　　**4.** $\begin{vmatrix} 6 & 4 \\ -5 & 2 \end{vmatrix}$

5. $\begin{vmatrix} 0 & -3 \\ -5 & 2 \end{vmatrix}$　　　**6.** $\begin{vmatrix} 4 & 3 \\ -1 & 0 \end{vmatrix}$

7. $\begin{vmatrix} 8 & -10 \\ 4 & 5 \end{vmatrix}$　　　**8.** $\begin{vmatrix} -3 & 2 \\ 6 & 4 \end{vmatrix}$

9. $\begin{vmatrix} \dfrac{1}{3} & \dfrac{1}{2} \\ -\dfrac{8}{5} & \dfrac{3}{5} \end{vmatrix}$　　　**10.** $\begin{vmatrix} 8 & \dfrac{2}{3} \\ -12 & \dfrac{3}{4} \end{vmatrix}$

11. $\begin{vmatrix} a & -a \\ -a & a \end{vmatrix}$　　　**12.** $\begin{vmatrix} -1 & -6 \\ a & 6a \end{vmatrix}$

13. $\begin{vmatrix} 3 & -a \\ -b & 2b \end{vmatrix}$　　　**14.** $\begin{vmatrix} 3b & -2 \\ ab & -a \end{vmatrix}$

Using Cramer's rule, solve the following two by two linear systems.

15. $6x - y = 4$
$7x - 3y = -10$

16. $x + 5y = 6$
$5x + 9y = -2$

17. $5x - 3y = 18$
$-9x + 4y = 21$

18. $-8x + 9y = -10$
$11x - 12y = 13$

19. $\dfrac{1}{2}x - \dfrac{2}{3}y = -2$
$-\dfrac{5}{4}x + y = -1$

20. $\dfrac{3}{5}x - 2y = -15$
$\dfrac{1}{3}x - \dfrac{5}{6}y = -5$

21. $13x - 12y = 8$
$-15x = 25$

22. $-9y = 15$
$11x - 12y = 13$

23. $9x = 11y - 17$
$4x + 3y = 1$

24. $9y + 3x = 7$
$8x = 14 - 3y$

25. $2x - 8y = -4$
$20y = 10 + 5x$

26. $3x - 12y = 6$
$4x + 8 = 16y$

27. $12(x - y) = 2x - y$
$-7(x - y) = 9 + y$

28. $-5(x + y) = 9x - 2y$
$9(x + y) = x - 17$

29. A 100-lb weight is supported by two cables as shown in Fig. 10.12. By using trigonometry (see Chapter 11) and the force law of equilibrium, it can be shown that the system $.643T_1 + .819T_2 = 100$ and $.766T_1 - .574T_2 = 0$ describes the tensions T_1 and T_2 of the cables. Using Cramer's rule, determine the tensions T_1 and T_2.

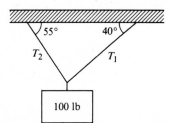

Figure 10.12

30. A 1500-lb weight is supported by two cables as shown in Fig. 10.13. By using trigonometry (see Chapter 11) and the force law of equilibrium, it can be shown that the system $.743T_1 + .423T_2 = 1500$ and $.669T_1 - .906T_2 = 0$ describes the tensions T_1 and T_2 of the cables. Using Cramer's rule, determine the tensions T_1 and T_2.

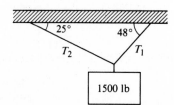

Figure 10.13

Chapter Review

REVIEW EXERCISES The following review exercises are grouped according to the objectives that should have been mastered in Chapter 10. Work each problem carefully. If any weaknesses appear, immediately refer to and read the subsection that matches that objective.

10.1 SOLVING SYSTEMS OF LINEAR EQUATIONS BY THE THE GRAPHING METHOD

Objectives

⊡ To determine if a given ordered pair is a solution of a two by two linear system.

Verify.

1. Is $(-3, 2)$ a solution for $\begin{array}{l} 3x - y = -11 \\ 4x + 2y = 2 \end{array}$?

2. Is $(-2, 5)$ a solution for $\begin{array}{l} 3x - y = -11 \\ 4x + 2y = 2 \end{array}$?

⊡ To solve a two by two linear system by the graphing method.

Solve by the graphing method.

3. $\begin{array}{l} 5x - 3y = 10 \\ y = -5 \end{array}$

4. $\begin{array}{l} 3x - 2y = 18 \\ y = 9 - 3x \end{array}$

⊡ To determine if a two by two linear system has exactly one solution, no solution, or an infinite number of solutions.

State the number of solutions for each system.

5. a) $\begin{array}{l} x + 3y = 1 \\ x - 3y = -1 \end{array}$

b) $\begin{array}{l} 2x - 6y = 2 \\ -3x + 9y = 3 \end{array}$

c) $\begin{array}{l} -5x - 15y = -5 \\ 4x + 12y = 4 \end{array}$

10.2 SOLVING SYSTEMS OF LINEAR EQUATIONS BY THE SUBSTITUTION METHOD

Objectives

⊡ To solve a two by two linear system by using the substitution method when one of the equations contains an isolated variable.

Solve by the substitution method.

6. $\begin{array}{l} 8x - y = 5 \\ y = 6x \end{array}$

7. $\begin{array}{l} x = 3y - 4 \\ -2x + y = 3 \end{array}$

⊡ To solve a two by two linear system by using the substitution method when neither of the equations contains an isolated variable.

Solve by the substitution method.

8. $\begin{array}{l} 5x + 4y = 1 \\ x + 2y = 5 \end{array}$

9. $\begin{array}{l} 2x - 7y = 8 \\ 6x - 5y = -8 \end{array}$

10. $\begin{array}{l} 4x - 2y = 8 \\ -6x + 3y = -12 \end{array}$

10.3 SOLVING SYSTEMS OF LINEAR EQUATIONS BY THE ADDITION METHOD

Objectives

⊡ To solve a two by two linear system by using the addition method.

Solve by the addition method.

11. $\begin{array}{l} 2x - 7y = 5 \\ 3x + 7y = 10 \end{array}$

12. $\begin{array}{l} 6x + y = 4 \\ -6x - 3y = 8 \end{array}$

$\boxdot$ To solve a two by two linear system by using the addition method when the multiplication property of equality must be applied first.

Solve by the addition method.

13. $2x - 3y = 6$
      $5x - y = -11$

14. $8x + 12y = 16$
      $20x - 9y = 53$

15. $x + 4y = 3$
      $8y = 5 - 2x$

10.4 VERBAL PROBLEMS CONTAINING TWO UNKNOWNS

Objectives

$\boxdot$ To solve various types of verbal problems that contain two unknowns by using a two by two linear system.

Solve.

16. The perimeter of a rectangle is 87.0 cm. The length is 12.1 cm more than the width. What are the dimensions of the rectangle?

17. A dry mixture of concrete is made up of cement and gravel in the ratio 3:2 by volume. How many cubic yards of each must be used to make 60 yd^3 of concrete?

18. A chemist has a 10% hydrochloric acid solution and a 15% hydrochloric acid solution. How many liters of each must be used to make a 25-L solution which is 12% hydrochloric acid?

19. An airplane flew 1600 mi in 4 h with a tail wind and returned in 5 h against a head wind. What was the speed of the airplane and the speed of the wind?

10.5 AN INTRODUCTION TO DETERMINANTS

Objectives

$\boxdot$ To evaluate a two by two determinant.

Evaluate.

20. a) $\begin{vmatrix} -3 & 4 \\ 5 & 6 \end{vmatrix}$
    b) $\begin{vmatrix} 0 & 8 \\ -1 & -3 \end{vmatrix}$
    c) $\begin{vmatrix} 9 & 6 \\ -6 & -4 \end{vmatrix}$

$\boxdot$ To use Cramer's rule to solve a two by two linear system.

Solve using Cramer's rule.

21. $5x + 6y = 2$
      $x - y = 7$

22. $15x = 13y + 7$
      $17y - 13 = 75x$

CHAPTER TEST
Time
50 minutes

Score
15, 14 excellent
13, 12 good
11, 10, 9 fair
below 8 poor

If you have worked through the Review Exercises and corrected any weaknesses, you are ready to take the following Chapter Test.

Determine whether the following linear systems have exactly one solution, no solution, or an infinite number of solutions.

1. $2y = 4 - 3x$
     $6x + 4y = 3$

2. $3x + 2y = -1$
     $2y = 3x - 1$

3. $6x - 2y = 1$
     $9x = 3y + \dfrac{3}{2}$

Evaluate the following determinants.

4. $\begin{vmatrix} 9 & 2 \\ -4 & 3 \end{vmatrix}$

5. $\begin{vmatrix} a & -2 \\ 3a & 0 \end{vmatrix}$

Solve the following linear systems by the indicated method.

6. $2x - 6y = 10$ (graphing
   $\quad\quad x = -1$ method)

7. $5x - y = 10$ (graphing
   $\quad\quad y = 11 - 2x$ method)

8. $3x + 5y = 4$ (substitution
   $\quad x = 3y - 8$ method)

9. $6x - y = -9$ (substitution
   $\quad 4x + 3y = 16$ method)

10. $2y = 9x - 5$ (addition
    $\quad 3x + 2y = 3$ method)

11. $5x + 4y = 16$ (addition
    $\quad 3x + 8y = 18$ method)

12. $\quad 7x - 2y = 3$ (Cramer's
    $\quad -x + y = 6$ rule)

13. $\quad\quad 18y = 15x$ (Cramer's
    $\quad 12x - 25y = -42$ rule)

Solve each of the following verbal problems by using a two by two linear system.

14. A bartender has a coffee brandy containing 20% alcohol and a vodka containing 50% alcohol. How many liters of each must be used to make a 10-L mixture of black russians that is 30% alcohol?

15. In an automobile, the rear axle ratio is the number of teeth in the ring gear compared to the number of teeth in the pinion gear. If the rear axle ratio of an automobile is 9:2 and the number of teeth in the pinion gear is 15 less than half the number of teeth in the ring gear, then how many teeth are there in each gear?

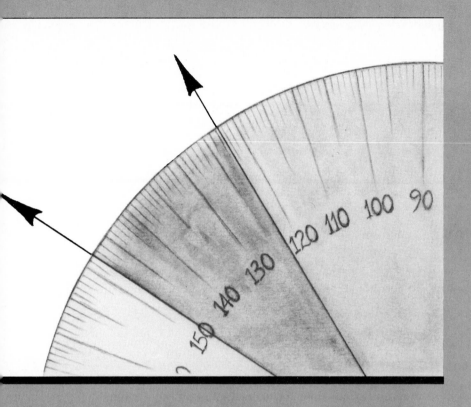

**GEOMETRY
AND
RIGHT
TRIANGLE
TRIGONOMETRY**

11.1 ANGLES

Objectives

☐ To determine the measure of a given angle by using a protractor.

☐ To convert angles from degrees, minutes, and seconds to degrees only or vice versa.

☐ Given a geometric figure containing adjacent, complementary, or supplementary angles, to determine the measure of the missing angle(s).

☐ Angular Measure

Two points, *A* and *B*, determine four basic geometric figures (see Fig. 11.1). The arrowheads on the end of a figure imply that the figure continues indefinitely in that direction. The geometric figure formed by two **rays** with a common endpoint

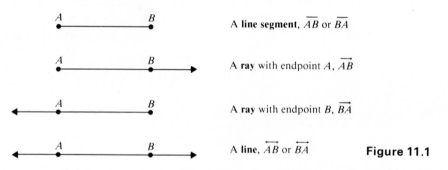

A **line segment**, $\overline{AB}$ or $\overline{BA}$

A **ray** with endpoint *A*, $\overrightarrow{AB}$

A **ray** with endpoint *B*, $\overrightarrow{BA}$

A **line**, $\overleftrightarrow{AB}$ or $\overleftrightarrow{BA}$

Figure 11.1

is called an **angle**. The rays form the **sides** of the angle and the common endpoint is called the **vertex** of the angle (see Fig. 11.2). The mathematical symbol for an angle is ∡. Sometimes angles are named by using three capital letters along with

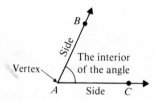

An **angle**, ∡*BAC*, ∡*CAB*, or ∡*A*   **Figure 11.2**

the symbol ∡. If angles are designated this way, the capital letter in the middle must correspond to the vertex of the angle. Thus, the angle in Fig. 11.2 can be designated ∡*BAC* or ∡*CAB*. If there is only one angle stemming from the vertex, a single capital letter may also be used to name an angle. Thus, the angle in Fig. 11.2 can also be designated ∡*A*.

Another method of naming an angle is to place a number, a lowercase letter, or a Greek letter in the interior of the angle (see Fig. 11.3). The most common way

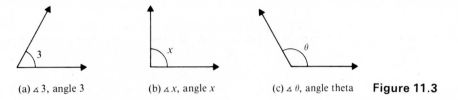

(a) ∡ 3, angle 3   (b) ∡ x, angle x   (c) ∡ θ, angle theta   **Figure 11.3**

to determine the size of an angle is to use a unit of measure called a **degree** (°). One instrument used to find the measure of an angle is a **protractor** (see Fig. 11.4).

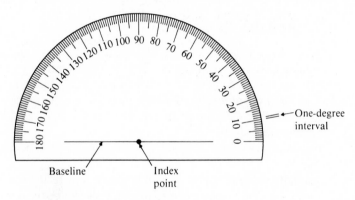

Figure 11.4

To use a protractor, proceed as follows:

1. place the vertex of the angle on the **index point**,
2. locate one side of the angle along the **baseline**, and
3. read the measure where the other side of the angle crosses the markings of the protractor.

EXAMPLE 1

Using the protractor in Fig. 11.5, find the measure of the following angles.

a) ∡ *AOB*   **b)** ∡ *AOC*   **c)** ∡ *COD*   **d** ∡ *BOC*

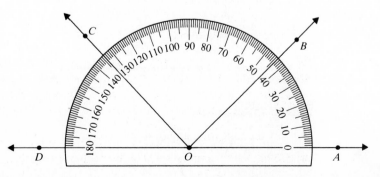

Figure 11.5

Solution:

a) $\sphericalangle AOB = 44°$ ■

b) $\sphericalangle AOC = 134°$ ■

c) $\sphericalangle COD = 180° - 134° = 46°$ ■

d) $\sphericalangle BOC = 134° - 44° = 90°$ ■

Note: In this chapter, we will only need to work with angles between 0° and 180°. In Chapter 12 we will discuss the idea of a *directed angle*. This will allow us to extend the meaning of an angle to include angles greater than 180° and less than 0°.

⬚ Angular Conversion

Many times engineers, surveyors, and astronomers must measure an angle with a smaller unit of measure than a degree. When greater precision is needed to measure an angle, each degree is divided into sixty equal parts. Each of these smaller units of measure is called a **minute** ('). If still greater precision is needed to measure an angle, then each minute is also divided into sixty equal parts. Each of these still smaller units of measure is called a **second** (″).

Transits, which are used by surveyors to measure angles for property lines, highways, and so on, are capable of measuring angles accurately to minutes and seconds. An angle measured on a transit and read by a surveyor as six degrees, twenty-one minutes, and thirty seconds, would be written 6° 21′ 30″.

On many calculators it is necessary to convert an angle written in degrees, minutes, and seconds to decimal parts of a degree before the calculator can be used to evalute the trigonometric ratios (see Section 11.5) of the angle.

Listed in Table 11.1 are the conversion factors for angular measure.

TABLE 11.1

Equal measures	Conversion factors
1 degree = 60 minutes $(1° = 60')$	$\dfrac{1°}{60'}$ or $\dfrac{60'}{1°}$
1 minute = 60 seconds $(1' = 60'')$	$\dfrac{1'}{60''}$ or $\dfrac{60''}{1'}$
1 degree = 3600 seconds $(1° = 3600'')$	$\dfrac{1°}{3600''}$ or $\dfrac{3600''}{1°}$

EXAMPLE 2

Convert the following angles to degrees only.

a) 57° 14′   **b)** 123° 34′ 27″

Solution:

a) For 57° 14′, we must convert 14′ to degrees.

$$14' = \frac{14'}{1}\left(\frac{1°}{60'}\right) = \frac{14°}{60} = .23° \quad \text{(retain two significant digits)}$$

$$57° \, 14' = 57.23° \quad \blacksquare$$

b) For 123° 34′ 27″, first change 34′ to seconds.

$$34' = \frac{34'}{1}\left(\frac{60''}{1'}\right) = 2040''$$

$$34' \, 27'' = 2040'' + 27'' = 2067''$$

Now convert 2067″ to degrees.

$$2067'' = \frac{2067''}{1}\left(\frac{1°}{3600''}\right) = \frac{2067°}{3600} = .5742 \quad \text{(retain four significant digits)}$$

$$123° \, 34' \, 27'' = 123.5742° \quad \blacksquare$$

EXAMPLE 3

Convert the following angles to degrees, minutes, and seconds where appropriate.

a) 32.8° **b)** 106.431°

Solution:

a) For 32.8°, we must convert .8° to minutes.

$$.8° = \frac{.8°}{1}\left(\frac{60'}{1°}\right) = 50' \quad \text{(retain one significant digit)}$$

$$32.8° = 32° \, 50' \quad \blacksquare$$

b. For 106.431°, first change .431° to minutes.

$$.431° = \frac{.431°}{1}\left(\frac{60'}{1°}\right) = 25.9' \quad \text{(retain three significant digits)}$$

$$106.431° = 106° \, 25.9'$$

Now, we convert .9′ to seconds.

$$.9' = \frac{.9'}{1}\left(\frac{60''}{1'}\right) = 50'' \quad \text{(retain one significant digit)}$$

$$106.431° = 106° \, 25.9' = 106° \, 25' \, 50'' \quad \blacksquare$$

⠒ Types of Angles

When two angles have a common vertex and share a common side, the angles are called **adjacent angles**. In Fig. 11.6, $\angle AOB$ and $\angle BOC$ are adjacent angles.

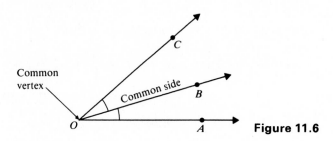

Figure 11.6

EXAMPLE 4

Referring to Fig. 11.6, if $\angle AOB$ measures $16°51'$ and $\angle BOC$ measures $25°42'$, find the measure of $\angle AOC$.

Solution:

$$\angle AOC = \angle AOB + \angle BOC$$
$$= 16°\ 51' + 25°\ 42'$$
$$= \boxed{41°\ 93'}$$

Note: $\left.\begin{array}{l} 93' = 60' + 33' \\ = 1° + 33' \end{array}\right\}$

$$= \boxed{42°\ 33'} \text{ or } 42.55° \quad \blacksquare$$

Listed in Table 11.2 are the names of several other types of angles that you will be working with throughout this chapter.

EXAMPLE 5

Given that $\angle DOF$ is a right angle, find the measure of $\angle DOE$ (see Fig. 11.7).

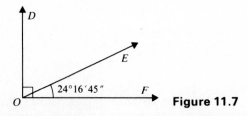

Figure 11.7

TABLE 11.2

Name and definition	Figure
ACUTE ANGLE: An angle whose measure is between 0° and 90°.	
OBTUSE ANGLE: An angle whose measure is between 90° and 180°.	
RIGHT ANGLE: An angle whose measure is 90°.	
STRAIGHT ANGLE: An angle whose measure is 180°.	
COMPLEMENTARY ANGLES: Two angles whose sum is 90°.	
SUPPLEMENTARY ANGLES: Two angles whose sum is 180°.	

Solution: If $\measuredangle DOF$ is a right angle, then the measure of $\measuredangle DOF$ is 90°. Therefore, $\measuredangle DOE$ and $\measuredangle EOF$ are complementary angles. Thus,

$$\measuredangle DOE + \measuredangle EOF = 90°.$$
$$\measuredangle DOE = 90° - \measuredangle EOF$$
$$= 90° - 24° \, 16' \, 45''$$

Note: This subtraction requires borrowing, so think of 90° as 89°59′60″.

$$= 89° \, 59' \, 60'' - 24° \, 16' \, 45''$$
$$\measuredangle DOE = 65° \, 43' \, 15'' \text{ or } 65.7208° \quad \blacksquare$$

EXAMPLE 6

Given that $\measuredangle COD$ and $\measuredangle AOB$ are straight angles, find the measure of $\measuredangle a$, $\measuredangle b$, and $\measuredangle c$ (see Fig. 11.8).

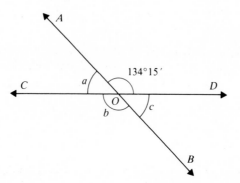

Figure 11.8

Solution: If $\measuredangle COD$ is a straight angle, then the measure of $\measuredangle COD$ is 180°. Therefore, $\measuredangle a$ and $\measuredangle AOD$ are supplementary angles. Thus,

$$\measuredangle a + \measuredangle AOD = 180°.$$

$$\measuredangle a = 180° - \measuredangle AOD$$

$$= \boxed{180°} - 134°15'$$

Note: This subtraction requires borrowing, so think of 180° as 179°60′.

$$= \boxed{179° \, 60'} - 134° \, 15'$$

$$\measuredangle a = 45° \, 45' \text{ or } 45.75° \quad \blacksquare$$

If $\measuredangle AOB$ is a straight angle, then the measure of $\measuredangle AOB$ is 180°. Therefore, $\measuredangle b$ and $\measuredangle a$ are supplementary angles. Thus,

$$\measuredangle a + \measuredangle b = 180°.$$

$$\measuredangle b = 180° - \measuredangle a$$

$$\measuredangle b = 179° \, 60' - 45° \, 45'$$

$$\measuredangle b = \underbrace{134° \, 15'}_{} \text{ or } 134.25° \quad \blacksquare$$

Note: $\measuredangle b$ and $\measuredangle AOD$ have the same measure.

Finally, ∢b and ∢c are also supplementary angles. Thus,

$$∢b + ∢c = 180°$$
$$∢c = 180° - ∢b$$
$$∢c = 179° 60' - 134° 15'$$
$$∢c = \underbrace{45° 45'}_{} \text{ or } 45.75° \quad ∎$$

Note: ∢c and ∢a have the same measure.

Note: We refer to ∢AOD and ∢b as *vertical angles*. Similarly ∢a and ∢c form a pair of vertical angles. Notice that two pairs of vertical angles are always formed when two lines intersect, and that a pair of **vertical angles** *always has the same measure*.

EXERCISES 11.1

Using the protractor in Fig 11.9, find the measure of the following angles.

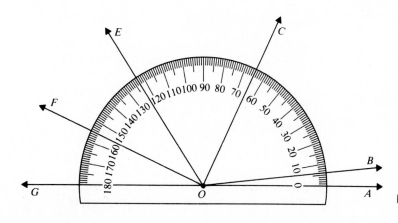

Figure 11.9

1. ∢AOB	**2.** ∢AOC	**3.** ∢AOE	**4.** ∢AOF	**5.** ∢FOG
6. ∢COG	**7.** ∢BOC	**8.** ∢EOF	**9.** ∢BOF	**10.** ∢BOE

Convert the following angles to degrees only.

11. 32°30′   **12.** 47°5$\bar{0}$′

13. 158°32′   **14.** 116°27′

15. 86°55′2$\bar{0}$″   **16.** 44°16′40″

17. 165°06′25″   **18.** 103°17′09″

Convert the following angles to degrees, minutes, and seconds when appropriate.

19. 82.7°   **20.** 15.3°

21. 112.80°   **22.** 157.92°

23. 13.043°

24. 8.630°

25. 144.2548°

26. 119.0065°

27. Referring to Fig. 11.10, find the measure of ∢*AOC*.

28. Referring to Fig. 11.11, find the measure of ∢*AOD*.

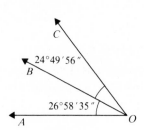

Figure 11.10

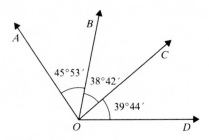

Figure 11.11

29. Referring to Fig. 11.12, find the measure of ∢*AOB*, given that ∢*AOC* is a right angle.

30. Referring to Fig. 11.13, find the measure of ∢*AOB*, given that ∢*AOC* is a straight angle.

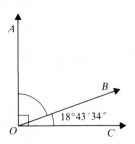

Figure 11.12

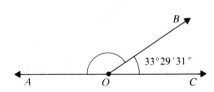

Figure 11.13

31. Referring to Fig. 11.14, find the measure of ∢*BOC*, given that ∢*AOD* is a straight angle.

32. Referring to Fig. 11.15, find the measure of ∢*COD*, given that ∢*AOD* is a right angle.

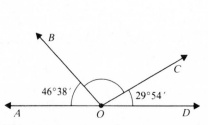

Figure 11.14

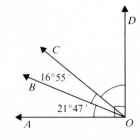

Figure 11.15

33. Three lines intersect as shown in Fig. 11.16. If ∢3 is complementary to ∢5 and the measure of ∢8 is 118°48′, find the measure of all the other angles.

34. Three lines intersect as shown in Fig. 11.16. If ∢3 is supplementary to ∢8 and the measure of ∢1 is 22° 16′, find the measure of all the other angles.

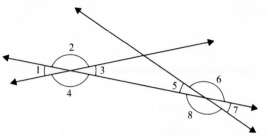

Figure 11.16

35. Referring to Fig. 11.17, find the measure of each angle, given that ∢AOB is a straight angle.

36. Referring to Fig. 11.18, find the measure of each angle, given that ∢AOB is a right angle.

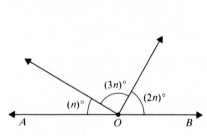

Figure 11.17

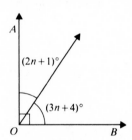

Figure 11.18

Solve the following verbal problems.

37. Two angles are complementary angles. One is 20° less than three times the other. What is the measure of each angle?

38. Two angles are supplementary angles. One is 17° more than twice the other. What is the measure of each angle?

39. Two angles are supplementary angles. The ratio of their measures is 5:7. What is the measure of each angle?

40. Two angles are complementary angles. The ratio of their measures is 7:29. What is the measure of each angle?

11.2 INTERIOR ANGLES OF TRIANGLES AND QUADRILATERALS

Objectives

⊡ Given a geometric figure containing two parallel lines cut by a transversal, to determine the measure of any missing angle.

⊡ Given a parallelogram, to determine the measure of any missing interior angle.

⊡ Given a triangle, to determine the measure of any missing interior angle.

⊡ Given a quadrilateral, to determine the measure of any missing interior angle.

⊡ Parallel Lines

A line that intersects two or more lines is called a **transversal**. In Fig. 11.19, $\overleftrightarrow{CD}$ and $\overleftrightarrow{EF}$ are cut by the transversal $\overleftrightarrow{AB}$. Notice that when two lines are cut by a

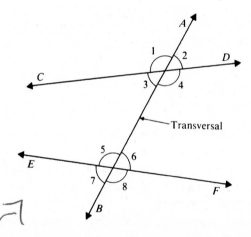

Figure 11.19

transversal, eight angles are formed. These angles are referred to by the following names:

1. ∡3, ∡4, ∡5, and ∡6 are called **interior** angles.
2. ∡1, ∡2, ∡7, and ∡8 are called **exterior** angles.
3. The pair ∡3, ∡6 and the pair ∡4, ∡5 are called **alternate interior** angles.
4. The pair ∡2, ∡7 and the pair ∡1, ∡8 are called **alternate exterior** angles.
5. The pairs ∡1, ∡5; ∡2, ∡6; ∡3, ∡7; and ∡4, ∡8 are called **corresponding angles**.

*If two **parallel** lines are cut by a transversal, then the alternate interior angles that are formed have the same measure.* Conversely, if a pair of alternate interior angles has the same measure, then the two lines that were cut by the transversal must be parallel.

EXAMPLE 1

In Fig. 11.20, $\overleftrightarrow{CD}$ and $\overleftrightarrow{EF}$ are two *parallel* lines cut by the transversal $\overleftrightarrow{AB}$, and the measure of $\angle 1$ is 110°. Find the measure of each of the other numbered angles.

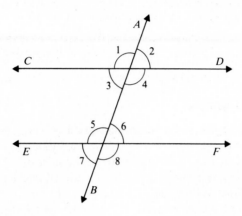

Figure 11.20

Solution:

1. $\angle 1 = 110°$ (given)
2. $\angle 2$ and $\angle 1$ are *supplementary angles*. Thus, $\angle 2 = 180° - \angle 1$; $\angle 2 = 70°$.
3. $\angle 3$ and $\angle 2$ are *vertical angles*. Thus, $\angle 3 = \angle 2$; $\angle 3 = 70°$.
4. $\angle 4$ and $\angle 1$ are *vertical angles*. Thus, $\angle 4 = \angle 1$; $\angle 4 = 110°$.
5. $\angle 5$ and $\angle 4$ are *alternate interior angles*. Since $\overleftrightarrow{CD}$ and $\overleftrightarrow{EF}$ are parallel, the alternate interior angles must be equal. Thus, $\angle 5 = \angle 4$; $\angle 5 = 110°$.
6. $\angle 6$ and $\angle 5$ are *supplementary angles*. Thus, $\angle 6 = 180° - \angle 5$; $\angle 6 = 70°$.
7. $\angle 7$ and $\angle 6$ are *vertical angles*. Thus, $\angle 7 = \angle 6$; $\angle 7 = 70°$.
8. $\angle 8$ and $\angle 5$ are *vertical angles*. Thus, $\angle 8 = \angle 5$; $\angle 8 = 110°$.

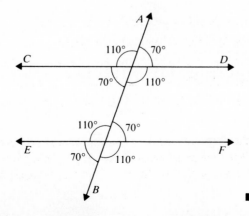

■

Note: We know that if two *parallel* lines are cut by a transversal, then the *alternate interior* angles have the same measure. The following observations are also true:

1. The *alternate exterior* angles have the same measure.
2. The *corresponding angles* have the same measure.
3. The *interior* angles on the *same* side of the transversal are supplementary angles.
4. The *exterior* angles on the *same* side of the transversal are supplementary angles.

⬚ Parallelograms

Recall from Chapter 3 that a **parallelogram** is a quadrilateral whose opposite sides are parallel and equal in length. The sides of a parallelogram determine four *interior angles* (see Fig. 11.21). If we extend the sides of the parallelogram as shown in Fig. 11.22, then ∡1 and ∡2 can be considered *interior angles* on the *same* side of the transversal $\overleftrightarrow{AB}$. Thus, ∡1 and ∡2 are supplementary angles (that is, ∡1 + ∡2 = 180°), and ∡1 and ∡2 are referred to as **consecutive angles** in the parallelogram. Similarly, the pairs ∡3, ∡4; ∡1, ∡3; and ∡2, ∡4 are consecutive angles.

In a parallelogram, any two consecutive angles are supplementary.

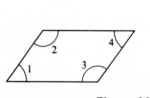

Figure 11.21

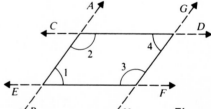

Figure 11.22

EXAMPLE 2

Given the parallelogram in Fig. 11.23, determine the measure of ∡A, ∡B, and ∡C.

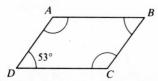

Figure 11.23

Solution:

1. $\measuredangle A$ and $\measuredangle D$ are *consecutive angles* and are therefore *supplementary*. Thus, $\measuredangle A = 180° - \measuredangle D$; $\measuredangle A = 127°$.

2. $\measuredangle C$ and $\measuredangle D$ are *consecutive angles* and are therefore *supplementary*. Thus, $\measuredangle C = 180° - \measuredangle D$; $\measuredangle C = 127°$.

3. $\measuredangle B$ and $\measuredangle C$ are *consecutive angles* and are therefore *supplementary*. Thus, $\measuredangle B = 180° - \measuredangle C$; $\measuredangle B = 53°$.

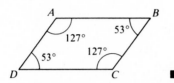

■

Note: There are other important observations.

1. $\measuredangle D$ and $\measuredangle B$ are called **opposite angles** in the parallelogram. Similarly, $\measuredangle A$ and $\measuredangle C$ are *opposite angles*. *In a parallelogram any two opposite angles have the same measure.*

2. The *sum* of the measures of the interior angles is 360°.

$$\measuredangle A = 127°$$
$$\measuredangle B = 53°$$
$$\measuredangle C = 127°$$
$$+\ \measuredangle D = 53°$$

$$\measuredangle A + \measuredangle B + \measuredangle C + \measuredangle D = 360°$$

Later in this section you will discover that the sum of the measures of the interior angles in *any quadrilateral* is 360°.

Triangles

Listed in Table 11.3 are several different types of triangles that you will be working with throughout this chapter.

TABLE 11.3

Name and definition	Figure
ACUTE TRIANGLE: A triangle in which *all* of the interior angles are acute angles.	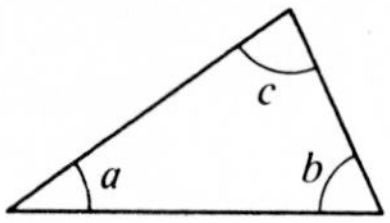
OBTUSE TRIANGLE: A triangle in which *one* of the interior angles is an obtuse angle.	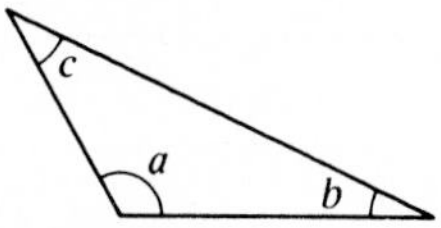
RIGHT TRIANGLE: A triangle in which *one* of the interior angles is a right angle.	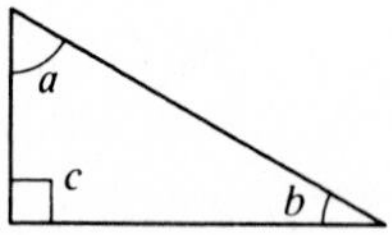
ISOSCELES TRIANGLE: A triangle in which *two* sides have the same length and the *two* interior angles opposite these sides have the same measure.	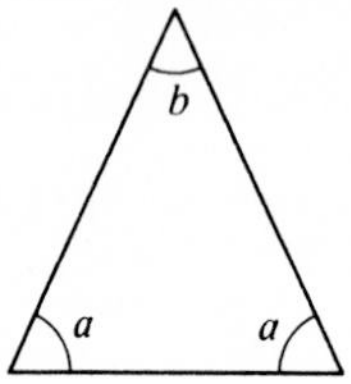
EQUILATERAL TRIANGLE: A triangle in which *all* three sides have the same length and *all* three interior angles have the same measure.	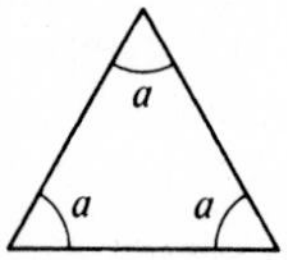

The sides of every triangle determine three *interior angles* as shown in Fig. 11.24.

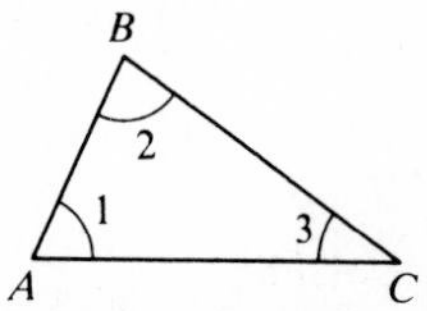

Figure 11.24

Suppose we draw a line through the vertex B that is *parallel* to $\overline{AC}$ (see Fig. 11.25).

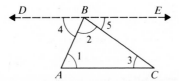

D B E

Figure 11.25

Since $\measuredangle DBE$ is a straight angle,

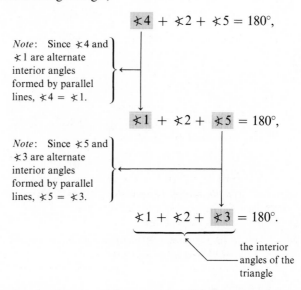

$$\measuredangle 4 + \measuredangle 2 + \measuredangle 5 = 180°,$$

Note: Since $\measuredangle 4$ and $\measuredangle 1$ are alternate interior angles formed by parallel lines, $\measuredangle 4 = \measuredangle 1$.

$$\measuredangle 1 + \measuredangle 2 + \measuredangle 5 = 180°,$$

Note: Since $\measuredangle 5$ and $\measuredangle 3$ are alternate interior angles formed by parallel lines, $\measuredangle 5 = \measuredangle 3$.

$$\measuredangle 1 + \measuredangle 2 + \measuredangle 3 = 180°.$$

the interior angles of the triangle

Thus, in any triangle, the sum of the measures of the interior angles is $180°$.

EXAMPLE 3

In Fig. 11.26, triangle ABC is isosceles with $\overline{AB} = \overline{AC}$ and $\measuredangle ABD$ is a straight angle. Find the measure of the numbered angles.

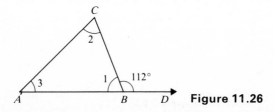

Figure 11.26

Solution:

1. $\not\lt 1$ and $\not\lt CBD$ are supplementary angles. Thus, $\not\lt 1 = 180° - \not\lt CBD$; $\not\lt 1 = 68°$.
2. Since triangle ABC is isosceles, with $\overline{AB} = \overline{AC}$, the angles *opposite* these sides must be equal. Thus, $\not\lt 2 = \not\lt 1$; $\not\lt 2 = 68°$.
3. The sum of the interior angles in any triangle is 180°.

$$\not\lt 1 + \not\lt 2 + \not\lt 3 = 180°$$
$$\not\lt 3 = 180° - \not\lt 1 - \not\lt 2$$
$$\not\lt 3 = 44°$$

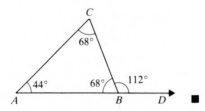

Note: $\not\lt CBD$ is an *exterior* angle of the triangle, and $\not\lt A$ and $\not\lt C$ are the *remote interior* angles for this exterior angle. Notice that

$$\not\lt A + \not\lt C = \not\lt CBD.$$

In any triangle, the measure of an exterior angle is the sum of the measures of its two remote interior angles.

⸬ Quadrilaterals

A quadrilateral is a four-sided polygon. The sides of a quadrilateral determine four interior angles (see Fig. 11.27). Note in Fig. 11.28 that a quadrilateral can be split into *two* triangular regions. We know that the sum of the measures of the interior angles in *each* triangular region is 180°.

$$\not\lt 1 + \not\lt 4 + \not\lt D = 180°$$
$$\not\lt 2 + \not\lt 3 + \not\lt B = 180°$$

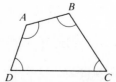

Figure 11.27

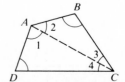

Figure 11.28

Thus, the sum of the measures of the interior angles in *both* triangular regions must be 180° + 180°, or 360°.

$$\angle 1 + \angle 2 + \angle B + \angle 3 + \angle 4 + \angle D = 360°$$

$$\angle A \quad + \angle B + \quad \angle C \quad + \angle D = 360°$$

the interior
angles of the
quadrilateral

In any quadrilateral, the sum of the measures of the interior angles is 360°.

EXAMPLE 4

Using a transit, a surveyor has measured three of the four interior angles for the piece of land shown in Fig. 11.29. What should be the measure of $\angle D$?

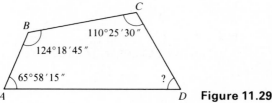

B
C
110°25′30″
124°18′45″
65°58′15″
?
A   *D*   **Figure 11.29**

Solution:

$$\angle A + \angle B + \angle C + \angle D = 360°$$

$$\angle D = 360° \quad - \quad (\angle A + \angle B + \angle C)$$

Note:
$$\angle A = 65° \ 58′15″$$
$$\angle B = 124° \ 18′45″$$
$$+ \ \angle C = 110° \ 25′30″$$

$$\angle A + \angle B + \angle C = 299°101′90″$$
$$= 299°102′30″$$
$$= 300° \ 42′30″$$

$$\angle D = \quad 360° \quad - \quad 300° \ 42′ \ 30″$$

$$\angle D = 359° \ 59′ \ 60″ - 300° \ 42′ \ 30″$$

$$\angle D = 59° \ 17′ \ 30″ \quad ■$$

Application

TRUSSES

A structural frame composed of triangles is called a **truss**. Trusses are most frequently used in the building of roofs and bridges. Before selecting a truss to carry a particular load, it is necessary to determine the measures of the angles between the members of the truss. Trigonometry can then be used to analyze the forces (compression and tension) in each member of the truss.

EXAMPLE 5

Shown in Fig. 11.30 is a four-panel Pratt truss. Determine general formulas for the measures of the angles between the members of the truss.

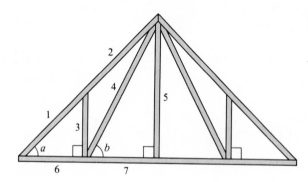

Figure 11.30

Solution:

Members	Angle between members	Remark
1/6	a	given
4/7	b	given
3/6	90	Every right angle measures 90°.
5/7	90	Every right angle measures 90°.
1/3	$90 - a$	The sum of the interior angles in a triangle is 180°.
4/5	$90 - b$	The sum of the interior angles in a triangle is 180°.
3/4	$90 - b$	complementary angle
2/3	$90 + a$	supplementary angle
2/4	$b - a$	The sum of the interior angles in a triangle is 180°. ∎

EXERCISES 11.2

Four straight roadways intersect as shown in Fig. 11.31. If $\overline{ED}$ is parallel to $\overrightarrow{AC}$, determine the measure of the following angles.

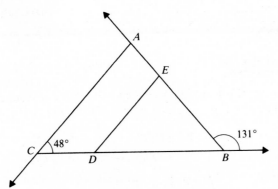

Figure 11.31

1. $\angle EDB$ **2.** $\angle EBD$ **3.** $\angle DEB$

4. $\angle AED$ **5.** $\angle CAE$ **6.** $\angle CDE$

Four straight roadways intersect as shown in Fig. 11.32. Determine the measure of the following angles.

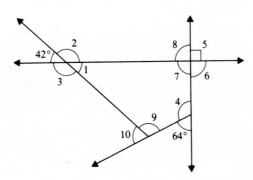

Figure 11.32

7. $\angle 1$ **8.** $\angle 2$ **9.** $\angle 3$

10. $\angle 4$ **11.** $\angle 5$ **12.** $\angle 6$

13. $\angle 7$ **14.** $\angle 8$ **15.** $\angle 9$

16. $\angle 10$

In the bridge truss shown in Fig. 11.33, *ARQP* and *BRPQ* are parallelograms. Determine the measure of the following angles.

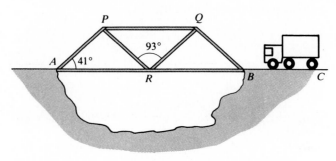

Figure 11.33

17. ⊀ *PQR*	**18.** ⊀ *QPR*	**19.** ⊀ *BQR*	**20.** ⊀ *APR*
21. ⊀ *ARP*	**22.** ⊀ *BRQ*	**23.** ⊀ *QBR*	**24.** ⊀ *CBQ*

In the wood frame shown in Fig. 11.34, triangle *ABC* is equilateral and triangle *CDE* is isosceles, with $\overline{CD} = \overline{ED}$. Find the measure of the following angles.

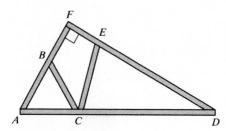

Figure 11.34

25. ⊀ *A*	**26.** ⊀ *ABC*	**27.** ⊀ *BCA*	**28.** ⊀ *F*	**29.** ⊀ *D*
30. ⊀ *DEC*	**31.** ⊀ *DCE*	**32.** ⊀ *FBC*	**33.** ⊀ *FEC*	**34.** ⊀ *ECB*

In the tract of land shown in Fig. 11.35, *ABCD* and *CEFG* are parallelograms. Find the measure of the following angles.

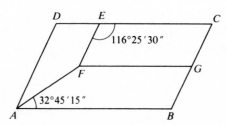

Figure 11.35

35. ⊀ *EFG*	**36.** ⊀ *CGF*	**37.** ⊀ *C*	**38.** ⊀ *FAD*	**39.** ⊀ *D*
40. ⊀ *B*	**41.** ⊀ *DEF*	**42.** ⊀ *BGF*	**43.** ⊀ *AFE*	**44.** ⊀ *AFG*

Determine the value of x that will make $\overleftrightarrow{AB}$ parallel to $\overleftrightarrow{CD}$.

45.

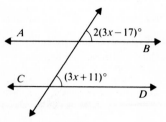

46.

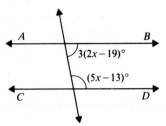

Determine the value of x that will make $ABCD$ a parallelogram.

47.

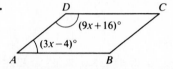

48.

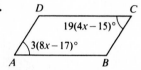

In the following table, the interior angles of the polygons are marked in numbers and the exterior angles are marked in letters. Fill in the table; then do exercises 49 and 50.

Figure	Name	Sum of the interior angles	Sum of the exterior angles
	triangle (three sides)		
	quadrilateral (four sides)		
	pentagon (five sides)		
	hexagon (six sides)		

49. Determine a general formula that can be used to find the sum of the measures of the *interior angles* for any *n*-sided polygon.

50. Determine the sum of the measures of the *exterior angles* for any *n*-sided polygon.

51. Shown in Fig. 11.36 is a six-panel Fink truss. Determine general formulas for the measures of the angles between the members of the truss.

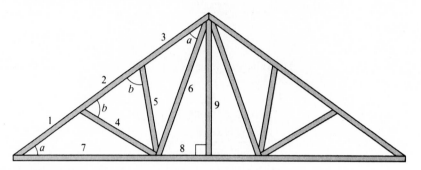

Figure 11.36

52. Shown in Fig. 11.37 is an eight-panel Howe truss. Determine general formulas for the measures of the angles between the members of the truss.

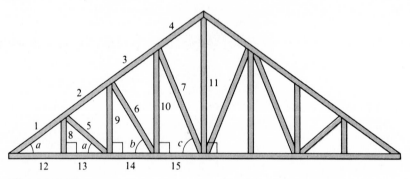

Figure 11.37

11.3 SIMILARITIES IN TRIANGLES

Objectives

- Given two similar triangles, to determine unknown angles and sides.
- To determine whether two triangles are similar.
- Given a right triangle and the altitude to its hypotenuse, to determine the length of an unknown side.

⊡ **Similar Triangles**

Two triangles are said to be **similar** if they have exactly the same shape but not necessarily the same size. In Fig. 11.38, triangle *ABC* is similar to triangle *DEF*,

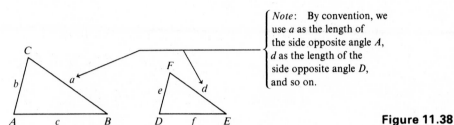

Note: By convention, we use *a* as the length of the side opposite angle *A*, *d* as the length of the side opposite angle *D*, and so on.

Figure 11.38

which is written △*ABC* ~ △*DEF*. If two triangles are similar, it is possible to match their *corresponding parts* in order to determine the measure of any unknown angle or the length of any unknown side.

For example, in Fig. 11.38, if triangle $\boxed{A}$ $\boxed{B}$ $\boxed{C}$ is similar to triangle $\boxed{D}$ $\boxed{E}$ $\boxed{F}$, then

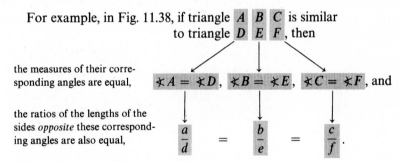

the measures of their corresponding angles are equal,

$\angle A = \angle D$, $\angle B = \angle E$, $\angle C = \angle F$, and

the ratios of the lengths of the sides *opposite* these corresponding angles are also equal,

$$\frac{a}{d} = \frac{b}{e} = \frac{c}{f}.$$

EXAMPLE 1

In Fig. 11.39, if triangle *ABC* is similar to triangle *DEF*, find the measure of

a) $\angle D$ **b)** $\angle B$ **c)** $\angle C$ **d)** $\angle F$

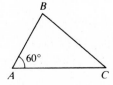

Figure 11.39

Solution:

a) If $\triangle ABC \sim \triangle DEF$, then $\angle A = \angle D$, so $\angle D = 60°$. ■

b) If $\triangle ABC \sim \triangle DEF$, then $\angle B = \angle E$, so $\angle B = 80°$. ■

c) Since the sum of the interior angles in a triangle is 180°,

$$\angle C = 180° - (\angle A + \angle B)$$
$$\angle C = 40°. \quad ■$$

d) If $\triangle ABC \sim \triangle DEF$, then $\angle C = \angle F$, so $\angle F = 40°$. ■

EXAMPLE 2

In Fig. 11.40, if triangle PQR is similar to triangle XYZ, find the length of

a) x **b)** r

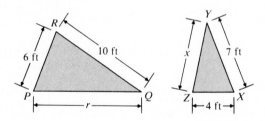

Figure 11.40

Solution:

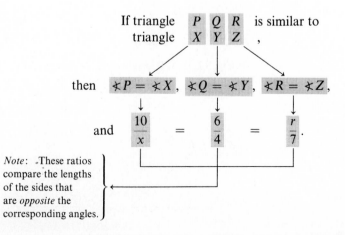

If triangle P Q R is similar to
triangle X Y Z ,

then $\angle P = \angle X$, $\angle Q = \angle Y$, $\angle R = \angle Z$,

and $\dfrac{10}{x} = \dfrac{6}{4} = \dfrac{r}{7}$.

Note: These ratios
compare the lengths
of the sides that
are *opposite* the
corresponding angles.

a) To find x, solve $\dfrac{10}{x} = \dfrac{6}{4}$.

$$(6)(x) = (10)(4) \qquad \text{cross product property}$$

$$6x = 40 \qquad \text{simplifying}$$

$$x = 6\frac{2}{3}\,\text{ft} \quad \blacksquare \qquad \text{dividing each side by 6}$$

b) To find r, solve $\dfrac{6}{4} = \dfrac{r}{7}$.

$$(4)(r) = (6)(7) \qquad \text{cross product property}$$

$$4r = 42 \qquad \text{simplifying}$$

$$r = 10\frac{1}{2}\,\text{ft} \quad \blacksquare \qquad \text{dividing each side by 4}$$

⚬ Conditions for Similarity

When given a geometric figure that contains two or more triangles, it is up to you to determine first whether any of the triangles are similar. Only then can you match corresponding parts of the triangles in order to find any unknown quantities.

There are three ways to determine whether two triangles are similar:

1. If two pairs of corresponding angles have the same measure, then the triangles are similar.
2. If corresponding sides are proportional, then the triangles are similar.
3. If one pair of corresponding sides is proportional to another pair of corresponding sides and the angles included between these sides are equal in measure, then the triangles are similar.

EXAMPLE 3

Five drill holes are located in an aluminum plate as shown in Fig. 11.41. Determine whether the triangles are similar. If they are similar, find the distance x.

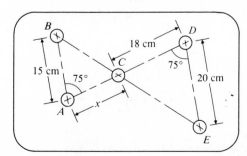

Figure 11.41

Solution:

1. $\angle A = \angle D$ given (both are 75°)
2. $\angle ACB = \angle DCE$ vertical angles
3. $\triangle ACB \sim \triangle DCE$ since two pairs of corresponding angles have the same measure

Now, since corresponding sides of similar triangles are proportional, we can write

$$\frac{15}{20} = \frac{x}{18}$$

$20x = 270$ cross product property

$x = 13\frac{1}{2}$ cm. ■   dividing each side by 20

EXAMPLE 4

A tract of land is subdivided into two triangular regions as shown in Fig. 11.42. Determine if the triangles are similar. If they are similar, find the measure of $\angle BAD$.

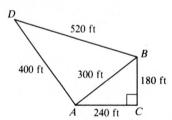

Figure 11.42

Solution: Are the corresponding sides proportional?

Does

$$\frac{180}{300} = \frac{240}{400} = \frac{300}{520}?$$

ratio of the smallest sides   ratio of the largest sides

ratio of the medium sides

Note that

$$\frac{180}{300} = \frac{240}{400}$$

since

$$(180)(400) = (300)(240)$$
$$72{,}000 = 72{,}000.$$

However,

$$\frac{240}{400} \neq \frac{300}{520}$$

since

$$(240)(520) \neq (400)(300)$$
$$124{,}800 \neq 120{,}000.$$

Thus, the triangles are *not* similar. We can't determine the measure of $\angle BAD$ by using corresponding parts of similar triangles. ■

Note: If the 520-ft distance were changed to 500 ft, $\triangle ACB \sim \triangle DAB$ (see if you can verify this), and $\angle BAD$ would also be a right angle.

EXAMPLE 5

Four roadways intersect as shown in Fig. 11.43. Determine whether the triangles are similar. If they are similar, find the distance x.

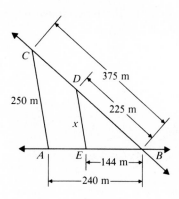

Figure 11.43

Solution:

1. $\dfrac{144}{240} = \dfrac{225}{375}$ since $(144)(375) = (240)(225)$
    $54{,}000 = 54{,}000$

2. $\angle DBE = \angle CBA$ Any angle is equal to itself.

3. $\triangle DBE \sim \triangle CBA$ since a pair of corresponding sides is proportional and the angle between these sides has the same measure

Now, since corresponding sides of similar triangles are proportional, we can write

$$\frac{144}{240} = \frac{x}{250}$$

$$240x = 36,000 \qquad \text{cross product property}$$

$$x = 150 \text{ m.} \quad \blacksquare \qquad \text{dividing each side by 240}$$

Note: Since $\triangle DBE \sim \triangle CBA$, it follows that $\measuredangle CAB = \measuredangle DEB$. Thus, $\overline{AC}$ is parallel to $\overline{DE}$.

⠒ Similar Right Triangles

A right triangle is a triangle with one right angle and two acute angles. The side opposite the right angle is called the **hypotenuse** and the other two sides are called the **legs** (see Fig. 11.44). Note that *the acute angles of a right triangle are complementary* since

$$\measuredangle A + \measuredangle B + \boxed{\measuredangle C} = 180° \qquad \begin{array}{l}\text{The sum of the interior angles}\\\text{in a triangle is }180°.\end{array}$$
$$\downarrow$$
$$\measuredangle A + \measuredangle B + \boxed{90°} = 180° \qquad \text{Every right angle measures }90°.$$
$$\measuredangle A + \measuredangle B = 90°. \qquad \begin{array}{l}\text{subtracting }90°\text{ from each side}\\\text{of the equation}\end{array}$$

Suppose we draw the altitude to the hypotenuse of the right triangle (see Fig. 11.45). Do any similar triangles appear?

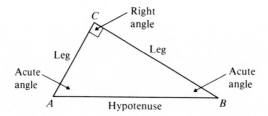

Figure 11.44

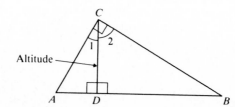

Figure 11.45

Note that

1. $\measuredangle ACB = \measuredangle ADC$ since both are right angles
2. $\measuredangle A = \measuredangle A$ Any angle is equal to itself.
3. $\triangle ACB \sim \triangle ADC$.   since two pairs of corresponding angles have the same measure

Also,

1. $\measuredangle ACB = \measuredangle CDB$ since both are right angles
2. $\measuredangle B = \measuredangle B$ Any angle is equal to itself.
3. $\triangle ACB \sim \triangle CDB$.   since two pairs of corresponding angles have the same measure

Finally,

1. $\measuredangle ADC = \measuredangle CDB$ since both are right angles
2. $\measuredangle 1 + \measuredangle 2 = 90°$ complementary angles
    or $\measuredangle 2 = 90° - \measuredangle 1$
3. $\measuredangle 1 + \measuredangle A = 90°$ The acute angles of a right triangle
    or $\measuredangle A = 90° - \measuredangle 1$ are complementary.
4. $\measuredangle 2 = \measuredangle A$ since both are equal to $90° - \measuredangle 1$
5. $\triangle ADC \sim \triangle CDB$.   since two pairs of corresponding angles have the same measure

When the altitude is drawn to the hypotenuse of a right triangle, the original triangle is separated into two triangles that are similar to each other and to the original triangle.

EXAMPLE 6

Referring to Fig. 11.46, find the distances.

a) x **b)** y **c)** z

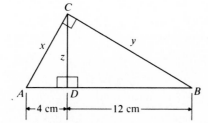

Figure 11.46

Solution:

a) $\triangle ACB \sim \triangle ADC$; thus

$$\frac{x}{16} = \frac{4}{x}.$$

ratio of the ⟋ ⟍ ratio of the
hypotenuses   shorter legs

$$x^2 = 64 \qquad \text{cross product property}$$
$$x = \pm 8 \qquad \text{square root property}$$
$$x = 8 \text{ cm} \quad \blacksquare \qquad \text{distances can only be positive}$$

b) $\triangle ACB \sim \triangle CDB$; thus

$$\frac{y}{16} = \frac{12}{y}.$$

ratio of the ⟋ ⟍ ratio of the
hypotenuses   longer legs

$$y^2 = 192 \qquad \text{cross product property}$$
$$y = \pm\sqrt{192} \qquad \text{square root property}$$
$$y = \sqrt{192} \qquad \text{distances can only be positive}$$
$$y = 8\sqrt{3} \text{ cm} \qquad \text{simplifying radicals}$$
$$y \approx 13.9 \text{ cm} \quad \blacksquare$$

c) $\triangle ADC \sim \triangle CDB$; thus

$$\frac{z}{12} = \frac{4}{z}.$$

ratio of the ⟋ ⟍ ratio of the
longer legs   shorter legs

$$z^2 = 48 \qquad \text{cross product property}$$
$$z = \pm\sqrt{48} \qquad \text{square root property}$$
$$z = \sqrt{48} \qquad \text{distances can only be positive}$$
$$z = 4\sqrt{3} \text{ cm} \qquad \text{simplifying radicals}$$
$$z \approx 6.92 \text{ cm} \quad \blacksquare$$

EXERCISES 11.3

Referring to Fig. 11.47, if $\triangle ABC \sim \triangle DEF$, find the following missing parts.

1. $\angle C$ 2. $\angle D$ 3. $\angle B$

4. $\angle E$ 5. d 6. f

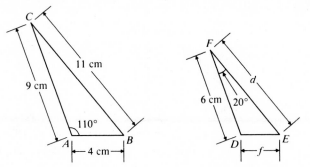

Figure 11.47

Referring to Fig. 11.48, if $\triangle PQR \sim \triangle XYZ$, find the following missing parts.

7. $\measuredangle P$ 8. $\measuredangle Z$

9. $\measuredangle Q$ 10. $\measuredangle Y$

11. r 12. y

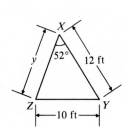

Figure 11.48

Each of the following exercises contains a pair of triangles. a) Determine whether the triangles are similar. b) If they are similar, find the value of x.

13.

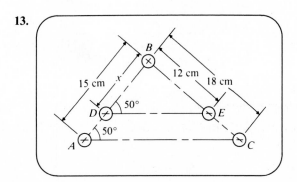

14.

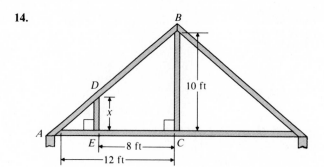

15.

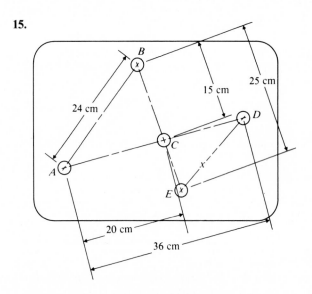

16.

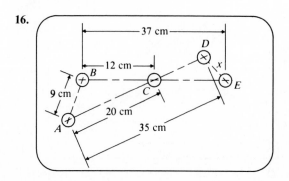

17.

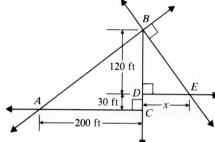

18.

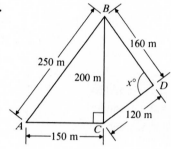

19.

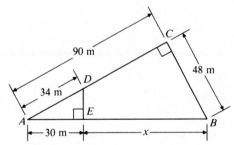

20.

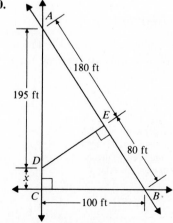

For exercises 21–26, refer to Fig. 11.49. Express all irrational answers in simplified radical form.

y
B
D
x
a
z
A
C
b

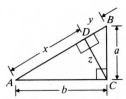

Figure 11.49

21. If $x = 5$ and $y = 4$, find a, b, and z.

22. If $x = 9$ and $y = 3$, find a, b, and z.

23. If $x = 9$ and $z = 6$, find y, a, and b.

24. If $x = 16$ and $z = 8$, find y, a, and b.

25. If $a = 12$ and $y = 9$, find x, z, and b.

26. If $a = 10$ and $y = 4$, find x, z, and b.

11.4 RIGHT TRIANGLES

Objectives

- ⚀ To determine the length of an unknown side of a right triangle by applying the Pythagorean theorem.
- ⚁ To determine the length of an unknown side of a 45°–90° triangle.
- ⚂ To determine the length of an unknown side of a 30°–60°–90° triangle.

⚀ Pythagorean Theorem

Consider the right triangle with legs a and b and hypotenuse c as shown in Fig. 11.50. Is there any special relationship between the lengths of these sides?

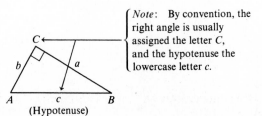

C
b
a
A
c
B
(Hypotenuse)
Note: By convention, the right angle is usually assigned the letter *C*, and the hypotenuse the lowercase letter *c*.

Figure 11.50

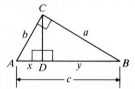

Figure 11.51

We know from Section 11.3 that when the altitude to the hypotenuse is drawn, the original right triangle is separated into two triangles that are similar to each other and to the original triangle. Thus, referring to Fig. 11.51 we can write

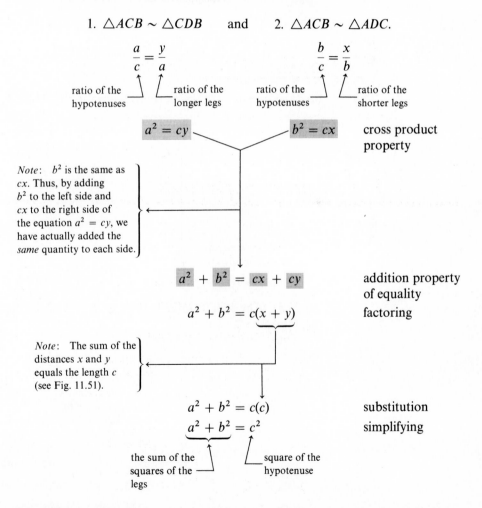

1. $\triangle ACB \sim \triangle CDB$ and   2. $\triangle ACB \sim \triangle ADC.$

$$\frac{a}{c} = \frac{y}{a} \qquad\qquad \frac{b}{c} = \frac{x}{b}$$

ratio of the hypotenuses — ratio of the longer legs   ratio of the hypotenuses — ratio of the shorter legs

$a^2 = cy$ $b^2 = cx$ cross product property

Note: b^2 is the same as cx. Thus, by adding b^2 to the left side and cx to the right side of the equation $a^2 = cy$, we have actually added the *same* quantity to each side.

$$a^2 + b^2 = cx + cy$$ addition property of equality

$$a^2 + b^2 = c(x + y)$$ factoring

Note: The sum of the distances x and y equals the length c (see Fig. 11.51).

$$a^2 + b^2 = c(c)$$ substitution

$$a^2 + b^2 = c^2$$ simplifying

the sum of the squares of the legs — square of the hypotenuse

This fundamental relationship between the sides of a right triangle is called the Pythagorean theorem.

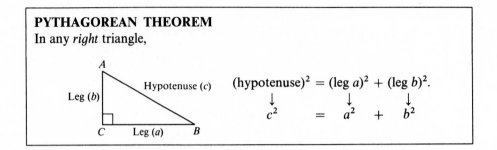

PYTHAGOREAN THEOREM
In any *right* triangle,

$$(\text{hypotenuse})^2 = (\text{leg } a)^2 + (\text{leg } b)^2.$$
$$c^2 = a^2 + b^2$$

 The Pythagorean theorem applies to *right triangles* only.

EXAMPLE 1

Find the length of the stair stringer shown in Fig. 11.52.

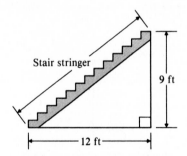

Stair stringer

9 ft

12 ft

Figure 11.52

Solution: The stair stringer is the hypotenuse of the right triangle. Let x be the length of the stair stringer.
Thus,

$$x^2 = \underbrace{12^2 + 9^2}$$ Pythagorean theorem

square of the hypotenuse ⟍ ⟋ the sum of the squares of the legs

$$x^2 = 144 + 81$$ squaring
$$x^2 = 225$$ simplifying
$$x = \pm\sqrt{225}$$ square root property
$$x = \sqrt{225}$$ distances are positive
$$x = 15 \text{ ft} \quad \blacksquare$$ simplifying radicals

Note: The ratio of the smaller leg to the larger leg is $9:12 = \dfrac{9}{12} = \dfrac{3}{4} = 3:4$ when reduced. The ratio of the larger leg to the hypotenuse is $12:15 = \dfrac{12}{15} = \dfrac{4}{5} = 4:5$ when reduced.

Thus, shorter leg:longer leg:hypotenuse = 3:4:5. *Any triangle whose sides are in the ratio 3:4:5 is a right triangle.* Builders often use this fact to determine the "squareness" of a wall or floor.

For some problems, you must first form a right triangle before the Pythagorean theorem can be applied.

EXAMPLE 2

Two drill holes are located in a steel plate as shown in Fig. 11.53. Find the distance *x*.

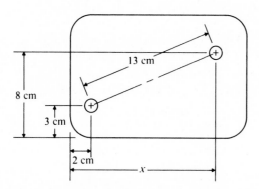

Figure 11.53

Solution: First, form a right triangle from the information that is given, as in Fig. 11.54.

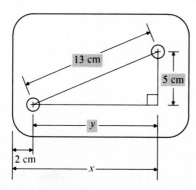

Figure 11.54

The length y can be determined by using the Pythagorean theorem as follows:

$$13^2 = 5^2 + y^2 \qquad \text{Pythagorean theorem}$$
$$169 = 25 + y^2 \qquad \text{squaring}$$
$$144 = y^2 \qquad \text{subtracting 25 from each side}$$
$$12 = y \qquad \text{square root property}$$

Now, $x = \boxed{y} + 2$ cm. So, $x = \boxed{12 \text{ cm}} + 2$ cm, or 14 cm. ∎

The nonequal side of an isosceles triangle is called its **base** and the angle opposite the base is called the **vertex angle** (see Fig. 11.55). If an altitude is drawn from the vertex angle to the base of an isosceles triangle, the vertex angle and the base are both cut in half and two identical right triangles are formed (see Fig. 11.56).

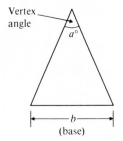

Figure 11.55

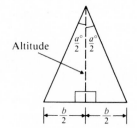

Figure 11.56

EXAMPLE 3

Find the area of the isosceles triangle shown in Fig. 11.57.

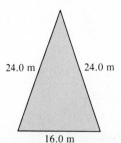

Figure 11.57

Solution: Recall from Chapter 3 that the area of a triangle is given by $A = \dfrac{bh}{2}$, where the height h is the perpendicular distance to the base b.

First, construct the altitude from the vertex angle to the base, as shown in Fig. 11.58. Next, apply the Pythagorean theorem to the right triangle in order to

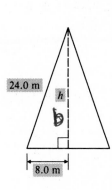

24.0 m

8.0 m

Figure 11.58

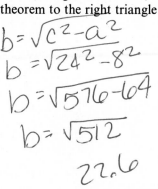

$b = \sqrt{c^2 - a^2}$
$b = \sqrt{24^2 - 8^2}$
$b = \sqrt{576 - 64}$
$b = \sqrt{512}$
22.6

$A = \dfrac{bh}{2}$
$\dfrac{22.6}{8}$
181

find the height h as follows:

$$(24.0)^2 = h^2 + (8.0)^2 \qquad \text{Pythagorean theorem}$$
$$576 = h^2 + 64 \qquad \text{squaring}$$
$$512 = h^2 \qquad \text{subtracting 64 from each side}$$
$$\sqrt{512} = h \qquad \text{square root property}$$
$$16\sqrt{2} = h \qquad \text{simplifying radicals}$$

Now,

$$A = \frac{bh}{2} = \frac{(16.0 \text{ m})(16\sqrt{2}\text{ m})}{2} = 128\sqrt{2} \text{ m}^2$$

$$= 181 \text{ m}^2. \quad \text{(three significant digits)} \ \blacksquare$$

⊡ 45°–90° Triangle

A right triangle whose legs are equal in length is called an **isosceles right triangle**. An isosceles right triangle is also referred to as a 45°–90° triangle since the angles opposite the equal sides have measures of 45° (see Fig. 11.59). *In an isosceles right*

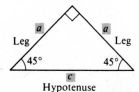

Leg   *a*   *a*   Leg

45°   45°

c
Hypotenuse

Figure 11.59

triangle, the hypotenuse is always $\sqrt{2}$ times as large as one of the legs. To show this, refer to Fig. 11.59 and apply the Pythagorean theorem as follows:

$$c^2 = a^2 + a^2 \qquad \text{Pythagorean theorem}$$
$$c^2 = 2a^2 \qquad \text{simplifying}$$
$$c = \pm\sqrt{2a^2} \qquad \text{square root property}$$
$$c = \sqrt{2a^2} \qquad \text{distances are positive}$$
$$c = a\sqrt{2} \qquad \text{simplifying radicals}$$

the hypotenuse ⎯⎤ ⎡⎯ $\sqrt{2}$ times as large
as the leg (a)

Note that the ratio of the legs is $a : a = \dfrac{a}{a} = \dfrac{1}{1} = 1 : 1$. Also, the ratio of the leg to the hypotenuse is $a : a\sqrt{2} = \dfrac{a}{a\sqrt{2}} = \dfrac{1}{\sqrt{2}} = 1 : \sqrt{2}$. Thus, in any 45°–90° triangle, the ratio of the sides is $1 : 1 : \sqrt{2}$.

45°–90° TRIANGLE (ISOSCELES RIGHT TRIANGLE)
The ratio of the sides is

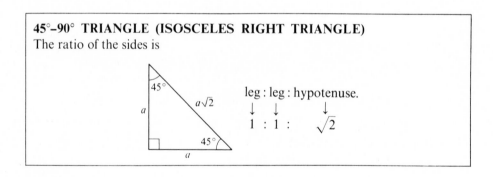

leg : leg : hypotenuse.
↓ ↓ ↓
1 : 1 : $\sqrt{2}$

EXAMPLE 4

The rafter shown in Fig. 11.60 includes a 3.0-ft overhang. Find the length of the rafter.

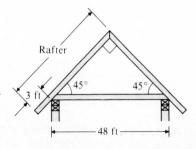

Figure 11.60

Solution: First, form an isosceles right triangle from the given information (Fig. 11.61).

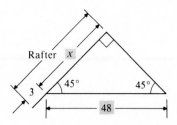

Figure 11.61

In a 45°–90° triangle,

$$\text{leg : hypotenuse} = 1 : \sqrt{2}$$
$$\quad\downarrow\qquad\qquad\downarrow$$
$$x\ :\qquad 48\qquad = 1 : \sqrt{2}$$

$$\frac{x}{48} = \frac{1}{\sqrt{2}}$$

$$x = \frac{48}{\sqrt{2}}\qquad\qquad \text{cross product property}$$

$$x = 24\sqrt{2}\qquad\quad \text{rationalizing the denominator}$$

$$x = 34 \text{ ft}\qquad\quad \text{two significant digits}$$

Therefore, the length of the rafter = x + overhang = 34 ft + 3 ft = 37 ft. ∎

∴ 30°–60°–90° Triangle

In an equilateral triangle, all interior angles measure 60° (see Fig. 11.62). When an altitude is drawn to one of its sides, the vertex angle and the base are both cut in half and two identical 30°–60°–90° triangles are formed (see Fig. 11.63). In a

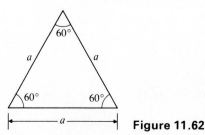

Figure 11.62

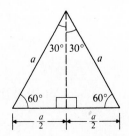

Figure 11.63

30°–60°–90° triangle, *the leg opposite the 30° angle* (the shorter leg) *is half the hypotenuse* (see Fig. 11.64). Thus, the ratio of the leg opposite the 30° angle to

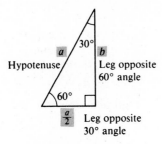

Hypotenuse

Leg opposite 60° angle

Leg opposite 30° angle

Figure 11.64

the hypotenuse is $1:2$. To find the length of the leg opposite the 60° angle (the longer leg), we can use the Pythagorean theorem as follows:

$$a^2 = \left(\frac{a}{2}\right)^2 + b^2 \qquad \text{Pythagorean theorem}$$

$$a^2 = \frac{a^2}{4} + b^2 \qquad \text{squaring}$$

$$a^2 - \frac{a^2}{4} = b^2 \qquad \text{subtracting } \frac{a^2}{4} \text{ from each side}$$

$$\frac{3a^2}{4} = b^2 \qquad \text{simplifying}$$

$$\pm\sqrt{\frac{3a^2}{4}} = b \qquad \text{square root property}$$

$$\sqrt{\frac{3a^2}{4}} = b \qquad \text{distances are positive}$$

$$\frac{a}{2}\sqrt{3} = b \qquad \text{simplifying radicals}$$

$\sqrt{3}$ times as large as the leg opposite the 30° angle

the leg opposite the 60° angle

Note that the ratio of the legs is $\dfrac{a}{2}:\dfrac{a}{2}\sqrt{3} = \dfrac{\dfrac{a}{2}}{\dfrac{a}{2}\sqrt{3}} = \dfrac{1}{\sqrt{3}} = 1:\sqrt{3}$ when reduced.

30°–60°–90° TRIANGLE
The ratio of the sides is

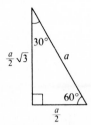

$$\text{leg opposite} \atop 30° \text{ angle} : \text{leg opposite} \atop 60° \text{ angle} : \text{hypotenuse.}$$

$$1 \quad : \quad \sqrt{3} \quad : \quad 2$$

EXAMPLE 5

Three roadways intersect as shown in Fig. 11.65. Determine the lengths of x and y.

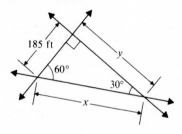

Figure 11.65

Solution: In a 30°–60°–90° triangle,

$$\frac{\text{leg opposite}}{30° \text{angle}} : \text{hypotenuse} = 1:2.$$

$$185 \quad : \quad x \quad = 1:2$$

$$\frac{185}{x} = \frac{1}{2}$$

$$x = 370 \text{ ft} \quad \blacksquare$$

$$\frac{\text{leg opposite}}{30° \text{ angle}} : \frac{\text{leg opposite}}{60° \text{ angle}} = 1:\sqrt{3}$$

$$185 \quad : \quad y \quad = 1:\sqrt{3}$$

$$\frac{185}{y} = \frac{1}{\sqrt{3}}$$

$$y = 185\sqrt{3} \text{ ft}$$

$$y = 32\overline{0} \text{ ft} \quad \text{(three significant digits)} \quad \blacksquare$$

EXAMPLE 6

Find the area of the triangular piece of land shown in Fig. 11.66.

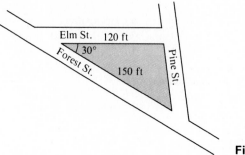

Figure 11.66

Solution: First, form a 30°–60°–90° triangle by drawing the altitude to Forest St. as shown in Fig. 11.67.

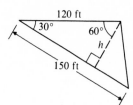

Figure 11.67

In a 30°–60°–90° triangle,

$$\frac{\text{leg opposite}}{30° \text{ angle}} : \text{hypotenuse} = 1:2.$$

$$\downarrow \qquad \downarrow$$

$$h \quad : \quad 120 \quad = 1:2$$

$$\frac{h}{120} = \frac{1}{2}$$

$$h = 60 \text{ ft}$$

$$A = \frac{bh}{2} = \frac{(150 \text{ ft})(60 \text{ ft})}{2} = 4500 \text{ sq ft} \quad \blacksquare$$

EXERCISES 11.4

In each of the following problems, find the value of *x*. Express all irrational answers to three significant digits.

1.

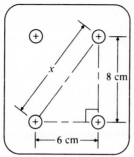

2.

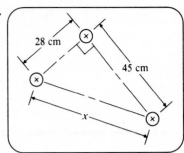

3.

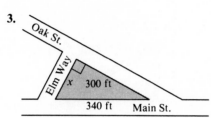

4.

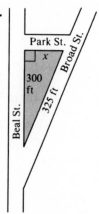

5.

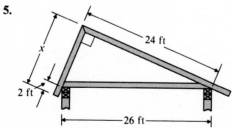

6.

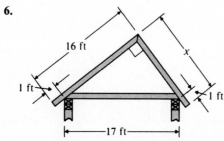

7.

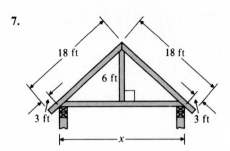

8.

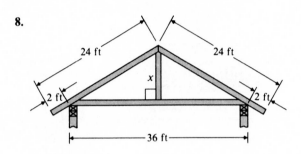

9.

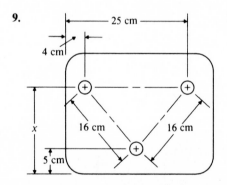

10.

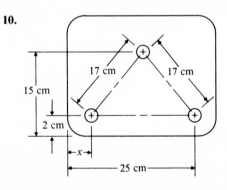

11.

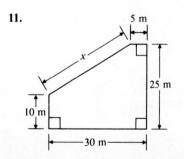

12.

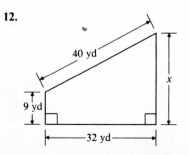

In each of the following, find the value of x and y. Express all irrational answers to three significant digits.

13.

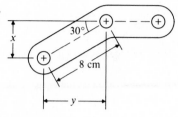

14.

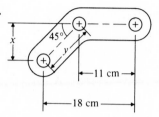

15.

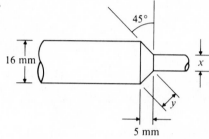

16.

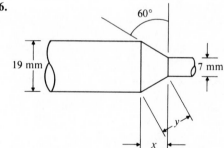

17.

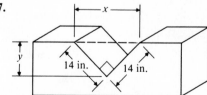

18.

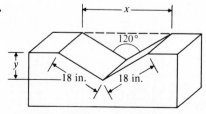

19.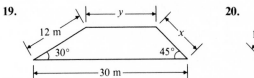

20.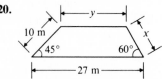

Find the area of each shaded region. Express all irrational answers to three significant digits.

21.

22.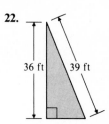

23.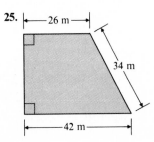

24.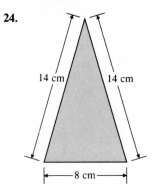

25.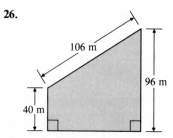

26.

27.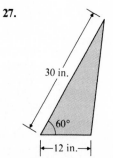

28.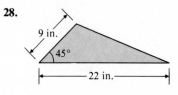

29.

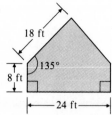

30.

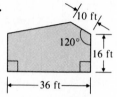

Determine a formula that describes the area of each shaded region. Express all irrational answers in simplified radical form.

31.

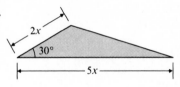

32.

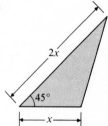

33.

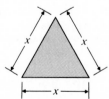

34.

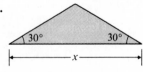

35.

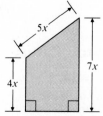

36.

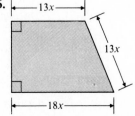

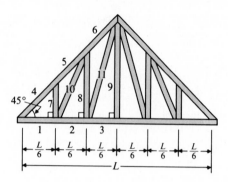

Figure 11.68

37. Referring to Fig. 11.68, determine general formulas that describe the lengths of each member of the six-panel Pratt truss. Express all irrational answers in simplified radical form.

38. Referring to Fig. 11.69, determine general formulas that describe the lengths of each member of the six-panel Howe truss. Express all irrational answers in simplified radical form.

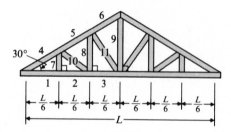

Figure 11.69

11.5 THE TRIGONOMETRIC RATIOS

Objectives

☐ Given a right triangle and the length of at least two of its sides, to determine the trigonometric ratios for an acute angle of the triangle.

☐ To determine the trigonometric ratios for a given acute angle by using a calculator.

☐ Given the value of a trigonometric ratio, to determine the measure of the acute angle by using a calculator.

☐ Defining the Trigonometric Ratios

In technical mathematics, we often use the Greek letter θ (theta) to represent one of the acute angles in a right triangle. The leg of the right triangle directly across from θ is then referred to as the **opposite side**, and the leg that (with the hypotenuse)

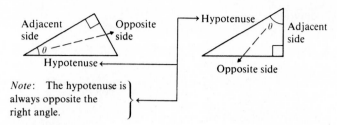

Figure 11.70

forms the angle θ is called the **adjacent side** (see Fig. 11.70). The word "trigonom-etry" comes from the Greek word meaning "triangle measure." In right triangle trigonometry, we are concerned with the ratios that compare the lengths of the sides of the right triangle. These ratios are called the **trigonometric ratios** and are defined as follows:

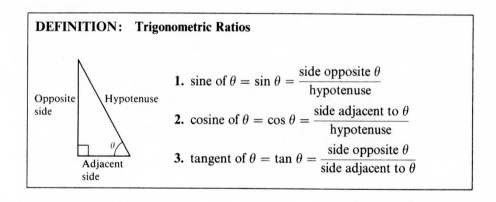

DEFINITION: Trigonometric Ratios

1. sine of $\theta = \sin\theta = \dfrac{\text{side opposite }\theta}{\text{hypotenuse}}$

2. cosine of $\theta = \cos\theta = \dfrac{\text{side adjacent to }\theta}{\text{hypotenuse}}$

3. tangent of $\theta = \tan\theta = \dfrac{\text{side opposite }\theta}{\text{side adjacent to }\theta}$

EXAMPLE 1

Determine the values of $\sin\theta$, $\cos\theta$, and $\tan\theta$ for each triangle. Write the ratios as decimal fractions.

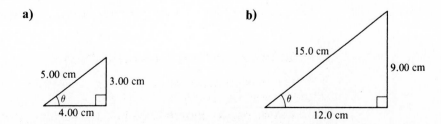

a)

b)

Solution:

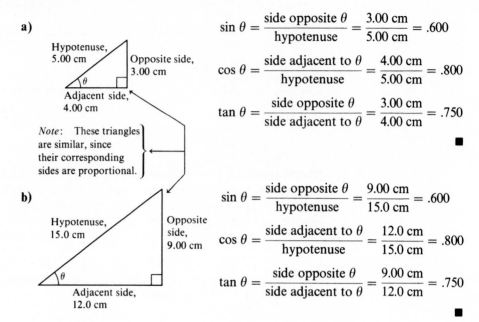

a)

$$\sin \theta = \frac{\text{side opposite } \theta}{\text{hypotenuse}} = \frac{3.00 \text{ cm}}{5.00 \text{ cm}} = .600$$

$$\cos \theta = \frac{\text{side adjacent to } \theta}{\text{hypotenuse}} = \frac{4.00 \text{ cm}}{5.00 \text{ cm}} = .800$$

$$\tan \theta = \frac{\text{side opposite } \theta}{\text{side adjacent to } \theta} = \frac{3.00 \text{ cm}}{4.00 \text{ cm}} = .750$$

■

b)

$$\sin \theta = \frac{\text{side opposite } \theta}{\text{hypotenuse}} = \frac{9.00 \text{ cm}}{15.0 \text{ cm}} = .600$$

$$\cos \theta = \frac{\text{side adjacent to } \theta}{\text{hypotenuse}} = \frac{12.0 \text{ cm}}{15.0 \text{ cm}} = .800$$

$$\tan \theta = \frac{\text{side opposite } \theta}{\text{side adjacent to } \theta} = \frac{9.00 \text{ cm}}{12.0 \text{ cm}} = .750$$

■

Note: Since these triangles are similar, we know that the measure of angle θ in both triangles has the same constant value. This is why the trigonometric ratios for θ in each of these triangles have the same corresponding values. It is important to note that the trigonometric ratios for a fixed angle θ are independent of the lengths of the sides of the triangle. The trigonometric ratios depend only on the measure of the angle θ, not on the size of the triangle.

EXAMPLE 2

Referring to the right triangle in Fig. 11.71, find

a) $\sin A$　　　　　　　　　　　　　　　**b)** $\cos B$.

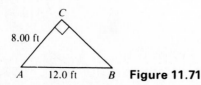

Figure 11.71

Solution: First, apply the Pythagorean theorem to find the length of leg $\overline{BC}$. Let x equal the length of $\overline{BC}$.

$$
\begin{aligned}
(12.0)^2 &= x^2 + (8.00)^2 && \text{Pythagorean theorem} \\
144 &= x^2 + 64.0 && \text{squaring} \\
80 &= x^2 && \text{subtracting 64.0 from each side} \\
\sqrt{80} &= x && \text{square root property} \\
8.94 &= x && \text{three significant digits}
\end{aligned}
$$

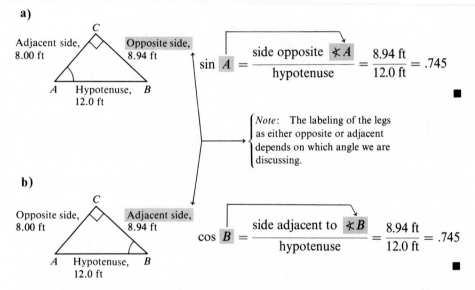

a)

Adjacent side, 8.00 ft

C

Opposite side, 8.94 ft

A   Hypotenuse,   *B*
           12.0 ft

$$\sin A = \frac{\text{side opposite } \angle A}{\text{hypotenuse}} = \frac{8.94 \text{ ft}}{12.0 \text{ ft}} = .745$$

∎

Note: The labeling of the legs as either opposite or adjacent depends on which angle we are discussing.

b)

Opposite side, 8.00 ft

C

Adjacent side, 8.94 ft

A   Hypotenuse,   *B*
           12.0 ft

$$\cos B = \frac{\text{side adjacent to } \angle B}{\text{hypotenuse}} = \frac{8.94 \text{ ft}}{12.0 \text{ ft}} = .745$$

∎

Note: Notice that if A and B are the acute angles of a right triangle, then $\sin A = \cos B$. In Section 11.3 you saw that the acute angles of a right triangle are complementary. Therefore, we can also state that

$$\sin A = \cos (90° - A).$$

complementary angles

⊡ From Angle to Trigonometric Ratio

In Example 1 you saw that the trigonometric ratios for a fixed angle θ are independent of the lengths of the sides of the triangle. *Once we know the measure of angle θ, the trigonometric ratios are determined.* For example, to find the tangent of 40°, we could use a protractor and two number lines as shown in Fig. 11.72. Referring to Fig. 11.72,

$$\tan 40° = \frac{\text{leg opposite 40° angle}}{\text{leg adjacent to 40° angle}} = \frac{8.4}{10} = .84 \text{ (two significant digits).}$$

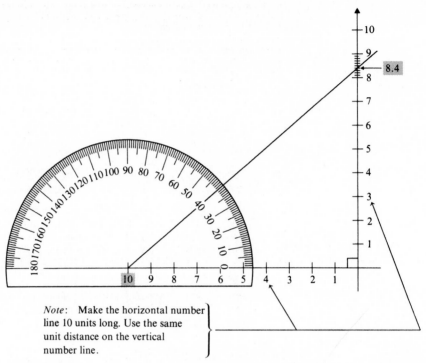

Note: Make the horizontal number line 10 units long. Use the same unit distance on the vertical number line.

Figure 11.72

We could find the trigonometric ratios for other angles by using this graphical approach, but the results would only be crude approximations at best. Fortunately, through more advanced mathematics, tables have been developed that record the values of the trigonometric ratios to several decimal places.

Notice from Table 11.4 that

$$\tan 40° = .839 \qquad \text{(three significant digits),}$$
$$\sin 40° = .643 \qquad \text{(three significant digits),}$$

and

$$\cos 40° = .766 \qquad \text{(three significant digits).}$$

By using the $\boxed{\sin}$, $\boxed{\cos}$, and $\boxed{\tan}$ keys on your calculator, it is also possible to find the trigonometric ratios for a given acute angle.

To find tan 40°, enter

$$\boxed{40} \ \boxed{\tan}. \ \underbrace{.83909963}_{\text{display}} \ \text{(eight significant digits)}$$

TABLE 11.4   Table of Trigonometric Ratios

θ	$\sin \theta$	$\cos \theta$	$\tan \theta$	θ	$\sin \theta$	$\cos \theta$	$\tan \theta$
1°	.017	1.000	.017	46°	.719	.695	1.035
2°	.035	.999	.035	47°	.731	.682	1.072
3°	.052	.999	.052	48°	.743	.669	1.111
4°	.070	.998	.070	49°	.755	.656	1.150
5°	.087	.996	.087	50°	.766	.643	1.192
6°	.105	.995	.105	51°	.777	.629	1.235
7°	.122	.993	.123	52°	.788	.616	1.280
8°	.139	.990	.141	53°	.799	.602	1.327
9°	.156	.988	.158	54°	.809	.588	1.376
10°	.174	.985	.176	55°	.819	.574	1.428
11°	.191	.982	.194	56°	.829	.559	1.483
12°	.208	.978	.213	57°	.839	.545	1.540
13°	.225	.974	.231	58°	.848	.530	1.600
14°	.242	.970	.249	59°	.857	.515	1.664
15°	.259	.966	.268	60°	.866	.5	1.732
16°	.276	.961	.287	61°	.875	.485	1.804
17°	.292	.956	.306	62°	.883	.469	1.881
18°	.309	.951	.325	63°	.891	.454	1.963
19°	.326	.946	.344	64°	.899	.438	2.050
20°	.342	.940	.364	65°	.906	.423	2.145
21°	.358	.934	.384	66°	.914	.407	2.246
22°	.375	.927	.404	67°	.921	.391	2.356
23°	.391	.921	.424	68°	.927	.375	2.475
24°	.407	.914	.445	69°	.934	.358	2.605
25°	.423	.906	.466	70°	.940	.342	2.747
26°	.438	.899	.488	71°	.946	.326	2.904
27°	.454	.891	.510	72°	.951	.309	3.078
28°	.469	.883	.532	73°	.956	.292	3.271
29°	.485	.875	.554	74°	.961	.276	3.487
30°	.5	.866	.577	75°	.966	.259	3.732
31°	.515	.857	.601	76°	.970	.242	4.011
32°	.530	.848	.625	77°	.974	.225	4.331
33°	.545	.839	.649	78°	.978	.208	4.705
34°	.559	.829	.675	79°	.982	.191	5.145
35°	.574	.819	.700	80°	.985	.174	5.671
36°	.588	.809	.727	81°	.988	.156	6.314
37°	.602	.799	.754	82°	.990	.139	7.115
38°	.616	.788	.781	83°	.993	.122	8.144
39°	.629	.777	.810	84°	.995	.105	9.514
40°	.643	.766	.839	85°	.996	.087	11.430
41°	.656	.755	.869	86°	.998	.070	14.301
42°	.669	.743	.900	87°	.999	.052	19.081
43°	.682	.731	.933	88°	.999	.035	28.636
44°	.695	.719	.966	89°	1.000	.017	57.290
45°	.707	.707	1.				

TABLE 11.5

Accuracy of the angle	Accuracy of the trigonometric ratios
Nearest degree	two significant digits
Nearest .1 degree or nearest 10 minutes	three significant digits
Nearest .01 degree or nearest minute	four significant digits
Nearest .001 degree or nearest 10 seconds	five significant digits
Nearest .001 degree or nearest second	six significant digits

To find sin 40°, enter

$$\boxed{40}\ \boxed{\sin}.\ \underbrace{.64278761}_{\text{display}}\ \text{(eight significant digits)}$$

To find cos 40°, enter

$$\boxed{40}\ \boxed{\cos}.\ \underbrace{.76604445}_{\text{display}}\ \text{(eight significant digits)}$$

We round the trigonometric ratios according to Table 11.5.

EXAMPLE 3

Using your calculator, find

a) sin 62° 15′ **b)** cos 62° 15′ **c)** tan 62° 15′.

Solution: On many calculators it is necessary to convert an angle written in degrees, minutes, and seconds to decimal parts of a degree before the calculator can be used to evaluate the trigonometric ratios of a particular angle.

For 62° 15′, we must convert 15′ to degrees.

$$15' = \frac{15^{x}}{1}\left(\frac{1°}{60^{x}}\right) = \frac{15°}{60} = .25° \text{ (see Section 11.1)}$$

Thus, 62° 15′ = 62.25°.

a) For sin 62° 15′, enter $\boxed{62.25}\ \boxed{\sin}.\ \underbrace{.88498764}_{\text{display}}$ (eight significant digits)

Thus, sin 62° 15′ = .8850 (four significant digits). ■

b) For cos 62° 15′, enter $\boxed{62.25}$ $\boxed{\cos}$. $\underset{\text{display}}{\underline{.46561452}}$ (eight significant digits)

Thus, cos 62° 15′ = .4656 (four significant digits).   ∎

c) For tan 62° 15′, enter $\boxed{62.25}$ $\boxed{\tan}$. $\underset{\text{display}}{\underline{1.9006874}}$ (eight significant digits)

Thus, tan 62° 15′ = 1.901 (four significant digits).   ∎

Note: Since 62° 15′ is accurate to the nearest minute, the trigonometric ratios are rounded to four significant digits, as suggested in Table 11.5.

⋰ From Trigonometric Ratio to Angle

Reversing the process, if the value of a trigonometric ratio is known, the angle can be determined by using the $\boxed{\text{inv}}$ or $\boxed{\text{arc}}$ keys of your calculator in conjunction with the $\boxed{\sin}$, $\boxed{\cos}$, and $\boxed{\tan}$ keys.

EXAMPLE 4

Find the acute angle θ.

a) sin θ = .935   **b)** cos θ = .935   **c)** tan θ = .935

Solution:

a) For sin θ = .935, enter $\boxed{.935}$ or $\begin{array}{c}\boxed{\text{inv}}\\\boxed{\text{arc}}\end{array}$ $\boxed{\sin}$. $\underset{\text{display}}{\underline{69.228145}}$

Thus, θ = 69.2° or 69° 10′.   ∎

b) For cos θ = .935, enter $\boxed{.935}$ or $\begin{array}{c}\boxed{\text{inv}}\\\boxed{\text{arc}}\end{array}$ $\boxed{\cos}$. $\underset{\text{display}}{\underline{20.771855}}$

Thus, θ = 20.8° or 20° 50′.   ∎

c) For tan θ = .935, enter $\boxed{.935}$ or $\begin{array}{c}\boxed{\text{inv}}\\\boxed{\text{arc}}\end{array}$ $\boxed{\tan}$. $\underset{\text{display}}{\underline{43.076059}}$

Thus, θ = 43.1° or 43° 00′.   ∎

Note: Since .935 is accurate to three significant digits, the angles are rounded to the nearest .1 degree or the nearest 10 minutes as suggested in Table 11.5.

EXERCISES 11.5

Referring to Fig. 11.73, determine the following trigonometric ratios. Write the ratios as decimal fractions.

1. sin *A*   **2.** cos *A*   **3.** tan *A*

4. sin *B*   **5.** cos *B*   **6.** tan *B*

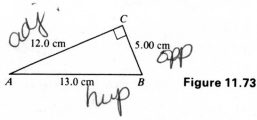

Figure 11.73

Referring to Fig. 11.74, determine the following trigonometric ratios. Write the ratios as decimal fractions.

7. tan *B*   **8.** cos *B*   **9.** sin *B*

10. tan *A*   **11.** cos *A*   **12.** sin *A*

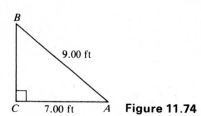

Figure 11.74

Referring to Fig. 11.75, determine the following trigonometric ratios. Write the ratios in reduced fractional form.

13. a) sin *x*   **14. a)** sin *y*   **15. a)** sin *z*   **16. a)** sin *A*

      b) cos *x*   **b)** cos *y*   **b)** cos *z*   **b)** cos *A*

      c) tan *x*   **c)** tan *y*   **c)** tan *z*   **c)** tan *A*

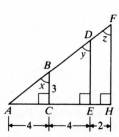

Figure 11.75

Using your calculator, evaluate the following. Round off your results in accordance with Table 11.5.

17. sin 82°

18. cos 65°

19. tan 34.9°

20. sin 6.3°

21. cos 18° 20′

22. tan 58° 50′

23. sin 49° 55′

24. cos 18° 43′

25. tan 16° 19′ 30″

26. sin 27° 06′ 10″

27. cos 6° 48′ 45″

28. tan 72° 15′ 15″

Using you calculator, find the measure of the acute angle θ. Round off your results in accordance with Table 11.5.

29. $\sin \theta = .26$

30. $\cos \theta = .89$

31. $\tan \theta = 1.64$

32. $\sin \theta = .234$

33. $\cos \theta = .0751$

34. $\tan \theta = .0530$

35. $\sin \theta = .6410$

36. $\cos \theta = .9123$

37. $\tan \theta = 2.7542$

38. $\sin \theta = .64107$

39. $\cos \theta = .925743$

40. $\tan \theta = 3.65420$

Using the fact that in a 30°–60°–90° triangle $\dfrac{\text{leg opposite}}{30° \text{ angle}} : \dfrac{\text{leg opposite}}{60° \text{ angle}} : \text{hypotenuse} = 1 : \sqrt{3} : 2$, and in a 45°–90° triangle leg : leg : hypotenuse = $1 : 1 : \sqrt{2}$, fill in the table below. Write the ratios in simplified radical form.

41.

θ	$\sin \theta$	$\cos \theta$	$\tan \theta$
30°			
60°			
45°			

11.6 RIGHT TRIANGLE TRIGONOMETRY

Objectives

- ⊡ Given the length of one side and the measure of an acute angle of a right triangle, to determine the lengths of the other sides.
- ⊡ Given the lengths of two sides of a right triangle, to determine the measure of the acute angles.

⊡ Determining the Unknown Sides of a Right Triangle

In a right triangle, when the length of one side and the measure of one acute angle are known, the trigonometric ratios can be used to determine the lengths of the other sides of the right triangle.

EXAMPLE 1

Three drill holes are located in a steel plate as shown in Fig. 11.76. Find the distances x and y.

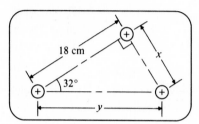

Figure 11.76

Solution: In relation to the given angle (32°), we know that

x is the *opposite* side,

18 cm is the *adjacent* side, and

y is the *hypotenuse*.

Now, to find x, use the fact that

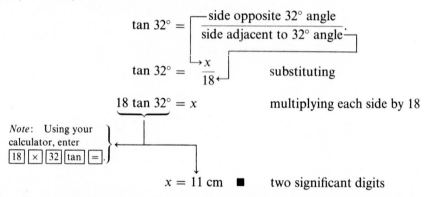

$$\tan 32° = \frac{\text{side opposite 32° angle}}{\text{side adjacent to 32° angle}}.$$

$$\tan 32° = \frac{x}{18} \qquad \text{substituting}$$

$$18 \tan 32° = x \qquad \text{multiplying each side by 18}$$

Note: Using your calculator, enter
[18] [×] [32] [tan] [=].

$$x = 11 \text{ cm} \quad \blacksquare \qquad \text{two significant digits}$$

To find y, use the fact that

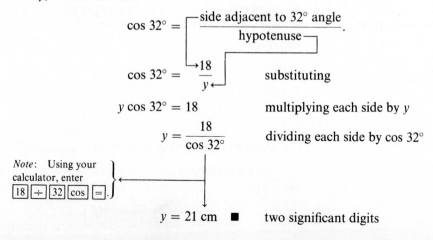

$$\cos 32° = \frac{\text{side adjacent to 32° angle}}{\text{hypotenuse}}.$$

$$\cos 32° = \frac{18}{y} \qquad \text{substituting}$$

$$y \cos 32° = 18 \qquad \text{multiplying each side by } y$$

$$y = \frac{18}{\cos 32°} \qquad \text{dividing each side by } \cos 32°$$

Note: Using your calculator, enter
[18] [÷] [32] [cos] [=].

$$y = 21 \text{ cm} \quad \blacksquare \qquad \text{two significant digits}$$

Note: To check, use the fact that

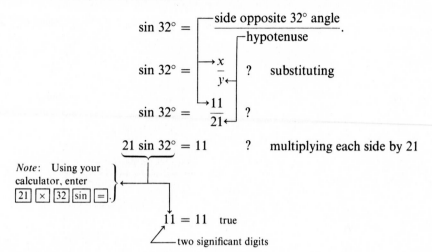

$$\sin 32° = \left[\begin{array}{l}\text{—side opposite } 32° \text{ angle} \\ \text{—hypotenuse}\end{array}\right].$$

$$\sin 32° = \left[\begin{array}{l}\dfrac{x}{y}\end{array}\right] \quad ? \quad \text{substituting}$$

$$\sin 32° = \dfrac{11}{21} \quad ?$$

$$21 \sin 32° = 11 \quad\quad ? \quad \text{multiplying each side by 21}$$

Note: Using your
calculator, enter
$\boxed{21}\ \boxed{\times}\ \boxed{32}\ \boxed{\sin}\ \boxed{=}$.

$$11 = 11 \quad \text{true}$$

— two significant digits

EXAMPLE 2

Find the area of the triangular piece of land shown in Fig. 11.77.

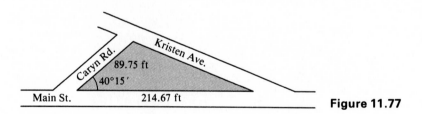

Figure 11.77

Solution: First, form a right triangle by drawing the altitude to Main St. as shown in Fig. 11.78.

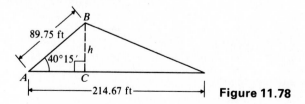

Figure 11.78

Working with $\triangle ABC$, in relation to the given angle (40° 15′) we know that

h is the *opposite* side and

89.75 ft is the *hypotenuse*.

Now, to find h, use the fact that

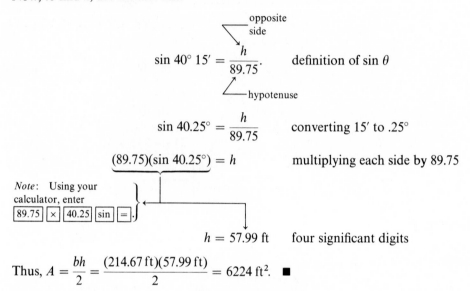

$$\sin 40° \, 15' = \frac{h}{89.75}.$$ definition of $\sin \theta$

$$\sin 40.25° = \frac{h}{89.75}$$ converting $15'$ to $.25°$

$$(89.75)(\sin 40.25°) = h$$ multiplying each side by 89.75

Note: Using your calculator, enter

$$\boxed{89.75} \; \boxed{\times} \; \boxed{40.25} \; \boxed{\sin} \; \boxed{=}.$$

$$h = 57.99 \text{ ft}$$ four significant digits

Thus, $A = \dfrac{bh}{2} = \dfrac{(214.67 \text{ ft})(57.99 \text{ ft})}{2} = 6224 \text{ ft}^2.$ ∎

Note: In a triangle (Fig. 11.79), if two sides (a and b) and the included angle (θ) between these sides are known, we can develop the following area formula for a triangle.

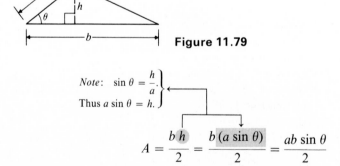

Figure 11.79

Note: $\sin \theta = \dfrac{h}{a}.$

Thus $a \sin \theta = h.$

$$A = \frac{bh}{2} = \frac{b\,(a \sin \theta)}{2} = \frac{ab \sin \theta}{2}$$

Using $A = \dfrac{ab \sin \theta}{2}$, find the area of the triangular piece of land in Fig. 11.77. See if you obtain the same result as above.

⊡ Determining the Acute Angles of a Right Triangle

In a right triangle, when the lengths of two sides are known, the definitions of the trigonometric ratios can be used to determine the measure of the acute angles of the right triangle.

EXAMPLE 3

Determine the measure of angles α (alpha) and β (beta) for the roof truss shown in Fig. 11.80.

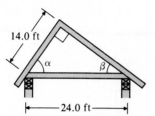

Figure 11.80

Solution: In relation to angle α, we know that

14.0 ft is the *adjacent* side and
24.0 ft is the *hypotenuse*.

Thus, to find α, use the fact that

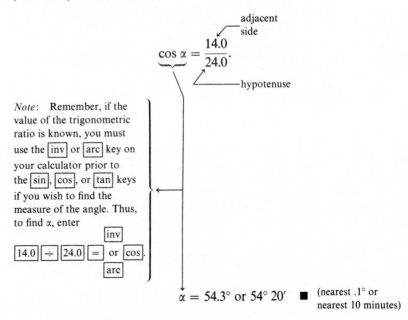

$$\cos \alpha = \frac{14.0}{24.0}.$$

adjacent side

hypotenuse

Note: Remember, if the value of the trigonometric ratio is known, you must use the $\boxed{\text{inv}}$ or $\boxed{\text{arc}}$ key on your calculator prior to the $\boxed{\text{sin}}$, $\boxed{\text{cos}}$, or $\boxed{\text{tan}}$ keys if you wish to find the measure of the angle. Thus, to find α, enter

$$\boxed{14.0}\ \boxed{\div}\ \boxed{24.0}\ \boxed{=}\ \text{or}\ \boxed{\text{cos}}\ \begin{array}{l}\boxed{\text{inv}}\\ \boxed{\text{arc}}\end{array}$$

$\alpha = 54.3°$ or $54°\,20'$ ■ (nearest .1° or nearest 10 minutes)

In relation to angle β, we know that

14.0 ft is the *opposite* side and
24.0 ft is the *hypotenuse*.

Thus, to find β, use the fact that

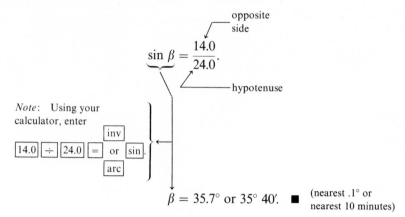

$$\underbrace{\sin \beta}_{} = \frac{14.0}{24.0}.$$

Note: Using your calculator, enter

$\boxed{14.0} \ \boxed{\div} \ \boxed{24.0} \ \boxed{=} \ \text{or} \ \boxed{\begin{array}{c}\text{inv}\\\text{sin}\\\text{arc}\end{array}}$.

$\beta = 35.7°$ or $35° \ 40'$. ■ (nearest .1° or nearest 10 minutes)

Note: To check, make sure that the acute angles of the right triangle are complementary.

$$\alpha + \beta = 90°?$$
$$54.3° + 35.7° = 90°?$$
$$90° = 90° \quad \text{true}$$

EXAMPLE 4

Find the measure of angle θ for the drill bit shown in Fig. 11.81(a).

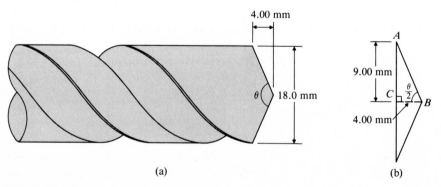

(a)   (b)

Figure 11.81

Solution: The end of the drill bit forms an isosceles triangle. If the altitude is drawn from the vertex angle to the base of an isosceles triangle, the vertex angle (θ) and the base (18.0 cm) are both cut in half and two identical right triangles are formed (Fig. 11.81b).

Working with $\triangle ABC$ in relation to the angle $\dfrac{\theta}{2}$, we know that

> 9.00 mm is the *opposite* side and
>
> 4.00 mm is the *adjacent* side.

Thus, to find $\dfrac{\theta}{2}$, use the fact that

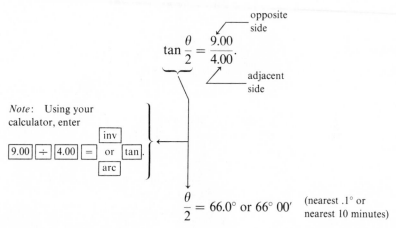

$$\tan \frac{\theta}{2} = \frac{9.00}{4.00}.$$

Note: Using your calculator, enter

$$\boxed{9.00}\ \boxed{\div}\ \boxed{4.00}\ \boxed{=}\ \text{or}\ \boxed{\begin{array}{c}\text{inv}\\ \text{tan}\\ \text{arc}\end{array}}.$$

$$\frac{\theta}{2} = 66.0° \text{ or } 66° \ 00'$$ (nearest .1° or nearest 10 minutes)

Now, if $\dfrac{\theta}{2} = 66.0°$, then $\theta = 2(66.0°) = 132°$. ∎

Application

SURVEYING

Shown in Fig. 11.82 is a **transit**, which is used by surveyors to measure angles. The instrument consists of a *telescope*, which the surveyor looks through to determine the line of sight. The telescope can be moved up or down in order to measure vertical angles. When sighting a point that is either above or below the horizontal, scale A is used to measure the vertical angle. If the point sighted is above the horizontal, the angle measured on scale A is called an **angle of elevation** (see Fig. 11.83). If the point sighted is below the horizontal, the angle measured on scale A is called the **angle of depression** (see Fig. 11.84). Suppose a surveyor must find the horizontal distance between points C and D (see Figs. 11.83 and 11.84). Suppose also that the terrain is too steep for this horizontal distance to be measured directly. By using a transit and trigonometry, the horizontal distance from C to D can be found as follows:

1. With the transit properly set up at point C, sight on point D and measure the vertical angle (angle of depression or angle of elevation) on scale A of the transit.

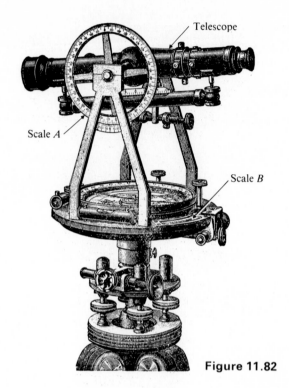

Figure 11.82

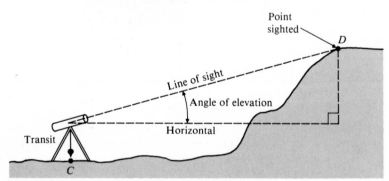

Figure 11.83

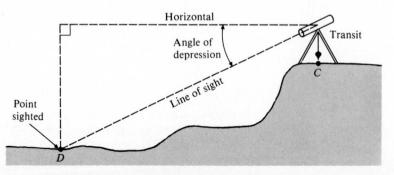

Figure 11.84

2. Measure the distance along the line of sight.

3. Use trigonometry to find the horizontal distance from point C to D.

EXAMPLE 5

Referring to Fig. 11.84, suppose the angle of depression measures 24° 51′ and the distance along the line of sight measures 87.65 ft. Find the horizontal distance from C to D.

Solution: Let x equal the horizontal distance from C to D (Fig. 11.85). In relation

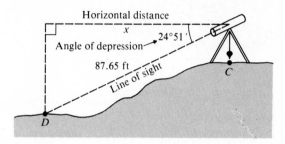

Figure 11.85

to the 24° 51′ angle, we know that

$$x \quad \text{is the } \textit{adjacent side} \text{ and}$$
$$87.65 \text{ ft is the } \textit{hypotenuse}.$$

Thus, to find x, use the fact that

$$\cos 24° 51' = \frac{x}{87.65}. \qquad \qquad \text{definition of } \cos \theta$$

$$\cos 24.85° = \frac{x}{87.65} \qquad \qquad \text{converting } 51' \text{ to } .85°$$

$$(87.65)(\cos 24.85°) = x \qquad \qquad \begin{array}{l}\text{multiplying each side}\\\text{by } 87.65\end{array}$$

Note: Using your calculator, enter
$\boxed{87.65} \boxed{\times} \boxed{24.85} \boxed{\cos} \boxed{=}.$

$$x = 79.53 \text{ ft} \quad \blacksquare \qquad \text{four significant digits}$$

   The telescope on the transit may also be moved to the left or right in order to measure horizontal angles. The horizontal angles are measured on scale B of the transit.

Suppose a surveyor must find the horizontal distance from point C to point D, which are on opposite sides of a river (see Fig. 11.86). Suppose also that the

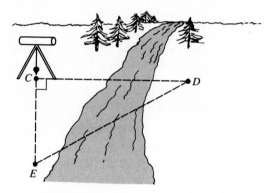

Figure 11.86

river is too wide for this horizontal distance to be measured directly. By using a transit and trigonometry, the horizontal distance from C to D can be found as follows:

1. With the transit properly set up at point C, sight on point D and turn the telescope 90° 00′ 00″ to some point E on the same side of the river as point C.
2. Measure the distance from C to E.
3. Set up the transit at point E and sight back to point C.
4. Turn the telescope to point D and read the measure of the horizontal angle ($\measuredangle CED$) on scale B of the transit.
5. Use trigonometry to find the distance from point C to point D.

EXAMPLE 6

Referring to Fig. 11.86, suppose the distance from point C to point E measures 100.00 ft and the measure of the horizontal angle ($\measuredangle CED$) is 72° 54′ 30″. Find the horizontal distance from C to D.

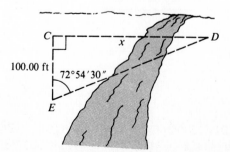

Figure 11.87

Solution: Let *x* equal the horizontal distance from *C* to *D* (Fig. 11.87). In relation to the 72° 54′ 30″ angle, we know that

$$x \qquad \text{is the } \textit{opposite} \text{ side and}$$
$$100.00 \text{ ft is the } \textit{adjacent} \text{ side.}$$

Thus, to find *x*, use the fact that

$$\tan 72° 54′ 30″ = \frac{x}{100.00}. \qquad \text{definition of tan } \theta$$

opposite side / adjacent side

$$\tan 72.908° = \frac{x}{100.00} \qquad \text{converting } 54′ 30″ \text{ to } .908°$$

$$(100.00)(\tan 72.908°) = x \qquad \text{multiplying each side by 100.00}$$

Note: Using your calculator, enter

| 100.00 | × | 72.908 | tan | = |

$$x = 325.22 \text{ ft} \quad \blacksquare \qquad \text{five significant digits}$$

EXERCISES 11.6

In each of the following problems, find the value of *x* and *y*.

1.

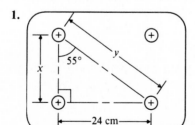

2.

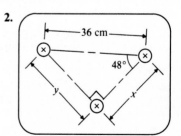

3.

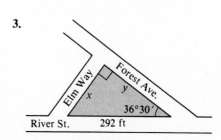

4.

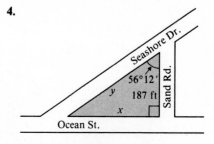

5.

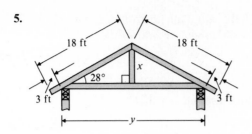

6.

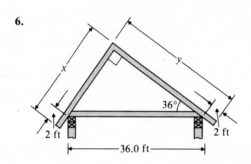

7.

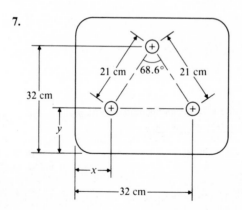

8.

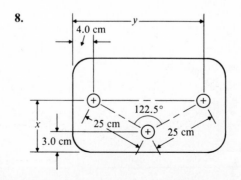

9.

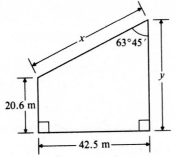

10.

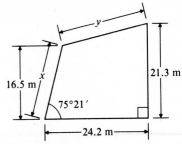

11.

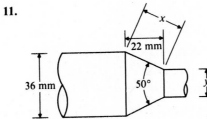

12.

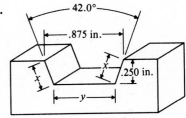

In each of the following problems, find the values of α and β.

13.

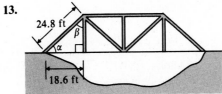

14.

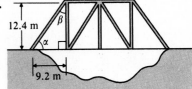

15.

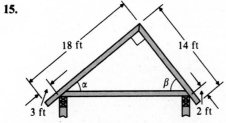

16.

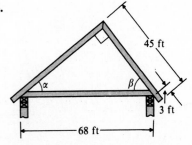

17.

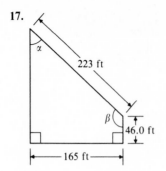

223 ft

α

β

46.0 ft

165 ft

18.

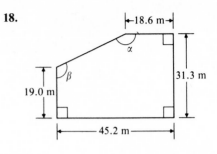

18.6 m

α

β

19.0 m

31.3 m

45.2 m

19.

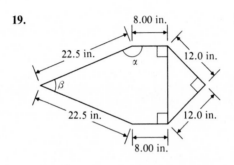

8.00 in.

22.5 in.

α

β

12.0 in.

22.5 in.

12.0 in.

8.00 in.

20.

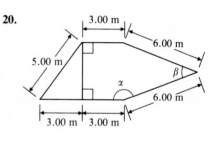

3.00 m

6.00 m

5.00 m

β

α

6.00 m

3.00 m 3.00 m

21.

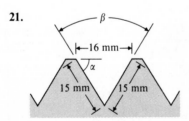

β

16 mm

α

15 mm 15 mm

22.

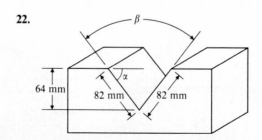

β

α

64 mm

82 mm 82 mm

23.

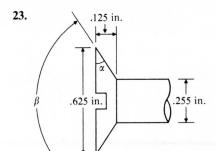

24.

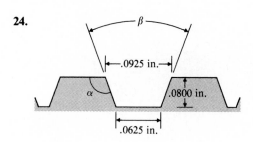

Find the area of each shaded region.

25.

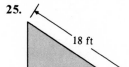

26.

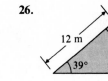

27.

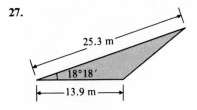

28.

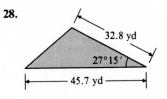

29.

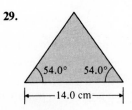

30.

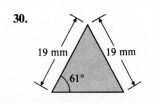

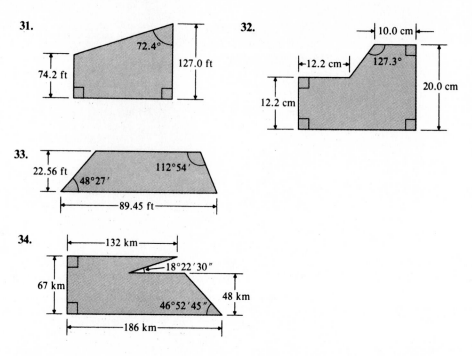

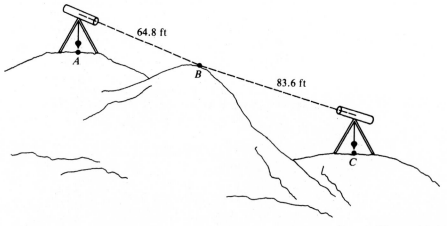

Figure 11.88

Referring to Fig. 11.88, the angle of depression from point A to point B is $21° \, 33'$, and the angle of elevation from point C to point B is $16° \, 09'$. Assume the points A, B, and C lie in the same vertical plane.

35. Determine the horizontal distance from point A to point C.

36. Determine the vertical distance from point A to point C.

Referring to Fig. 11.89, a plot plan calls for the length of a property line to be 227.63 ft. When surveying this property, it is found that a large tree is directly on line making the distance of

227.63 ft impossible to measure directly. From the property line the surveyor turns the transit 90° 00′ 00″ to point C and measures 20.00 ft. Setting up the transit at point C, she sights back to point A.

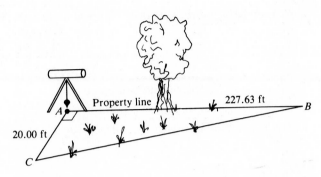

Property line   227.63 ft   B

20.00 ft

A

C

Figure 11.89

37. Determine the measure of the horizontal angle ($\angle ACB$) to which she must turn the transit to locate point B correctly. (Record your answer to the nearest 10″).

38. Determine the distance to be measured from point C to locate point B correctly.

Chapter Review

REVIEW EXERCISES

The following review exercises are grouped according to the objectives that should have been mastered in Chapter 11. Work each problem carefully. If any weaknesses appear, immediately refer to and read the subsection that matches that objective.

11.1 ANGLES

Objectives

⊡ To determine the measure of a given angle by using a protractor.

Measure.

1. Using the protractor in Fig. 11.90, find the measure of
a) $\angle AOB$ **b)** $\angle AOC$ **c)** $\angle COD$ **d)** $\angle BOC$

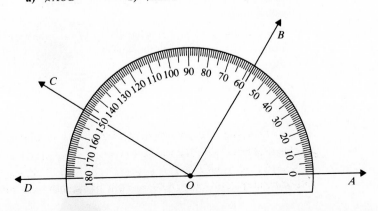

Figure 11.90

☐ To convert angles from degrees, minutes, and seconds to degrees only or vice versa.

Convert.

2. Convert the following angles to degrees only.

    a) 42° 25′   **b)** 143° 15′ 15″

3. Convert the following angles to degrees, minutes, and seconds where appropriate.

    a) 14.7°   **b)** 102.643°

☐ Given a geometric figure containing adjacent, complementary or supplementary angles, to determine the measure of the missing angle(s).

Find the measure of the designated angles.

4. Referring to Fig. 11.91, find the measure of ∢AOC.

5. Referring to Fig. 11.92, find the measure of ∢DOE.

6. Referring to Fig. 11.93, find the measure of ∢a, ∢b, and ∢c.

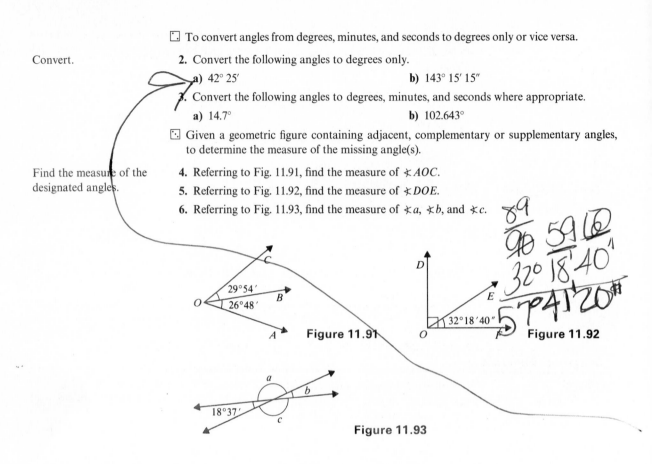

Figure 11.91

Figure 11.92

Figure 11.93

11.2 INTERIOR ANGLES OF TRIANGLES AND QUADRILATERALS

Objectives

☐ Given a geometric figure containing two parallel lines cut by a transversal, to determine the measure of any missing angle.

Find the measure of the designated angles.

7. Referring to Fig. 11.94, $\overleftrightarrow{CD}$ is parallel to $\overleftrightarrow{EF}$, and the measure of ∢1 is 106°. Determine the measure of each of the other angles.

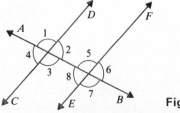

Figure 11.94

⊡ Given a parallelogram, to determine the measure of any missing interior angle.

Find the measure of the designated angles.

8. Referring to the parallelogram in Fig. 11.95, determine the measure of ⊀A, ⊀B, and ⊀C.

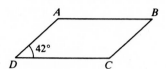

Figure 11.95

⊡ Given a triangle, to determine the measure of any missing interior angle.

Find the measure of the designated angles.

9. The triangle in Fig. 11.96 is isosceles, with $\overline{AB} = \overline{AC}$. Determine the measure of the numbered angles.

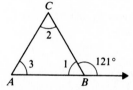

Figure 11.96

⊡ Given a quadrilateral, to determine the measure of any missing interior angle.

Find the measure of the designated angle.

10. Referring to the tract of land in Fig. 11.97, find the measure of ⊀D.

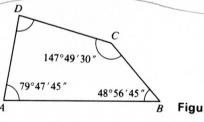

Figure 11.97

Handwritten note:
60 min = 1 deg.
60 sec. = 1 min.
45°25′ (25)(1/60)
↳ or
25/60

11.3 SIMILARITIES IN TRIANGLES

Objectives

⊡ Given two similar triangles, to determine unknown angles and sides.

Solve by using similar triangles.

11. Referring to Fig. 11.98, if △ABC ~ △DEF, find the measure of

a) ⊀B   **b)** ⊀F   **c)** ⊀A   **d)** ⊀D

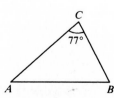

Figure 11.98

12. Referring to Fig. 11.99, if triangle $\triangle PQR \sim \triangle XYZ$, find the length of

a) y **b)** r

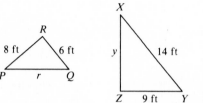

Figure 11.99

⊡ To determine whether two triangles are similar.

Solve (if possible) by similar triangles.

13. Determine whether the triangles formed by the drill holes in Fig. 11.100 are similar. If they are similar, find the distance x.

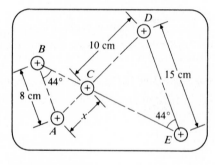

Figure 11.100

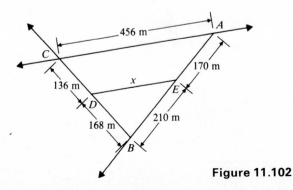

Figure 11.101

14. Determine whether the triangles formed by the subdivision in Fig. 11.101 are similar. If they are similar, find the measure of $\angle ABD$.

15. Determine whether the triangles formed by the roadways in Fig. 11.102 are similar. If they are similar, find the distance x.

Figure 11.102

⊡ Given a right triangle and the altitude to its hypotenuse, to determine the length of an unknown side.

Solve by using similar triangles.

16. Referring to Fig. 11.103, find the distances.

a) x **b)** y **c)** z

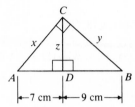

Figure 11.103

11.4 RIGHT TRIANGLES

Objectives

⊡ To determine the length of an unknown side of a right triangle by applying the Pythagorean theorem.

Solve by using the Pythagorean theorem.

17. Referring to Fig. 11.104, find the length of the stair stringer.

18. Referring to the drill holes in Fig. 11.105, determine the distance x.

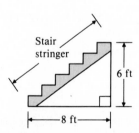

Figure 11.104

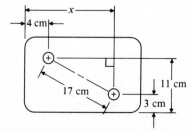

Figure 11.105

19. Find the area of the isosceles triangle in Fig. 11.106.

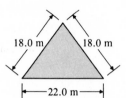

Figure 11.106

☐ To determine the length of an unknown side of a 45°–90° triangle.

Solve by using the ratio of the sides of a 45°–90° triangle.

20. Referring to Fig. 11.107, find the length of the rafter. Write the answer in simplified radical form.

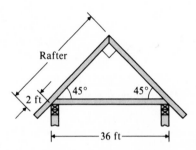

Figure 11.107

☐ To determine the length of an unknown side of a 30°–60°–90° triangle.

Solve by using the ratio of the sides of a 30°–60°–90° triangle.

21. Referring to the intersecting roadways in Fig. 11.108, find the distances x and y. Write the answers in simplified radical form.

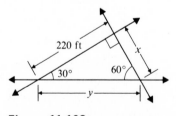

Figure 11.108

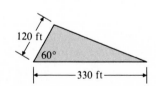

Figure 11.109

22. Find the area of the triangular piece of land shown in Fig. 11.109. Write the answer in simplified radical form.

11.5 THE TRIGONOMETRIC RATIOS

Objectives

☐ Given a right triangle and the length of at least two of its sides, to determine the trigonometric ratios for an acute angle of the triangle.

Determine the
trigonometric ratios.

23. Determine the values of sin θ, cos θ, and tan θ for each of the following triangles. Write the ratios as decimal fractions.

a)
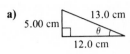

b)
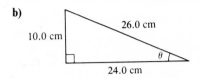

24. Referring to the right triangle in Fig. 11.110, find

a) sin A   **b)** cos B

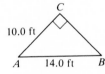

Figure 11.110

⊡ To determine the trigonometric ratios for a given acute angle by using a calculator.

Evaluate.

25. Using your calculator, evaluate

a) sin 43° 27′   **b)** cos 43° 27′   **c)** tan 43° 27′

⊡ Given the value of a trigonometric ratio, to determine the measure of the acute angle by using a calculator.

Solve for θ.

26. Using your calculator, find the acute angle θ.

a) sin θ = .621   **b)** cos θ = .621   **c)** tan θ = .621

11.6 RIGHT TRIANGLE TRIGONOMETRY

Objectives

☐ Given the length of one side and the measure of an acute angle of a right triangle, to determine the lengths of the other sides.

Solve by using the
trigonometric ratios.

27. Referring to the drill holes in Fig. 11.111, find the distances x and y.

28. Find the area of the tract of land in Fig. 11.112.

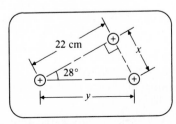

Figure 11.111

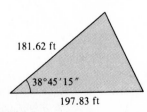

Figure 11.112

☐ Given the lengths of two sides of a right triangle, to determine the measure of the acute angles.

Solve by using the trigonometric ratios.

29. Referring to the roof truss in Fig. 11.113, find the measure of angles α and β.

30. Referring to the drill bit in Fig. 11.114, find the measure of angle θ.

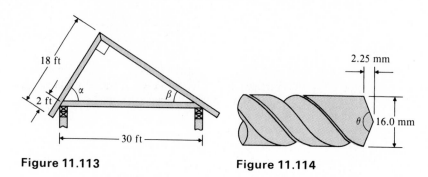

Figure 11.113   **Figure 11.114**

CHAPTER TEST
Time
50 minutes

Score
15, 14 excellent
13, 12 good
11, 10, 9 fair
below 8 poor

If you have worked through the Review Exercises and corrected any weaknesses, you are ready to take the following Chapter Test.

1. Convert 67° 24′ 30″ to degrees only.

2. Convert 87.36° to degrees, minutes, and seconds if appropriate.

3. Referring to Fig. 1.115, find the measure of $\angle BOC$.

4. Four straight roadways intersect as shown in Fig. 11.116. If $\overrightarrow{AC}$ is parallel to $\overrightarrow{ED}$, find the measures of the numbered angles.

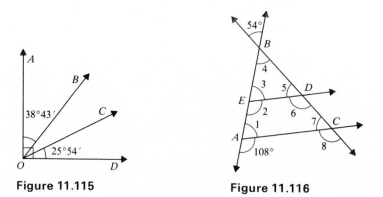

Figure 11.115   **Figure 11.116**

5. The tract of land shown in Fig. 11.117 is subdivided into a triangle and a parallelogram. Find the measures of the numbered angles.

6. Determine whether the triangles formed by the drill holes in Fig. 11.118 are similar. If they are similar, determine the distance *x*.

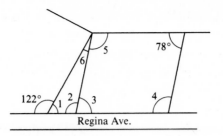

Figure 11.117

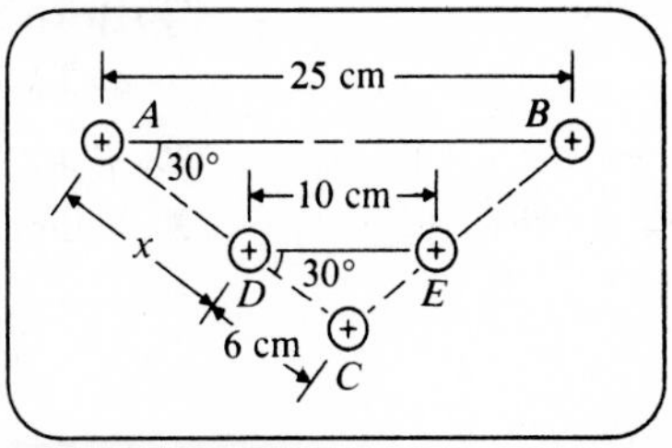

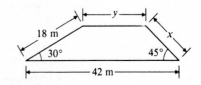

Figure 11.118

7. Referring to Fig. 11.119, if $x = 16$ and $z = 12$, find y, a, and b. Express all irrational answers in simplified radical form.

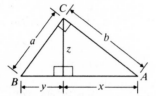

Figure 11.119

8. Referring to the roof truss in Fig. 11.120, find the distance x.

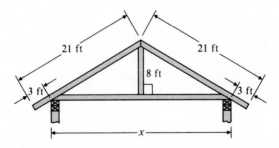

Figure 11.120

9. Find the area of the isosceles triangle shown in Fig. 11.121.

10. Referring to Fig. 11.122, find the distances x and y. Express all irrational answers in simplified radical form.

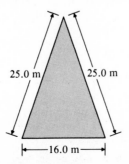

Figure 11.121

Figure 11.122

11. Referring to Fig. 11.123, determine

a) tan *A*, tan *B* **b)** sin *A*, sin *B* **c)** cos *A*, cos *B*

12. Referring to Fig. 11.124, find the distances *x* and *y*.

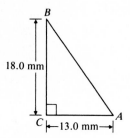

Figure 11.123

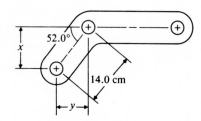

Figure 11.124

13. Referring to the bridge truss in Fig. 11.125, find the measures of α and β.

14. Find the area of the tract of land shown in Fig. 11.126.

15. Referring to Fig. 11.127, determine *x* and θ.

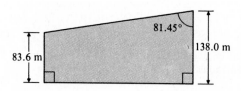

Figure 11.125

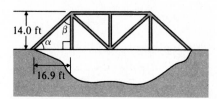

Figure 11.126

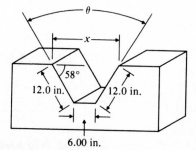

Figure 11.127

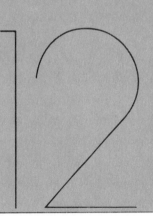

12

TRIGONOMETRY

12.1 ANGLES IN THE RECTANGULAR COORDINATE SYSTEM

Objectives

☐ To determine the smallest positive coterminal angle for a given directed angle.

☐ To determine the reference angle for a given directed angle.

☐ To determine whether the trigonometric ratios of a given directed angle are positive or negative.

☐ Coterminal Angles

For geometry and right triangle trigonometry (see Chapter 11), it was sufficient to work with angles between 0° and 180°. For most trigonometry, however, we need to extend the meaning of an angle to include angles greater than 180° and less than 0°.

(a) Initial position   (b) Terminal position   **Figure 12.1**

Consider a gear rotating *counterclockwise* from an initial position to a terminal position as shown in Fig. 12.1. We can think of this rotation as forming an acute angle ($\sphericalangle AOB$) with ray $\overrightarrow{OA}$ the *initial side* and ray $\overrightarrow{OB}$ the *terminal side* of this angle (see Fig. 12.2). Now, consider a gear rotating *clockwise* from an initial position to a terminal position as shown in Fig. 12.3. Again, we can think of this rotation as forming an acute angle ($\sphericalangle AOD$) with ray $\overrightarrow{OA}$ the *initial side* and ray $\overrightarrow{OD}$ the *terminal side* of this angle (see Fig. 12.4).

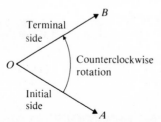

Figure 12.2

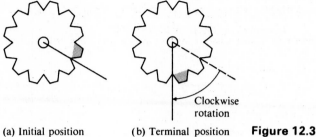

(a) Initial position   (b) Terminal position   **Figure 12.3**

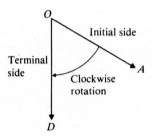

Figure 12.4

Angles generated by clockwise and counterclockwise rotation are referred to as **directed angles**. In the study of trigonometry, directed angles are placed in the rectangular coordinate system using standard position. **Standard position** means that the vertex of the directed angle is at the origin and its initial side is along the positive x-axis (see Fig. 12.5). One degree ($1°$) is defined as $\dfrac{1}{360}$ of a full counterclockwise rotation. By convention, we consider angles formed by counterclockwise rotation **positive angles** and angles formed by clockwise rotation **negative angles**.

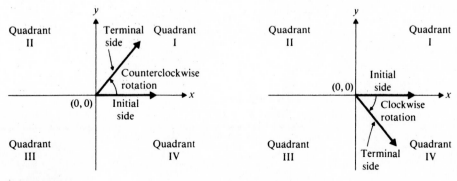

Figure 12.5

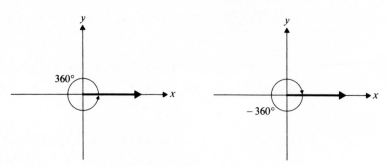

Figure 12.6

Thus, there are 360° in one full counterclockwise rotation and −360° in one full clockwise rotation (see Fig. 12.6). In two counterclockwise rotations there are 2(360°) = 720°, and in two clockwise rotations there are 2(−360°) = −720° (see Fig. 12.7). In one-half of a counterclockwise rotation there are $\frac{1}{2}$ (360°) = 180°, and in one-half of a clockwise rotation there are $\frac{1}{2}$ (−360°) = −180° (see Fig. 12.8). In one-fourth of a counterclockwise rotation there are $\frac{1}{4}$ (360°) = 90°, and in one-fourth of a clockwise rotation there are $\frac{1}{4}$ (−360°) = −90° (see Fig. 12.9).

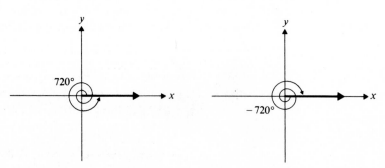

Figure 12.7

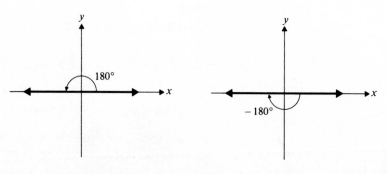

Figure 12.8

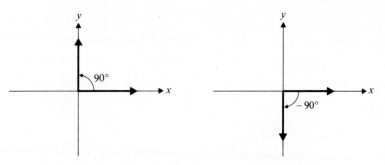

Figure 12.9

In three-fourths of a counterclockwise rotation, there are $\frac{3}{4}(360°) = 270°$ and in three-fourths of a clockwise rotation there are $\frac{3}{4}(-360°) = -270°$ (see Fig. 12.10). If an angle has no rotation at all, then its measure is defined as $0°$ (see Fig. 12.11). You may think that $0°$, $360°$, $-360°$, $720°$, and $-720°$ are the same angle. This is not quite true, since these angles differ in the amount or direction of their rotation. These angles are closely related in that they have the same terminal side. Angles in standard position that have the same terminal side are called **coterminal angles**.

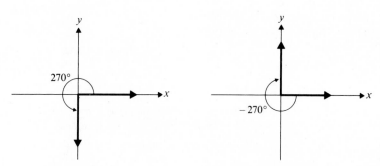

Figure 12.10

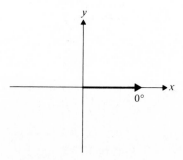

Figure 12.11

Thus, 0°, 360°, −360°, 720°, and −720° are coterminal angles. Similarly,

90° and −270° are coterminal angles;

−90° and 270° are coterminal angles; and

180° and −180° are coterminal angles.

EXAMPLE 1

Determine the smallest positive coterminal angle for

a) −290°   **b)** 600°   **c)** −780°

Solution:

a) A −290° angle is between −270° and −360°. Therefore, −290° is located in the first quadrant (Fig. 12.12). The smallest positive coterminal angle for −290° is 360° − 290°, or 70°. ■

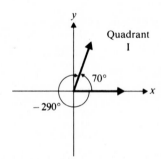

Figure 12.12

b) Think of 600° as 360° + 240° (Fig. 12.13). The smallest positive coterminal

one full counter-_____ ∖ ∠__a third
clockwise rotation   quadrant angle

angle for 600° is 600° − 360°, or 240°. ■

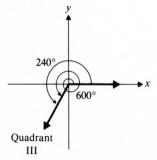

Figure 12.13

c) Think of $-780°$ as $-720° + (-60°)$ (Fig. 12.14). Now, the smallest positive

two full clock-⟍   ⎿ a fourth
wise rotations   quadrant angle

coterminal angle for $-60°$ is $360° - 60°$, or $300°$. Thus, the smallest positive coterminal angle for $-780°$ is also $300°$. ∎

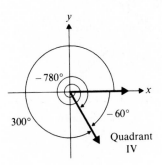

Figure 12.14

⬚ Reference Angles

In Section 12.2 you will be given the value of a trigonometric ratio and asked to find the directed angles between $0°$ and $360°$ that have that ratio. In order to do this, you will need to work with reference angles. A **reference angle** is a positive acute angle that is formed with the terminal side of a directed angle and the x-axis. Shown in Fig. 12.15 are the reference angles for angles whose terminal sides are in the first, second, third, and fourth quadrants.

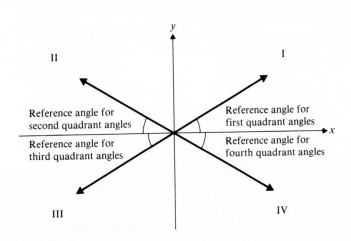

Figure 12.15

A reference angle is always formed with respect to the x-axis, *never* with respect to the y-axis (Fig. 12.16).

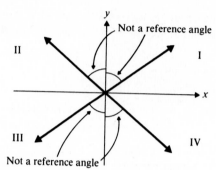

Figure 12.16

EXAMPLE 2

Find the reference angle for

a) 20°   **b)** 120°   **c)** 225°   **d)** 350°   **e)** 600°

Solution:

a) The reference angle for any positive acute angle is itself. Thus, the reference angle for 20° is 20° (Fig. 12.17). ∎

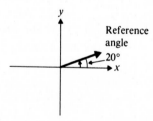

Figure 12.17

b) A 120° angle is located in the second quadrant. The reference angle for 120° is 180° − 120°, or 60° (Fig. 12.18). ∎

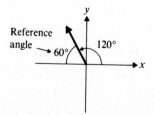

Figure 12.18

c) A 225° angle is located in the third quadrant. The reference angle for 225° is 225° − 180°, or 45° (Fig. 12.19).   ■

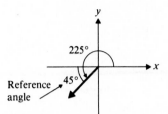

Figure 12.19

d) A 350° angle is located in the fourth quadrant. The reference angle for 350° is 360° − 350°, or 10° (Fig. 12.20).   ■

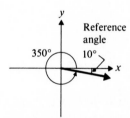

Figure 12.20

e) The smallest positive coterminal angle for 600° is 240° (see Example 1b). Since 240° is located in the third quadrant, the reference angle for 240° is 240° − 180°, or 60°. Thus, the reference angle for 600° is also 60° (Fig. 12.21).   ■

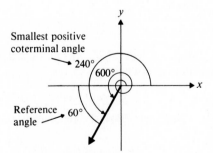

Figure 12.21

⠗ Algebraic Signs of the Trigonometric Ratios

Suppose we draw an acute angle θ in standard position and let (x, y) be a point on its terminal side (see Fig. 12.22). If we draw a perpendicular line from the point (x, y) to the positive x-axis as shown in Fig. 12.23, a right triangle is formed.

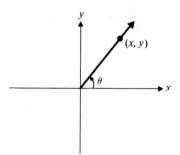

Figure 12.22

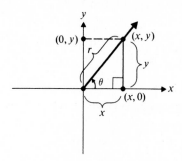

Figure 12.23

The value of the y-coordinate can be considered the *opposite side* and the value of the x-coordinate can be considered the *adjacent side* of the right triangle. Let r be the *hypotenuse* of the right triangle; then

$$r^2 = x^2 + y^2 \qquad \text{Pythagorean theorem}$$

$$\text{or} \quad r = \sqrt{x^2 + y^2}. \qquad \text{square root property}$$

Note: The hypotenuse is the positive square root of the sum of the squares of the coordinates x and y.

Using the right triangle in Fig. 12.23, we can now define the three basic trigonometric ratios by using the coordinates (x, y).

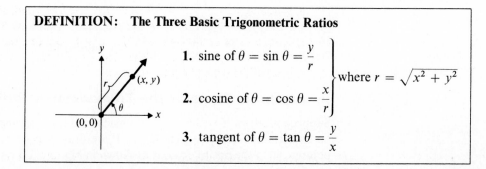

DEFINITION: The Three Basic Trigonometric Ratios

1. sine of $\theta = \sin \theta = \dfrac{y}{r}$

2. cosine of $\theta = \cos \theta = \dfrac{x}{r}$

where $r = \sqrt{x^2 + y^2}$

3. tangent of $\theta = \tan \theta = \dfrac{y}{x}$

At this time we will also define three additional trigonometric ratios.

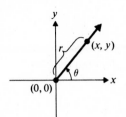

DEFINITION: The Three Additional Trigonometric Ratios

1. cosecant of $\theta = \csc \theta = \dfrac{r}{y}$

2. secant of $\theta = \sec \theta = \dfrac{r}{x}$

3. cotangent of $\theta = \cot \theta = \dfrac{x}{y}$

where $r = \sqrt{x^2 + y^2}$

Recall that two numbers whose product equals one are said to be reciprocals of each other. Observe that

1. $(\sin \theta)(\csc \theta) = \left(\dfrac{y}{r}\right)\left(\dfrac{r}{y}\right) = 1$

2. $(\cos \theta)(\sec \theta) = \left(\dfrac{x}{r}\right)\left(\dfrac{r}{x}\right) = 1$

3. $(\tan \theta)(\cot \theta) = \left(\dfrac{y}{x}\right)\left(\dfrac{x}{y}\right) = 1$

Thus, we have the three basic reciprocal relations.

1. $\csc \theta = \dfrac{1}{\sin \theta}$

2. $\sec \theta = \dfrac{1}{\cos \theta}$

3. $\cot \theta = \dfrac{1}{\tan \theta}$

If the terminal side of a directed angle is in the first quadrant, the values of x and y are both positive. Therefore the trigonometric ratios are also positive. However, *if the terminal side of an angle is in the second quadrant, the value of x is negative, while the value of y remains positive.* Thus, some of the trigonometric ratios will be positive while others will be negative.

EXAMPLE 3

Determine whether the trigonometric ratios for 140° are positive or negative.

Solution: A 140° angle is an angle whose terminal side is in the second quadrant (Fig. 12.24). In the second quadrant, *y is positive* and *x is negative*, while *r is always positive* since it is defined as a positive square root.

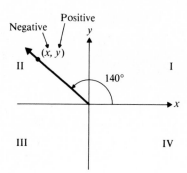

Figure 12.24

Thus,

$$\sin 140° = \frac{y}{r} = \frac{+}{+} = +(\text{positive})$$

$$\cos 140° = \frac{x}{r} = \frac{-}{+} = -(\text{negative})$$

$$\tan 140° = \frac{y}{x} = \frac{+}{-} = -(\text{negative})$$

$$\csc 140° = \frac{r}{y} = \frac{+}{+} = +(\text{positive})$$

$$\sec 140° = \frac{r}{x} = \frac{+}{-} = -(\text{negative})$$

$$\cot 140° = \frac{x}{y} = \frac{-}{+} = -(\text{negative}). \quad \blacksquare$$

Note: Notice that for any angle whose terminal side is in the second quadrant, only the *sine* and *cosecant* are positive.

∠—reciprocals—⌐

If the terminal side of an angle is in the third quadrant, the values of both x and y are negative.

EXAMPLE 4

Determine whether the trigonometric ratios for 580° are positive or negative.

Solution: The smallest positive coterminal angle for 580° is 580° − 360°, or 220°.

third quadrant ⟍
angle

Thus, 580° is an angle whose terminal side is in the third quadrant (Fig. 12.25). In the third quadrant, *both x and y are negative*, while *r is always positive*, since it is defined as a positive square root.

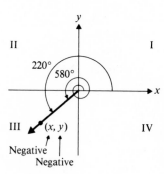

Figure 12.25

Thus,

$$\sin 580° = \frac{y}{r} = \frac{-}{+} = -(\text{negative})$$

$$\cos 580° = \frac{x}{r} = \frac{-}{+} = -(\text{negative})$$

$$\tan 580° = \frac{y}{x} = \frac{-}{-} = +(\text{positive})$$

$$\csc 580° = \frac{r}{y} = \frac{+}{-} = -(\text{negative})$$

$$\sec 580° = \frac{r}{x} = \frac{+}{-} = -(\text{negative})$$

$$\cot 580° = \frac{x}{y} = \frac{-}{-} = +(\text{positive}). \quad ■$$

Note: Notice that for any angle whose terminal side is in the third quadrant, only the *tangent* and *cotangent* are positive.

∠—reciprocals—⟍

If the terminal side of an angle is in the fourth quadrant, the value of x is positive, while the value of y is negative.

EXAMPLE 5

Determine whether the trigonometric ratios for $-40°$ are positive or negative.

Solution: A $-40°$ angle is an angle whose terminal side is in the fourth quadrant (Fig. 12.26). In the fourth quadrant, *x is positive* and *y is negative*, while *r is always positive* since it is defined as a positive square root.

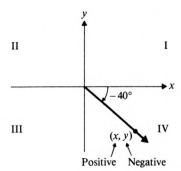

Figure 12.26

Thus,

$$\sin(-40°) = \frac{y}{r} = \frac{-}{+} = -(\text{negative})$$

$$\cos(-40°) = \frac{x}{r} = \frac{+}{+} = +(\text{positive})$$

$$\tan(-40°) = \frac{y}{x} = \frac{-}{+} = -(\text{negative})$$

$$\csc(-40°) = \frac{r}{y} = \frac{+}{-} = -(\text{negative})$$

$$\sec(-40°) = \frac{r}{x} = \frac{+}{+} = +(\text{positive})$$

$$\cot(-40°) = \frac{x}{y} = \frac{+}{-} = -(\text{negative}) \quad \blacksquare$$

Note: Notice that for any angle whose terminal side is in the fourth quadrant, only the *cosine* and *secant* are positive.

⌐ reciprocals ⌐

Whether the trigonometric ratios are positive or negative does not depend on whether the angle is positive or negative, but instead depends on the quadrant in which the terminal side of the angle lies. It is important before beginning Section 12.2 that you memorize in which quadrants the trigonometric ratios are positive. You can use the following phrase to help you: A̲ll S̲tudents T̲ake C̲ourses.

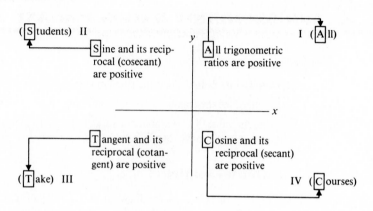

EXERCISES 12.1

Determine the smallest positive coterminal angle for the following directed angles.

1. $-60°$ **2.** $-265°$

3. $-130°$ **4.** $-310°$

5. $580°$ **6.** $695°$

7. $-1824°$ **8.** $1200°$

9. $1000°$ **10.** $-906°$

Determine the reference angle for the following directed angles.

11. $35°$ **12.** $-75°$

13. $-242°$ **14.** $173°$

15. $191°$ **16.** $-164°$

17. $-60°$ **18.** $296°$

19. $843°$ **20.** $-427°$

21. $-527°$ **22.** $1030°$

23. $132° \ 15'$ **24.** $92° \ 42' \ 30''$

Determine whether the following trigonometric ratios are positive or negative.

25. $\sin 82°$ **26.** $\cos 6°$ **27.** $\tan(-320°)$

28. $\csc 418°$ **29.** $\sec 132°$ **30.** $\cot 98°$

31. cos 483°   32. tan(−216°)   33. csc 196°

34. sec 253°   35. cot(−127°)   36. sin 551°

37. tan 305°   38. csc 281°   39. sec 995°

40. cot(−16°)   41. sin(−42°)   42. cos(−425°)

12.2 TRIGONOMETRIC RATIOS FOR ANY ANGLE

Objectives

☐ To use a unit point to find the trigonometric ratios for a quadrantal angle.

☐ To determine the trigonometric ratios for any angle by using a calculator.

☐ Given the value of a trigonometric ratio, to use a calculator to find all angles between 0° and 360° that have that value.

☐ Quadrantal Angles

In Chapter 11 you saw that the trigonometric ratios for a fixed acute angle were independent of the lengths of the sides of the right triangle. Similarly, the trigonometric ratios for a fixed directed angle are independent of the point (x, y) on its terminal side.

For convenience, let's choose a point (x, y) on the terminal side of a directed angle such that the value of r is always positive one ($r = 1$). Thus, the point (x, y) will lie on a circle whose radius is 1 and whose center is at the origin $(0, 0)$. We refer to this circle as the **unit circle** (see Fig. 12.27). The points $(1, 0)$, $(0, 1)$, $(−1, 0)$,

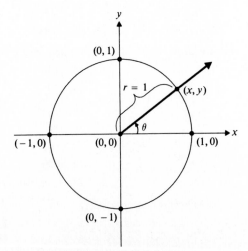

Figure 12.27

and $(0, -1)$ are called the **unit points**. The unit points can be used to find the trigonometric ratios of any quadrantal angle. A **quadrantal angle** is a directed angle that has its terminal side on either the x- or y-axis. The non-negative quadrantal angles that are less than $360°$ are $0°$, $90°$, $180°$, and $270°$.

EXAMPLE 1

Use a unit point to find the trigonometric ratios for $0°$.

Solution: The unit point associated with $0°$ is $(1, 0)$ (Fig. 12.28).

$$x\text{-value} \quad \quad y\text{-value}$$

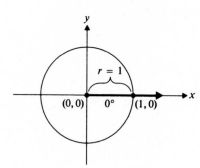

Figure 12.28

Using $x = 1$, $y = 0$, and $r = 1$, we have the following:

$$\sin 0° = \frac{y}{r} = \frac{0}{1} = 0$$

$$\cos 0° = \frac{x}{r} = \frac{1}{1} = 1$$

$$\tan 0° = \frac{y}{x} = \frac{0}{1} = 0$$

$$\csc 0° = \frac{r}{y} = \frac{1}{0} = \text{undefined}$$

$$\sec 0° = \frac{r}{x} = \frac{1}{1} = 1$$

$$\cot 0° = \frac{x}{y} = \frac{1}{0} = \text{undefined} \quad ■$$

Note: Remember, division by zero is undefined.

Note: Any angle coterminal with 0° will have the same values. For example,

$$\sin 0° = \sin 360° = \sin 720° = 0,$$
$$\cos 0° = \cos 360° = \cos 720° = 1,$$
$$\tan 0° = \tan 360° = \tan 720° = 0, \text{ and so on.}$$

EXAMPLE 2

Use a unit point to find the trigonometric ratios for $-90°$.

Solution: The unit point associated with $-90°$ is $(0, -1)$ (Fig. 12.29).

x-value —⌐ ⌐— y-value

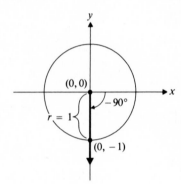

Figure 12.29

Using $x = 0$, $y = -1$, and $r = 1$, we have the following:

$$\sin(-90°) = \frac{y}{r} = \frac{-1}{1} = -1$$

$$\cos(-90°) = \frac{x}{r} = \frac{0}{1} = 0$$

$$\tan(-90°) = \frac{y}{x} = \frac{-1}{0} = \text{undefined}$$

$$\csc(-90°) = \frac{r}{y} = \frac{1}{-1} = -1$$

$$\sec(-90°) = \frac{r}{x} = \frac{1}{0} = \text{undefined}$$

$$\cot(-90°) = \frac{x}{y} = \frac{0}{-1} = 0 \quad ■$$

Note: On the unit circle, since $\sin \theta = \dfrac{y}{r}$ and $r = 1$, we can state that $\sin \theta = y$. The maximum value of y will occur at 90° when $y = 1$. The minimum value of y will occur at $-90°$ (or 270°) when $y = -1$. Thus, we know that the value of $\sin \theta$ is always between -1 and 1. We can state this by writing $-1 \leq \sin \theta \leq 1$.

From Angle to Trigonometric Ratio

To find the trigonometric ratios for other angles, use a protractor to mark off 1° increments on the unit circle and mark off the distances from the origin to the unit points in .1 unit intervals, as shown in Fig. 12.30. Note in Fig. 12.30 that the

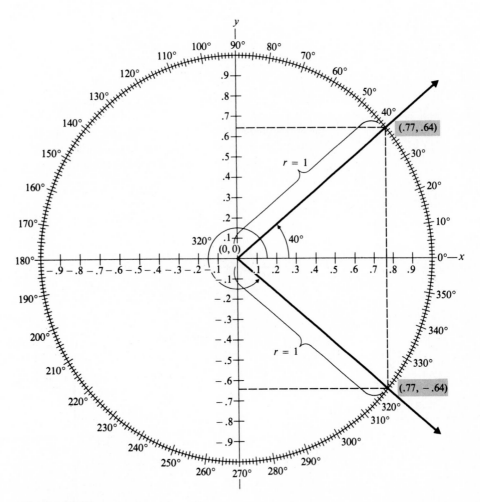

Figure 12.30

point associated with 40° is approximately (.77, .64). Thus, using $x = .77$, $y = .64$,

x-value ⤴ ⤴ y-value

and $r = 1$, we have the following:

$$\sin 40° = \frac{y}{r} = \frac{.64}{1} = .64$$

$$\cos 40° = \frac{x}{r} = \frac{.77}{1} = .77$$

$$\tan 40° = \frac{y}{x} = \frac{.64}{.77} = .83$$

→ *Note*: Retaining two significant digits.

$$\csc 40° = \frac{r}{y} = \frac{1}{.64} = 1.6$$

$$\sec 40° = \frac{r}{x} = \frac{1}{.77} = 1.3$$

$$\cot 40° = \frac{x}{y} = \frac{.77}{.64} = 1.2$$

These graphical values of sin 40°, cos 40°, and tan 40° can be verified by using a trig table for acute angles (see Table 11.4, page 595).

θ	$\sin \theta$	$\cos \theta$	$\tan \theta$
39°	.629	.777	.810
from Appendix A → 40°	.643	.766	.839
41°	.656	.755	.869

The graphical values of csc 40°, sec 40°, and cot 40° can be verified by taking the reciprocals of the table values for sin 40°, cos 40°, and tan 40°, respectively.

$$\csc 40° = \frac{1}{\sin 40°} = \frac{1}{.643} = 1.56$$

$$\sec 40° = \frac{1}{\cos 40°} = \frac{1}{.766} = 1.31$$

Note: Retaining three significant digits.

$$\cot 40° = \frac{1}{\tan 40°} = \frac{1}{.839} = 1.19$$

Note also in Fig. 12.30 that the point (x, y) associated with 320° is approximately $(.77, -.64)$. These are the same values of x and y that were used for 40°, except

— y-value

— x-value

that y is now *negative*. The trigonometric ratios for 320° will be the same as for 40°, except for some of the algebraic signs. Thus, retaining two significant digits, we have the following:

$$\sin 320° = \frac{y}{r} = \frac{-.64}{1} = -.64$$

$$\cos 320° = \frac{x}{r} = \frac{.77}{1} = .77$$

$$\tan 320° = \frac{y}{x} = \frac{-.64}{.77} = -.83$$

Note: In the fourth quadrant, only the cosine and secant are positive.

$$\csc 320° = \frac{r}{y} = \frac{1}{-.64} = -1.6$$

$$\sec 320° = \frac{r}{x} = \frac{1}{.77} = 1.3$$

$$\cot 320° = \frac{x}{y} = \frac{.77}{-.64} = -1.2$$

Observe that the reference angle for both 40° and 320° is 40°. *If two angles are in different quadrants but have the same reference angle, then the values of their corresponding trigonometric ratios will differ by sign only.*

The graphical values of sin 320°, cos 320°, and tan 320° can also be verified by using a trig table for acute angles (see Table 11.4), provided that you work with the reference angle and attach the proper algebraic sign to these ratios. Recall from Section 12.1 that in the fourth quadrant, $\cos \theta$ and $\sec \theta$ are positive, while $\sin \theta$, $\tan \theta$, $\csc \theta$, and $\cot \theta$ are negative.

$$\sin 320° = -(\sin 40°) = -.643$$
$$\cos 320° = \cos 40° = .766$$
$$\tan 320° = -(\tan 40°) = -.839$$

Note: Values from Table 11.4

Note: In the fourth quadrant, only the cosine and secant and positive.

$$\csc 320° = \frac{1}{\sin 320°} = \frac{1}{-(\sin 40°)} = \frac{1}{-.643} = -1.56$$

$$\sec 320° = \frac{1}{\cos 320°} = \frac{1}{\cos 40°} = \frac{1}{.766} = 1.31$$

$$\cot 320° = \frac{1}{\tan 320°} = \frac{1}{-(\tan 40°)} = \frac{1}{-.839} = -1.19$$

Note: retaining three significant digits.

We could find the trigonometric ratios for other directed angles by using the graphical approach (Fig. 12.30) or by using a trig table for acute angles (see Table 11.4). However, your calculator can be used to quickly find the trig ratios for any angle.

To find sin 320° enter $\boxed{320}$ $\boxed{\text{sin}}$. $\underbrace{-.64278761}_{\text{display}}$ ←┐

Note: Your calculator will automatically record the correct algebraic sign for each ratio.

To find cos 320° enter $\boxed{320}$ $\boxed{\text{cos}}$. $\underbrace{.76604445}_{\text{display}}$ ←┤

To find tan 320° enter $\boxed{320}$ $\boxed{\text{tan}}$. $\underbrace{-.83909963}_{\text{display}}$ ←┘

To find csc 320°, remember that $\csc \theta = \dfrac{1}{\sin \theta}$.

Enter $\boxed{320}$ $\boxed{\text{sin}}$ $\boxed{\dfrac{1}{x}}$. $\underbrace{-1.5557238}_{\text{display}}$

To find sec 320°, remember that $\sec \theta = \dfrac{1}{\cos \theta}$.

Enter $\boxed{320}$ $\boxed{\text{cos}}$ $\boxed{\dfrac{1}{x}}$. $\underbrace{1.3054073}_{\text{display}}$

To find cot 320°, remember that $\cot \theta = \dfrac{1}{\tan \theta}$.

Enter $\boxed{320}$ $\boxed{\text{tan}}$ $\boxed{\dfrac{1}{x}}$. $\underbrace{-1.1917536}_{\text{display}}$

EXAMPLE 3

Using your calculator, determine the trigonometric ratios for $-95°$. Retain four significant digits.

Solution:

1. For sin$(-95°)$, enter $\boxed{95}$ $\boxed{+/-}$ $\boxed{\text{sin}}$. $\underbrace{-.9961947}_{\text{display}}$

   Thus, sin$(-95°) = -.9962$ (four significant digits). ■

2. For cos$(-95°)$, enter $\boxed{95}$ $\boxed{+/-}$ $\boxed{\text{cos}}$. $\underbrace{-.08715574}_{\text{display}}$

   Thus, cos$(-95°) = -.08716$ (four significant digits). ■

3. For $\tan(-95°)$, enter $\boxed{95}$ $\boxed{+/-}$ $\boxed{\tan}$. $\underbrace{11.430052}_{\text{display}}$

Thus, $\tan(-95°) = 11.43$ (four significant digits).   ■

4. For $\csc(-95°)$, enter $\boxed{95}$ $\boxed{+/-}$ $\boxed{\sin}$ $\boxed{\dfrac{1}{x}}$. $\underbrace{-1.0038198}_{\text{display}}$

Thus, $\csc(-95°) = -1.004$ (four significant digits).   ■

5. For $\sec(-95°)$, enter $\boxed{95}$ $\boxed{+/-}$ $\boxed{\cos}$ $\boxed{\dfrac{1}{x}}$. $\underbrace{-11.473713}_{\text{display}}$

Thus, $\sec(-95°) = -11.47$ (four significant digits).   ■

6. For $\cot(-95°)$, enter $\boxed{95}$ $\boxed{+/-}$ $\boxed{\tan}$ $\boxed{\dfrac{1}{x}}$. $\underbrace{.08748866}_{\text{display}}$

Thus, $\cot(-95°) = .08749$ (four significant digits).   ■

⠂ From Trigonometric Ratio to Angle

Recall from Chapter 11 that when the value of a trigonometric ratio is known, you can use your calculator and the $\boxed{\text{inv}}$ or $\boxed{\text{arc}}$ keys to find the acute angle θ. For example, to find θ for $\sin \theta = .5$

$$\text{enter } \boxed{.5} \text{ or } \begin{array}{c}\boxed{\text{inv}}\\ \boxed{\sin}\\ \boxed{\text{arc}}\end{array}. \quad \underbrace{30}_{\text{display}}$$

Thus, if $\sin \theta = .5$, then $\theta = 30°$.

/ acute
angle

When working with directed angles, however, any angle coterminal wih 30° can also be an answer. Thus, other possible values of θ for which $\sin \theta = .5$ are $\theta = 390°$, $\theta = -330°$, and so on (see Fig. 12.31). Remember also that the sine is positive in

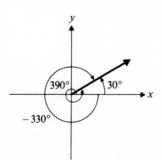

Figure 12.31

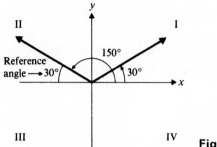

Figure 12.32

the second quadrant as well as in the first quadrant. Thus, using a reference angle of 30° in the second quadrant, we also have $\theta = 150°$ as another possible value of θ for which $\sin\theta = .5$ (see Fig. 12.32). Any angle coterminal with 150° can also be an answer. Thus, still other possible values of θ for which $\sin\theta = .5$ are $\theta = 510°$, $\theta = -210°$, and so on (see Fig. 12.33). As you can see, for a given value of a trigonometric ratio, there are an infinite number of angles in standard position that have that value. In this section, we will restrict the values of θ such that $0° \le \theta < 360°$. Under this restriction, there will be *two* angles that have the same trigonometric ratio. For $\sin\theta = .5$, with $0° \le \theta < 360°$, we have only $\theta = 30°$ and $\theta = 150°$.

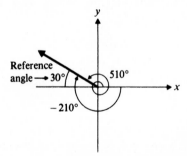

Figure 12.33

When given the value of a trigonometric ratio and asked to find all angles between 0° and 360° that have that value, it is best to proceed as follows:

1. Find the reference angle on your calculator by entering only the *magnitude* of the given trigonometric ratio.
2. Look at the algebraic sign of the given ratio and determine in which two quadrants to place the reference angle.
3. Record the values of θ such that $0° \le \theta < 360°$.

EXAMPLE 4

Find all θ with $0° \le \theta < 360°$ such that $\tan\theta = .6250$. Record your answers to the nearest tenth of a degree.

Solution: First, find the reference angle, α.

Enter $\boxed{.6250}$ or $\boxed{\begin{array}{c}\text{inv}\\ \text{arc}\end{array}}$ $\boxed{\text{tan}}$.   32.005383
                                                              display

Thus, α = 32.0°.

reference
angle

Next, note that for tan θ = .6250, the ratio is *positive*. Recall that the *tangent is positive* in the *first* and *third* quadrants. Thus, place a reference angle of 32.0° in the *first* and *third* quadrants (Fig. 12.34).

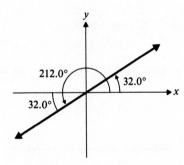

Figure 12.34

Therefore, θ = 32.0° and θ = 180° + 32.0° = 212.0°. ∎

EXAMPLE 5

Find all θ with 0° ≤ θ < 360° such that cos θ = −.5592. Record your answers to the nearest tenth of a degree.

Solution: First, find the reference angle, α.

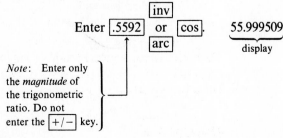

Enter $\boxed{.5592}$ or $\boxed{\begin{array}{c}\text{inv}\\ \text{arc}\end{array}}$ $\boxed{\text{cos}}$.   55.999509
                                                              display

Note: Enter only the *magnitude* of the trigonometric ratio. Do not enter the $\boxed{+/-}$ key.

Thus, α = 56.0°.

reference
angle

Next, note that for cos θ = −.5592, the ratio is *negative*. Recall that the cosine is positive in the first and fourth quadrants. Thus, it is *negative* in the *second* and *third* quadrants. Therefore, place a reference angle of 56.0° in the *second* and *third* quadrants (Fig. 12.35).

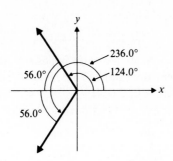

Figure 12.35

Therefore, θ = 180° − 56.0° = 124.0° and θ = 180° + 56.0° = 236.0°. ∎

EXAMPLE 6

Find all θ with 0° ≤ θ < 360° such that csc θ = −1.642. Record your answer to the nearest minute.

Solution: If csc θ = −1.642, then $\sin θ = \dfrac{1}{-1.642}$

reciprocals

To find the reference angle, enter 1.642 $\frac{1}{x}$ $\boxed{\text{inv}}$ or $\boxed{\sin}$.   37.518199

display

Note: Enter only the *magnitude* of the trigonometric ratio. Do not enter the $\boxed{+/-}$ key.

$.518199° = \dfrac{.518199°}{1}\left(\dfrac{60'}{1°}\right) = 31'$ (nearest minute)

Thus, α = 37° 31′.

reference angle

Next, note that for csc θ = −1.642, the ratio is *negative*. Recall that the cosecant (like its reciprocal the sine) is positive in the first and second quadrants.

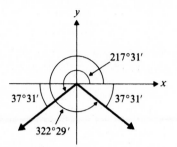

Figure 12.36

Thus, it is *negative* in the *third* and *fourth* quadrants. Therefore, place a reference angle of 37° 31′ in the *third* and *fourth* quadrants (Fig. 12.36). So

$$\theta = 180° + 37° \; 31′ = 217° \; 31′$$

and

$$\theta = 360° - 37° \; 31′ = 322° \; 29′. \quad ■$$

EXERCISES 12.2

Use a unit point to find the value of the following trigonometric ratios.

1. cos 90°	**2.** sin 90°	**3.** tan 180°
4. csc 180°	**5.** sec 270°	**6.** cot 270°
7. sin 360°	**8.** cos 360°	**9.** csc 630°
10. tan 720°	**11.** cot(−630°)	**12.** sec(−180°)

Using your calculator, determine the value of the following trigonometric ratios. Retain four significant digits.

13. tan 127°	**14.** csc 158°	**15.** sec 254°
16. cot 198°	**17.** sin 303°	**18.** cos 290°
19. cos(−22°)	**20.** sin(−73°)	**21.** csc 394°
22. tan 468°	**23.** cot(−475°)	**24.** sec(−820°)

For each of the following, find all θ such that $0° \le \theta < 360°$. Record your answers to the nearest tenth of a degree.

25. sin θ = .5299	**26.** cos θ = .6947	**27.** tan θ = 2.145
28. csc θ = 1.236	**29.** sec θ = 9.107	**30.** cot θ = 8.636
31. cos θ = −.9573	**32.** sin θ = −.2181	**33.** csc θ = −1.203
34. tan θ = −.3428	**35.** cot θ = −.0695	**36.** sec θ = −2.673

For each of the following, find all θ such that $0° \le \theta < 360°$. Record your answers to the nearest minute.

37. sin θ = −.8620	**38.** cos θ = −.2461	**39.** tan θ = 3.024
40. csc θ = 1.095	**41.** sec θ = −1.623	**42.** cot θ = −5.425

For each of the following, use the *unit circle* to find all θ such that $0° \leq \theta < 360°$.

43. $\cos \theta = 0$ **44.** $\cos \theta = 1$ **45.** $\cos \theta = -1$

46. $\sin \theta = 0$ **47.** $\sin \theta = 1$ **48.** $\sin \theta = -1$

49. $\tan \theta = 0$ **50.** $\tan \theta = 1$ **51.** $\tan \theta = -1$

52. $\cot \theta = 0$ **53.** $\sec \theta = 1$ **54.** $\csc \theta = -1$

55. On the unit circle, since $\cos \theta = \dfrac{x}{r}$ and $r = 1$, we can state that $\cos \theta = x$. Find the maximum and minimum values of $\cos \theta$. At what angles do they occur?

56. Does $\tan \theta$ have a maximum or minimum value? Explain.

12.3 AN INTRODUCTION TO VECTORS

Objectives

☐ To translate a given vector into standard position.

☐ Given a vector in polar form, to express it in rectangular form or vice versa.

☐ To find the sum of two vectors.

☐ Vector Notation

There are many quantities that can be fully described by using a number (called the **magnitude** of the quantity) and a unit of measure. For example, the *length* of a steel beam is fully described by expressing it as

$$16.0 \text{ ft}$$

magnitude ⟋ ⟋ unit of measure

Quantities that can be completely described by magnitude are called **scalar** quantities. Examples of scalar quantities are length, area, volume, time, electrical resistance, and so on.

There are also some quantities that *cannot* be fully described by magnitude alone. For example, the *force* acting at the center of a horizontal beam is not fully described by expressing it as 100 lb. We need to know in what **direction** the force is acting. Upward? Downward? At an angle? (see Fig. 12.37). *For a force to be*

Figure 12.37

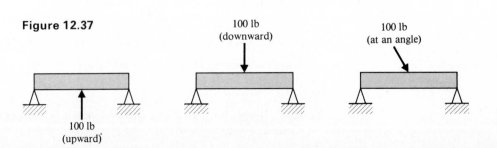

100 lb (upward)

100 lb (downward)

100 lb (at an angle)

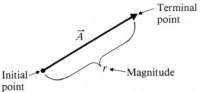

Figure 12.38

fully described, we need to know both its magnitude and direction. Quantities that need to be described by magnitude and direction are called **vector** quantities. Examples of vector quantities are force, velocity, acceleration, electrical current, displacement, and so on.

In technical mathematics, vectors are represented by arrows. The arrowhead represents the **terminal point** of the vector and the opposite end represents its **initial point**. Shown in Fig. 12.38 is vector A ($\vec{A}$). The *magnitude* of a vector is indicated by the length of the arrow drawn to scale. The magnitude of the vector shown in Fig. 12.38 is r units. The *direction* of a vector is indicated by the arrow's orientation in the plane. In order to have some sense of direction, we will place all vectors in the rectangular coordinate system with the initial point of the vector at the origin (0, 0). This is called **standard position** of a vector. When a vector is in standard position, it can be easily identified by using either rectangular form or polar form.

To represent a vector in **rectangular form**, we use the notation $\langle x, y \rangle$, where x and y are the coordinates of the terminal point of the vector. To represent a vector in **polar form**, we use the notation $r\underline{/\theta}$, where r is the magnitude (length) of the vector and θ is the measure of the directed angle (see Fig. 12.39).

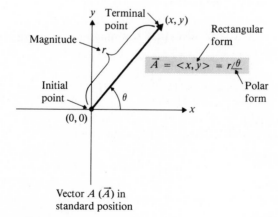

Vector A ($\vec{A}$) in standard position

Figure 12.39

EXAMPLE 1

Translate the vector shown in Fig. 12.40 into standard position; then identify the vector using rectangular form.

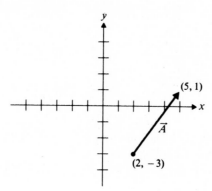

Figure 12.40

Solution: To be in standard position, the vector must have its initial point at the origin, (0, 0). Therefore, the initial point must be adjusted by moving *2 units left* and *3 units up*. To preserve the magnitude and direction of the vector, a similar adjustment must be made to the terminal point (see Fig. 12.41). The notation

$$\overset{\substack{\text{rectangular} \\ \text{form}}}{\overbrace{\vec{A} = \langle 3, 4 \rangle}}$$

can now be used to identify this vector since its terminal point is (3, 4) when in standard position. ∎

Note: It is important to note that the original vector and the vector in standard position are *equal* since they have the same magnitude and the same direction. The translation of a vector from one position to another graphically does not change its value, provided that the magnitude and direction of the vector are unchanged.

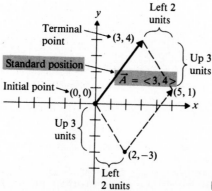

Figure 12.41

EXAMPLE 2

Translate the vector shown in Fig. 12.42 into standard position; then identify the vector using polar form.

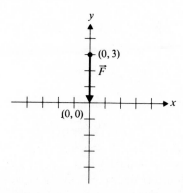

Figure 12.42

Solution: To be in standard position, the vector must have its initial point at the origin, (0, 0). Therefore, the initial point must be adjusted by moving *3 units down*. To preserve the magnitude and direction of the vector, a similar adjustment must be made to the terminal point (see Fig. 12.43).

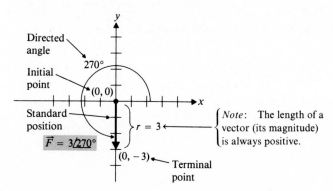

Figure 12.43

The notation $\vec{F} = 3\underline{/270°}$ can now be used to identify this vector since its magnitude is 3 when in standard position and its directed angle is 270°. ■

Note: Vector *F* can also be represented by using a negative angle or by using rectangular form.

$$\vec{F} = 3\ \underline{/270°} = 3\ \underline{/-90°} = \langle 0, -3 \rangle$$

polar form ⟋   polar form
            negative angle

rectangular
form

⊡ Changing between Polar and Rectangular Form

When working with vectors in polar form, it is often necessary to express them in rectangular form or vice versa. The coordinates x and y can be expressed in terms of r and θ as follows:

$$\cos \theta = \frac{x}{r} \text{ thus } \boxed{x = r \cos \theta}$$

$$\sin \theta = \frac{y}{r} \text{ thus } \boxed{y = r \sin \theta}$$

The values of r and θ can be expressed in terms of x and y as follows:

$$\boxed{r = \sqrt{x^2 + y^2}}$$

$$\boxed{\tan \theta = \frac{y}{x}}$$

EXAMPLE 3

Express $\vec{A}$ in rectangular form if $\vec{A} = 8.0\underline{/116°}$.

Solution: First, draw the vector in standard position using $r = 8.0$ and $\theta = 116°$ (Fig. 12.44). Now determine the values of x and y (see Fig. 12.45).

$$x = r \cos \theta = (8.0)(\cos 116°) = -3.5 \quad \text{(two significant digits)}$$
$$y = r \sin \theta = (8.0)(\sin 116°) = 7.2 \quad \text{(two significant digits)}$$

Thus, $\vec{A} = 8.0\underline{/116°} = \langle -3.5, 7.2 \rangle$ ■

polar ⟍ ⟋ rectangular
form   form

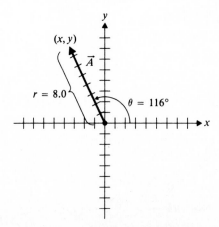

Figure 12.44

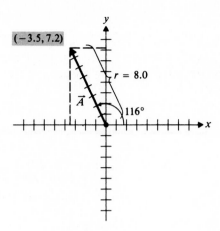

Figure 12.45

EXAMPLE 4

Express $\vec{B}$ in polar form if $\vec{B} = \langle 525, -386 \rangle$.

Solution: First, draw the vector in standard position using $(525, -386)$ as the terminal point (Fig. 12.46). Now determine the values of r and θ (see Fig. 12.47).

$$r = \sqrt{x^2 + y^2} = \sqrt{(525)^2 + (-386)^2} = 652 \quad \text{(three significant digits)}$$

$$\tan \theta = \frac{y}{x} = \frac{-386}{525}$$

Note: Place a reference angle of $36.3°$ in the fourth quadrant, since y is negative and x is positive.

$\begin{cases} \textit{Note}: & \text{Using your calculator,} \\ & \text{enter } \boxed{386}\,\boxed{\div}\,\boxed{525}\,\boxed{=}\,\boxed{\text{inv}}\,\boxed{\text{tan}}. \end{cases}$

$$\theta = 323.7° \quad \text{(nearest .1°)}$$

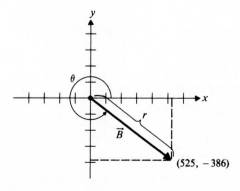

Figure 12.46

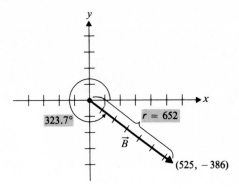

Figure 12.47

Thus, $\vec{B} = \langle 525, -386 \rangle = 652/323.7°$. ∎

rectangular form ⟍ ⟋ polar form

⸫ The Sum of Two Vectors

If $\vec{A}_1$ and $\vec{A}_2$ are in standard position, with $\vec{A}_1 = \langle x_1, y_1 \rangle$ and $\vec{A}_2 = \langle x_2, y_2 \rangle$, the *sum* of these vectors is defined as

$$\vec{A}_1 + \vec{A}_2 = \langle x_1 + x_2, y_1 + y_2 \rangle \text{ (see Fig. 12.48)}$$

sum of the x-components ⟍ ⟋ sum of the y-components

The sum of the vectors $\vec{A}_1 + \vec{A}_2$ can be found graphically by translating $\vec{A}_2$ such that its initial point corresponds to the terminal point of $\vec{A}_1$ (see Fig. 12.49), or by translating $\vec{A}_1$ such that its initial point corresponds to the terminal point of $\vec{A}_2$ (see Fig. 12.50). It is important to observe that geometrically $\vec{A}_1 + \vec{A}_2$ represents the diagonal of a parallelogram formed by the vectors $\vec{A}_1$ and $\vec{A}_2$.

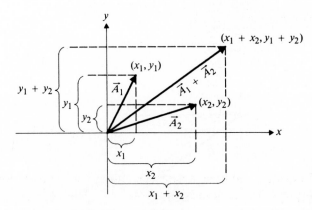

Figure 12.48

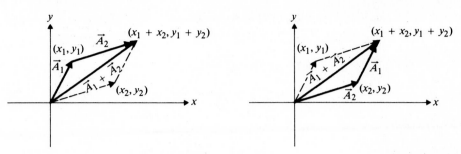

Figure 12.49   **Figure 12.50**

EXAMPLE 5

Find $\vec{A}_1 + \vec{A}_2$ if $\vec{A}_1 = \langle 2, 5 \rangle$ and $\vec{A}_2 = \langle -8, -3 \rangle$.

Solution:

Add the x-components.

$$\vec{A}_1 + \vec{A}_2 = \langle 2, 5 \rangle + \langle -8, -3 \rangle = \langle \boxed{2 + (-8)}, \boxed{5 + (-3)} \rangle = \langle -6, 2 \rangle \quad \blacksquare$$

Add the y-components.

Note: As a check, see if $\vec{A}_1 + \vec{A}_2 = \langle -6, 2 \rangle$ represents the diagonal of the parallelogram formed by $\vec{A}_1$ and $\vec{A}_2$ (see Fig. 12.51).

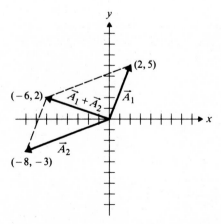

Figure 12.51

In order to add two vectors that are given in polar form, it is first necessary to express them in rectangular form.

EXAMPLE 6

Find $\vec{A}_1 + \vec{A}_2$ if $\vec{A}_1 = 25\underline{/160°}$ and $\vec{A}_2 = 35\underline{/295°}$. Express the sum in polar form.

Solution: First, express $\vec{A}_1$ and $\vec{A}_2$ in rectangular form. To help organize this work, set up a table as follows:

Polar form $r\underline{/\theta}$	x-Component $x = r \cos \theta$	y-Component $y = r \sin \theta$	Rectangular form $\langle x, y \rangle$
$\vec{A}_1 = 25\underline{/160°}$	$x_1 = 25 \cos 160°$ $x_1 = -23.5$	$y_1 = 25 \sin 160°$ $y_1 = 8.55$	$\langle -23.5, 8.55 \rangle$
$\vec{A}_2 = 35\underline{/295°}$	$x_2 = 35 \cos 295°$ $x_2 = 14.8$	$y_2 = 35 \sin 295°$ $y_2 = -31.7$	$\langle 14.8, -31.7 \rangle$

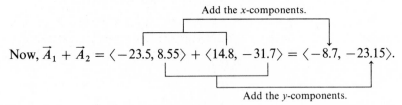

Add the *x*-components.

Now, $\vec{A}_1 + \vec{A}_2 = \langle -23.5, 8.55 \rangle + \langle 14.8, -31.7 \rangle = \langle -8.7, -23.15 \rangle.$

Add the *y*-components.

$\vec{A}_1 + \vec{A}_2 = \langle -8.7, -23.15 \rangle$ should be the diagonal of the parallelogram formed by $\vec{A}_1$ and $\vec{A}_2$. Be sure to check that this is reasonable before expressing the sum in polar form (Fig. 12.52).

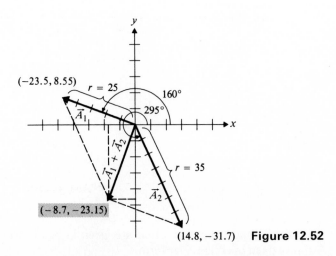

Figure 12.52

Expressing the sum in polar form, we have

$$r = \sqrt{x^2 + y^2} = \sqrt{(-8.7)^2 + (-23.15)^2} = 25 \quad \text{(two significant digits)}.$$

$$\tan \theta = \frac{y}{x} = \frac{-23.15}{-8.7}$$

Note: Place the reference angle of 69° in the third quadrant, since both x and y are negative.

Note: Using your calculator, enter 23.15 ÷ 8.7 = inv tan.

$$\theta = 249° \quad \text{(nearest degree)}$$

Thus, $\vec{A}_1 + \vec{A}_2 = \langle -8.7, -23.15 \rangle = 25\underline{/249°}.$ ∎

EXERCISES 12.3

Translate the given vectors into standard position; then identify each vector using rectangular form.

1.

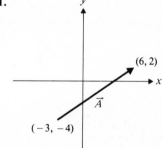

2.

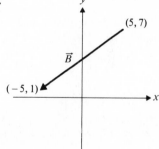

3.

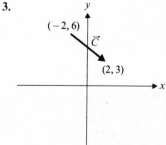

4.

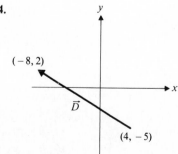

Translate the given vectors into standard position, then identify each vector using polar and rectangular form.

5.

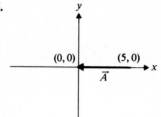

6.

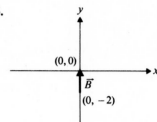

7.

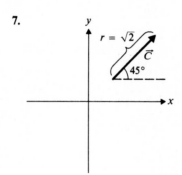

8.

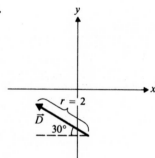

9.

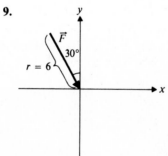

10.

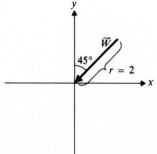

Express each of the following vectors in rectangular form.

11. $\vec{A} = 7.2\underline{/32°}$ **12.** $\vec{A} = 9.3\underline{/57°}$ **13.** $\vec{B} = 19.5\underline{/125.6°}$

14. $\vec{B} = 72.6\underline{/154.8°}$ **15.** $\vec{C} = 16\underline{/224°}$ **16.** $\vec{C} = 35\underline{/192°}$

17. $\vec{D} = 585\underline{/341.5°}$ **18.** $\vec{D} = 625\underline{/-32.5°}$ **19.** $\vec{F} = 120\underline{/0°}$

20. $\vec{F} = 320\underline{/180°}$ **21.** $\vec{W} = 1500\underline{/-90°}$ **22.** $\vec{W} = 7300\underline{/90°}$

Express each of the following vectors in polar form.

23. $\vec{A} = \langle 87, 72 \rangle$ **24.** $\vec{A} = \langle 19, 32 \rangle$ **25.** $\vec{B} = \langle -32.6, 45.6 \rangle$

26. $\vec{B} = \langle -92.1, 37.9 \rangle$ **27.** $\vec{C} = \langle -120, -140 \rangle$ **28.** $\vec{C} = \langle -780, -950 \rangle$

29. $\vec{D} = \langle 582, -327 \rangle$ **30.** $\vec{D} = \langle 621, -125 \rangle$ **31.** $\vec{F} = \langle 0, 36 \rangle$

32. $\vec{F} = \langle 0, -48 \rangle$ **33.** $\vec{W} = \langle -92.6, 0 \rangle$ **34.** $\vec{W} = \langle 843, 0 \rangle$

Given that $\vec{A}_1 = \langle 16, 18 \rangle$, $\vec{A}_2 = \langle -13, 16 \rangle$, $\vec{A}_3 = \langle 0, -15 \rangle$, $\vec{B}_1 = 15\underline{/20°}$, $\vec{B}_2 = 23\underline{/195°}$, and $\vec{B}_3 = \underline{/-17°}$, find the following vector sums. Express each sum in polar form.

35. $\vec{A}_1 + \vec{A}_2$ **36.** $\vec{A}_1 + \vec{A}_3$

37. $\vec{A}_2 + \vec{A}_3$ **38.** $\vec{A}_1 + \vec{B}_1$

39. $\vec{A}_1 + \vec{B}_2$ **40.** $\vec{A}_2 + \vec{B}_3$

41. $\vec{B}_1 + \vec{B}_2$ **42.** $\vec{B}_2 + \vec{B}_3$

43. $\vec{B}_1 + \vec{B}_2 + \vec{B}_3$ **44.** $\vec{A}_1 + \vec{A}_2 + \vec{A}_3 + \vec{B}_1 + \vec{B}_2 + \vec{B}_3$

12.4 ANALYSIS OF A FORCE SYSTEM

Objectives

⊡ To find the resultant force of a force system.

⊡ Given a force system, to determine the unknown force necessary to produce equilibrium.

⊡ Resultant Force

A **force system** is a group of forces acting on a particular object. In engineering technology, we use vectors to represent the forces in the system. In doing so, we form what is called a **free body diagram**. Shown in Fig. 12.53 is a pictorial diagram

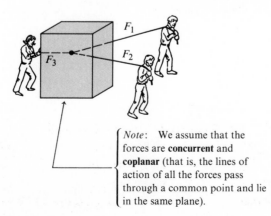

Note: We assume that the forces are **concurrent** and **coplanar** (that is, the lines of action of all the forces pass through a common point and lie in the same plane).

Figure 12.53

of three forces acting on a large box (two men pulling and one woman pushing). Shown in Fig. 12.54 is the free body diagram associated with these forces.

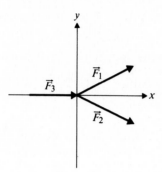

Figure 12.54

A **resultant force** is a single force that can replace the force system and have the same physical effect upon the object. *The resultant force is simply the vector sum of all the forces in the system.*

EXAMPLE 1

Suppose the forces acting on the box in Fig. 12.53 are given by the free body diagram shown in Fig. 12.55. Find the resultant force.

Solution: First, place all vector forces in standard position and express each of them in polar form (Fig. 12.56).

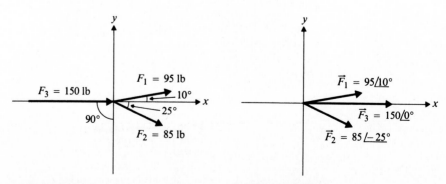

Figure 12.55

Figure 12.56

Next, use a table to express each vector force in rectangular form.

Polar form $r\underline{/\theta}$	x-Component $x = r\cos\theta$	y-Component $y = r\sin\theta$	Rectangular form $\langle x, y\rangle$
$\vec{F}_1 = 95\underline{/10°}$	$x_1 = 95\cos 10°$ $x_1 = 93.6$	$y_1 = 95\sin 10°$ $y_1 = 16.5$	$\langle 93.6, 16.5\rangle$
$\vec{F}_2 = 85\underline{/-25°}$	$x_2 = 85\cos(-25°)$ $x_2 = 77.0$	$y_2 = 85\sin(-25°)$ $y_2 = -35.9$	$\langle 77.0, -35.9\rangle$
$\vec{F}_3 = 150\underline{/0°}$	$x_3 = 150\cos 0°$ $x_3 = 150$	$y_3 = 150\sin 0°$ $y_3 = 0$	$\langle 150, 0\rangle$

Now, the resultant force $\vec{F}_r$ is simply the vector sum of all these forces (Fig. 12.57).

$$\vec{F}_r = \qquad \vec{F}_1 \qquad + \qquad \vec{F}_2 \qquad + \quad \vec{F}_3$$
$$\vec{F}_r = \langle 93.6, 16.5\rangle + \langle 77.0, -35.9\rangle + \langle 150, 0\rangle$$
$$\vec{F}_r = \langle 320.6, -19.4\rangle$$

sum of the $\diagup$ $\diagdown$ sum of the
x-components   y-components

Finally, expressing $\vec{F}_r$ in polar form, we have

$$r = \sqrt{x^2 + y^2} = \sqrt{320.6^2 + (-19.4)^2} = 320. \quad \text{(two significant digits)}$$

$$\tan\theta = \frac{y}{x} = \frac{-19.4}{320.6}$$

Note: Place the reference angle of 3° in the fourth quadrant, since y is negative and x is positive.

Note: Using your calculator, enter $\boxed{19.4}$ $\boxed{\div}$ $\boxed{320.6}$ $\boxed{=}$ $\boxed{\text{inv}}$ $\boxed{\text{tan}}$.

$$\theta = 357° \quad \text{(nearest degree)}$$

Therefore, $\vec{F}_r = 320$ lb $\underline{/357°}$. ∎

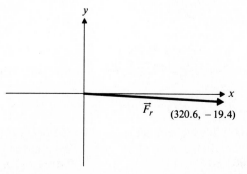

Figure 12.57

Note: A single force of 320 lb directed at an angle of 357° would have the same physical effect upon the box as would the three forces F_1, F_2, and F_3.

⋰ Forces in Equilibrium

Tension is a force that causes an object to be stretched while **compression** is a force that causes an object to be crushed (see Fig. 12.58). Consider a weight being lifted by the derrick shown in Fig. 12.59. The *cable* is being stretched since the

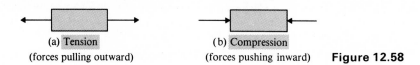

(a) Tension — (forces pulling outward)   (b) Compression — (forces pushing inward)   **Figure 12.58**

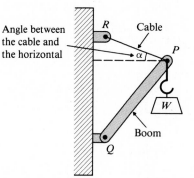

Figure 12.59

forces at *P* and *R* are pulling outward. Thus, there is tension in the cable. The *boom* is being crushed since the forces at *P* and *Q* are pushing inward. Thus, there is compression in the boom. If the derrick is *in equilibrium* (at rest), *the vector sum of all the forces acting through the common point P must be zero.*

$$\vec{W} + \vec{T} + \vec{C} = \langle 0, 0 \rangle$$

weight being lifted ⟋   tension in the cable ↑   compression in the boom ⟋

EXAMPLE 2

Suppose a 1$\overline{0}$00-lb weight is being lifted by the derrick shown in Fig. 12.59. If the structure is in equilibrium, find the compression in the boom if the tension in the cable is 750 lb and $\alpha = 30°$.

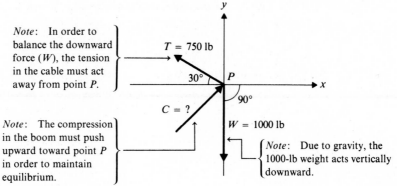

Note: In order to balance the downward force (*W*), the tension in the cable must act away from point *P*.

Note: The compression in the boom must push upward toward point *P* in order to maintain equilibrium.

Note: Due to gravity, the 1000-lb weight acts vertically downward.

Figure 12.60

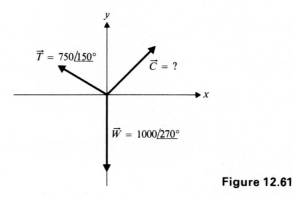

Figure 12.61

Solution: Since all forces act through point *P*, choose point *P* as the origin $(0, 0)$ and draw the free body diagram as in Fig. 12.60. Now, place all vector forces in standard position and express the known forces in polar form (Fig. 12.61). Next, use a table to express these known forces in rectangular form.

Polar form $r \underline{/\theta}$	x-Component $x = r \cos \theta$	y-Component $y = r \sin \theta$	Rectangular form $\langle x, y \rangle$
$\vec{W} = 1000 \underline{/270°}$	$x_W = 1000 \cos 270°$ $x_W = 0$	$y_W = 1000 \sin 270°$ $y_W = -1000$	$\langle 0, -1000 \rangle$
$\vec{T} = 750 \underline{/150°}$	$x_T = 750 \cos 150°$ $x_T = -650$	$y_T = 750 \sin 150°$ $y_T = 375$	$\langle -650, 375 \rangle$

For equilibrium,

$$\vec{W} \quad + \quad \vec{T} \quad + \quad \vec{C} \quad = \langle 0, 0 \rangle$$
$$\langle 0, -1000 \rangle + \langle -650, 375 \rangle + \langle x_C, y_C \rangle = \langle 0, 0 \rangle$$

Thus, the sum of the x-components must be zero,

$$0 + (-650) + x_C = 0$$
$$x_C = 650,$$

and the sum of the y-components must be zero,

$$-1000 + 375 + y_C = 0$$
$$y_C = 625.$$

Therefore, $\vec{C} = \langle 650, 625 \rangle$ (Fig. 12.62). Finally, expressing $\vec{C}$ in polar form, we have

$$r = \sqrt{x^2 + y^2} = \sqrt{(650)^2 + (625)^2} = 9\overline{0}0 \quad \text{(two significant digits)}$$

$$\tan \theta = \frac{y}{x} = \frac{625}{650} \longleftarrow$$

Note: Place the reference angle of 44° in the first quadrant, since both x and y are positive.

$\left\{\begin{array}{l}\textit{Note:} \text{ Using your calculator,} \\ \text{enter } \boxed{625} \boxed{\div} \boxed{650} \boxed{=} \boxed{\text{inv}} \boxed{\text{tan}}.\end{array}\right.$

$$\theta = 44° \quad \text{(nearest degree)}$$

Therefore, $\vec{C} = 9\overline{0}0$ lb $\underline{/44°}$. ∎

Note: The compression in the boom is $9\overline{0}0$ lb directed at an angle of 44° above the horizontal. Thus, the angle between the weight W and the boom is $90° - 44°$, or 46° (Fig. 12.63).

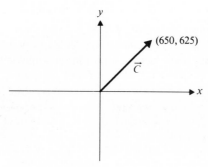

Figure 12.62

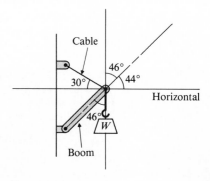

Figure 12.63

EXERCISES 12.4

Determine the resultant force, $\vec{F}_r$, for each of the following force systems.

1.

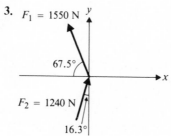

$F_2 = 130$ lb
$F_1 = 210$ lb
$55°$
$20°$

2.

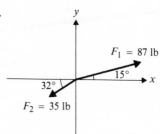

$F_1 = 87$ lb
$15°$
$32°$
$F_2 = 35$ lb

3. $F_1 = 1550$ N

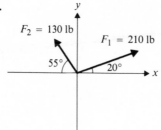

$67.5°$
$F_2 = 1240$ N
$16.3°$

4.

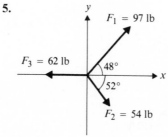

$40.5°$
$62.8°$
$F_2 = 175$ N
$F_1 = 125$ N

5.

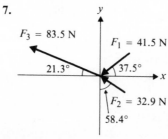

$F_1 = 97$ lb
$F_3 = 62$ lb
$48°$
$52°$
$F_2 = 54$ lb

6.

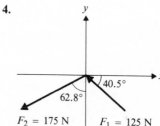

$F_1 = 470$ lb
$37°$
$78°$
$F_2 = 420$ lb
$F_3 = 260$ lb

7.

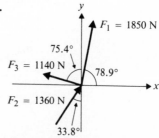

$F_3 = 83.5$ N
$F_1 = 41.5$ N
$21.3°$
$37.5°$
$F_2 = 32.9$ N
$58.4°$

8.

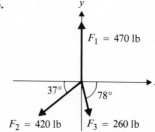

$F_1 = 1850$ N
$75.4°$
$F_3 = 1140$ N
$78.9°$
$F_2 = 1360$ N
$33.8°$

Determine the unknown forces necessary to produce equilibrium in each of the following force systems.

9.

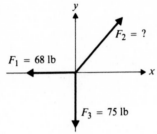

10.

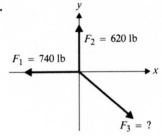

11.

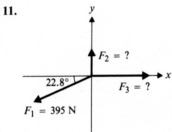

12.

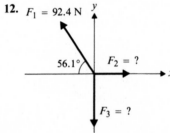

13.

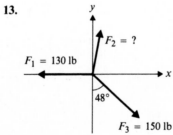

14.

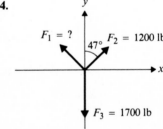

15.

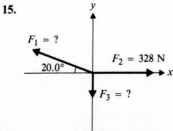

16.

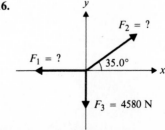

For exercises 17–22, refer to the derrick in Fig. 12.64 and assume that the system is in equilibrium.

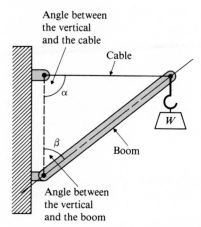

Figure 12.64

17. Determine the tension T in the cable and the compression C in the boom if $\alpha = 90°$, $\beta = 35°$, and $W = 650$ lb.

18. Determine the tension T in the cable and the compression C in the boom if $\alpha = 90°$, $\beta = 65°$, and $W = 2500$ N.

19. Determine the compression C in the boom if the tension T in the cable is 950 lb, $\alpha = 115°$, and $W = 1200$ lb.

20. Determine the tension T in the cable if the compression C in the boom is 4500 N, $\beta = 48°$, and $W = 5400$ N.

21. Determine the tension T in the cable and the weight W being supported if the compression C in the boom is 870 lb, $\alpha = 60°$, and $\beta = 90°$.

22. Determine the compression C in the boom and the weight W being supported if the tension T in the cable is 1400 N, $\alpha = 90°$, and $\beta = 75°$.

12.5 THE LAW OF SINES

Objectives

☐ To find the missing parts of an oblique triangle when given the measure of two angles and the length of one side.

☐ To find the missing parts of an oblique triangle when given the lengths of two sides and the measure of the angle opposite one of them.

☐ Given Two Angles and One Side of an Oblique Triangle

Having developed the trigonometric ratios for any angle, it is now possible to solve various types of problems that involve oblique triangles. An **oblique triangle** is a triangle that contains *no* right angle (see Fig. 12.65). Note that the oblique triangle in Fig. 12.65 is separated into two right triangles when the altitude *h* is drawn from angle *C* to side *c* (see Fig. 12.66).

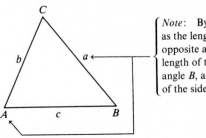

Note: By convention, we use *a* as the length of the side opposite angle *A*, *b* as the length of the side opposite angle *B*, and *c* as the length of the side opposite angle *C*.

Figure 12.65

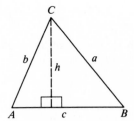

Figure 12.66

Referring to Fig. 12.66 and using the right triangle on the left, we can state that

$$\sin A = \frac{h}{b},$$

with *h* labeled opposite side and *b* labeled hypotenuse

or $h = b \sin A$.

Using the right triangle on the right, we can also state that

$$\sin B = \frac{h}{a},$$

opposite side

hypotenuse

or $h = a \sin B$.

Equating the values of h, we have

$$a \sin B = b \sin A.$$

Finally, dividing each side by $(\sin A)(\sin B)$, we have

$$\frac{a}{\sin A} = \frac{b}{\sin B}.$$

In a given triangle, the ratio of the length of a side to the sine of the angle opposite that side has the same constant value.

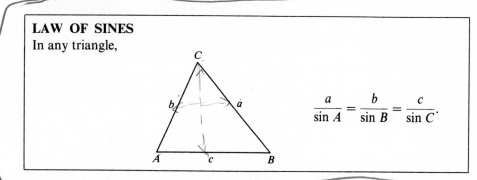

LAW OF SINES
In any triangle,

$$\frac{a}{\sin A} = \frac{b}{\sin B} = \frac{c}{\sin C}.$$

If two angles and the length of one side of a triangle are known, the law of sines can be used to find the lengths of the remaining sides.

EXAMPLE 1

Three drill holes are located in a steel plate as shown in Fig. 12.67. Find the distances x and y.

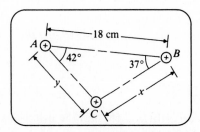

Figure 12.67

Solution: Since the sum of the interior angles in any triangle is 180°, we know that

$$\angle C = 180° - (42° + 37°)$$
$$\angle C = 101°.$$

Now, using the law of sines we have

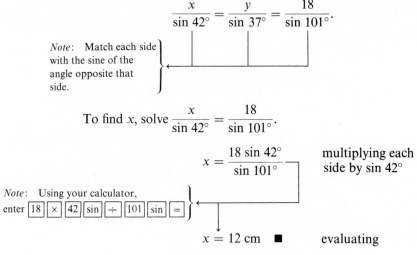

$$\frac{x}{\sin 42°} = \frac{y}{\sin 37°} = \frac{18}{\sin 101°}.$$

Note: Match each side with the sine of the angle opposite that side.

To find *x*, solve $\dfrac{x}{\sin 42°} = \dfrac{18}{\sin 101°}$.

$$x = \frac{18 \sin 42°}{\sin 101°} \qquad \text{multiplying each side by } \sin 42°$$

Note: Using your calculator, enter ⌜18⌝ ⌜×⌝ ⌜42⌝ ⌜sin⌝ ⌜÷⌝ ⌜101⌝ ⌜sin⌝ ⌜=⌝

$$x = 12 \text{ cm} \quad \blacksquare \qquad \text{evaluating}$$

To find *y*, solve

$$\frac{y}{\sin 37°} = \frac{18}{\sin 101°}$$

$$y = \frac{18 \sin 37°}{\sin 101°} \qquad \text{multiplying each side by } \sin 37°$$

$$y = 11 \text{ cm} \quad \blacksquare \qquad \text{evaluating}$$

Note: As a rough check, make sure the longest side of the triangle is opposite the largest angle and the shortest side is opposite the smallest angle. The Pythagorean theorem cannot be used to check since the triangle is not a right triangle.

⌞⌟ Given Two Sides and the Angle Opposite One of Them

In an oblique triangle, *if two sides and the measure of the angle opposite one of them are known, the law of sines can be used to find the remaining side and angles.*

EXAMPLE 2

For the roof truss shown in Fig. 12.68, find the distance *x*.

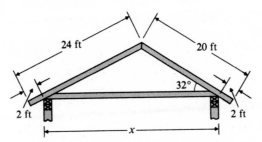

Figure 12.68

Solution: First, draw and label the oblique triangle that is formed (Fig. 12.69).

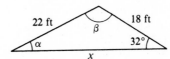

Figure 12.69

Thus, according to the law of sines,

$$\frac{22}{\sin 32°} = \frac{18}{\sin \alpha} = \frac{x}{\sin \beta}.$$

Note: Match each side with the sine of the angle opposite that side.

Next, find angle α by solving

$$\frac{22}{\sin 32°} = \frac{18}{\sin \alpha}.$$

$$22 \sin \alpha = 18 \sin 32° \qquad \text{cross product property}$$

$$\sin \alpha = \frac{18 \sin 32°}{22} \qquad \text{dividing each side by 22}$$

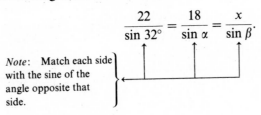

Note: Enter

18 × 32 sin ÷ 22 = inv sin .

Recall that the sine is positive in the first and second quadrants. At this point, be sure to record both angles.

$$\alpha = 26° \text{ or } \alpha = 154° \qquad \text{nearest degree}$$

Since the sum of the interior angles in a triangle must be 180°, we can exclude $\alpha = 154°$ as a possible value since when used with the given angle of 32° it would violate this condition. Thus, $\alpha = 26°$ only. Therefore, $\beta = 180° - (32° + 26°) = 122°$. Finally, to find x, solve

$$\frac{22}{\sin 32°} = \frac{x}{\sin 122°}.$$ replacing β with $122°$

$$x \sin 32° = 22 \sin 122°$$ cross product property

$$x = \frac{22 \sin 122°}{\sin 32°}$$ dividing each side by $\sin 32°$

$$x = 35 \text{ ft} \quad \blacksquare$$ evaluating

If the given angle is acute ($\theta < 90°$) and the side opposite this angle is smaller than the other given side ($a < b$), then there are two possible oblique triangles that can be constructed (see Fig. 12.70).

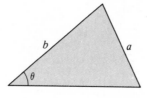 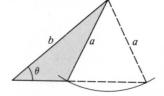

Figure 12.70

In most technical situations a fairly accurate picture of the problem is usually known before you attempt to solve for any missing parts. Using the picture, you can observe which solution is really desired. However, if you don't have a picture of the oblique triangle, you must include both triangles as possible solutions.

EXAMPLE 3

A plot plan of a triangular piece of land is shown in Fig. 12.71. Determine the measure of α and β and find the distance x.

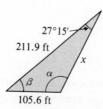

Figure 12.71

Solution: According to the law of sines,

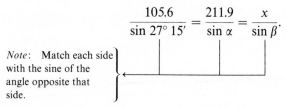

$$\frac{105.6}{\sin 27° \; 15'} = \frac{211.9}{\sin \alpha} = \frac{x}{\sin \beta}.$$

Note: Match each side with the sine of the angle opposite that side.

First, find angle α by solving

$$\frac{105.6}{\sin 27.25°} = \frac{211.9}{\sin \alpha}$$ converting 15′ to degrees

$$(105.6)(\sin \alpha) = (211.9)(\sin 27.25°)$$ cross product property

$$\sin \alpha = \frac{(211.9)(\sin 27.25°)}{105.6}$$ dividing each side by 105.6

Note: Enter $\boxed{211.9}\;\boxed{\times}\;\boxed{27.25}\;\boxed{\sin}\;\boxed{\div}\;\boxed{105.6}\;\boxed{=}\;\boxed{\text{inv}}\;\boxed{\sin}$.
The sine is positive in the first and second quadrants. At this point, be sure to record both angles.

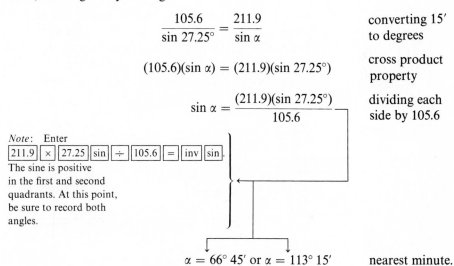

$$\alpha = 66° \; 45' \text{ or } \alpha = 113° \; 15'$$ nearest minute.

Both 66° 45′ and 113° 15′ are perfectly acceptable values for α. Neither violates the fact that the sum of the interior angles in the triangle is 180°. However, since the plot plan shows that α is obtuse, we will retain 113° 15′ as the measure of α. Then $\beta = 180° - (27° \; 15' + 113° \; 15') = 39° \; 30'$. Finally, to find x, solve

$$\frac{105.6}{\sin 27.25°} = \frac{x}{\sin 39.5°}.$$ replacing β with 39° 30′ = 39.5°

$$(x)(\sin 27.25°) = (105.6)(\sin 39.5°)$$ cross product property

$$x = \frac{(105.6)(\sin 39.5°)}{\sin 27.25°}$$ dividing each side by sin 27.25°

$$x = 146.7 \text{ ft } \blacksquare$$ evaluating

Note: If the picture of the triangle had not been shown, we would not know whether $\alpha = 66° \; 45'$ or $\alpha = 113° \; 15'$. Because of this ambiguity, mathematicians refer to this as the **ambiguous case** for oblique triangles. The ambiguous case will occur only when the given angle is acute and the side opposite this angle is smaller than the other given side.

In summary, the law of sines can be used to find the missing parts of an oblique triangle provided one of the ratios $\dfrac{a}{\sin A}$, $\dfrac{b}{\sin B}$, or $\dfrac{c}{\sin C}$ is known and one additional measurement is given.

Application

FORCE TRIANGLE

Consider a load W being supported by two cables as shown in Fig. 12.72, and assume that the structure is in equilibrium. The load W is acting vertically downward and the tension in the cables is acting away from point P as shown in the free body diagram (Fig. 12.72). By translating the vectors T_1 and T_2, we can form a **force triangle** as shown in Fig. 12.73. Using the force triangle, we can apply the law of sines to determine the tension in the cables.

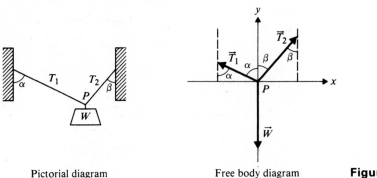

Pictorial diagram   Free body diagram   **Figure 12.72**

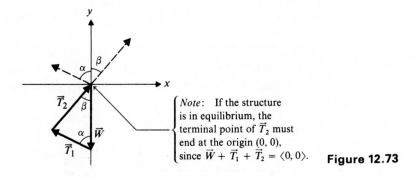

Note: If the structure is in equilibrium, the terminal point of $\vec{T}_2$ must end at the origin $(0, 0)$, since $\vec{W} + \vec{T}_1 + \vec{T}_2 = \langle 0, 0 \rangle$.   **Figure 12.73**

EXAMPLE 4

Referring to the load being supported by the cables in Fig. 12.72, find the tension in the cables if $W = 80\bar{0}$ lb, $\alpha = 65°$, and $\beta = 42°$.

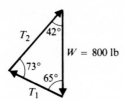

Figure 12.74

Solution: First, draw the force triangle (Fig. 12.74). Using the law of sines, we have

$$\frac{800}{\sin 73°} = \frac{T_1}{\sin 42°} = \frac{T_2}{\sin 65°}.$$

Find T_1 by solving

$$\frac{800}{\sin 73°} = \frac{T_1}{\sin 42°}.$$

$$T_1 = \frac{800 \sin 42°}{\sin 73°}$$

$$T_1 = 56\overline{0} \text{ lb}\ \ \text{(three significant digits)} \quad \blacksquare$$

Find T_2 by solving

$$\frac{800}{\sin 73°} = \frac{T_2}{\sin 65°}$$

$$T_2 = \frac{800 \sin 65°}{\sin 73°}$$

$$T_2 = 758 \text{ lb}\ \ \text{(three significant digits)} \quad \blacksquare$$

EXERCISES 12.5

1. Referring to the bridge truss in Fig. 12.75, determine the length x.

2. Referring to the bridge truss in Fig. 12.75, determine the length y.

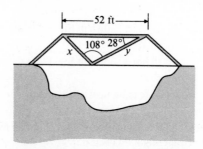

Figure 12.75

3. Referring to the roof in Fig. 12.76, determine the rafter length x.
4. Referring to the roof in Fig. 12.76, determine the rafter length y.

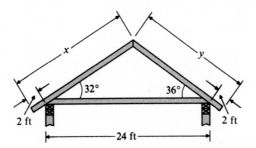

Figure 12.76

5. From the survey notes shown in Fig. 12.77, determine the length of the pond.
6. Referring to the pulleys in Fig. 12.78, determine the distance x between pulley B and pulley C.

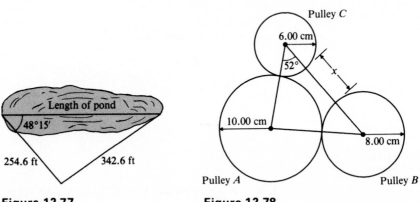

Figure 12.77   **Figure 12.78**

7. Referring to the three drill holes shown in Fig. 12.79, determine the length x. Assume that angle C is acute.
8. Referring to the three drill holes shown in Fig. 12.79, determine the length x. Assume that angle C is obtuse.
9. Determine the amount of fencing needed to enclose the triangular piece of land shown in Fig. 12.80. Assume that angle C is obtuse.
10. Determine the amount of fencing needed to enclose the triangular piece of land shown in Fig. 12.80. Assume that angle C is acute.

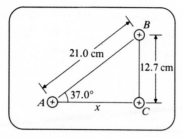

Figure 12.79

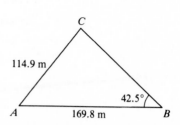

Figure 12.80

11. From the survey notes shown in Fig. 12.81, determine the width of the river.

12. Determine the depth of the canyon shown in Fig. 12.82.

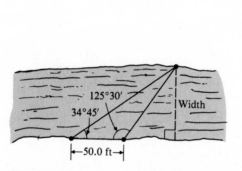

Figure 12.81

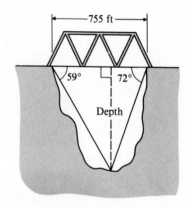

Figure 12.82

For the roof truss shown in Fig. 12.83, determine the following:

13. The length of member 1.

14. The length of member 2.

15. The length of member 3.

16. The length of member 4.

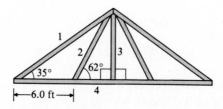

Figure 12.83

For the roof truss shown in Fig. 12.84, determine the following:

17. The measure of angle α. **18.** The measure of angle β.

19. The length of member 1. **20.** The measure of angle γ.

21. The length of member 2. **22.** The length of member 3.

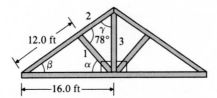

Figure 12.84

For the dormer addition shown in Fig. 12.85, determine the following:

23. The measure of angle α. **24.** The measure of angle β.

25. The length x. **26.** The measure of angle γ.

27. The length y. **28.** The length z.

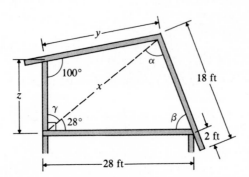

Figure 12.85

Using a force triangle and the law of sines, determine the tension in the cables.

29.

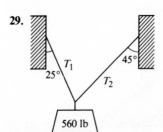

30.

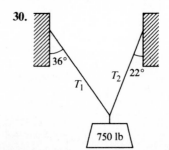

31.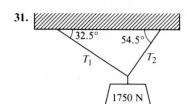

32.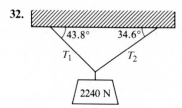

33.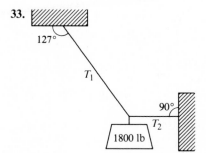

34.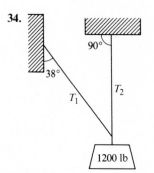

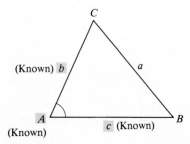

12.6   THE LAW OF COSINES

Objectives

⊡ To find the missing parts of an oblique triangle when given the lengths of two sides and the measure of the angle between them.

⊡ To find the missing parts of an oblique triangle when given all three sides.

⊡ Given Two Sides and the Angle Included between Them

Suppose in an oblique triangle that sides b and c and the angle between them, ⦨A, are known (see Fig. 12.86). It is important to note that the law of sines *cannot* be used to find the length of side a since none of the ratios $\dfrac{a}{\sin A}$, $\dfrac{b}{\sin B}$, or $\dfrac{c}{\sin C}$

(Known) b a

A c (Known) B

(Known)

Figure 12.86

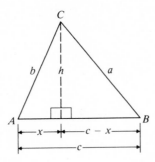

Figure 12.87

can be determined. To find the length of side a, we must develop a new law as follows:

First, draw an altitude h from angle C to side c. The altitude separates the oblique triangle into two right triangles and separates side c into two parts, x and $c - x$ (see Fig. 12.87). Referring to Fig. 12.87 and using the right triangle on the left, we can state that

$$\cos A = \frac{x}{b}$$

adjacent side

hypotenuse

or $\boxed{x = b \cos A}$.

Also, by the Pythagorean theorem,

$$b^2 = h^2 + x^2$$

or $\boxed{h^2 = b^2 - x^2}$.

Using the right triangle on the right and the Pythagorean theorem, we can state that

$$a^2 = h^2 + (c - x)^2$$

or $\boxed{h^2 = a^2 - (c - x)^2}$

Now, equating the values of h^2 we have

$a^2 - (c - x)^2 = b^2 - x^2.$

$$\begin{aligned}
a^2 &= b^2 - x^2 + (c - x)^2 & \text{adding } (c - x)^2 \text{ to each side} \\
a^2 &= b^2 - x^2 + c^2 - 2cx + x^2 & \text{expanding} \\
a^2 &= b^2 + c^2 - 2cx & \text{simplifying} \\
a^2 &= b^2 + c^2 - 2bc \cos A & \text{replacing } x \text{ with } b \cos A
\end{aligned}$$

The square of the unknown side is the sum of the squares of the given sides minus twice the product of those sides and the cosine of the angle *opposite* the unknown side.

LAW OF COSINES
In any triangle,

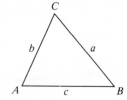

$$a^2 = b^2 + c^2 - 2bc \cos A,$$
$$b^2 = a^2 + c^2 - 2ac \cos B,$$
$$c^2 = a^2 + b^2 - 2ab \cos C.$$

If two sides and the angle between the two sides are known, the law of cosines can be used to find the length of the remaining side. Once the unknown side is determined by the law of cosines, one of the ratios $\dfrac{a}{\sin A}$, $\dfrac{b}{\sin B}$, or $\dfrac{c}{\sin C}$ can be found. The law of sines can then be used to find the missing angles.

EXAMPLE 1

Three meshed gears are shown in Fig. 12.88. Determine the distance x and the measure of angles α and β.

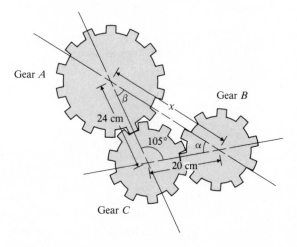

Figure 12.88

Solution: Using the law of cosines, we have

$$x^2 = 24^2 + 20^2 - 2(24)(20) \cos 105°,$$

Note: Match the unknown side with the cosine of the angle opposite that side.

so $x = \sqrt{24^2 + 20^2 - 2(24)(20) \cos 105°}$.

To find the value of x, use your calculator as follows:

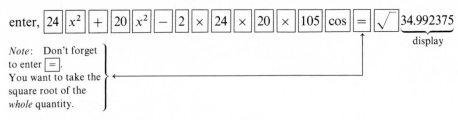

enter, $\boxed{24}\ \boxed{x^2}\ \boxed{+}\ \boxed{20}\ \boxed{x^2}\ \boxed{-}\ \boxed{2}\ \boxed{\times}\ \boxed{24}\ \boxed{\times}\ \boxed{20}\ \boxed{\times}\ \boxed{105}\ \boxed{\cos}\ \boxed{=}\ \boxed{\sqrt{\ }}\ 34.992375$

display

Note: Don't forget to enter $\boxed{=}$. You want to take the square root of the *whole* quantity.

Therefore, $x = 35$ cm. ■

Since we now know the length of a side ($x = 35$ cm) and the measure of the angle opposite that side (105°), we can use the law of sines to find α and β.

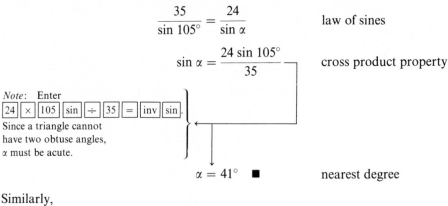

$$\frac{35}{\sin 105°} = \frac{24}{\sin \alpha} \qquad \text{law of sines}$$

$$\sin \alpha = \frac{24 \sin 105°}{35} \qquad \text{cross product property}$$

Note: Enter
$\boxed{24}\ \boxed{\times}\ \boxed{105}\ \boxed{\sin}\ \boxed{\div}\ \boxed{35}\ \boxed{=}\ \boxed{\text{inv}}\ \boxed{\sin}$.
Since a triangle cannot have two obtuse angles, α must be acute.

$$\alpha = 41° \quad ■ \qquad \text{nearest degree}$$

Similarly,

$$\frac{35}{\sin 105°} = \frac{20}{\sin \beta} \qquad \text{law of sines}$$

$$\sin \beta = \frac{20 \sin 105°}{35} \qquad \text{cross product property}$$

$$\beta = 34° \quad ■ \qquad \text{nearest degree}$$

Note: To check, be sure the longest side is opposite the largest angle and the shortest side is opposite the smallest angle. Also, be sure the sum of the interior angles is 180°.

⚬ Given Three Sides

Suppose that in an oblique triangle the three sides are known (see Fig. 12.89). It is important to note that the law of sines *cannot* be used to find the measure of any of the interior angles, since none of the ratios $\dfrac{a}{\sin A}$, $\dfrac{b}{\sin B}$, or $\dfrac{c}{\sin C}$ can be

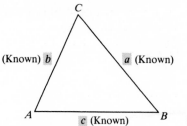

Figure 12.89

determined. However, *the law of cosines can be used to find an unknown angle of an oblique triangle when all three sides are known.*

EXAMPLE 2

For the bracket in Fig. 12.90, determine the measure of angle θ.

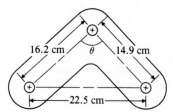

Figure 12.90

Solution: Using the law of cosines, we have

$$(22.5)^2 = (16.2)^2 + (14.9)^2 - 2(16.2)(14.9) \cos \theta$$

Note: Match the cosine of the unknown angle with the side opposite that angle.

Now, solve for $\cos \theta$ as follows:

$$(22.5)^2 - (16.2)^2 - (14.9)^2 = -2(16.2)(14.9)\cos \theta \qquad \text{subtracting } (16.2)^2 \text{ and } (14.9)^2 \text{ from each side}$$

$$\frac{(22.5)^2 - (16.2)^2 - (14.9)^2}{-2(16.2)(14.9)} = \cos \theta \qquad \text{dividing each side by } -2(16.2)(14.9)$$

To find the value of cos θ, use your calculator as follows:

enter, $($ $\boxed{22.5}$ $\boxed{x^2}$ $\boxed{-}$ $\boxed{16.2}$ $\boxed{x^2}$ $\boxed{-}$ $\boxed{14.9}$ $\boxed{x^2}$ $)$ $\boxed{\div}$ $($ $\boxed{2}$ $\boxed{+/-}$ $\boxed{\times}$ $\boxed{16.2}$ $\boxed{\times}$ $\boxed{14.9}$ $)$ $\boxed{=}$ $-.045157$

display

Note: Be sure to enter these parentheses in order to preserve the order of operations.

$\cos \theta = -.045157$

Note: Recall, the cosine is negative in the second and third quadrants. The third quadrant angle must be excluded since it would be greater than 180°. Thus, only the obtuse angle is retained.

$\theta = 92.6°$ ■ (nearest .1 degree)

Application

DISPLACEMENT USING BEARINGS

The **displacement** between two points is the length and direction of the straight line between the two points. Since displacement is fully described by magnitude (length) and direction, it is considered a vector quantity.

When working with displacements, the direction is usually given with respect to a compass reading, called a **bearing**. Referring to Fig. 12.91, the direction from point A to point B is given by the bearing S–20°–E, and the direction from point A to point C is given by the bearing N–65°–W. The law of cosines can be used to help solve displacement problems.

Note: North (N) or South (S) is always used as a reference line when describing a bearing.

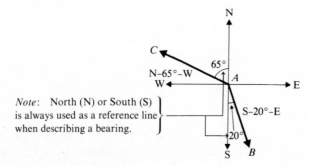

Figure 12.91

EXAMPLE 3

The displacement from an airport to an old landing strip is known to be $26\bar{0}$ mi S–85° 00′–W. A small aircraft located $20\bar{0}$ mi N–88° 00′–W of the airport is having engine trouble and needs to reach the landing strip. What is the airplane's displacement to the old landing strip?

Solution: First, draw the displacements and form an oblique triangle (Fig. 12.92).

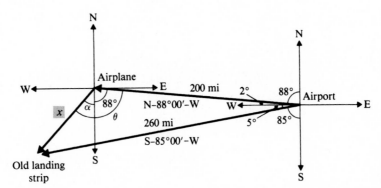

Figure 12.92

Now, using the law of cosines, we have

$$x^2 = 260^2 + 200^2 - 2(260)(200)\cos 7°.$$

$$\angle\!\!\!\!\!\!\!\!\!\!{}(2° + 5°)$$

$$x = \sqrt{260^2 + 200^2 - 2(260)(200)\cos 7°}$$

$$x = 66.1 \text{ mi} \text{ (three significant digits)}$$

Using the law of sines, we have

$$\frac{66.1}{\sin 7°} = \frac{260}{\sin \theta}.$$

$$\sin \theta = \frac{260 \sin 7°}{66.1}$$

$$\theta = 151.36° \longleftarrow \begin{cases} Note: \text{ Since } \theta \text{ is obtuse, place the} \\ \text{reference angle of } 28.64° \text{ in the} \\ \text{second quadrant.} \end{cases}$$

$$\theta = 151° \ 20' \text{ (nearest 10 minutes)}$$

Now, $\alpha = 151° \ 20' - 88° \ 00' = 63° \ 20'$. Therefore the airplane should fly 66.1 mi S–63° 20′–W in order to reach the old landing strip. ■

EXERCISES 12.6

1. Referring to the bridge truss in Fig. 12.93, determine the measure of angle α.

2. Referring to the bridge truss in Fig. 12.93, determine the measure of angle β.

3. Referring to the roof in Fig. 12.94, determine the measure of angle α.

4. Referring to the roof in Fig. 12.94, determine the measure of angle β.

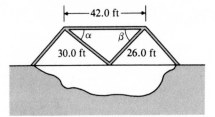

Figure 12.93

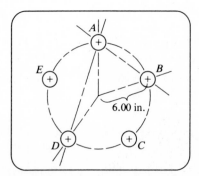

Figure 12.94

5. Referring to Fig. 12.95, five drill holes are equally spaced around a circle whose radius is 6.00 in. Find the distance between the centers of holes A and B.

6. Referring to Fig. 12.95, find the distance between the centers of holes A and D.

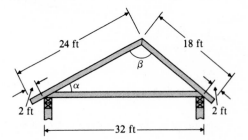

Figure 12.95

7. Referring to the pulleys in Fig. 12.96, determine the distance x.

8. Referring to the four drill holes in Fig. 12.97, determine the distance x.

9. Referring to Fig. 12.96, find the measure of angle θ.

10. Referring to Fig. 12.97, find the measure of angle θ.

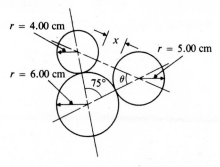

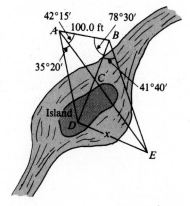

Figure 12.96   **Figure 12.97**

11. From the survey notes shown in Fig. 12.98, find the distance x.

12. From the survey notes shown in Fig. 12.99, find the distance x.

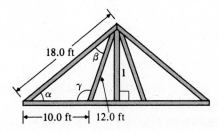

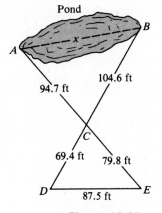

Figure 12.98   **Figure 12.99**

For the roof truss shown in Fig. 12.100, determine the following:

13. The measure of angle α.   **14.** The measure of angle β.

15. The measure of angle γ.   **16.** The length of member 1.

Figure 12.100

For the roof truss shown in Fig. 12.101, determine the following:

17. The length of member 1.   **18.** The measure of angle α.

19. The measure of angle β.   **20.** The measure of angle γ.

21. The length of member 2.   **22.** The length of member 3.

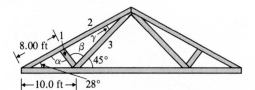

Figure 12.101

Use the law of cosines to find the following displacements.

23. A small aircraft leaves airport A and flies 87 mi N–32°00′–E. It then changes course and flies 109 mi N–79° 00′–E to land at airport B. What is the displacement from airport A to airport B?

24. A sailboat leaves island A and sails 32 mi S–11° 00′–E. It then changes course and sails 22 mi S–68° 00′–E to dock at island B. What is the displacement from island A to island B?

25. The displacement from a harbor on the mainland to a small island is known to be 72 mi N–42° 30′–E. A small sailboat located 87 mi N–38° 15′–E of the harbor learns of an impending tropical storm and needs to reach the island. What is the sailboat's displacement to the island?

26. The displacement from an airport to an old landing strip is known to be 127 mi N–7° 30′–E. A small aircraft located 152 mi N–10° 20′–W of the airport is having engine trouble and needs to reach the landing strip. What is the airplane's displacement to the old landing strip?

12.7 RADIAN MEASURE

Objectives

☐ To place an angle in standard position when the measure of the angle is given in radians.

☐ To convert from degree measure to radian measure or vice versa.

☐ To determine the measure of a central angle or the length of the arc subtended by a central angle.

☐ To determine the angular speed of a rotating object.

⋅ Definition of a Radian

Thus far in our study of geometry and trigonometry, it has been sufficient to work with angles measured in degrees. For many branches of engineering technology, however, we need to specify the measure of an angle by using **radian measure**.

Recall from Chapter 3 that the circumference C of a circle is given by the formula $C = 2\pi r$, where r is the radius of the circle. Since $2\pi \approx 6.28$, we can state that $C \approx 6.28r$. *This means that in any circle the radius r can be marked off 6.28 times along its circumference* (see Fig. 12.102).

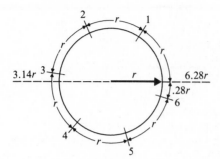

Figure 12.102

Now, **one radian** (rad) is formed when we rotate the radius counterclockwise such that the length along the circumference is equal to the radius of the circle (see Fig. 12.103). Similarly, an angle of *two radians* is formed when we rotate the radius counterclockwise such that the length along the circumference is equal to two radii of the circle, and so on (see Fig. 12.104).

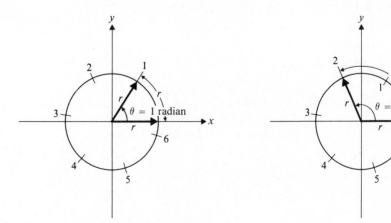

Figure 12.103   **Figure 12.104**

EXAMPLE 1

Place the following angles in standard position.

a) 5.4 rad   **b)** 7 rad

Solution:

a) An angle of 5.4 rad is located in the fourth quadrant and is four-tenths of the way between 5 rad and 6 rad (Fig. 12.105). ∎

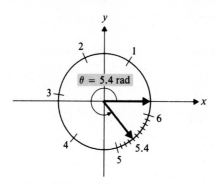

Figure 12.105

b) Since there are approximately 6.28 rad in one full revolution, 7 rad is slightly more than one full revolution (Fig. 12.106). ∎

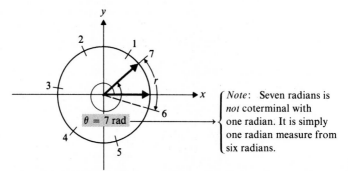

Note: Seven radians is *not* coterminal with one radian. It is simply one radian measure from six radians.

Figure 12.106

⠢ Converting between Degree Measure and Radian Measure

In one complete revolution there are approximately 6.28 rad. In one complete revolution there are also 360°.

$$6.28 \text{ radians} \approx 360 \text{ degrees} \qquad \text{one full revolution}$$

$$2\pi \text{ radians} = 360 \text{ degrees} \qquad \text{replacing 6.28 with } 2\pi$$

$$\pi \text{ radians} = 180 \text{ degrees} \qquad \text{dividing each side by 2}$$

dividing each side by π   dividing each side by 180

$$1 \text{ radian} = \frac{180}{\pi} \text{ degrees} \qquad 1 \text{ degree} = \frac{\pi}{180} \text{ radians}$$

$$1 \text{ radian} \approx 57.296 \text{ degrees} \qquad 1 \text{ degree} \approx .01745 \text{ radians}$$

We will use the fact that π rad and $180°$ are equal measures in order to convert from degree measure to radian measure or vice versa (see Table 12.1).

TABLE 12.1

Equal measures	Conversion factors
π radians $= 180$ degrees	$\dfrac{\pi \text{ rad}}{180°}$ or $\dfrac{180°}{\pi \text{ rad}}$
$(\pi \text{ rad} = 180°)$	

EXAMPLE 2

Convert the following angles to radians.

a) $54°$ **b)** $240°$

Solution:

a) $54° = \dfrac{54°}{1}\left(\dfrac{\pi \text{ rad}}{180°}\right) = \dfrac{54\pi}{180}\text{ rad} = \dfrac{3\pi}{10}\text{ rad} = .942 \text{ rad}$ ■

(fractions reduced: $54/180 = 3/10$, in terms of π, to three significant digits)

b) $240° = \dfrac{240°}{1}\left(\dfrac{\pi \text{ rad}}{180°}\right) = \dfrac{240\pi}{180}\text{ rad} = \dfrac{4\pi}{3}\text{ rad} = 4.19 \text{ rad}$ ■

(fractions reduced: $240/180 = 4/3$, in terms of π, to three significant digits.)

EXAMPLE 3

Convert the following angles to degrees.

a) 1.57 rad **b)** $\dfrac{9\pi}{4}$ rad

Solution:

to three significant digits

a) $1.57 \text{ rad} = \dfrac{1.57 \text{ rad}}{1}\left(\dfrac{180°}{\pi \text{ rad}}\right) = \dfrac{(1.57)(180)°}{\pi} = 90.0°$ ■

b) $\dfrac{9\pi}{4}\text{ rad} = \dfrac{9\pi \text{ rad}}{4}\left(\dfrac{180°}{\pi \text{ rad}}\right) = \dfrac{(9)(180)°}{4} = 405°$ ■

(exact)

⸬ Arc Length and Central Angles

Shown in Fig. 12.107 is a circle of *radius r* with a *central angle* θ that subtends an *arc* of length *s*. For a central angle of $360° = 2\pi$ radians, the arc length is the

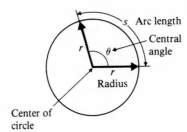

Figure 12.107

circumference of the entire circle. Since the arc length *s* is directly proportional to the measure of its central angle θ, we have

$$\frac{s}{2\pi r} = \frac{\theta}{2\pi}$$

circumference of the   central angle (in radians)
entire circle   for one full revolution

$$\frac{2\pi s}{2\pi r} = \theta \qquad \text{multiplying each side by } 2\pi$$

$$\frac{s}{r} = \theta \qquad \text{cancelling}$$

Note: The ratio of the arc length to the radius of the circle is equal to the central angle in radians.

FORMULA   Central Angle

$\theta = \dfrac{s}{r}$, where θ is the central angle in radians
        s is the length of the subtended arc
        r is the radius of the circle

When using the central angle formula, be sure that θ is in radians, and that *s* and *r* have the same units of measure.

EXAMPLE 4

Referring to Fig. 12.108, through what angle has the gear rotated?

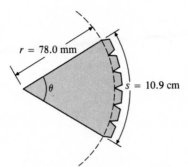

$r = 78.0$ mm

$s = 10.9$ cm

θ

Figure 12.108

Solution: Before the formula can be applied, s and r must have the same units of measure. Let's convert 10.9 cm to mm.

$$10.9 \text{ cm} = \frac{10.9 \text{ cm}}{1}\left(\frac{10 \text{ mm}}{1 \text{ cm}}\right) = 109 \text{ mm}$$

Thus,

$$\theta = \frac{s}{r} = \frac{109 \text{ mm}}{78.0 \text{ mm}} = 1.40. \quad \text{(three significant digits)}$$

Note: The units of measure (mm) cancel, leaving a dimensionless value of θ. When an angle is dimensionless, it is understood to be a radian measure.

Therefore, $\theta = 1.40$ rad ∎

Note: If you want to express the angle in degrees, convert as follows:

$$1.40 \text{ rad} = \frac{1.40 \text{ rad}}{1}\left(\frac{180°}{\pi \text{ rad}}\right) = 8\overline{0}° \quad \text{(two significant digits)}$$

EXAMPLE 5

Referring to the intersecting highways in Fig. 12.109, determine the arc length of the curve from the point of tangency (*PT*) to the point of curvature (*PC*).

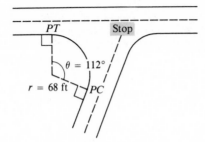

Figure 12.109

Solution: Before the formula can be applied, we must convert 112° to radians.

$$112° = \frac{112°}{1}\left(\frac{\pi \text{ rad}}{180°}\right) = 1.95 \text{ rad}$$

Since

$$\theta = \frac{s}{r}, s = \theta r.$$

$$s = \theta r = (1.95)(68 \text{ ft}) = 130 \text{ ft} \quad \blacksquare$$

Note: We need a dimension-less angle in order to preserve a proper unit of measure (ft) for the arc length.

⸬ Linear and Angular Speed

Linear speed v is defined as the linear distance s traveled per unit of time t.

$$\text{linear speed } v = \frac{s}{t}$$

For example, if it takes an automobile 2 h to travel a linear distance of 80 mi, its linear speed is

$$v = \frac{s}{t} = \frac{80 \text{ mi}}{2 \text{ h}} = 40 \text{ mph}.$$

In a similar way, we define **angular speed** ω (the Greek letter omega) as the amount of rotation θ per unit of time t.

$$\text{angular speed } \omega = \frac{\theta}{t}$$

For example, if it takes .5 s for a wheel to rotate 2 rad, its angular speed is

$$\omega = \frac{2 \text{ rad}}{.5 \text{ s}} = 4 \frac{\text{rad}}{\text{s}}.$$

For many applications in engineering technology, it is necessary to know the relationship between linear speed and angular speed. Recall that

$$\theta = \frac{s}{r}$$

$$s = \theta r.$$

Dividing each side by t, we have

$$\frac{s}{t} = \frac{\theta r}{t}$$

$$\boxed{\frac{s}{t} = r\,\frac{\theta}{t}}$$

$$v = r\,\omega$$

FORMULA   Linear Speed in Terms of Angular Speed

$$v = r\omega,$$

where v is the linear speed
ω is the angular speed in radians per unit of time
r is the radius of rotation

When using the formula $v = r\omega$, be sure that ω is in radians per unit of time and that the unit of length of v and r is the same.

EXAMPLE 6

An automobile is traveling at a speed of 60.0 mph. The tires have a radius of 15.0 in. Find the number of revolutions the tire makes in one minute (rpm).

Solution: Before the formula $v = r\omega$ can be applied, the unit of length of v and r must be the same. Let's convert 60.0 mph to *feet* per second and 15.0 in. to *feet*.

$$60.0 \text{ mph} = \frac{60.0 \cancel{\text{ mi}}}{\cancel{h}} \left(\frac{5280 \text{ ft}}{1 \cancel{\text{ mi}}}\right)\left(\frac{1\cancel{h}}{3600 \text{ s}}\right) = 88.0\,\frac{\text{ft}}{\text{s}}$$

$$15.0 \text{ in.} = \frac{15.0 \cancel{\text{ in.}}}{1}\left(\frac{1 \text{ ft}}{12 \cancel{\text{ in.}}}\right) = 1.25 \text{ ft}$$

Since $v = r\omega$, then $\omega = \dfrac{v}{r}$.

Thus, $\omega = \dfrac{v}{r} = \dfrac{88.0 \text{ } \cancel{\text{ft}}/\text{s}}{1.25 \text{ } \cancel{\text{ft}}} = 70.4\,\dfrac{\text{rad}}{\text{s}}$

Therefore, the angle through which the tire rotates in one minute (60 s) is

$$\theta = \omega t = \left(70.4 \, \frac{\text{rad}}{\text{s}} \right)(60\,\text{s}) = 4224 \text{ rad}.$$

Since every 2π rad equals one revolution, we have

$$\theta = 4224 \text{ rad} = \frac{4224 \text{ rad}}{1} \left(\frac{1 \text{ rev}}{2\pi \text{ rad}} \right) = 672 \text{ rev}. \quad \text{(three significant digits)}$$

Thus, the tire is rotating at 672 rpm when the automobile travels at 60 mph (Fig. 12.110). ■

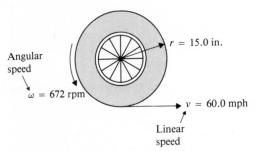

Angular speed

$\omega = 672$ rpm

$r = 15.0$ in.

$v = 60.0$ mph

Linear speed

Figure 12.110

Application

SIMPLE HARMONIC MOTION

Assume that the spring shown in Fig. 12.111(a) is in equilibrium. If the spring is pulled down and then released as shown in Fig. 12.111(b), it will oscillate back and

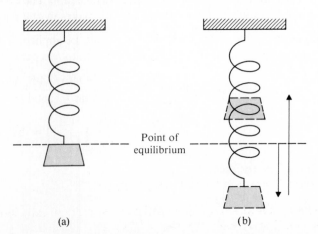

Point of equilibrium

(a)

(b)

Figure 12.111

forth through the point of equilibrium. If we ignore frictional forces, this oscillating motion will repeat itself over and over again. Motion of this type is called **simple harmonic motion**. The mathematical models that describe simple harmonic motion are

$$d = A \sin \omega t$$

or

$$d = A \cos \omega t ,$$

where A is the maximum displacement,
    ω is the angular speed in rad/s, and
    d is the displacement after a time of t seconds.

EXAMPLE 7

A spring is pulled down 15 cm and then released as shown in Fig. 12.111. Suppose its displacement is described by $d = 15 \cos \pi t$. Find its displacement after $\frac{1}{2}$ second.

Solution:

$$d = 15 \cos \pi t$$

$$d = 15 \cos\left(\pi \, \frac{\text{rad}}{\cancel{s}} \, \frac{1}{2} \, \cancel{s}\right) \qquad \text{replacing } t \text{ with } \tfrac{1}{2} \text{ s}$$

Note: Angular speed ω is in rad/s.

$$d = 15 \cos \frac{\pi}{2} \qquad \text{simplifying the angle in radians}$$

$$d = 15 \underline{\cos 90^\circ} \qquad \text{converting to degrees}$$
$$\downarrow$$

$$d = 15 \quad (0) \qquad \text{evaluating } \cos 90^\circ$$

$$d = 0 \text{ cm} \quad \blacksquare \qquad \text{simplifying}$$

— at the point of equilibrium

Note: Shown in the following table are the displacements after 0 s, $\frac{1}{2}$ s, 1 s, $\frac{3}{2}$ s, and 2 s. Plotting these points allows us to graphically look at the motion of the spring (Fig. 12.112).

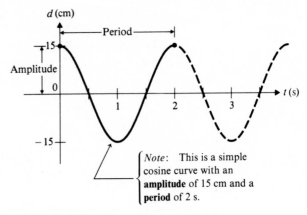

t (s)	d (cm)
0	15
$\dfrac{1}{2}$	0
1	-15
$\dfrac{3}{2}$	0
2	15

Figure 12.112

Electrical current and voltage are also described by the models for harmonic motion.

EXAMPLE 8

The voltage v in an electrical circuit is given by $v = 20 \sin 4t$. Find the times when $v = 20$ V.

Solution: The voltage $v = 20$ V only when $\sin 4t = 1$. Recall from the unit circle that $\sin \theta$ will equal 1 only when

$$\theta = 90° \text{ or } \frac{\pi}{2} \text{ rad},$$

$$\theta = (90° + 360°) = 450° \text{ or } \frac{5\pi}{2} \text{ rad},$$

one revolution

$$\theta = (90° + 720°) = 810° \text{ or } \frac{9\pi}{2} \text{ rad, and so on.}$$

two revolutions

Now, since $\theta = 4t$, we can find the times when $v = 20$ V by solving

$$4t = \frac{\pi}{2} \rightarrow t = \frac{\pi}{8} \text{ s}$$

$$4t = \frac{5\pi}{2} \rightarrow t = \frac{5\pi}{8} \text{ s}$$

$$4t = \frac{9\pi}{2} \rightarrow t = \frac{9\pi}{8} \text{ s, and so on.} \quad \blacksquare$$

Note: Shown in the following table are the voltages after $0\,s$, $\dfrac{\pi}{8}\,s$, $\dfrac{\pi}{4}\,s$, $\dfrac{3\pi}{8}\,s$, and $\dfrac{\pi}{2}\,s$. Plotting these points allows us to graphically look at this *alternating* voltage (Fig. 12.113).

t (s)	v (V)
0	0
$\dfrac{\pi}{8}$	20
$\dfrac{\pi}{4}$	0
$\dfrac{3\pi}{8}$	-20
$\dfrac{\pi}{2}$	0

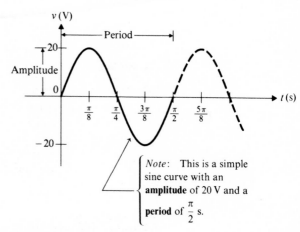

Note: This is a simple sine curve with an **amplitude** of 20 V and a **period** of $\dfrac{\pi}{2}$ s.

Figure 12.113

EXERCISES 12.7

Place the following angles in standard position.

1. 4 rad
2. 6 rad
3. .3 rad
4. .8 rad
5. 1.6 rad
6. 3.1 rad
7. 8 rad
8. 9 rad

Convert the following angles to radians. Write your answers in terms of π (reduced) and as decimal numbers accurate to three significant digits.

9. 90°
10. 270°
11. 180°
12. 360°
13. 30°
14. 60°
15. 45°
16. 225°
17. 300°
18. 150°
19. 72°
20. 198°
21. 375°
22. 520°

Convert the following angles to degrees.

23. .646 rad
24. .209 rad
25. 3.75 rad
26. 6.37 rad
27. $\dfrac{5\pi}{4}$ rad
28. $\dfrac{7\pi}{6}$ rad

29. $\dfrac{5\pi}{3}$ rad

30. $\dfrac{\pi}{18}$ rad

31. $\dfrac{3\pi}{8}$ rad

32. $\dfrac{11\pi}{8}$ rad

For the gear shown in Fig. 12.114, fill in the table below.

| | | Central angle (θ) | |
Arc length s	Radius r	radians	degrees
33. 17.0 in.	1.00 ft		
34. 26.2 cm	90.0 mm		
35.	8.5 in.		95°
36. 9.3 mm			215°

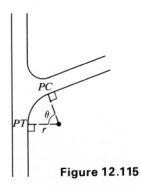

Figure 12.114

For the intersecting highways shown in Fig. 12.115, fill in the table below.

| | | Central angle (θ) | |
Arc length (*PT* to *PC*)	Radius (*r*)	radians	degrees
37.	42.0 ft		85°
38.	118 m		130°
39. 195 ft			72°
40. 82.0 m	49.0 m		

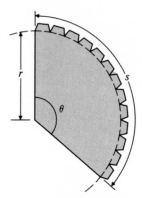

Figure 12.115

41. The pulley for the tuning dial of the radio shown in Fig. 12.116 has a radius of 54 mm. How far will the frequency marker move when the pulley is moved one-half revolution?

42. The pulley for the tuning dial of the radio shown in Fig. 12.116 has a diameter of 84 mm. If the frequency marker moves 66 mm, what part of a revolution has the pulley moved?

43. An automobile is traveling at a speed of 45.0 mph. The tires have a radius of 14.0 in. Find the number of revolutions the tire makes in one minute (rpm).

44. An automobile is traveling at a speed of 30.0 mph. The tires have a radius of 13.0 in. Find the number of revolutions the tire makes in one minute (rpm).

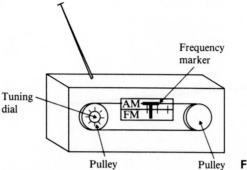

Frequency marker

Tuning dial

AM
FM

Pulley   Pulley   **Figure 12.116**

45. A pulley of diameter 8.00 in. has an angular speed of 725 rpm. What is the speed of the belt connected to the pulley? (Record your answer in feet per second.)

46. A pulley of radius 6.4 cm has an angular speed of 1800 rpm. What is the speed of the belt connected to the pulley? (Record your answer in meters per second.)

47. The displacement (in centimeters) of an oscillating spring is given by $d = 12 \cos \frac{\pi}{2} t$. Fill in the table; then graph the motion of the spring.

t (s)	d (cm)
0	
1	
2	
3	
4	

48. The voltage (in volts) of an electric circuit is given by $v = 120 \sin 6\pi t$. Fill in the table; then graph the alternating voltage.

t (s)	v (V)
0	
$\frac{1}{12}$	
$\frac{1}{6}$	
$\frac{1}{4}$	
$\frac{1}{3}$	

49. The current (in amperes) in an electric circuit is given by $i = 5 \sin 3t$. Determine the times when $i = 5$ A; then graph one period of this alternating current.

50. The displacement (in inches) of an oscillating spring is given by $d = 16 \cos 5t$. Determine the times when $d = 0$ cm; then graph one period of this harmonic motion.

Chapter Review

REVIEW EXERCISES

The following review exercises are grouped according to the objectives that should have been mastered in Chapter 12. Work each problem carefully. If any weaknesses appear, immediately refer to and read the subsection that matches that objective.

12.1 ANGLES IN THE RECTANGULAR COORDINATE SYSTEM

Objectives

Find the smallest positive coterminal angle.

▫ To determine the smallest positive coterminal angle for a given directed angle.

1. a) $-230°$ **b)** $570°$ **c)** $-810°$

▫ To determine the reference angle for a given directed angle.

Find the reference angle.

2. a) $30°$ **b)** $140°$ **c)** $260°$
d) $300°$ **e)** $500°$

▫ To determine whether the trigonometric ratios of a given directed angle are positive or negative.

Determine whether the ratios are positive or or negative.

3. Determine whether the trigonometric ratios for $150°$ are positive or negative.

4. Determine whether the trigonometric ratios for $550°$ are positive or negative.

5. Determine whether the trigonometric ratios for $-25°$ are positive or negative.

12.2 TRIGONOMETRIC RATIOS FOR ANY ANGLE

Objectives

▫ To use a unit point to find the trigonometric ratios for a quadrantal angle.

Find the trigonometric ratios.

6. Use a unit point to find the trigonometric ratios for $90°$.

7. Use a unit point to find the trigonometric ratios for $-180°$.

▫ To determine the trigonometric ratios for any angle by using a calculator.

Find the trigonometric ratios.

8. Using your calculator, determine the trigonometric ratios for $-10°$. Retain four significant digits.

▫ Given the value of a trigonometric ratio, to use a calculator to find all angles between $0°$ and $360°$ that have that value.

Find all θ, with
$0° \le \theta < 360°$.

9. $\tan \theta = 1.150$ Record your answers to the nearest .1 degree.

10. $\cos \theta = -.9613$ Record your answers to the nearest .1 degree.

11. $\csc \theta = -2.875$ Record your answer to the nearest minute.

12.3 AN INTRODUCTION TO VECTORS

Objectives

☐ To translate a given vector into standard position.

Translate into standard
position and identify
using rectangular form.

12.

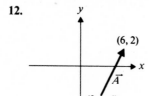

Translate into standard
position and identify
using polar form.

13.

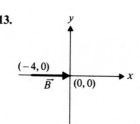

☐ Given a vector in polar form, to express it in rectangular form or vice versa.

Express in rectangular
form.

14. $9.2\underline{/127°}$

Express in polar form.

15. $\langle -285, -358 \rangle$

☐ To find the sum of two vectors.

Find the sum $\vec{A_1} + \vec{A_2}$.

16. $\vec{A_1} = \langle 4, -6 \rangle$, $\vec{A_2} = \langle -9, 5 \rangle$

17. $\vec{A_1} = 25\underline{/210°}$, $\vec{A_2} = 15\underline{/130°}$

12.4 ANALYSIS OF A FORCE SYSTEM

Objectives

☐ To find the resultant force of a force system.

Find the resultant force.   **18.**

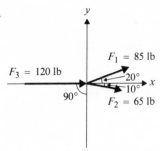

☐ Given a force system, to determine the unknown force necessary to produce equilibrium.

Find the force F_1
necessary to produce
equilibrium.   **19.**

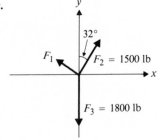

12.5 THE LAW OF SINES

Objectives

⊡ To find the missing parts of an oblique triangle when given the measure of two angles and the length of one side.

Find the distances
x and y.   **20.**

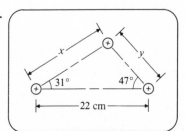

⊡ To find the missing parts of an oblique triangle when given the lengths of two sides and the measure of the angle opposite one of them.

Find the distance x.

21.

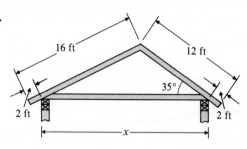

Find the measure of angles α and β and the distance x.

22.

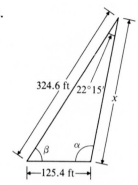

12.6 THE LAW OF COSINES

Objectives

☐ To find the missing parts of an oblique triangle when given the lengths of two sides and the measure of the angle included between them.

Find the distance x and the measure of angles α and β.

23.

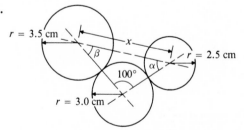

☐ To find the missing parts of an oblique triangle when given all three sides.

Find the measure of angle θ.

24.

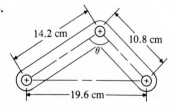

12.7 **RADIAN MEASURE**

Objectives

⊡ To place an angle in standard position when the measure of the angle is given in radians.

Place in standard position.

25. a) 4.3 rad

    b) 10 rad

⊡ To convert from degree measure to radian measure or vice versa.

Convert to radians.

26. a) 36°

    b) 210°

Convert to degrees.

27. a) 3.14 rad

    b) $\dfrac{13\pi}{6}$ rad

⊡ To determine the measure of a central angle or the length of the arc subtended by a central angle.

Find the angle θ.

28.

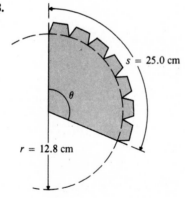

$s = 25.0$ cm

$r = 12.8$ cm

Find the arc length from the *PT* to the *PC*.

29.

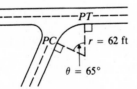

PT

PC $r = 62$ ft

$\theta = 65°$

⊡ To determine the angular speed of a rotating object.

Solve.

30. An automobile is traveling at a speed of 45.0 mph. The tires have a radius of 15.0 in. Find the number of revolutions the tire makes in one minute (rpm).

CHAPTER TEST
Time
50 minutes

Score
15, 14 excellent
13, 12 good
11, 10, 9 fair
below 8 poor

If you have worked through the Review Exercises and corrected any weaknesses, you are ready to take the following Chapter Test.

1. Find the smallest positive coterminal angle for $-920°$.

2. Determine the reference angle for $-920°$.

3. Using your calculator, determine all six trigonometric ratios for $230°$. Retain four significant digits.

4. If $\cos \theta = -.5736$, find all θ with $0° \leq \theta < 360°$. Record your answer to the nearest $.1°$.

5. If $\cot \theta = -2.567$, find all θ with $0° \leq \theta < 360°$. Record your answer to the nearest minute.

6. Express $\vec{A}_1 = 22.0 \underline{/-56°}$ in rectangular form.

7. Express $\vec{A}_2 = \langle -16.8, -12.5 \rangle$ in polar form.

8. Referring to exercises 6 and 7, find $\vec{A}_1 + \vec{A}_2$. Express the sum in polar form.

9. Convert $225°$ to radians. Leave your answer in terms of π.

10. Convert 3.62 rad to degrees.

11. From the survey notes shown in Fig. 12.117, determine the length of the pond.

12. For the roof truss shown in Fig. 12.118, determine the measure of angle θ.

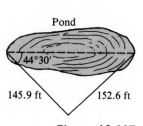

Figure 12.117

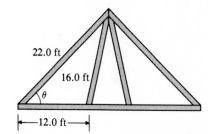

Figure 12.118

13. Determine the unknown forces necessary to produce equilibrium for the force system shown in Fig. 12.119.

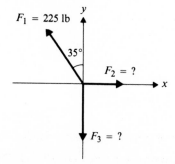

Figure 12.119

14. Find the angle θ through which the gear in Fig. 12.120 has rotated.

15. The water wheel shown in Fig. 12.121 has a radius of 8.0 ft and rotates at 19 rpm. What is the speed of the river? Record your answer in miles per hour.

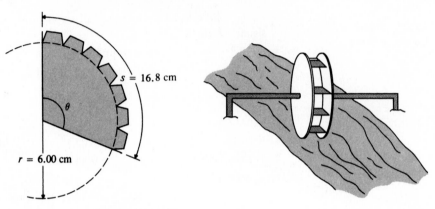

Figure 12.120   **Figure 12.121**

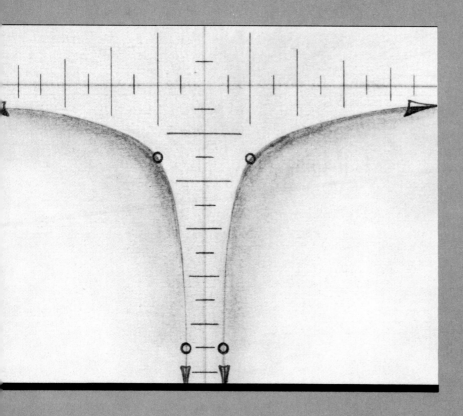

CHAPTER 1

Exercises 1.1 (pp. 8, 9)

1. $v - 8$ **3.** $\dfrac{H}{3}$ **5.** $2T_2 + 6$ **7.** $2w - 4$ **9.** $P_1 - 5 + P_2$ **11.** 213   **13.** 80   **15.** 277   **17.** 56

19. 732   **21.** $3q$ **23.** $3p + 16$ **25.** $4M + 2R + 14$ **27.** $h + h + h + h$ **29.** $K + K + K + S + S$

31. yes

Exercises 1.2 (pp. 13, 14)

1. $n = 6$ **3.** $m = 16$ **5.** $A = 5$ **7.** $x = 34$ **9.** $x = 3$ **11.** $v - 6 = 4; v = 10$

13. $T + 16 = 50; T = 34$ **15.** $2H - 4 = 30; H = 17$

	AB	*DE*
17.	10	5
19.	$3a + 2b$	$4m - 5n$

21. $x = z - y$ **23.** $x = 39$ in.   **25.** $R_t = R_1 + 2R_2 + 33$

Exercises 1.3 (pp. 17, 18)

1. -100 **3.** -500 **5.** -16 **7.** 16   **9.** $-x$ **11.** 4   **13.** 22   **15.** -7 **17.** -19 **19.** 9

21. $<$ **23.** $>$ **25.** $=$ **27.** $<$ **29.** $=$ **31.** false; $|-2| \neq -2$ **33.** true   **35.** true   **37.** $x = 8$

39. $t = -15$

Exercises 1.4 (pp. 24, 25)

1. 0   **3.** 8   **5.** -30 **7.** -437 **9.** -14 **11.** 65   **13.** -16 **15.** 0   **17.** $3 + 2n$ **19.** 8

21. $-3 + t + d$ **23.** 4   **25.** $3a$ **27.** $n = -5$ **29.** $R = -6$ **31.** $F_1 = 5$ **33.** $D = -1$ **35.** no

37. yes

39.	*a*	*b*	$|a|$	$|b|$	$|a| + |b|$	$|a + b|$	Conclusion:
	3	2	3	2	5	5	$|a| + |b| \neq |a + b|$
	-3	2	3	2	5	1	
	3	-2	3	2	5	1	
	-3	-2	3	2	5	5	

Exercises 1.5 (pp. 28, 29)

1. 10   **3.** -11 **5.** -21 **7.** -1 **9.** -68 **11.** 19   **13.** -10 **15.** $-y + a$ **17.** $-n + f - 3$

19. $x + y + 9$ **21.** $5 - x$ **23.** 8   **25.** 0   **27.** $2d - 10$ **29.** $x = -2$ **31.** $m = -10$ **33.** $a = -1$

35. $45°F$ **37.** $105°C$ **39.** $|2a|°F$

Exercises 1.6 (pp. 36, 37)

1. -48 **3.** -96 **5.** -60 **7.** 0   **9.** 25   **11.** 96   **13.** 90   **15.** 48   **17.** 0   **19.** -93 **21.** 9

23. -64 **25.** -16 **27.** 1   **29.** 100   **31.** $3 - 3y$ **33.** $6 - 3t$ **35.** $-18x$ **37.** n^2 **39.** 0

41. $5n^2d$ **43.** $-x^3y^2$ **45.** $-9x^3$ **47.** $-25x^4y^2$ **49.** $x = 16$ **51.** $x = -16$ **53.** $n = 25$

55. $y = -1$ **57.** $x = 3$ and $x = -3$ **59.** no solution

Exercises 1.7 (pp. 43, 44)

1. 8   **3.** -8 **5.** 6   **7.** -25 **9.** 1   **11.** undefined   **13.** -79 **15.** 0   **17.** -1 **19.** tv

21. $-3p$ **23.** 1   **25.** $n = -10$ **27.** $a = -6$ **29.** $x = 48$ **31.** $f = -9$ **33.** $t = 32$ **35.** $-5°F$

Exercises 1.8 (pp. 50–52)

1. 7   **3.** 48   **5.** 116   **7.** -24 **9.** 9   **11.** -9 **13.** 9   **15.** 32   **17.** 19   **19.** 36   **21.** 0

23. 3   **25.** $-10 - t$ **27.** $t - 7$ **29.** $q - p$ **31.** $19 + h$

33.

	$2x + 3x$	$5x$
$x = 3$	15	15
$x = 5$	25	25
$x = -3$	-15	-15
$x = -5$	-25	-25

Conclusion:
$2x + 3x = 5x$

35.

	$(x + y)^2$	$x^2 + y^2$
$x = 2$, $y = 3$	25	13
$x = -2$, $y = -3$	25	13
$x = -2$, $y = 3$	1	13
$x = 2$, $y = -3$	1	13

Conclusion:
$(x + y)^2 \neq x^2 + y^2$

37. $1220°F$ **39.** $2804°F$ **41.** $960°C$ **43.** boiling point $= 355°C$; freezing point $= -40°F$

Exercises 1.9 (pp. 58, 59)

1. $x + 13$ **3.** $y - 15$ **5.** $-4x - 2$ **7.** $26 - 2t$ **9.** $30t^2$ **11.** $-27a^3$ **13.** $6x^2y^2$ **15.** $16m^2n^2$

17. $-5x - 20$ **19.** $3p^2 - 6p$ **21.** $5v - v^2$ **23.** $-18x^2 - 12x^3$ **25.** $-12 - 18n$ **27.** $14 - 5w$

29. $45y^2 - 60y$ **31.** $9y$ **33.** $-6n$ **35.** $16a^3$ **37.** $4 - 2x$ **39.** $12y - 2y^2$ **41.** $1 - 9x$

43. $44n^3 - 33n^2$ **45.** $12x - y$ **47.** $14w - 5wf + f$ **49.** $y = 2x + 6$ **51.** $w = 5z + 2$ **53.** $z = 11 - 5y$

55. $x = 8 - 4y$

CHAPTER REVIEW (pp. 60–64)

Review Exercises

1. $5 + x$ **2.** $x - y$ **3.** x^2 **4.** $9x$ **5.** $\dfrac{8}{x}$ **6.** $4x - 1$ **7.** $5 - 2x$ **8.** $2x - 5$ **9.** 288   **10.** 28

11. $6y$ **12.** $4x + 3y$ **13.** $3t + 12$ **14.** $3V = 18$ **15.** $2F + 8 = 20$ **16.** $t = 11$ **17.** $x = 8$

18. $n = 5$ **19.** $x = 2y + 7$ **20.** $y = 6$ **21.** -600 **22.** -18 **23.** 8   **24.** 9   **25.** 14   **26.** $<$

27. $>$ **28.** $=$ **29.** -37 **30.** -4 **31.** 4   **32.** $2y + 5$ **33.** -1 **34.** -18 **35.** 6

36. $x - y + 4$ **37.** -1 **38.** $y - 1$ **39.** -24 **40.** -60 **41.** 72   **42.** 25   **43.** -25 **44.** 64

45. -32 **46.** $-2t - 3x$ **47.** x^2 **48.** 0   **49.** $16n^3$ **50.** $-18t^5$ **51.** undefined   **52.** 0   **53.** -7

54. -7 **55.** 7   **56a.** $-2xy$ **56b.** $2xy$ **57a.** 1   **57b.** -1 **58.** -51 **59.** 62   **60.** -125

61. $-x - 3$ **62.** $x - 3$ **63.** $x + 4$ **64.** 13   **65.** $13 - t$ **66.** $-27x$ **67.** $12a^3b^2$ **68.** 18

69. $5x + 15$ **70.** $6 - 8y$ **71.** $38 - 40y$ **72.** $12y$ **73.** $6x^2$ **74.** $13p^2 - 8p$ **75.** 3

Chapter Test

1. $2x - 7$ **2.** $2P_1 - 3 = 21; P_1 = 12$ **3.** 3   **4.** -16 **5.** -20 **6.** -17 **7.** -15 **8.** -36

9. 18   **10.** -64 **11.** 84   **12.** 81   **13.** 64   **14.** 9   **15.** -27 **16.** -8 **17.** 3   **18.** undefined

19. 0   **20.** -7 **21.** 12   **22.** 36   **23.** -7 **24.** -1 **25.** $3T$ **26.** 5   **27.** $b - a$ **28.** $2x - 13$

29. $-3p$ **30.** $20a^3$ **31.** $-4x^3$ **32.** $-25x^3y^3$ **33.** $3st$ **34.** -1 **35.** $x - 1$ **36.** $8 - 5y$

37. $6x^2 - 12x$ **38.** $3x - 4$ **39.** $-4x$ **40.** $14a$ **41.** $6s^3 - 6s^2$ **42.** $9h^2 - 4h$ **43.** $18f - 7f^2$

44. $3x^2 - 2xy - y^2$ **45.** $x = 4y - 2$

CHAPTER 2

Exercises 2.1 (pp. 73, 74)

1a. $\dfrac{3}{4}$ **1b.** $\dfrac{7}{4}$ **1c.** $-\dfrac{1}{4}$ **1d.** $-\dfrac{5}{4}$ **3.** 15   **5.** -44 **7.** -24 **9.** $3y$ **11.** $-20xy$ **13.** $21x$

15. $-8t^2u^2$ **17.** $30x$ **19.** $10x^2y^2$ **21.** $\dfrac{8}{9}$ **23.** $-\dfrac{9}{5}$ **25.** $\dfrac{2}{3}$ **27.** $\dfrac{3t}{4}$ **29.** $-\dfrac{2x^2}{3}$ **31.** $\dfrac{1}{2ab}$

33. -8 **35.** $\dfrac{y}{7z}$

Exercises 2.2 (pp. 82–84)

1. $-\dfrac{2}{15}$ **3.** $-\dfrac{10}{3}$ **5.** $\dfrac{65}{4}$ **7.** $\dfrac{1}{2}$ **9.** -4 **11.** $\dfrac{14}{15}$ **13.** $-\dfrac{1}{4}$ **15.** $\dfrac{3}{10}$ **17.** $\dfrac{4}{9}$ **19.** $-\dfrac{64}{27}$

21. $-\dfrac{15t^2}{32}$ **23.** $\dfrac{2p^2}{3q^2}$ **25.** $\dfrac{8z}{15}$ **27.** $-\dfrac{2}{3}$ **29.** $\dfrac{2x}{3y}$ **31.** $-\dfrac{3m^2}{10}$ **33.** $\dfrac{9x^2}{25}$ **35.** $-\dfrac{x^3}{8y^3}$ **37.** $\dfrac{4}{y}$

39. $-\dfrac{3}{yz}$ **41.** $P = \dfrac{1}{24}$ in.; $L = \dfrac{1}{12}$ in.   **43.** 36 turns

Exercises 2.3 (pp. 90, 91)

1. $\dfrac{3}{2}$ **3.** $-\dfrac{3}{5}$ **5.** $\dfrac{3}{11}$ **7.** 1   **9.** $\dfrac{4}{5}$ **11.** $\dfrac{1}{5}$ **13.** x **15.** $-\dfrac{y}{4}$ **17.** $-\dfrac{2}{5n}$ **19.** $\dfrac{x - 5y}{5y}$

21. $\dfrac{3m - n}{2n}$ **23.** $-\dfrac{w^2}{xy}$ **25.** $\dfrac{a}{2x}$ **27.** $\dfrac{2x}{y}$ **29.** $-\dfrac{x}{ab}$ **31.** 0   **33.** $\dfrac{4m}{n^2}$ **35.** x

Exercises 2.4 (pp. 102, 103)

1. $2^2 \cdot 7$ **3.** $3^3 \cdot 2$ **5.** $2^2 \cdot 5^2$ **7.** $2^4 \cdot 3^2$ **9.** $2^2 \cdot 3 \cdot 11$ **11.** LCD $= 24; \dfrac{13}{24}$ **13.** LCD $= 20; \dfrac{31}{20}$

15. LCD $= 100; -\dfrac{31}{100}$ **17.** LCD $= 189; -\dfrac{38}{189}$ **19.** LCD $= 42; \dfrac{1}{6}$ **21.** LCD $= 30; \dfrac{23x}{30}$

23. LCD $= 40; -\dfrac{13a^2}{40}$ **25.** LCD $= 36x; \dfrac{13y}{36x}$ **27.** LCD $= mn; \dfrac{3n + 4m}{mn}$ **29.** LCD $= 4x^2; \dfrac{1 - 6x}{4x^2}$

31. LCD $= m^2n^2; \dfrac{4n - 2m^2}{m^2n^2}$ **33.** LCD $= ab; \dfrac{a^2 + b^2}{ab}$ **35.** LCD $= 126x^2y; \dfrac{8 - 35x}{126x^2y}$

37. LCD $= 4a^2; \dfrac{b^2 - 4ac}{4a^2}$ **39.** LCD $= 36x^2; \dfrac{69x - 10}{36x^2}$ **41.** $\dfrac{1}{2}$ in.   **43.** $\dfrac{17}{32}$ in.

Exercises 2.5 (pp. 108, 109)

1. $\dfrac{29}{8}$ **3.** $-\dfrac{9}{2}$ **5.** $-\dfrac{25}{7}$ **7.** $\dfrac{4y+4}{y}$ **9.** $\dfrac{3-5m^2}{5m}$ **11.** $\dfrac{7x}{5}$ **13.** $\dfrac{5t^2}{4}$ **15.** $\dfrac{5x}{2}$ **17.** $1+\dfrac{1}{a}$

19. $x-\dfrac{5}{x}$ **21.** $\dfrac{4y^2}{3}+2xy$ **23.** $\dfrac{3y^2}{2}-3$ **25.** $\dfrac{6}{xy^2}+3$ **27.** $\dfrac{v}{10}$ **29.** $\dfrac{10v+1}{10}$ **31.** $\dfrac{2F-9}{3}$

33. $\dfrac{P_1+2P_2}{2}$ **35.** $L=\dfrac{13d_1}{12}$ **37.** $L=\dfrac{11d_2}{8}$

CHAPTER REVIEW (pp. 110–113)

Review Exercises

1.

2. 20 **3.** -18 **4.** $-9y^2$ **5.** $33x^2y$ **6.** $\dfrac{3}{7}$ **7.** $-\dfrac{9}{5}$ **8.** 3 **9.** $\dfrac{4y}{3x}$ **10.** $\dfrac{10}{27}$ **11.** $-\dfrac{7}{10}$ **12.** $6y^2$

13. $\dfrac{27t}{w}$ **14.** $-\dfrac{8y^3}{125}$ **15.** $\dfrac{21}{22}$ **16.** $-\dfrac{1}{6}$ **17.** $\dfrac{5}{2}$ **18.** $\dfrac{1}{12t}$ **19.** $\dfrac{4}{3}$ **20.** $\dfrac{1}{3}$ **21.** $-\dfrac{t}{2}$ **22.** $\dfrac{1}{2p^2}$

23. $\dfrac{1}{2}$ **24.** $\dfrac{4}{5}$ **25.** $-2x$ **26.** $\dfrac{6y-5}{6y}$ **27.** $\dfrac{3a^2}{7}$ **28.** $\dfrac{2mn}{p}$ **29.** $2\cdot3\cdot5$ **30.** $2^3\cdot3^2$ **31.** 198

32. 360 **33.** 36 **34.** $8x^3y^2$ **35.** mn **36.** $\dfrac{13}{99}$ **37.** $-\dfrac{67}{360}$ **38.** $\dfrac{x}{36}$ **39.** $\dfrac{4x-3y}{8x^3y^2}$ **40.** $\dfrac{9n+3m}{mn}$

41. $-\dfrac{19}{8}$ **42.** $\dfrac{4x+5}{4x}$ **43.** $\dfrac{21y^2-2}{3y}$ **44.** $\dfrac{3}{x}-x$ **45.** $\dfrac{10}{n}-3n$ **46.** $\dfrac{3n}{5}$ **47.** $\dfrac{y-27}{9}$ **48.** $\dfrac{5a+24}{6}$

Chapter Test

1. 18 **2.** $-15t^2$ **3.** $-\dfrac{3}{7}$ **4.** $\dfrac{1}{4x}$ **5.** $-\dfrac{1}{2}$ **6.** $\dfrac{14}{3}$ **7.** $\dfrac{3}{2}$ **8.** $-\dfrac{7}{5}$ **9.** $\dfrac{19}{48}$ **10.** $\dfrac{11}{196}$

11. $\dfrac{53}{234}$ **12.** $-\dfrac{71}{630}$ **13.** $\dfrac{8}{3}$ **14.** $-\dfrac{5}{2x}$ **15.** $\dfrac{2t}{3}$ **16.** $-\dfrac{x}{7y}$ **17.** $\dfrac{43x}{40}$ **18.** $\dfrac{9x+10y^2}{12x^2y^2}$ **19.** $\dfrac{5+6w}{5}$

20. $-\dfrac{x^3}{27}$ **21.** $\dfrac{2}{15}$ **22.** $-\dfrac{a^2}{2}$ **23.** $\dfrac{2}{t}$ **24.** $\dfrac{2y-3x}{xy}$ **25.** $\dfrac{4t^2}{15}$ **26.** $-\dfrac{29y}{4}$ **27.** $n-\dfrac{5}{4n}$ **28.** $\dfrac{9}{x}-5y$

29. $\dfrac{2V-18}{3}$ **30.** $\dfrac{4F_1+F_2}{2}$

CHAPTER 3

Exercises 3.1 (pp. 118, 119)

1. .05 **3.** .018 **5.** .616 **7.** 600.016 **9.** .00009 **11.** 9.100 **13.** 58 **15.** 2.7 **17.** 190

19. .0034 **21.** 27.70 **23.** 6800 **25.** .000400 **27.** $100\bar{0}$

Exercises 3.2 (pp. 123, 124)

1. 4 is exact; 5 is exact; 500 is approximate; 12.2 is approximate

Accuracy	Precision	g.p.e.	Lower limit	Upper limit
3. three significant digits	ones	.5	480.5	481.5
5. three significant digits	tenths	.05	9.95	10.05
7. four significant digits	hundredths	.005	19.995	20.005
9. one significant digit	hundredths	.005	.055	.065
11. two significant digits	thousandths	.0005	.0595	.0605
13. two significant digits	hundreds	50	4450	4550
15. four significant digits	ones	.5	4499.5	4500.5

17. 1.80 is more accurate and more precise   **19.** same accuracy and same precision

21. $15\bar{0}$ is more accurate and .35 is more precise   **23.** $256\bar{0}$ is more accurate and more precise

Exercises 3.3 (pp. 131, 132)
1. 16.2   **3.** 1.27   **5.** -44.4 **7.** -2.66 **9.** 200   **11.** $-53,600$ **13.** .20   **15.** .0018   **17.** 3380
19. -1.357 **21.** $-.011$ **23.** 3.36   **25.** 3.51 in.   **27.** 13 cards   **29.** $1500\,\Omega$ **31.** $84\,\Omega$

Exercises 3.4 (pp. 137, 138)
1. 16.0 ft   **3.** 2.57 yd   **5.** 172,000 ft   **7.** 3.7 cm   **9.** $75,\bar{0}00$ m   **11.** 505 ft   **13.** 1.609 km
15. 6.57 in.   **17.** .764 in.   **19.** 15.875 mm   **21.** 26.19375 mm   **23.** $\frac{5}{16}$ in.   **25.** 15 mm   **27.** $\frac{53}{64}$ in.
29. 19 mm

Exercises 3.5 (pp. 144, 145)
1. 34 ft, 0 in.   **3.** 42 ft, $11\frac{1}{2}$ in.   **5.** 55.8 cm   **7.** 20.34 m   **9.** 192π mm   **11.** $12x$ **13.** $4x + 2y$
15. $\frac{3x}{2}$ **17.** $\frac{5x}{2}$ **19.** 9.97 m   **21.** 51.4 ft   **23.** $22.57x$ **25.** $7.142x$

Exercises 3.6 (pp. 154–156)
1. 9 ft^2 **3.** 3.9 Ac   **5.** 351 ft^2 **7.** 5.75 m^2 **9.** .0063 km^2 **11.** $\frac{3}{4}$ yd^2 **13.** $64\bar{0}$ Ac   **15.** 25.2 ft^2
17. 120 m^2 **19.** 1520 in.^{2} **21.** $11\frac{1}{9}$ ft^2 **23.** 5.7 m^2 **25.** $40x^2$ **27.** $4xy$ **29.** $9\pi x^2$ **31.** $\frac{9y^2}{16}$
33. 175 ft^2 **35.** 195 m^2 **37.** $12x^2$ **39.** $8.429x^2$ **41.** 3.21

Exercises 3.7 (pp. 165–167)
1. 27 ft^3 **3.** 53,000 cm^3 **5.** 10.61 in.^{3} **7.** 101 ft^3 **9.** 305 in.^{3} **11.** 1.0 ft^3 **13.** 3.79 L
15. 1650 m^3 **17.** 8860 m^3 **19.** $2\bar{0}0$ in.^{3} **21.** 304 in^3.   **23.** .111 m^3 **25.** 679 m^3 **27.** $18x^3$
29. $3\pi x^3$ **31.** $\frac{y^3}{4}$ **33.** $180a^3$ **35.** $100\bar{0}$ gal

CHAPTER REVIEW (pp. 168–174)

Review Exercises

1a. .0100   **1b.** .110   **1c.** 100.010   **2a.** 479.8   **2b.** 479.85   **2c.** 480   **2d.** 48$\bar{0}$
3. 10,000 is approximate; 10 is exact; 35 is approximate; 25 is approximate; 750 is exact

Accuracy	Precision		Upper limit	Lower limit
4a. three significant digits	tenths	**5a.**	.0035	.0025
4b. four significant digits	hundredths	**5b.**	28,500	27,500
4c. four significant digits	hundredths			
4d. one significant digit	thousandths			
4e. two significant digits	thousands			
4f. four significant digits	tens			

6. 145.9   **7.** -9.0 **8.** .12   **9.** $-.00646$ **10.** 342   **11.** 1299   **12.** $-.35$ **13.** 6.75 cm   **14.** 2255 in.

15. 15,000 yd   **16.** 8730 km   **17.** 25.7 cm   **18.** 5740 ft   **19.** 1.191 mm   **20.** 73 ft, 8 in.   **21.** 23.3 in.

22. $\dfrac{13x}{3}$ **23.** 110 cm   **24.** 133 cm   **25.** 7.142x **26.** 62.0 yd^2 **27.** 77,800 ft^2 **28.** 4.66 Ac

29. 160 m^2 **30.** 43.8 m^2 **31.** $36\pi x^2$ **32.** $\dfrac{5x^2}{6}$ **33.** 190 cm^2 **34.** $10x^2$ **35.** 34 yd^3 **36.** 120 in.^{3}

37. 137 gal   **38.** 3950 in.^{3} **39.** 69.0 ft^3 **40.** $\dfrac{3x^3}{8}$ **41.** 25$\bar{0}$ gal   **42.** $25\pi x^3$

Chapter Test

1. .500   **2.** 5.00001   **3.** 80.30   **4.** 5740

Accuracy	Precision	g.p.e.	Lower limit	Upper limit
5. two significant digits	ten-thousandths	.00005	.00395	.00405
6. three significant digits	tens	5	8195	8205

7. -899.9 **8.** 1120   **9.** 1.194   **10.** -89 **11.** 7.144 mm   **12.** 6.13 Ac   **13.** 7$\bar{0}$ m^2 **14.** 4.151 ft^3

15. 234 ft, 10 in.   **16.** 3.34 m   **17.** 20$\bar{0}$ cm^2 **18.** 19.4 yd^2 **19.** 2331 m^3 **20.** 1$\bar{0}$,000 gal   **21.** 46.5 ft

22. 14$\bar{8}$ ft^3 **23.** 7$\bar{0}0$ in^3 **24.** $9\pi x^2$ **25.** $20x$ **26.** $\dfrac{50x^2}{3}$ **27.** $6x + 2\pi x \approx 12.284x$

28. $2x^2 + \pi x^2 \approx 5.142x^2$ **29.** $\dfrac{x^3}{2}$ **30.** $\dfrac{\pi x^3}{9}$

CHAPTER 4

Exercises 4.1 (pp. 183, 184)

1. $x = 47$ **3.** $z = 34$ **5.** $x = 15$ **7.** $x = -2\dfrac{1}{2}$ **9.** $y = 50$ **11.** $y = \dfrac{1}{2}$ **13.** $w = -\dfrac{1}{8}$ **15.** $h = \dfrac{37}{24}$

17. $y = -18$ **19.** $m = -2.15$ **21.** $x = -11.4$ **23.** $f = 6\frac{1}{2}$ **25.** $z = -.32$ **27.** $g = -15$ **29.** $z = 7\frac{1}{8}$

31. $t = -.09$ **33.** $p = -5.0$ **35.** $z = 4.8$ **37.** $y = \frac{12}{5}$ **39.** $n = 0$ **41.** $t = \frac{d}{r}$ **43.** $R_2 = R_t - R_1 - R_3$

45. $B = 180 - A$ **47.** $v_0 = v - at$ **49.** $s = \frac{SD}{d}$ **51.** $t = \frac{T}{r}$ **53.** $R_2 = 21.6\Omega$ **55.** $t = 15$ teeth

Exercises 4.2 (pp. 190, 191)

1. $x = 4$ **3.** $x = -1$ **5.** $p = -75$ **7.** $v = -25$ **9.** $x = 5$ **11.** $t = 4$ **13.** $n = -8$

15. $g = -\frac{14}{3}$ **17.** $w = -9$ **19.** $p = 16$ **21.** $y = -5$ **23.** $x = 2$ **25.** $x = 1$ **27.** $x = 3$

29. $x = 38$ **31.** $x = \frac{7}{6}$ **33.** $y = 4$ **35.** $x = 6$ **37.** $k = 81$ **39.** $z = 6$ **41.** $t = \frac{v - v_0}{a}$

43. $d_2 = \frac{M - F_1 d_1 - F_3 d_3}{F_2}$ **45.** $h = \frac{2A}{b_1 + b_2}$ **47.** $\alpha = \frac{R - R_0}{R_0 t}$ **49.** $E = IR_i + IR$ **51.** $l = \frac{6\delta EI + Px^3}{3Px^2}$

53. $I = \frac{5wl^4}{384\delta_{\max} E}$ **55.** $E = 8.7$ V

Exercises 4.3 (pp. 197, 198)

1. 32.8 m   **3.** 3.28 ft   **5.** 3.0 ft   **7.** 17 ft   **9.** 22.1 cm   **11.** equal sides are 25.0 cm; nonequal side is 9.6 cm
13. first exam score is 87 points; second exam score is 41 points   **15.** longer piece is 88 in.; shorter piece is 20 in.
17. first piece is 60 in.; second piece is 10 in.; third piece is 38 in.   **19.** length is 6.0 cm; width is 2.5 cm

Exercises 4.4 (pp. 202, 203)

1. $I_2 = .23$ A   **3.** $I_1 = .015$ A; $I_3 = .003$ A   **5.** $I_2 = .17$ A; $I_3 = .20$ A   **7.** $V_1 = 9.8$ V
9. $V_1 = 4.9$ V; $V_2 = 11.1$ V   **11.** $E = 212.8$ V; $V_2 = 90.6$ V   **13.** $I_1 = 18$ A; $I_2 = 6$ A; $I_3 = 3$ A; $I_4 = 9$ A

Exercises 4.5 (pp. 210, 211)

1a. $F_3 = 522$ lb   **1b.** $d_3 = 11.3$ ft   **3a.** $F_2 = 29.0$ lb; $F_3 = 203$ lb   **3b.** $d_2 = 10.67$ ft; $d_3 = 6.67$ ft
5a. $F_1 = 200$ N; $F_2 = 600$ N; $F_3 = 800$ N   **5b.** $d_1 = 16$ m; $d_2 = 28$ m; $d_3 = 25$ m
7a. $F_1 = 180$ lb; $F_2 = 400$ lb; $F_3 = 380$ lb; $F_4 = 200$ lb   **7b.** $d_1 = 12$ in.; $d_2 = 81$ in.; $d_3 = 72$ in.; $d_4 = 36$ in.
9. at least 79 lb

Exercises 4.6 (pp. 217, 218)

1. $\frac{5}{9}$ **3.** $19:6$

	a. Fractional form	b. Colon form	c. Decimal-to-one form
5.	$\dfrac{1}{4}$	1:4	.25:1
7.	$\dfrac{5}{8}$	5:8	.625:1
9.	$\dfrac{11}{40}$	11:40	.275:1
11.	$\dfrac{5}{32}$	5:32	.15625:1

13. $\dfrac{1}{4}$ L of oil, $3\dfrac{3}{4}$ L of gas   **15.** width is 120 in.; length is 160 in.

17. $10\dfrac{1}{2}$ yd^3 of cement; $15\dfrac{3}{4}$ yd^3 of sand; $15\dfrac{3}{4}$ yd^3 of gravel   **19.** 4:1   **21.** 16:1   **23.** 9.75 A

Exercises 4.7 (pp. 223–227)

1. $x = 68$ **3.** $p = -\dfrac{14}{3}$ **5.** $x = \dfrac{13}{2}$ **7.** $z = \dfrac{11}{4}$ **9.** $x = -2$ **11.** $t = 6$ **13.** 41.4 m   **15.** 52$\bar{0}$ rpm

17. 35.0 cm   **19.** 19 teeth   **21.** 10 V across 40 Ω resistor; 25 V across 100 Ω resistor   **23.** (inversely) 9 in.

25. (directly) 8 cm   **27.** 3.49 cm   **29.** 775 mm^2 **31.** 35 V   **33.** $\dfrac{1}{4}$ A

Exercises 4.8 (pp. 233, 234)

1. $92,000   **3.** $1054.63   **5.** 4.2%   **7.** $125,000   **9.** 3.5 lb   **11.** 5 lb   **13.** 25%   **15.** 1600 Ω

17. 10°C   **19.** $3640

CHAPTER REVIEW (pp. 234–239)

Review Exercises

1. $x = -\dfrac{10}{3}$ **2.** $y = -15$ **3.** $n = -26.6$ **4.** $t = -\dfrac{1}{4}$ **5.** $x = -51$ **6.** $m = \dfrac{11}{9}$ **7.** $p = 1.76$

8. $b = \dfrac{12I}{d^3}$ **9.** $b = P - 2a$ **10.** $x = 8$ **11.** $w = -9$ **12.** $x = -2$ **13.** $t = -\dfrac{5}{3}$ **14.** $h = \dfrac{A - 2\pi r^2}{2\pi r}$

15. $Z_s = \dfrac{I_p \omega M - I_s Z_L}{I_s}$ **16.** 6.9 ft   **17.** 75 points   **18.** equal sides are 30.1 ft; nonequal side is 2.3 ft

19. first score is 69 points; second score is 74 points; third score is 52 points   **20.** $I_3 = .023$ A; $I_2 = .138$ A

21. $V_1 = 8.25$ V; $V_2 = 3.75$ V   **22.** $F_1 = 475$ lb; $F_3 = 2150$ lb   **23.** $d_1 = 1.13$ ft; $d_2 = 6.27$ ft; $d_3 = 5.13$ ft

24a. $\dfrac{1}{4}$ **24b.** 1:4   **24c.** .25:1   **25.** 1575   **26.** 19 cm, 19 cm, 38 cm, 57 cm   **27.** $x = \dfrac{14}{3}$ **28.** $y = 7$

29. $V_p = 37\dfrac{1}{2}$ V   **30.** 360 lb   **31.** 30%   **32.** 8 gal   **33.** 12.9%   **34.** 300 rpm

Chapter Test

1. $y = 28$ **2.** $t = -2$ **3.** $h = 12$ **4.** $m = \dfrac{7}{3}$ **5.** $x = \dfrac{2}{5}$ **6.** $y = -16$ **7.** $p = 7$ **8.** $f = \dfrac{1}{2\pi X_c C}$

9. $H_2 = H_1 - H_1 E$ **10.** length is 16 cm; width is 8 cm   **11.** equal sides are 40 in.; nonequal side is 24 in.

12. $N_p = 3200$ turns   **13.** 2.1 L   **14.** 4.0 yd^3 **15.** 720 psi

CHAPTER 5

Exercises 5.1 (p. 246)

1. $2^6 = 64$ **3.** $2^8 = 256$ **5.** x^{11} **7.** x^{28} **9.** $m^7 n^4$ **11.** $-p^{10} q^{11}$ **13.** $20x^{11} y^{12}$ **15.** $24x^{15} y^8 z^2$

17. $(p - 4)^8$ **19.** $(p - 4)^6$ **21.** $64x^3 y^{12}$ **23.** $-108m^9 n^{19}$ **25.** $16m^{25} n^{31}$ **27.** $81x^4 y^4$ **29.** a^{5t}

31. a^{4t} **33.** $(2^{5t+1})(a^{4t})$

Exercises 5.2 (pp. 252, 253)

1. 1   **3.** $2^3 = 8$ **5.** $\dfrac{1}{2^5} = \dfrac{1}{32}$ **7.** $\dfrac{1}{3}$ **9.** $\dfrac{gh^4}{2}$ **11.** $\dfrac{-1}{7x^3 yz^7}$ **13.** $\dfrac{7t^9}{8r^2}$ **15.** $(y + 2)^3$ **17.** $\dfrac{64y^7 z}{9x^2}$

19. $\dfrac{-64}{z^{12}}$ **21.** $\dfrac{1}{32t^{25}}$ **23.** $\dfrac{16}{r^{16} s^4 t^4}$ **25.** $\dfrac{9q^{13} r}{4}$ **27.** a^{2t} **29.** $\dfrac{1}{a^t}$ **31.** $\dfrac{-1}{a^{9t}}$

Exercises 5.3 (pp. 259, 260)

1. false   **3.** true   **5.** false   **7.** true   **9.** true   **11.** true   **13.** false   **15.** true   **17.** true   **19.** $\dfrac{4}{xz}$

21. $\dfrac{y^2}{3x^2}$ **23.** $\dfrac{1}{x^4}$ **25.** $\dfrac{1}{x^2}$ **27.** $\dfrac{1}{y}$ **29.** $\dfrac{1}{9x^2 y^2 z^2}$ **31.** $\dfrac{1}{(1 - y)^6}$ **33.** $\dfrac{y^3}{27}$ **35.** 1   **37.** $(x + y)^6$

39. $\dfrac{1}{y^8}$ **41.** $\dfrac{256}{x^8 y^8}$ **43.** $\dfrac{8z^{15}}{x^{12} y^3}$ **45.** $\dfrac{x^3 y^6}{27}$ **47.** $\dfrac{4}{x^4 y^{20} z^4}$ **49.** $x(y - z)^7$

Exercises 5.4 (pp. 263, 264)

1. .0000065   **3.** 1,740,000   **5.** 7,600,000,000   **7.** .00000000000000000016   **9.** 625,000,000

11. 1.17×10^{-5} **13.** 1.44×10^5 **15.** 8.3×10^9 **17.** 6.24×10^{18} **19.** 1.00×10^{-3}

Exercises 5.5 (pp. 272–274)

1. $10^{-1} = .1$ **3.** $10^{-6} = .000001$ **5.** $10^2 = 100$ **7.** 110,000   **9.** 100   **11.** 1.5×10^3 **13.** 8.60×10^1

15. 1.065×10^8 **17.** 4.29×10^{-9} **19.** 1.54×10^{-15} **21.** 4.5×10^{-1} **23.** 3.41×10^{-14} **25.** 20.5 MΩ

27. 5.31 mw   **29.** 1.66 MΩ **31.** $\dfrac{1}{10}$ in.   **33.** 5.46 ft

Exercises 5.6 (p. 280)

1. 6   **3.** -7 **5.** ± 10 **7.** not a real number   **9.** .3   **11.** $\dfrac{1}{3}$ **13.** 4   **15.** -5

17. not a real number   **19.** -1 **21.** x^2 **23.** z^{18} **25.** $x + y$ **27.** x^3 **29.** z^{22} **31.** $x + y$

33. 1.732   **35.** -6.403 **37.** 3.803   **39.** -8.481 **41.** not a real number   **43.** -1.516

Exercises 5.7 (pp. 286, 287)

1. $3xy$ **3.** $10x^2y^4z^{50}$ **5.** $2x^5$ **7.** $3x^2y^9$ **9.** $2\sqrt{6}$ **11.** $3\sqrt{5}$ **13.** $2\sqrt[3]{5}$ **15.** $5\sqrt[3]{2}$ **17.** $x^4\sqrt{x}$

19. $x^4\sqrt[3]{x}$ **21.** $2x^2y^3\sqrt{3}$ **23.** $4xy^4\sqrt{2xz}$ **25.** $2y^2\sqrt[3]{2}$ **27.** $3x^3y^2\sqrt[3]{3y^2z}$ **29.** already simplified

31. already simplified   **33.** $3x\sqrt{25+y^2}$ **35.** $xz\sqrt[3]{8+y^3}$ **37.** $5x$ **39.** $7x^2y\sqrt{y}$ **41.** $3x\sqrt[3]{y^2}$ **43.** $5xy$

Exercises 5.8 (p. 295)

1. $\dfrac{2x^2}{3y}$ **3.** $\dfrac{z\sqrt{z}}{4xy^8}$ **5.** $\dfrac{2x}{y^3}$ **7.** $\dfrac{2y^2\sqrt[3]{y^2}}{5x^3}$ **9.** $\dfrac{\sqrt{5}}{10}$ **11.** $\dfrac{\sqrt{15}}{12}$ **13.** $\dfrac{\sqrt[3]{4}}{4}$ **15.** $\dfrac{2\sqrt[3]{9}}{9}$ **17.** $\dfrac{\sqrt{x}}{x^2}$ **19.** $\dfrac{\sqrt[3]{x}}{x^2}$

21. $\dfrac{x\sqrt{2}}{4y}$ **23.** $\dfrac{xy\sqrt{2z}}{10z^2}$ **25.** $\dfrac{x^2\sqrt[3]{4}}{4y}$ **27.** $\dfrac{xy^3\sqrt[3]{2z}}{4z^2}$ **29.** $\dfrac{\sqrt{4+x^2}}{2x}$ **31.** $\dfrac{3}{x^2}$ **33.** $\dfrac{2\sqrt{11x}}{x^3}$ **35.** $\dfrac{2\sqrt[3]{x}}{x^2}$

37. $\dfrac{2\sqrt[3]{11x}}{x^2}$ **39.** $3x$ **41.** $2x$ **43.** $995\,\text{kHz}$

Exercises 5.9 (pp. 300, 301)

1. ± 7 **3.** 4 **5.** $\pm 5\sqrt{3}$ **7.** $2\sqrt[3]{7}$ **9.** $\pm 4\sqrt{3}$ **11.** $3\sqrt[3]{2}$ **13.** no real solution   **15.** -5

17. ± 4 **19.** $\pm\dfrac{\sqrt{6}}{6}$ **21.** -2 **23.** $\dfrac{\sqrt[3]{4}}{2}$ **25.** no real solution   **27.** $s=\sqrt{A}$ **29.** $r=\sqrt[3]{\dfrac{3V}{4\pi}}$

31. $V=\sqrt{PR}$ **33.** $L=4\sqrt[3]{\dfrac{6EIy}{5W}}$ **35.** $15.4\,\text{ft}$

CHAPTER REVIEW (pp. 302–306)

Review Exercises

1. y^{13} **2.** x^4y^{11} **3.** $-6x^{11}y^2z^{10}$ **4.** y^{40} **5.** $(2-x)^8$ **6.** $64y^6$ **7.** $-27x^9y^6z^{15}$ **8.** $256x^{14}y^{14}$

9. $\dfrac{5}{4}$ **10.** x^8y^2 **11.** $\dfrac{9x^2z}{16}$ **12.** $\dfrac{1}{x^3y}$ **13.** $\dfrac{2y^4}{3x}$ **14.** $\dfrac{16}{y^8}$ **15.** $\dfrac{-x^{12}}{8y^9}$ **16.** 100 **17.** $\dfrac{82}{9}$ **18.** $\dfrac{1}{8y^3}$

19. $\dfrac{y^2}{3x^4}$ **20.** $\dfrac{3y^6}{4x^3}$ **21.** $\dfrac{1}{xy^5}$ **22.** $\dfrac{1}{(x-1)^6}$ **23.** $\dfrac{16x^8}{y^{12}}$ **24.** $(x+y)^6$ **25.** $\dfrac{x^3y^3}{16}$ **26.** $386{,}000{,}000$

27. $.0000730$ **28.** 2.2×10^8 **29.** 5.60×10^{-6} **30.** 3.58×10^2 **31.** 4.8×10^{-17} **32.** 1.56×10^{-8}

33. 7.345×10^9 **34a.** 8 **34b.** -5 **34c.** ± 6 **34d.** $.3$ **34e.** $\dfrac{1}{3}$ **35a.** 1.732 **35b.** -1.732 **36a.** x^2

36b. y^5 **36c.** z^{18} **37a.** -4 **37b.** not a real number   **38.** -2.520 **39a.** x^3 **39b.** x^{27} **40.** $4x^3$

41. already simplified   **42.** $5x^4yz^{32}$ **43.** $6\sqrt{3}$ **44.** $x^5\sqrt{x}$ **45.** $3x^5z^3\sqrt{3yz}$ **46.** $8x^2y^3$ **47.** $4x^4y$

48. $3\sqrt[3]{2}$ **49a.** $x^3\sqrt[3]{x}$ **49b.** $x^3\sqrt[3]{x^2}$ **50.** $5x^4y^4z^4\sqrt[3]{xz^2}$ **51.** $4xy^2\sqrt[3]{x}$ **52.** $\dfrac{6x}{7y^3}$ **53.** $\dfrac{z^2\sqrt{11z}}{3x^2y^6}$

54. $\dfrac{x^5\sqrt{7}}{14}$ **55.** $\dfrac{5x^2\sqrt{2xy}}{4y^4}$ **56.** $\dfrac{3x^2\sqrt{3xy^2}}{2z^3}$ **57.** $\dfrac{x^9\sqrt[3]{4}}{8}$ **58.** $\dfrac{x\sqrt[3]{xy}}{4y^3}$ **59.** $x=\pm3\sqrt{7}$ **60.** $x=\pm\dfrac{1}{4}$

61. no real solution   **62.** $I=\sqrt{\dfrac{P}{R}}$ **63.** $x=2\sqrt[3]{5}$ **64.** $x=-2$ **65.** $L=2\sqrt[3]{\dfrac{15EIy}{w}}$

Chapter Test

1. $\dfrac{1}{x}$ **2.** x^9 **3.** $\dfrac{1}{(x-1)^3}$ **4.** 1   **5.** $-8r^9s^3t^{12}$ **6.** $\dfrac{x^2}{9y^2}$ **7.** $\dfrac{-12x^2y^7}{z}$ **8.** $\dfrac{6y^2}{x^4}$ **9.** $\dfrac{-27y^{18}}{x^6}$

10. $\dfrac{64}{225}$ **11.** 2×10^8 **12.** 6.36×10^{12} **13.** $2\sqrt{15}$ **14.** $2\sqrt[3]{9}$ **15.** $\dfrac{\sqrt{5}}{10}$ **16.** $\dfrac{\sqrt[3]{36}}{12}$ **17.** $(x^3+8)^2$

18. $2h^2\sqrt{10h}$ **19.** $5vz^2\sqrt[3]{2v^2z}$ **20.** $6x^3$ **21.** $\dfrac{x^2}{2y^8}$ **22.** $\dfrac{x\sqrt[3]{2}}{4}$ **23.** $\dfrac{t^2\sqrt{30u}}{4u^5}$ **24.** $\dfrac{2}{x}$ **25.** $\dfrac{\sqrt[3]{27+x^3}}{3x}$

26. $\dfrac{\sqrt{13}}{6}$ **27.** $x = \pm\sqrt{15}$ **28.** $x = -\dfrac{1}{2}$ **29.** $d = \sqrt{\dfrac{\alpha l}{R}}$ **30.** $h = \sqrt[3]{\dfrac{12I}{b}}$

CHAPTER 6

Exercises 6.1 (p. 310)

1. binomial   **3.** monomial   **5.** trinomial   **7.** none (it is a five-term polynomial)
9. none (it is not a polynomial since it contains negative exponents)   **11.** $x^3 + 6x^2y + 5xy^2 + 2y^3$ **13.** $8 - 6x$
15. $6pq^4 - 7pq$ **17.** $-x^2y$ **19.** $x - 6$ **21.** $xy - 1$ **23.** -1

Exercises 6.2 (pp. 317, 318)

1. $8p^5$ **3.** $8p^2 + 4p^5$ **5.** $4 + 2p^2 + p^5$ **7.** $8 + 4p^3 + 2p^2 + p^5$ **9.** $-6t^4 + 24t^3 - 24t^2 + 6t$
11. $4xy^3 - 28xy + 7x^3y$ **13.** $16xy^3 - 4x^3y^3 - 28xy + 7x^3y$ **15.** $4p^5 - 5p^2$ **17.** $7p^5 - 14p^2$
19. $x^2 + 9x + 14$ **21.** $6x^2 - 8$ **23.** $8 + 2t - t^2$ **25.** $27a^3b^3$ **27.** 0   **29.** $6y^2 - y - 15$
31. $-4x^2 + 5x - 8$ **33.** $x^3 - 5x^2 + 6x - 8$ **35.** $-3x^2 + 5x - 2$ **37.** $-18t^2 - 11t - 2$
39. $-30m^2 + 6m^3 + 18m^4$ **41.** $x^3 - 5x^2 + 5x + 3$ **43.** $-x^2$ **45.** x^2 **47.** $x^4 + 4x^2 - x + 6$
49. $3x^4 - 7x^3 + 3x + 2$ **51.** x^3 **53.** 0

Exercises 6.3 (pp. 323, 324)

1. $x^2 - 1$ **3.** $x^2 + 2x + 1$ **5.** $4x^2 - 9$ **7.** $4x^2 - 12x + 9$ **9.** $16t^6 - y^2$ **11.** $16t^6 + 8t^3y + y^2$
13. $16t^6y^2$ **15.** $3x^3 - 75x$ **17.** $3x - x^2 + 25$ **19.** $32m^3 - 32m^2 + 8m$ **21.** $12m - 4m^2 - 1$
23. $-50t^3 + 40t^2 - 8t$ **25.** $25t^2 - 22t + 4$ **27.** $x^3 + 3x^2 + 3x + 1$ **29.** $27 - 54x + 36x^2 - 8x^3$
31. $8x^6 + 12x^4y + 6x^2y^2 + y^3$ **33.** $8x^6y^3$ **35.** $x^3 + x^2 - x - 1$ **37.** $x^3 - 27x - 54$
39. $x^4 + 4x^3y + 6x^2y^2 + 4xy^3 + y^4$ **41.** $540p^3 - 180p^2 + 15p$ **43.** $a^4 - 32a^2 + 256$ **45.** $4x^2 - 4x + 1$
47. $3x^2$ **49.** $x^3 + 8x^2 + 16x$

Exercises 6.4 (pp. 330, 331)

1. 5 and 2   **3.** -12 and -3 **5.** 8 and -6 **7.** -8 and 7   **9.** -5 and 5   **11.** $2x$ **13.** $9xy^2$
15. $-8t^3$ **17.** $-6x^2z^{20}$ **19.** 7   **21.** $3 - 5t$ **23.** $1 - t + t^2$ **25.** $2x(4y - 7x)$ **27.** $4y^2(4 - xy + 2x)$
29. $3x^2y^2(xy + 5x - 3y - 1)$ **31.** $6xy(1 - x - 2y + 3x^2)$ **33.** $(x + 4)(x^2 + 3)$ **35.** $(x + 2)(4y^3 - 3)$
37. $(5 - y)(3 - 2x^2)$ **39.** $(t + 2)(t^2 + 1)$ **41.** $3x(3x - 1)(x^2 - 6)$ **43.** $5y^2(9 + w)(1 + y^2)$ **45.** $\pi(R^2 - r^2)$
47. $4r^2(7 - \pi)$

Exercises 6.5 (pp. 339, 340)

1. $(2x + 3)(x + 2)$ **3.** $(3x - 5)(x - 2)$ **5.** $(x + 6)(x + 1)$ **7.** $(2t - 5)(3t - 2)$ **9.** $(y + 3)(2y - 5)$
11. $(x - 6)(x - 2)$ **13.** $(2x + 3)(2x + 1)$ **15.** $(y + 9)(y - 9)$ **17.** prime **19.** $(x - 8)(x + 7)$
21. $(2x + 5y)(2x - 5y)$ **23.** $(2d^3 + 5)(2d^3 - 5)$ **25.** prime **27.** $(2x + y)(5x + y)$ **29.** $(p + 6q)(p - 4q)$
31. $(5x - y)^2$ **33.** $2(2 + x)(2 - x)$ **35.** $(8 + x)(4 - x)$ **37.** $3(2x - 3)(4x + 1)$ **39.** $5y(x + 4)^2$
41. $3t^3(2t^2 + t + 1)$ **43.** $12x(x^2 + 4y^2)$ **45.** $-2x^2y(y - 5)(y + 1)$ **47.** $(x^4 + y^4)(x^2 + y^2)(x + y)(x - y)$
49. $(1 + x)(1 - x)(2 + y)(2 - y)$

CHAPTER REVIEW (pp. 340–342)

Review Exercises

1a. binomial **1b.** monomial **1c.** trinomial **1d.** a four-term polynomial **1e.** not a polynomial
2. $2x^5 - x^2$ **3.** $-5y$ **4.** $7xy - 4y$ **5.** $-21y^2 + 14y^5$ **6.** $2x^2y^6 - 5x^4y^3 + xy^4 - xy^3$ **7.** $24x - 9x^3y$
8. $3x^2 + 2xz + 9xy + 6yz$ **9.** $24x - 16xy^2 + 3y^3 - 2y^5$ **10.** $4x^2 + 25xy - 21y^2$ **11.** $20 - 9y^3 + y^6$
12. $-3x^2 + 19x + 25$ **13.** $3x^3 - 8x^2y + y^3$ **14.** $x^4 - 13x^2 + 4$ **15.** $4x^2 - 25$ **16.** $16x^4 - y^{14}$
17. $4x^2 + 20x + 25$ **18.** $16x^4 - 8x^2y^7 + y^{14}$ **19.** $2x^3 - 50x$ **20.** $2x^3 - 20x^2 + 50x$
21. $8x^3 - 12x^2 + 6x - 1$ **22.** -7 and -4 **23.** 16 and -4 **24a.** $6b$ **24b.** $-7x^2$ **24c.** $-11y^6z$
25. $8(2x^4 - 3y^3)$ **26.** $4x^3(2x - 3 - 5x^2)$ **27.** $3xy(5y^2 - 4x + 3xy + 1)$ **28.** $(2x^2 - 3)(4x + 3)$
29. $(3 - y)(4 - 3x^2)$ **30.** $3y(x + 1)(2x + 1)(2x - 1)$ **31.** $(2x + 3)(3x + 1)$ **32.** $(2x + 5)^2$
33. $3y(3y + 1)(y - 5)$ **34.** $(3x - y)(x + 2y)$ **35.** $(x - 7)(x + 6)$ **36.** $2y(x - 4)^2$ **37.** prime
38. $9(2x + y)(2x - y)$ **39.** $5x(x + 5)(x - 5)$ **40.** $(x^3 + 4)(x^3 - 4)$ **41.** $(y^2 + 9)(y + 3)(y - 3)$

Chapter Test

1. $2y^2 + 2xy$ **2.** $5xy^2 - x^2y$ **3.** $1 - 3xy^2 + 6x^4y^2 - 9x^3y^3$ **4.** $3x^3 + x^2y^2 + 6xy^3 + 2y^5$
5. $2x^2 + x - 21$ **6.** $x^3 - x - 6$ **7.** $8 - 25x^2$ **8.** $18x^3 - 24x^2 + 8x$ **9.** $3y^3 + 3y^2 - 18y$
10. $8 - 36x + 54x^2 - 27x^3$ **11.** $3x(3y - 6xy^3 - 2)$ **12.** $(x^2 + 3)(x - 2)$ **13.** $(2x + 3)(2x + 1)$
14. $(x - 3y)(5x - y)$ **15.** $(x + 9)(x - 4)$ **16.** $2xy(x - 8)(x + 7)$ **17.** $6x(1 + 2x)(1 - 2x)$
18. $(8 - x)(5 - 2x)$ **19.** prime **20.** $(3 - x)(2 + x)(2 - x)$

CHAPTER 7

Exercises 7.1 (pp. 348, 349)

1. $x = \dfrac{bc}{a}$ **3.** $x = \dfrac{c}{a - b}$ **5.** $x = \dfrac{a}{dc - b}$ **7.** $x = \dfrac{be}{cd - ae}$ **9.** $x = \pm\sqrt{\dfrac{c - b}{a}}$ **11.** $t = \dfrac{W}{I^2R_L + I^2R_i}$

13. $d = \dfrac{L - a}{n - 1}$ **15.** $L = \dfrac{Rd^2}{\alpha + \alpha t}$ **17.** $H_1 = \dfrac{H_2}{1 - e}$ **19.** $R_1 = \dfrac{R_2R_t}{R_2 - R_t}$ **21.** $I = \sqrt{\dfrac{P}{R_1 + R_2}}$

Exercises 7.2 (pp. 357, 358)

1. $x = 4; x = -3$ **3.** $x = -\dfrac{7}{2}; x = -\dfrac{1}{3}$ **5.** $t = 0; t = 3$ **7.** $y = 0; y = \dfrac{9}{2}; y = 4$

9. $m = 0; m = \dfrac{5}{7}; m = -2$ **11.** $R = 5$ **13.** $x = \dfrac{2\sqrt{3}}{3}; x = \dfrac{-2\sqrt{3}}{3}; x = -9$ **15.** $v = -1$

17. $x = -5; x = 2$ **19.** $x = \dfrac{1}{2}; x = -5$ **21.** $y = 8; y = 2$ **23.** $t = -5; t = \dfrac{3}{4}$ **25.** $w = 0; w = 5$

27. $t = -2; t = \dfrac{5}{3}$ **29.** $x = 0; x = -4; x = -1$ **31.** $h = 0; h = \dfrac{1}{2}$ **33.** $x = 2; x = -2; x = -\dfrac{3}{2}$

35. $x = 2$ **37.** $x = 0; x = 2\sqrt{2}; x = -2\sqrt{2}$ **39.** $x = 0$ **41.** $t = -4; t = 2$ **43.** $x = 2; x = -1$

45. $t = 8$ s   **47.** $I = 6$ A   **49.** $r = 3$ cm

Exercises 7.3 (pp. 365, 366)

1. $4; (x + 2)^2$ **3.** $121; (y - 11)^2$ **5.** $\dfrac{81}{4}; \left(t + \dfrac{9}{2}\right)^2$ **7.** $\dfrac{1}{4}; \left(x - \dfrac{1}{2}\right)^2$ **9.** $\dfrac{1}{16}; \left(y + \dfrac{1}{4}\right)^2$ **11.** $\dfrac{9}{100}; \left(t - \dfrac{3}{10}\right)^2$

13. $\dfrac{b^2}{4}, \left(x + \dfrac{b}{2}\right)^2$ **15.** $x = 3; x = -5$ **17.** $t = \dfrac{2}{3}; t = 0$ **19.** $x = -2 + \sqrt{17} \approx 2.12; x = -2 - \sqrt{17} \approx -6.12$

21. $n = 1 + 2\sqrt{5} \approx 5.47; n = 1 - 2\sqrt{5} \approx -3.47$ **23.** $x = \dfrac{-1 + 4\sqrt{3}}{3} \approx 1.98; x = \dfrac{-1 - 4\sqrt{3}}{3} \approx -2.64$

25. $x = -2; x = -5$ **27.** $y = 6; y = 2$ **29.** $t = -2 + \sqrt{10} \approx 1.16; t = -2 - \sqrt{10} \approx -5.16$

31. $m = 4 + 2\sqrt{5} \approx 8.47; m = 4 - 2\sqrt{5} \approx -.472$ **33.** $x = \dfrac{1 + \sqrt{37}}{2} \approx 3.54; x = \dfrac{1 - \sqrt{37}}{2} \approx -2.54$

35. no real solution   **37.** $t = -2 + \dfrac{5\sqrt{2}}{2} \approx 1.54; t = -2 - \dfrac{5\sqrt{2}}{2} \approx -5.54$ **39.** $y = 3; y = -\dfrac{7}{5}$

41. $x = 0; x = -b$

Exercises 7.4 (p. 373)

1. $x = 5; x = -2$ **3.** $y = -3; y = 2$ **5.** $x = -\dfrac{1}{3}; x = \dfrac{3}{2}$ **7.** $x = \dfrac{3}{2}$

9. $t = \dfrac{-3 + \sqrt{37}}{2} \approx 1.54; t = \dfrac{-3 - \sqrt{37}}{2} \approx -4.54$ **11.** no real solution

13. $m = \dfrac{3 + 3\sqrt{5}}{2} \approx 4.85; m = \dfrac{3 - 3\sqrt{5}}{2} \approx -1.85$ **15.** $x = \dfrac{-2 + \sqrt{10}}{2} \approx .581; x = \dfrac{-2 - \sqrt{10}}{2} \approx -2.58$

17. $w = 2 + \sqrt{15} \approx 5.87; w = 2 - \sqrt{15} \approx -1.87$ **19.** $x = -4 + 2\sqrt{3} \approx -.536; x = -4 - 2\sqrt{3} \approx -7.46$

21. no real solution   **23.** $x = 0; x = -\dfrac{5}{3}$ **25.** $y = \dfrac{\sqrt{6}}{6} \approx .408; y = -\dfrac{\sqrt{6}}{6} \approx -.408$

27. $x = \dfrac{3 + \sqrt{65}}{4} \approx 2.77; x = \dfrac{3 - \sqrt{65}}{4} \approx -1.27$ **29.** $t = \dfrac{-2 + \sqrt{7}}{3} \approx .215; t = \dfrac{-2 - \sqrt{7}}{3} \approx -1.55$

31. $x = \dfrac{-3 + \sqrt{33}}{2} \approx 1.37; x = \dfrac{-3 - \sqrt{33}}{2} \approx -4.37$ **33.** $6\dfrac{2}{3}$ ft and $3\dfrac{1}{3}$ ft

Exercises 7.5 (pp. 376–378)

1. 30 ft by 24 ft   **3.** base is 12 m; height is 8 m   **5.** 27.9 m by 25.9 m   **7.** 5.62 ft   **9.** 9 in. by 9 in.

11. 32 ft by 24 ft

CHAPTER REVIEW (pp. 379–381)

Review Exercises

1. $x = \dfrac{d - ac}{ab}$ **2.** $x = \dfrac{b}{cd - a}$ **3.** $L_0 = \dfrac{L}{1 + \alpha t}$ **4.** $C_2 = \dfrac{C_1 C_t}{C_1 - C_t}$ **5.** $x = 5; x = -3$

6. $x = 0; x = 4; x = \dfrac{4}{3}$ **7.** $y = -8; y = 3$ **8.** $t = \dfrac{2}{3}; t = -5$ **9.** $z = 0; z = 8$ **10.** $x = \dfrac{3}{2}; x = -4$

11. $x = 0; x = 4$ **12.** $v = 3$ **13a.** 49   **13b.** $\dfrac{25}{4}$ **14.** $x = 1; x = -7$

15. $p = 5 + 3\sqrt{2} \approx 9.24; p = 5 - 3\sqrt{2} \approx .757$ **16.** $x = 4 + 5\sqrt{2} \approx 11.1; x = 4 - 5\sqrt{2} \approx -3.07$

17. $y = \dfrac{-1 + 3\sqrt{5}}{2} \approx 2.85; y = \dfrac{-1 - 3\sqrt{5}}{2} \approx -3.85$ **18.** $t = \dfrac{5}{3}; t = 1$ **19.** no real solution

20. $x = 4; x = -7$ **21.** $y = \dfrac{9 + \sqrt{65}}{2} \approx 8.53; y = \dfrac{9 - \sqrt{65}}{2} \approx .469$

22. $t = \dfrac{5 + \sqrt{22}}{3} \approx 3.23; t = \dfrac{5 - \sqrt{22}}{3} \approx .103$ **23.** no real solution   **24.** 12 ft

Chapter Test

1. $P = \dfrac{A}{1 + rt}$ **2.** $x = 0; x = -\dfrac{3}{4}; x = 5$ **3.** $x = -6; x = 5$ **4.** $x = \dfrac{1}{3}$ **5.** $y = 0; y = 5$

6. $t = 0; t = \dfrac{1}{4}; t = -2$ **7.** 100   **8.** $\dfrac{49}{4}$ **9.** $x = 2 + 2\sqrt{5} \approx 6.47; x = 2 - 2\sqrt{5} \approx -2.47$

10. $m = -6 + 3\sqrt{2} \approx -1.76; m = -6 - 3\sqrt{2} \approx -10.2$ **11.** $x = \dfrac{-3 + \sqrt{53}}{2} \approx 2.14; x = \dfrac{-3 - \sqrt{53}}{2} \approx -5.14$

12. $n = 1 + 2\sqrt{2} \approx 3.83; n = 1 - 2\sqrt{2} \approx -1.83$ **13.** $x = \dfrac{4}{3}; x = -1$ **14.** no real solution

15. 32.6 m by 55.2 m

CHAPTER 8

Exercises 8.1 (pp. 391, 392)

1. $x \neq 0$ **3.** $y \neq -4$ **5.** $z \neq 3; z \neq -6$ **7.** none   **9.** $t \neq 0; t \neq 3; t \neq -4$ **11.** $28x^4 y^3$

13. $y^2 + y - 6$ **15.** $2z^2 - 18$ **17.** $6t^3 - 30t^2$ **19.** $7w^2 - 23w + 6$ **21.** $1 - 3x$ **23.** $\dfrac{3 - 2y + x}{x}$

25. $\dfrac{1}{y + 7}$ **27.** $-\dfrac{1}{3}$ **29.** $6 + 6t$ **31.** $\dfrac{3y - 3x}{2}$ **33.** $\dfrac{t - 2}{2t - 1}$ **35.** $\dfrac{-4 - w}{3w + 2}$ **37.** $\dfrac{6 - x}{x^2 + 6x}$

39. $2x^2 - 3x - 2$

Exercises 8.2 (p. 397)

1. $\dfrac{4xz^3}{y}$ **3.** $\dfrac{2m}{15n}$ **5.** $\dfrac{4}{x^2}$ **7.** $\dfrac{6n - 3m}{2}$ **9.** $y^3 - 7y^2$ **11.** $9w^2 + 6w + 1$ **13.** $\dfrac{t^4 - 2t^3}{2 - 8t^2}$ **15.** $\dfrac{3x^2 + 18x}{2x - 12}$

17. $\dfrac{3x-6}{4x^2}$　　**19.** $x+2$　　**21.** $\dfrac{y}{3x+3y}$　　**23.** $4-x^2$　　**25.** $\dfrac{6+t-t^2}{35-3t-2t^2}$　　**27.** $\dfrac{x+3y}{x+y}$　　**29.** $\dfrac{7-t}{3-t}$

31. $\dfrac{98}{x^4}$

Exercises 8.3 (pp. 404, 405)

1. $\dfrac{x}{y^4}$　　**3.** $\dfrac{1}{m^2 n^4}$　　**5.** 3　　**7.** $-\dfrac{1+2b}{b}$　　**9.** $\dfrac{11x-2y}{3x-y}$　　**11.** 1　　**13.** $2x-10$　　**15.** $-\dfrac{1}{5y^2}$　　**17.** 2

19. $\dfrac{4}{t-2}$　　**21.** $\dfrac{2}{2y-3}$　　**23.** $-\dfrac{1}{3x}$　　**25.** $\dfrac{6-2n}{n^2+6n+9}$　　**27.** $\dfrac{w-1}{w-3}$　　**29.** $\dfrac{2x-5}{x-3}$　　**31.** $\dfrac{3x^2-10x-3}{3x^2-5x-2}$

33. 0　　**35.** $6-x$

Exercises 8.4 (pp. 413, 414)

1. $\dfrac{20y+9x}{24x^2 y^2}$　　**3.** $\dfrac{19m-5n}{30m^5}$　　**5.** $\dfrac{b^2-a^2}{a^4 b^3}$　　**7.** $\dfrac{12x-3xy-7}{4xy}$　　**9.** $\dfrac{8-7y^2-21y}{10y(y+3)^2}$　　**11.** $\dfrac{9x+12}{x^2+3x}$

13. $\dfrac{5k^2+2k}{k^2-4}$　　**15.** $\dfrac{11p}{6p+18}$　　**17.** $\dfrac{1}{2x+8}$　　**19.** $\dfrac{8t-t^2}{t^2-t-2}$　　**21.** $\dfrac{2-2x}{6x-9}$　　**23.** $\dfrac{-3t-12}{(t+3)^2}$

25. $\dfrac{2p^2-5p+3}{4p^2-9}$　　**27.** $\dfrac{-4}{(x-2)(x+2)^2}$　　**29.** $\dfrac{x+2}{x+4}$　　**31.** $\dfrac{2t+9}{2t+1}$　　**33.** $\dfrac{5w^2+12w-2}{6w(w+2)^2}$　　**35.** $\dfrac{y-13}{y-3}$

Exercises 8.5 (pp. 421–423)

1. $\dfrac{6t^4+4v}{3t^3}$　　**3.** $\dfrac{11w^2-1}{5w}$　　**5.** $\dfrac{y^2+7y+5}{y}$　　**7.** $\dfrac{24x^2-7x+3}{5x}$　　**9.** $\dfrac{8y-y^2}{y-4}$　　**11.** $\dfrac{4+t^2}{2-t}$　　**13.** $\dfrac{5x^2-2}{x+1}$

15. $\dfrac{-a}{a+3}$　　**17.** $2m$　　**19.** $1+x$　　**21.** $\dfrac{1-y}{2+y}$　　**23.** $\dfrac{x^2+3x-10}{x^2+x-2}$　　**25.** $\dfrac{t+2}{t+9}$　　**27.** $\dfrac{3n^2+9n+6}{5n^2-9n-2}$

29a. $\dfrac{R_1 R_2}{R_1+R_2}$　　**29b.** $\dfrac{2R_1^2+3R_1 R_2}{R_1+R_2}$

Exercises 8.6 (pp. 432, 433)

1. $x=4$　　**3.** $y=0$　　**5.** $t=2$　　**7.** $x=-\dfrac{1}{2}$　　**9.** $y=2$　　**11.** $p=18$　　**13.** $w=-\dfrac{108}{19}$　　**15.** $x=5$

17. $y=15$　　**19.** $t=2$　　**21.** $w=5;\, w=-3$　　**23.** $x=10;\, x=-1$　　**25.** $x=3;\, x=1$　　**27.** $k=-5;\, k=1$

29. $m=5;\, m=-3$　　**31.** $x=-8$　　**33.** $x=6$　　**35.** no solution　　**37.** $x=\dfrac{abc}{a+b}$　　**39.** $x=\dfrac{bc+ac-ab}{a-c}$

41. 27 cm

CHAPTER REVIEW (pp. 433–436)

Review Exercises

1a. $t\neq 0$　　**1b.** $y\neq 3$　　**1c.** $x\neq -5;\, x\neq 4$　　**1d.** $z\neq \pm 1$　　**2.** $6x^4 y^2+18x^3 y^2$　　**3.** $2y^2-13y+6$

4. $2t^2-18$　　**5.** $1-2x$　　**6.** $\dfrac{1}{t+6}$　　**7.** $\dfrac{2y-1}{6-2y}$　　**8.** $\dfrac{5+x}{5-x}$　　**9.** $\dfrac{3}{2y}$　　**10.** $\dfrac{2y+1}{4y-2y^2}$　　**11.** $\dfrac{7-t}{3}$

12. $\dfrac{3x^2y}{x^2 - y^2}$ **13.** $\dfrac{-1}{4t^2 + 18t + 20}$ **14.** $-1 - x$ **15.** $\dfrac{2x + 2}{x}$ **16.** $\dfrac{-3}{3 + x}$ **17.** $\dfrac{y - 6x}{5y^2}$ **18.** $\dfrac{2t + 10}{t^2 + 3t - 40}$

19. $\dfrac{9x}{x - 2}$ **20.** $\dfrac{5}{w - 5}$ **21.** $30x^2y^3$ **22.** $3y(y + 3)^2$ **23.** $4 - y^2$ **24.** $t^2 + 2t - 15$ **25.** $\dfrac{25x + 3y^3 - 6y^2}{30x^2y^3}$

26. $\dfrac{3y^2 - 7y + 15}{3y(y + 3)^2}$ **27.** $\dfrac{1 - y}{2 + y}$ **28.** $\dfrac{16}{t^2 + 2t - 15}$ **29.** $\dfrac{15x^2y^7 + 3}{5x^2y^5}$ **30.** $\dfrac{3x^2 - 5x - 1}{x}$ **31.** $\dfrac{6t^2 + t - 7}{3t - 1}$

32. $\dfrac{2y - 1}{y}$ **33.** $\dfrac{y - 3}{y + 7}$ **34.** $t = 9$ **35.** $x = 5$ **36.** $y = -6$ **37.** $y = 1; y = 9$ **38.** $x = -4$

Chapter Test

1. $x \neq 2; x \neq -1$ **2.** $3y^2 + 11y + 6$ **3.** $4x - 1$ **4.** $-\dfrac{t + 6}{t + 3}$ **5.** $\dfrac{x + 1}{2x}$ **6.** $\dfrac{2 - w}{8 + 6w + w^2}$ **7.** $\dfrac{2}{2t + 5}$

8. $\dfrac{-2}{x + 8}$ **9.** $\dfrac{8t - 1}{2t - 1}$ **10.** $\dfrac{8y - 1}{9y^2 - 3y}$ **11.** $\dfrac{x^2 + 6x - 9}{3x^3 + 15x^2 + 18x}$ **12.** $\dfrac{k^2 - k - 9}{k + 2}$ **13.** $\dfrac{y^2 + 4y + 4}{2y + 8}$

14. $t = 1$ **15.** $x = -2$

CHAPTER 9

Exercises 9.1 (pp. 442–444)

1. $C = \pi d$ **3.** $d = \dfrac{C}{\pi}$ **5.** $l = \dfrac{A}{4}$ **7.** $w = \dfrac{P - 10}{2}$ **9.** $y = 20x$ **11.** $A = 15t$ **13.** $A = 45{,}000 - 15t$

15. $y = 3000 + 10x$

17. domain is all real numbers between 0 and 3000; range is all real numbers between 0 and 45,000

19. domain is all real numbers between 0 and 3000; range is all real numbers between 0 and 45,000

21. $(0, 0), (1, 1), (-1, -1), (2, 2), (-2, -2)$ **23.** $(0, 6), (3, 0), (-3, 12), \left(\dfrac{1}{2}, 5\right), \left(-\dfrac{1}{2}, 7\right)$

25. $(0, 0), (1, 1), (-1, 1), (3, 9), (-3, 9)$ **27.** $\left(3, \dfrac{1}{3}\right), \left(-3, -\dfrac{1}{3}\right), \left(\dfrac{1}{3}, 3\right), \left(-\dfrac{1}{3}, -3\right)$

29. $(1, 4), (4, 2), \left(9, \dfrac{4}{3}\right), (16, 1)$ **31.** $(0, 2), \left(\dfrac{1}{2}, 0\right), (2, 0), (1, -1), (-1, 9)$ **33.** only $y = x$

Exercises 9.2 (pp. 448–450)

1. $y = 8x$ **3.** $y = \dfrac{90}{x}$ **5.** $y = \dfrac{x^2}{4}$ **7.** $y = \dfrac{1000}{x^2}$ **9a.** $W = 80L$ **9b.** $W = 640$ lb   **11a.** $C = .103d$

11b. $C = 7.21$ ft   **13a.** $S = \dfrac{6500}{d}$ **13b.** $S = 500$ lb   **15a.** $I = \dfrac{100}{R}$ **15b.** $I = 5$ A   **17a.** $P = 5I^2$

17b. $P = 45$ watts   **19a.** $F = \dfrac{80{,}000}{d^2}$ **19b.** $F = 50$ N

x	y		x	y
21. 18	12	**23.** 18	12
12	8	12	27
9	6	9	48
3	2	3	432
1	$\frac{2}{3}$	1	3888

Exercises 9.3 (pp. 462–464)

1.

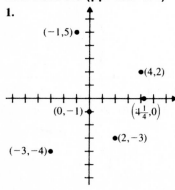

3. $A(0, 0)$; $B(3, 1)$; $C(5, -1)$ $D(-1, -4)$; $E(-3, 0)$; $F(-4, 4)$; $G(0, 3)$

5.

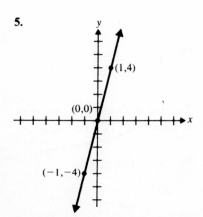

7.

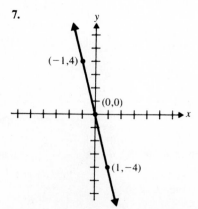

9.

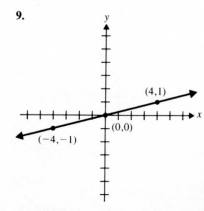

11.

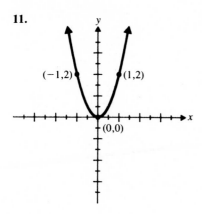

13.

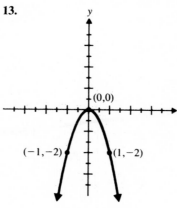

15.

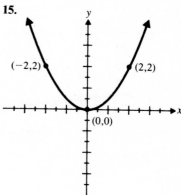

17.

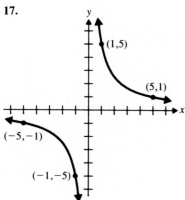

19.

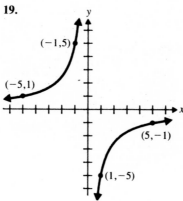

21.

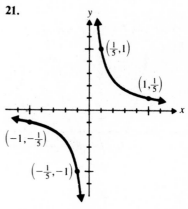

23.

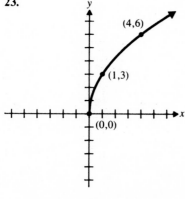

25.

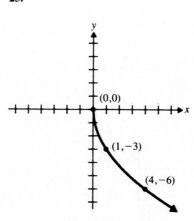

27.

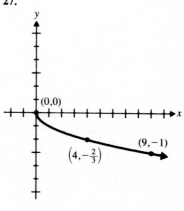

29.

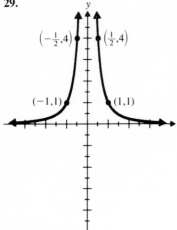

31.

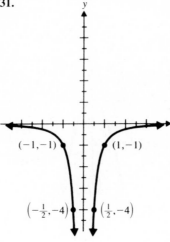

33.

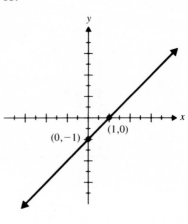

35.

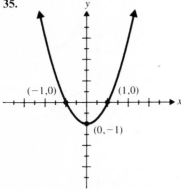

37.

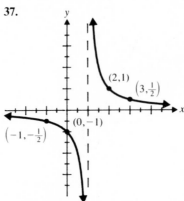

39.

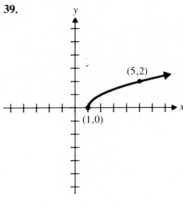

41.

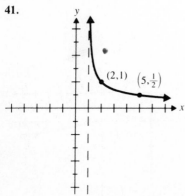

43.

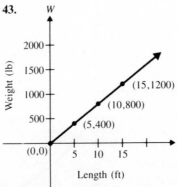

45.

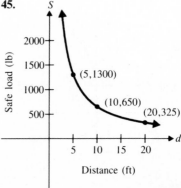

47.

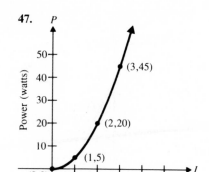

49.

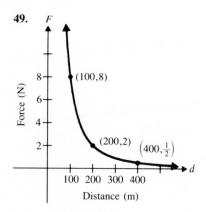

Exercises 9.4 (pp. 477, 478)

1.

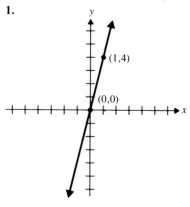

3.

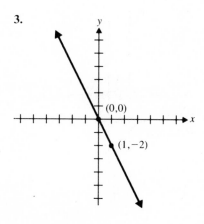

5.

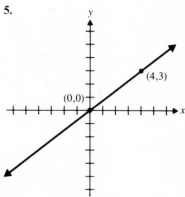

7.

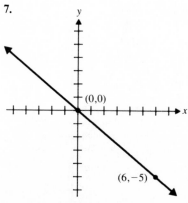

9.

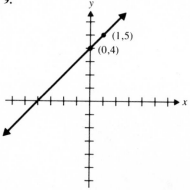

11.

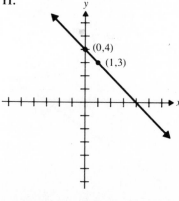

13.

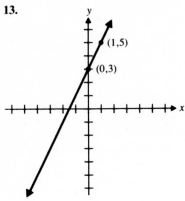

15.

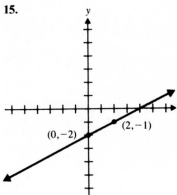

17.

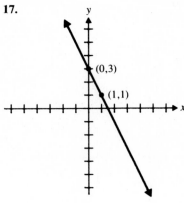

19.

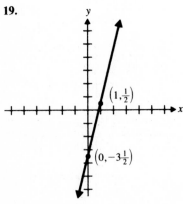

21.

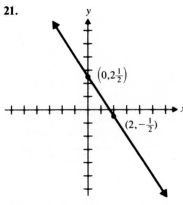

23.

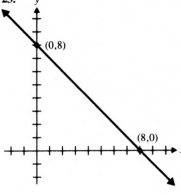

25.

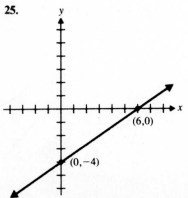

27.

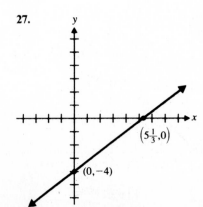

29.

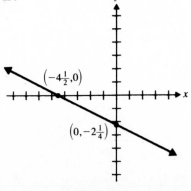

31.

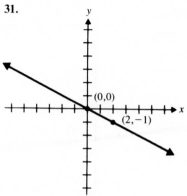

33.

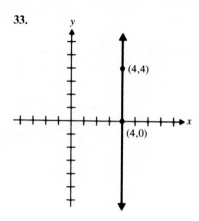

35.

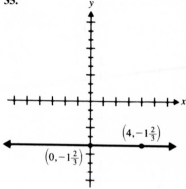

37.

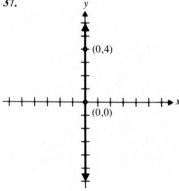

39.

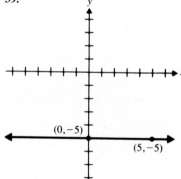

41a.

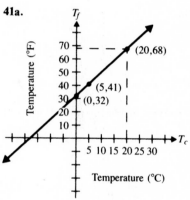

41b. 68°F (see graph in 41a)

43a.

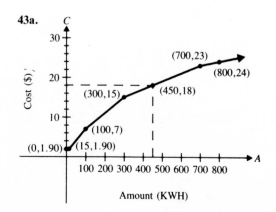

43b. $18 (see graph in 43a)

Exercises 9.5 (pp. 488)

1.

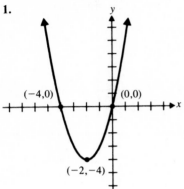

3.

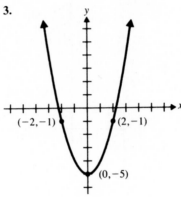

5.

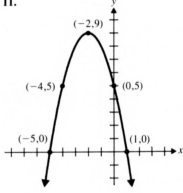

7.

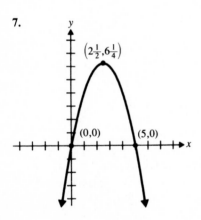

9.

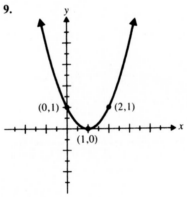

11.

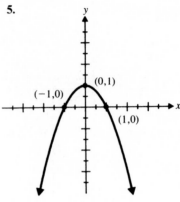

13.

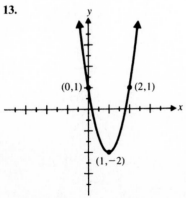

15.

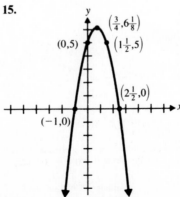

17.

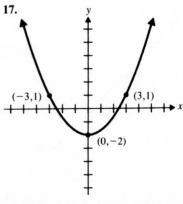

19.

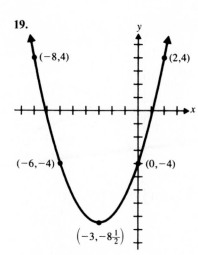

21.

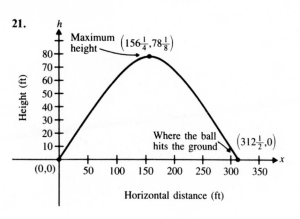

CHAPTER REVIEW (pp. 488-491)

Review Exercises

1. $s = \dfrac{P}{4}$ **2.** $y = 8 - 2x$ **3a.** $(0, 1)$ **3b.** $(2, -35)$ **3c.** $(-2, -35)$ **3d.** $\left(-\dfrac{1}{3}, 0\right)$ **4.** $y = 6x$

5. $d = \dfrac{F}{80}$ **6.** $y = \dfrac{24}{x}$ **7.** $P = \dfrac{120}{V}$ **8.** $A = .785d^2$ **9.** $t = \dfrac{32}{d^2}$ **10.**

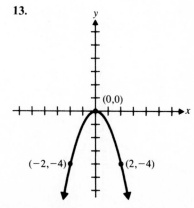

11.

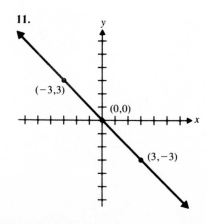

12.

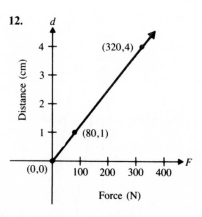

13.

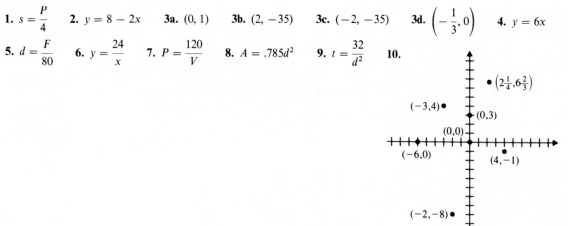

14.

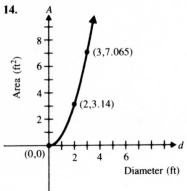

15.

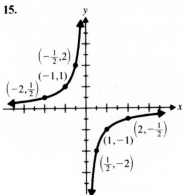

16.

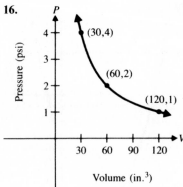

17.

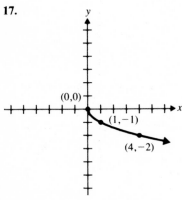

18.

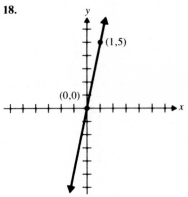

19.

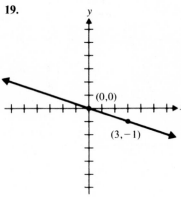

20.

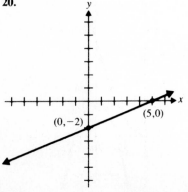

21.

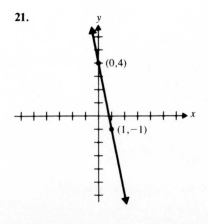

22.

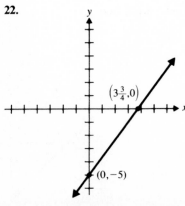

23a.

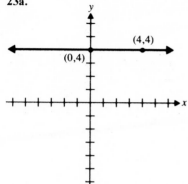

23b.

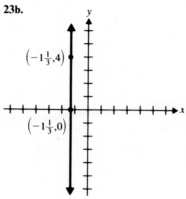

24.

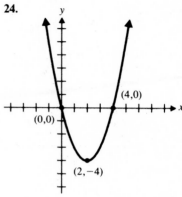

25.

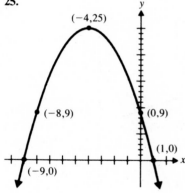

26.

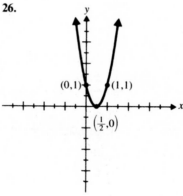

27.

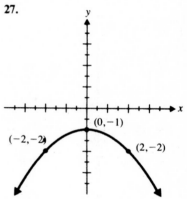

Chapter Test

1. $C = 10n + 400$ **2.**

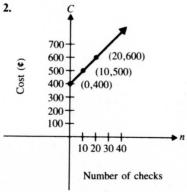

Number of checks

3. $L = \dfrac{v^2}{100}$ **4.**

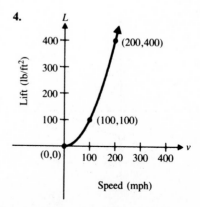

Speed (mph)

5. $F = \dfrac{120}{L}$

6.

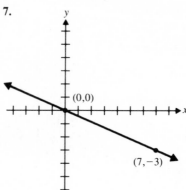

7.

8.

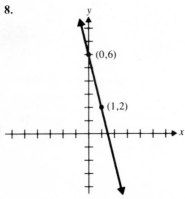

9.

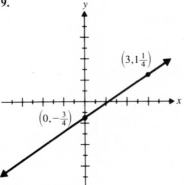

10.

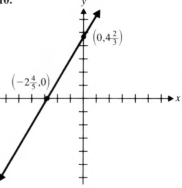

11.

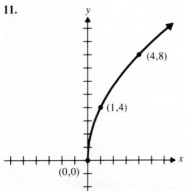

12.

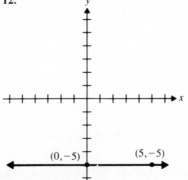

13.

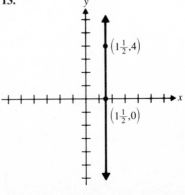

14.

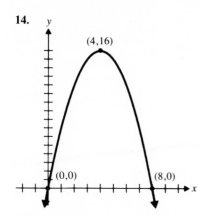

15.

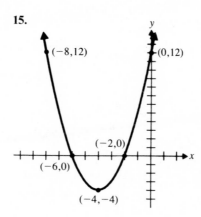

CHAPTER 10

Exercises 10.1 (pp. 500, 501)

1. no   **3.** yes   **5.** no   **7.** yes   **9.** no   **11.** $(5, -2)$ **13.** $(-2, 1)$ **15.** $(2, 0)$ **17.** $(3, 3)$

19. $(8, 4)$ **21.** no solution   **23.** infinite number of solutions   **25.** exactly one solution, $(0, 2)$

27. infinite number of solutions   **29.** no solution

Exercises 10.2 (pp. 510, 511)

1. $(5, 4)$ **3.** $\left(\dfrac{5}{2}, \dfrac{1}{2}\right)$ **5.** $(2, 1)$ **7.** $(3, 6)$ **9.** infinite number of solutions   **11.** $(5, 7)$ **13.** $\left(-\dfrac{13}{4}, \dfrac{11}{4}\right)$

15. $\left(-\dfrac{1}{2}, -1\right)$ **17.** $(-2, 3)$ **19.** $(0, 3)$ **21.** no solution   **23.** $(21, -7)$ **25.** $\left(2, -\dfrac{1}{2}\right)$

27. $F_1 = 250$ lb; $F_2 = 150$ lb   **29.** $F_1 = 480$ N; $F_2 = 320$ N

Exercises 10.3 (pp. 519, 520)

1. $(7, 2)$ **3.** $\left(2, -\dfrac{1}{3}\right)$ **5.** $\left(-\dfrac{9}{8}, -\dfrac{3}{2}\right)$ **7.** $\left(\dfrac{5}{4}, -\dfrac{43}{20}\right)$ **9.** $(-2, 13)$ **11.** $\left(\dfrac{1}{4}, -\dfrac{3}{2}\right)$ **13.** $\left(2, -\dfrac{4}{3}\right)$

15. $(-1, 3)$ **17.** $(4, 5)$ **19.** $\left(\dfrac{3}{2}, 5\right)$ **21.** no solution   **23.** $\left(2, -\dfrac{1}{2}\right)$ **25.** $\left(\dfrac{9}{2}, 0\right)$ **27.** $I_1 = 1$ A; $I_2 = 2$ A

29. $I_1 = \dfrac{5}{6}$ A; $I_2 = \dfrac{1}{6}$ A

Exercises 10.4 (pp. 527, 528)

1. equal sides are 25.0 cm; nonequal side is 9.6 cm   **3.** large piece is 6.4 ft; short piece is 1.6 ft

5. amount of gasoline is $3\dfrac{3}{4}$ L; amount of oil is $\dfrac{1}{4}$ L   **7.** amount of cement is $16\dfrac{4}{5}$ yd^3; amount of gravel is $25\dfrac{1}{5}$ yd^3

9. amount of vermouth is $\dfrac{2}{3}$ L; amount of gin is $3\dfrac{1}{3}$ L   **11.** speed of the airplane is 350 mph; speed of wind is 50 mph

13. amount of forest land is 5 Ac; amount of swamp is 35 Ac

15. drain $7\frac{1}{2}$ qt and replace with $7\frac{1}{2}$ qt of pure antifreeze   **17.** slope is $-\frac{1}{3}$; y-intercept is 3

Exercises 10.5 (pp. 533, 534)

1. -8 **3.** 4   **5.** -15 **7.** 80   **9.** 1   **11.** 0   **13.** $6b - ab$ **15.** $(2, 8)$ **17.** $\left(-\frac{135}{7}, -\frac{267}{7}\right)$

19. $(8, 9)$ **21.** $\left(-\frac{5}{3}, -\frac{89}{36}\right)$ **23.** $\left(-\frac{40}{71}, \frac{77}{71}\right)$ **25.** an infinite number of solutions   **27.** $\left(-\frac{99}{17}, -\frac{90}{17}\right)$

29. $T_1 = 57.6$ lb; $T_2 = 77.9$ lb

CHAPTER REVIEW (pp. 535–537)

Review Exercises

1. no   **2.** yes   **3.** $(-1, -5)$ **4.** $(4, -3)$ **5a.** exactly one solution, $\left(0, \frac{1}{3}\right)$ **5b.** no solution

5c. an infinite number of solutions   **6.** $\left(\frac{5}{2}, 15\right)$ **7.** $(-1, 1)$ **8.** $(-3, 4)$ **9.** $(-3, -2)$

10. an infinite number of solutions   **11.** $\left(3, \frac{1}{7}\right)$ **12.** $\left(\frac{5}{3}, -6\right)$ **13.** $(-3, -4)$ **14.** $\left(\frac{5}{2}, -\frac{1}{3}\right)$

15. no solution   **16.** length is 27.8 cm; width is 15.7 cm

17. amount of cement is 36 yd^3; amount of gravel is 24 yd^3

18. amount of 10% solution is 15 L; amount of 15% solution is 10 L

19. speed of airplane is 360 mph; speed of wind is 40 mph   **20a.** -38 **20b.** 8   **20c.** 0   **21.** $(4, -3)$

22. $\left(-\frac{2}{5}, -1\right)$

Chapter Test

1. no solution   **2.** exactly one solution, $\left(0, -\frac{1}{2}\right)$ **3.** an infinite number of solutions   **4.** 35   **5.** $6a$

6. $(-1, -2)$ **7.** $(3, 5)$ **8.** $(-2, 2)$ **9.** $\left(-\frac{1}{2}, 6\right)$ **10.** $\left(\frac{2}{3}, \frac{1}{2}\right)$ **11.** $\left(2, \frac{3}{2}\right)$ **12.** $(3, 9)$

13. $\left(\frac{252}{53}, \frac{210}{53}\right)$ **14.** amount of coffee brandy is $6\frac{2}{3}$ L; amount of vodka is $3\frac{1}{3}$ L

15. number of teeth in ring gear is 54; number of teeth in pinion gear is 12

CHAPTER 11

Exercises 11.1 (pp. 547–549)

1. $6°$ **3.** $123°$ **5.** $25°$ **7.** $59°$ **9.** $149°$ **11.** $32.5°$ **13.** $158.53°$ **15.** $86.9222°$ **17.** $165.1069°$
19. $82° \, 40'$ **21.** $112° \, 48'$ **23.** $13° \, 02' \, 30''$ **25.** $144° \, 15' \, 17''$ **27.** $51° \, 48' \, 31''$ **29.** $71° \, 16' \, 26''$
31. $103° \, 28'$ **33.** $\angle 1 = \angle 3 = 28° \, 48'$; $\angle 2 = \angle 4 = 151° \, 12'$; $\angle 5 = \angle 7 = 61° \, 12'$; $\angle 6 = 118° \, 48'$
35. $30°, 90°, 60°$ **37.** $27\frac{1}{2}°$ and $62\frac{1}{2}°$ **39.** $75°$ and $105°$

Exercises 11.2 (pp. 559–562)

1. $48°$ **3.** $83°$ **5.** $83°$ **7.** $42°$ **9.** $138°$ **11.** $90°$ **13.** $90°$ **15.** $112°$ **17.** $41°$ **19.** $93°$
21. $46°$ **23.** $46°$ **25.** $60°$ **27.** $60°$ **29.** $30°$ **31.** $75°$ **33.** $105°$ **35.** $63° \, 34' \, 30''$ **37.** $63° \, 34' \, 30''$
39. $116° \, 25' \, 30''$ **41.** $63° \, 34' \, 30''$ **43.** $149° \, 10' \, 45''$ **45.** $x = 15$ **47.** $x = 14$
49. $S = (n - 2)180°$, where S is the sum of the interior angles and n is the number of sides

Members	Angle
51. 1/7, 3/6	a
2/4, 2/5	b
1/4, 3/5	$180 - b$
4/7, 5/6	$b - a$
4/5	$180 - 2b$
6/8	$2a$
6/9	$90 - 2a$

Exercises 11.3 (pp. 570–574)

1. $20°$ **3.** $50°$ **5.** $7\frac{1}{3}$ cm **7.** $52°$ **9.** $57°$ **11.** 18 ft **13.** $\triangle BDE \sim \triangle BAC$; $x = 10$ cm
15. no similar triangles **17.** $\triangle ABC \sim \triangle BED$; $x = 90$ ft **19.** $\triangle ADE \sim \triangle ABC$; $x = 72$ m
21. $a = 6$; $b = 3\sqrt{5}$; $z = 2\sqrt{5}$ **23.** $y = 4$; $a = 2\sqrt{13}$; $b = 3\sqrt{13}$ **25.** $x = 7$; $z = 3\sqrt{7}$; $b = 4\sqrt{7}$

Exercises 11.4 (pp. 584–590)

1. 10 cm **3.** 160 ft **5.** 12 ft **7.** 27.5 ft **9.** 17.1 cm **11.** 29.2 m
13. $x = 4$ cm; $y = 4\sqrt{3}$ cm ≈ 6.93 cm **15.** $x = 6$ mm; $y = 5\sqrt{2}$ mm ≈ 7.07 mm
17. $x = 14\sqrt{2}$ in. ≈ 19.8 in.; $y = 7\sqrt{2}$ in. ≈ 9.90 in. **19.** $x = 6\sqrt{2}$ m ≈ 8.49 m; $y = (24 - 6\sqrt{3})$ m ≈ 13.6 m
21. 294 ft^2 **23.** 72.0 cm^2 **25.** 1020 m^2 **27.** $90\sqrt{3}$ in^2 ≈ 156 in^2 **29.** $(192 + 108\sqrt{2})$ft^2 ≈ 345 ft^2
31. $\dfrac{5x^2}{2}$ **33.** $\dfrac{x^2\sqrt{3}}{4}$ **35.** $22x^2$

Member	Length
37. 1, 2, 3, 7	$\dfrac{L}{6}$
4, 5, 6	$\dfrac{L\sqrt{2}}{6}$
8	$\dfrac{L}{3}$
9	$\dfrac{L}{2}$
10	$\dfrac{L\sqrt{5}}{6}$
11	$\dfrac{L\sqrt{10}}{6}$

Exercises 11.5 (pp. 598, 599)

1. .385   **3.** .417   **5.** .385   **7.** 1.24   **9.** .778   **11.** .778   **13a.** $\dfrac{4}{5}$ **13b.** $\dfrac{3}{5}$ **13c.** $\dfrac{4}{3}$ **15a.** $\dfrac{4}{5}$

15b. $\dfrac{3}{5}$ **15c.** $\dfrac{4}{3}$ **17.** .99   **19.** .698   **21.** .949   **23.** .7651   **25.** .29289   **27.** .992940   **29.** 15°

31. 58.6° or 58° 40′   **33.** 85.7° or 85° 40′   **35.** 39.87° or 39° 52′   **37.** 70.045° or 70° 02′ 40″

39. 22.2193° or 22° 13′ 1$\bar{0}$″

θ	sin θ	cos θ	tan θ
41. 30°	$\dfrac{1}{2}$	$\dfrac{\sqrt{3}}{2}$	$\dfrac{\sqrt{3}}{3}$
60°	$\dfrac{\sqrt{3}}{2}$	$\dfrac{1}{2}$	$\sqrt{3}$
45°	$\dfrac{\sqrt{2}}{2}$	$\dfrac{\sqrt{2}}{2}$	1

Exercises 11.6 (pp. 609–615)

1. $x = 17$ cm; $y = 29$ cm   **3.** $x = 174$ ft; $y = 235$ ft   **5.** $x = 7.0$ ft; $y = 26$ ft   **7.** $x = 8$ cm; $y = 15$ cm

9. $x = 47.4$ m; $y = 41.6$ m   **11.** $x = 24$ mm; $y = 16$ mm   **13.** $\alpha = 41.4°$; $\beta = 48.6°$ **15.** $\alpha = 39°$; $\beta = 51°$

17. $\alpha = 47.7°$; $\beta = 132.3°$ **19.** $\alpha = 157.8°$; $\beta = 44.3°$ **21.** $\alpha = 58°$; $\beta = 64°$ **23.** $\alpha = 34.0°$; $\beta = 112°$

25. 74 ft^2 **27.** 55.2 m^2 **29.** 67.4 cm^2 **31.** 16,700 ft^2 **33.** 1685 ft^2 **35.** 140.6 ft   **37.** 84° 58′ 40″

CHAPTER REVIEW (pp. 615–624)

Review Exercises

1a. $58°$ **1b.** $148°$ **1c.** $32°$ **1d.** $90°$ **2a.** $42.42°$ **2b.** $143.2542°$ **3a.** $14° \, 40'$ **3b.** $102° \, 38' \, 30''$

4. $56° \, 42'$ **5.** $57° \, 41' \, 20''$ **6.** $\measuredangle a = \measuredangle c = 161° \, 23'; \; \measuredangle b = 18° \, 37'$

7. $\measuredangle 3 = \measuredangle 5 = \measuredangle 7 = 106°; \; \measuredangle 2 = \measuredangle 4 = \measuredangle 6 = \measuredangle 8 = 74°$ **8.** $\measuredangle A = \measuredangle C = 138°; \; \measuredangle B = 42°$

9. $\measuredangle 1 = \measuredangle 2 = 59°; \; \measuredangle 3 = 62°$ **10.** $83° \, 26' \, 00''$ **11a.** $62°$ **11b.** $77°$ **11c.** $41°$ **11d.** $41°$

12a. $y = 12 \text{ ft}$ **12b.** $r = 9\frac{1}{3} \text{ ft}$ **13.** $\triangle ABC \sim \triangle DEC; \; x = 5\frac{1}{3} \text{ cm}$ **14.** $\triangle ABC \sim \triangle ADB; \; \measuredangle ABD = 90°$

15. $\triangle ABC \sim \triangle EBD; \; x = 252 \text{ m}$ **16a.** $x = 4\sqrt{7} \text{ cm} \approx 10.6 \text{ cm}$ **16b.** $y = 12 \text{ cm}$

16c. $z = 3\sqrt{7} \text{ cm} \approx 7.94 \text{ cm}$ **17.** 10 ft **18.** $x = 19 \text{ cm}$ **19.** 157 m^2 **20.** $(2 + 18\sqrt{2}) \text{ft} \approx 27.5 \text{ ft}$

21. $x = \dfrac{220\sqrt{3}}{3} \text{ ft} \approx 127 \text{ ft}; \; y = \dfrac{440\sqrt{3}}{3} \text{ ft} \approx 254 \text{ ft}$ **22.** $17{,}000 \text{ ft}^2$

23a. $\sin \theta = .385; \cos \theta = .923; \tan \theta = .417$ **23b.** $\sin \theta = .385; \cos \theta = .923; \tan \theta = .417$ **24a.** $\dfrac{2\sqrt{6}}{7} \approx .700$

24b. $\dfrac{2\sqrt{6}}{7} \approx .700$ **25a.** $.6877$ **25b.** $.7260$ **25c.** $.9473$ **26a.** $38.4°$ or $38° \, 20'$ **26b.** $51.6°$ or $51° \, 40'$

26c. $31.8°$ or $31° \, 50'$ **27.** $x = 12 \text{ cm}; y = 25 \text{ cm}$ **28.** $11{,}246 \text{ ft}^2$ **29.** $\alpha = 58°; \beta = 32°$ **30.** $\theta = 148.6°$

Chapter Test

1. $67.408°$ **2.** $87° \, 22'$ **3.** $25° \, 23'$

4. $\measuredangle 1 = \measuredangle 3 = 72°; \; \measuredangle 2 = 108°; \; \measuredangle 4 = \measuredangle 5 = \measuredangle 7 = 54°; \; \measuredangle 6 = \measuredangle 8 = 126°$

5. $\measuredangle 1 = 58°; \; \measuredangle 3 = 78°; \; \measuredangle 2 = \measuredangle 4 = \measuredangle 5 = 102°; \; \measuredangle 6 = 20°$ **6.** $\triangle ABC \sim \triangle DEC; \; x = 9 \text{ cm}$

7. $y = 9; a = 15; b = 20$ **8.** $x = 32 \text{ ft}$ **9.** 189 m^2 **10.** $x = 9\sqrt{2} \text{ m} \approx 12.7 \text{ m}; y = (33 - 9\sqrt{3}) \text{m} \approx 17.4 \text{ m}$

11a. $\tan A = 1.38; \tan B = .722$ **11b.** $\sin A = .811; \sin B = .586$ **11c.** $\cos A = .586; \cos B = .811$

12. $x = 11.0 \text{ cm}; y = 8.62 \text{ cm}$ **13.** $\alpha = 39.6°; \beta = 50.4°$ **14.** $40{,}100 \text{ m}^2$ **15.** $\theta = 64°; x = 18.7 \text{ in.}$

CHAPTER 12

Exercises 12.1 (pp. 639, 640)

1. $300°$ **3.** $230°$ **5.** $220°$ **7.** $336°$ **9.** $280°$ **11.** $35°$ **13.** $62°$ **15.** $11°$ **17.** $60°$ **19.** $57°$

21. $13°$ **23.** $47° \, 45'$ **25.** positive   **27.** positive   **29.** negative   **31.** negative   **33.** negative

35. positive   **37.** negative   **39.** positive   **41.** negative

Exercises 12.2 (pp. 651, 652)

1. 0 **3.** 0 **5.** undefined   **7.** 0 **9.** -1 **11.** 0 **13.** -1.327 **15.** -3.628 **17.** $-.8387$

19. $.9272$ **21.** 1.788 **23.** $.4663$ **25.** $32.0°$ and $148.0°$ **27.** $65.0°$ and $245.0°$ **29.** $83.7°$ and $276.3°$

31. $163.2°$ and $196.8°$ **33.** $236.2°$ and $303.8°$ **35.** $94.0°$ and $274.0°$ **37.** $239° \, 33'$ and $300° \, 27'$

39. $71° \, 42'$ and $251° \, 42'$ **41.** $128° \, 02'$ and $231° \, 58'$ **43.** $90°$ and $270°$ **45.** $180°$ **47.** $90°$ **49.** $0°$ and $180°$

51. $135°$ and $315°$ **53.** $0°$

55. The maximum value of x occurs at $0°$ when $x = 1$; the minimum value of x occurs at $180°$ when $x = -1$. Thus, $-1 \le \cos \theta \le 1$.

Exercises 12.3 (pp. 661–663)

1.

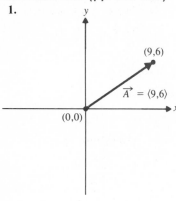

3.

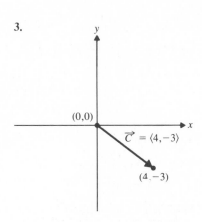

5.

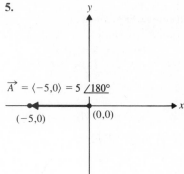

7.

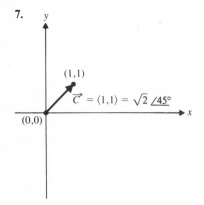

9.

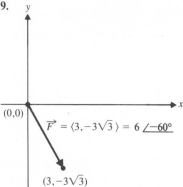

11. $\vec{A} = \langle 6.1, 3.8 \rangle$ **13.** $\vec{B} = \langle -11.4, 15.9 \rangle$ **15.** $\vec{C} = \langle -12, -11 \rangle$ **17.** $\vec{D} = \langle 555, -186 \rangle$

19. $\vec{F} = \langle 120, 0 \rangle$ **21.** $\vec{W} = \langle 0, -1500 \rangle$ **23.** $\vec{A} = 110\underline{/40^\circ}$ **25.** $\vec{B} = 56.1\underline{/125.6^\circ}$ **27.** $\vec{C} = 180\underline{/229^\circ}$

29. $\vec{D} = 668\underline{/330.7^\circ}$ **31.** $\vec{F} = 36\underline{/90^\circ}$ **33.** $\vec{W} = 92.6\underline{/180^\circ}$ **35.** $34\underline{/85^\circ}$ **37.** $13\underline{/176^\circ}$ **39.** $14\underline{/117^\circ}$

41. $8\underline{/186^\circ}$ **43.** $25\underline{/336^\circ}$

Exercises 12.4 (pp. 669–671)

1. $220\,\text{lb}\underline{/55^\circ}$ **3.** $2630\,\text{N}\underline{/95.3^\circ}$ **5.** $47\,\text{lb}\underline{/39^\circ}$ **7.** $140.5\,\text{N}\underline{/170.9^\circ}$ **9.** $\vec{F}_2 = 101\,\text{lb}\underline{/48^\circ}$

11. $\vec{F}_2 = 153\,\text{N}\underline{/90^\circ}; \vec{F}_3 = 364\,\text{N}\underline{/0^\circ}$ **13.** $\vec{F}_2 = 102\,\text{lb}\underline{/80^\circ}$ **15.** $\vec{F}_1 = 349\,\text{N}\underline{/160.0^\circ}; \vec{F}_3 = 119\,\text{N}\underline{/270.0^\circ}$

17. $\vec{C} = 790\,\text{lb}\underline{/55^\circ}; \vec{T} = 460\,\text{lb}\underline{/180^\circ}$ **19.** $\vec{C} = 1800\,\text{lb}\underline{/62^\circ}$ **21.** $\vec{T} = 10\overline{0}0\,\text{lb}\underline{/150^\circ}; \vec{W} = 5\overline{0}0\,\text{lb}\underline{/270^\circ}$

Exercises 12.5 (pp. 679–683)

1. $x = 26\,\text{ft}$ **3.** $x = 17\,\text{ft}$ **5.** $454.7\,\text{ft}$ **7.** $x = 18.0\,\text{cm}$ **9.** $403.4\,\text{m}$ **11.** $68.7\,\text{ft}$ **13.** $11.7\,\text{ft}$

15. $6.7\,\text{ft}$ **17.** 47° **19.** $8.4\,\text{ft}$ **21.** $6.6\,\text{ft}$ **23.** 125° **25.** $15.6\,\text{ft}$ **27.** $14\,\text{ft}$

29. $T_1 = 420\,\text{lb}; T_2 = 250\,\text{lb}$ **31.** $T_1 = 1020\,\text{N}; T_2 = 1480\,\text{N}$ **33.** $T_1 = 2300\,\text{lb}; T_2 = 1400\,\text{lb}$

Exercises 12.6 (pp. 690–692)

1. $\alpha = 37.9°$ **3.** $\alpha = 27°$ **5.** 7.05 in. **7.** $x = 3.81$ cm **9.** $\theta = 49°$ **11.** $x = 116.6$ ft **13.** 38.9°

15. 109.5° **17.** 4.77 ft **19.** 83° **21.** 16.2 ft **23.** $18\bar{0}$ mi, N–58° 20′–E **25.** 16 mi, S–18° 50′–W

Exercises 12.7 (pp. 703–706)

1.

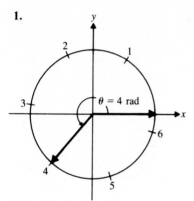

3.

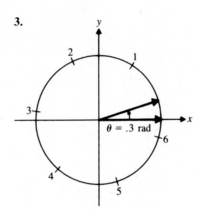

5.

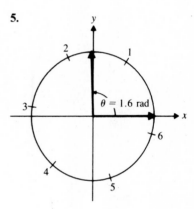

7.

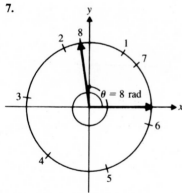

9. $\dfrac{\pi}{2}$ rad ≈ 1.57 rad **11.** π rad ≈ 3.14 rad **13.** $\dfrac{\pi}{6}$ rad $\approx .524$ rad **15.** $\dfrac{\pi}{4}$ rad $\approx .785$ rad

17. $\dfrac{5\pi}{3}$ rad ≈ 5.24 rad **19.** $\dfrac{2\pi}{5}$ rad ≈ 1.26 rad **21.** $\dfrac{25\pi}{12}$ rad ≈ 6.54 rad **23.** 37.0° **25.** 215° **27.** 225°

29. 300° **31.** $67\dfrac{1}{2}°$

	Radians	Degrees		Arc length	Radians		Arc length	Radians		Radius	Radians
33.	1.42 rad	81.2°	**35.**	14 in.	1.66 rad	**37.**	62.3 ft	1.48 rad	**39.**	155 ft	1.26 rad

41. 17 cm   **43.** 540 rpm   **45.** $25.3 \dfrac{\text{ft}}{\text{sec}}$

t(s)	d(cm)
47. 0	12
1	0
2	-12
3	0
4	12

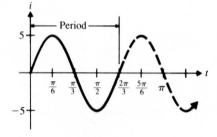

49. $t = \dfrac{\pi}{6}$ s; $t = \dfrac{5\pi}{6}$ s; $t = \dfrac{3\pi}{2}$ s; and so on

CHAPTER REVIEW (pp. 706–712)

Review Exercises

1a. 130°   **1b.** 210°   **1c.** 270°   **2a.** 30°   **2b.** 40°   **2c.** 80°   **2d.** 60°   **2e.** 40°

θ	sin θ	cos θ	tan θ	csc θ	sec θ	cot θ
3. 150°	$+$	$-$	$-$	$+$	$-$	$-$
4. 550°	$-$	$-$	$+$	$-$	$-$	$+$
5. $-25°$	$-$	$+$	$-$	$-$	$+$	$-$

θ	sin θ	cos θ	tan θ	csc θ	sec θ	cot θ
6. 90°	1	0	undefined	1	undefined	0
7. $-180°$	0	-1	0	undefined	-1	undefined
8. $-10°$	$-.1736$	$.9898$	$-.1763$	-5.759	1.015	-5.671

9. 49.0° and 229.0°   **10.** 164.0° and 196.0°   **11.** 200° 21′ and 339° 39′

12.

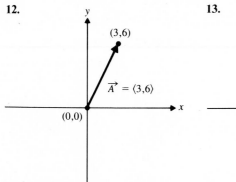

13.

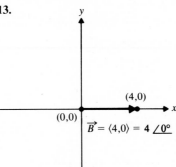

14. $\langle -5.5, 7.3 \rangle$ **15.** $458\underline{/231.5°}$ **16.** $\langle -5, -1 \rangle$ **17.** $31\underline{/182°}$ **18.** 265 lb$\underline{/4°}$ **19.** 950 lb$\underline{/146°}$

20. $x = 16$ cm; $y = 12$ cm   **21.** $x = 21$ ft   **22.** $\alpha = 101° \ 26'$; $\beta = 56° \ 19'$; $x = 275.6$ ft

23. $x = 9.2$ cm; $\alpha = 44°$; $\beta = 36°$ **24.** $102.4°$

25a.

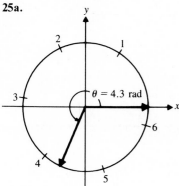

25b.

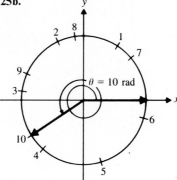

26a. $\dfrac{\pi}{5}$ rad $\approx .628$ rad   **26b.** $\dfrac{7\pi}{6}$ rad ≈ 3.67 rad   **27a.** $18\overline{0}°$ **27b.** $390°$ **28.** $111.9°$ **29.** $7\overline{0}$ ft

30. 504 rpm

Chapter Test

1. $160°$ **2.** $20°$

θ	sin θ	cos θ	tan θ	csc θ	sec θ	cot θ
3. 230°	$-.7660$	$-.6428$	1.192	-1.305	-1.556	.8391

4. $125.0°$ and $235.0°$ **5.** $158° \ 43'$ and $338° \ 43'$ **6.** $\langle 12.3, -18.2 \rangle$ **7.** $20.9\underline{/216.7°}$ **8.** $31.0\underline{/261.7°}$

9. $\dfrac{5\pi}{4}$ rad   **10.** $207°$ **11.** 217.3 ft   **12.** $45.2°$ **13.** $\vec{F}_2 = 129$ lb$\underline{/0°}$; $\vec{F}_3 = 184$ lb$\underline{/270°}$ **14.** $16.4°$

15. 11 mph

INDEX